Skylab

NEWS REFERENCE

MARCH 1973

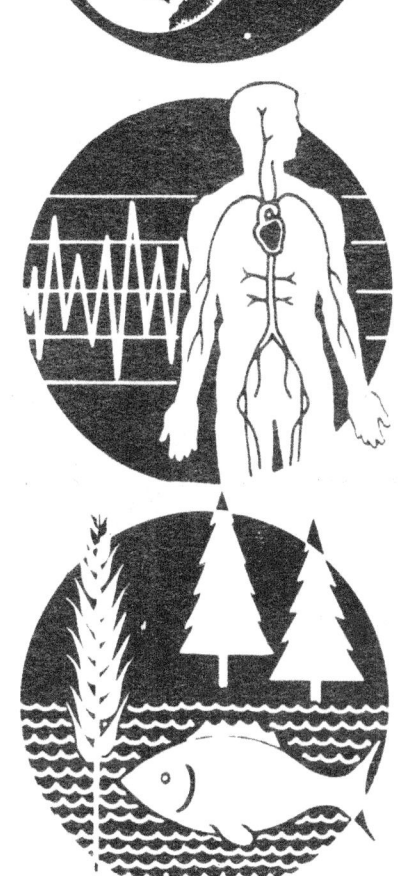

©2012 Periscope Film LLC
All Rights Reserved
ISBN #978-1-937684-84-6
www.PeriscopeFilm.com

NATIONAL AERONAUTICS AND SPACE ADMINISTRATION
OFFICE OF PUBLIC AFFAIRS WASHINGTON, D.C. 20546

TABLE OF CONTENTS

Section	Subject	Page
	Foreword	i
	List of Illustrations	ii-1
	Introduction	iii-1
I	Hardware	I-1
I-1	Orbital Workshop	I-1
I-2	Multiple Docking Adapter	I-15
I-3	Airlock Module	I-22
I-4	Apollo Telescope Mount	I-27
I-5	Payload Shroud	I-32
I-6	Command and Service Module	I-34
I-7	Saturn V	I-37
I-8	Saturn IB	I-40
I-9	Restraints and Mobility Aids	I-43
I-10	Backup Hardware	I-46
II	Systems	II-1
II-1	Life Support	II-1
II-2	Thermal Control	II-37
II-3	Electrical Power	II-40
II-4	Attitude and Pointing Control	II-54
II-5	Communications	II-61
II-6	Crew Equipment	II-80
II-7	Provisions and Stowage	II-89
III	Experiments	III-1
III-1	Life Sciences	III-1
III-2	Solar Physics	III-21
III-3	Earth Observations	III-29
III-4	Astrophysics	III-39
III-5	Materials Science and Manufacturing in Space	III-47
III-6	Engineering and Technology Experiments	III-52
III-7	Student Experiments	III-63
IV	Missions	IV-1
IV-1	SL-1 Launch	IV-2
IV-2	SL-2 Launch	IV-3
IV-3	SL-3 Launch	IV-5
IV-4	SL-4 Launch	IV-6
IV-5	Skylab Rescue Mission and Rescue Kit	IV-6
IV-6	Crew Biographies	IV-8
V	Launch Facilities and Operations	V-1
V-1	Launch Complex 39	V-1
V-2	Skylab Launch Operations	V-7

TABLE OF CONTENTS (Concluded)

Section	Subject	Page
VI	Flight Operations	VI-1
VI-1	Operations – Mission Control Center	VI-1
VI-2	Operations Support – Marshall Space Flight Center	VI-6
VI-3	Recovery	VI-15
VII	Crew Training	VII-1
VIII	Crew Operations	VIII-1
VIII-1	Crew Schedule	VIII-1
VIII-2	Extravehicular Activities	VIII-3
VIII-3	Scheduled Maintenance	VIII-7
IX	Spaceflight Tracking and Data Network	IX-1
X	Manufacturing and Testing Facilities, Transportation	X-1
X-1	Contractor Facilities	X-1
X-2	NASA Facilities	X-6
X-3	Transportation	X-7
XI	Management	XI-1
XII	Contracts	XII-1
	Appendix A – Acronyms and Abbreviations	A-1
	Appendix B – Glossary	B-1
	Index	

FOREWORD

The purpose of this document is to compile the best basic information available on all elements of the Skylab Program.

Subsequent press kits — three of them — will furnish more specific information on how each of the three manned Skylab visits will be conducted, how communications and data will be handled and what can be anticipated in terms of preliminary or early investigative results.

> This reference document should be retained for the duration of the Skylab Program. It will not be updated or reprinted.

LIST OF ILLUSTRATIONS

Figure	Title	Page
I-1.	Skylab	I-3
I-2.	Launch Configuration	I-5
I-3.	OWS Major Assemblies	I-5
I-4.	OWS Habitation Area	I-6
I-5.	OWS Solar Array Wing Assembly	I-6
I-6.	TACS Spheres, Meteoroid Shield	I-7
I-7.	OWS Forward Compartment, Launch Configuration	I-8
I-8.	OWS Forward Compartment, On-Orbit Configuration	I-8
I-9.	OWS Stowage Ring	I-9
I-10.	Scientific Airlock	I-10
I-11.	Wardroom Arrangement	I-10
I-12.	WMC Arrangement	I-11
I-13.	Crewman Color Coding	I-12
I-14.	Sleep Compartment	I-13
I-15.	Experiment Compartment	I-14
I-16.	OWS Maintenance Equipment	I-15
I-17.	OWS Waste Tank	I-16
I-18.	ATM Control and Display Console	I-17
I-19.	M512 Facility	I-17
I-20.	Multiple Docking Adapter	I-18
I-21.	MDA Internal View, +Y	I-19
I-22.	MDA Internal View, -Y	I-20
I-23.	MDA Hatch	I-21
I-24.	MDA Docking Mechanism	I-21
I-25.	Airlock Module	I-23
I-26.	AM Structural Elements	I-24

LIST OF ILLUSTRATIONS (Continued)

Figure	Title	Page
I-27.	AM Interior, -Y+Z	I-24
I-28.	AM Interior, +Y-Z	I-25
I-29.	Fixed Airlock Shroud, Deployment Assembly	I-26
I-30.	Operational ATM	I-27
I-31.	ATM Major Structural Components	I-28
I-32.	ATM Center Work Station	I-29
I-33.	ATM Sun End	I-29
I-34.	ATM Rack Mounted Equipment	I-30
I-35.	ATM Spar, Sun End View	I-31
I-36.	ATM Deployment Assembly	I-33
I-37.	PS Separation Bellows	I-34
I-38.	PS Separation	I-34
I-39.	CSM With LES	I-35
I-40.	Command Module	I-36
I-41.	Launch Escape System	I-37
I-42.	SL-1 Vehicle	I-38
I-43.	SL-2 Vehicle	I-41
I-44.	SL-2 on Launch Pedestal	I-42
I-45.	External Restraints and Mobility Aids	I-44
I-46.	Equipment Restraints	I-45
I-47.	Hand Restraints	I-46
I-48.	Grid Type Restraints	I-47
I-49.	Crewman Restraints	I-48
II-1.	Pressurization and Gas Distribution	II-2
II-2.	Atmosphere Control Diagram	II-3
II-3.	OWS Atmosphere Control Equipment	II-3

LIST OF ILLUSTRATIONS (Continued)

Figure	Title	Page
II-4.	OWS Fan-Muffler	II-4
II-5.	Molecular Sieve Bed, Adsorb Mode	II-4
II-6.	OWS Water Systems	II-5
II-7.	Water Tank	II-7
II-8.	Water Tank Cutaway	II-7
II-9.	Food Table Pedestal	II-8
II-10.	Portable Water Tank Cutaway	II-9
II-11.	OWS Food Table	II-12
II-12.	Food Tray	II-12
II-13.	Non-Refrigerated Food Supply – Daily	II-13
II-14.	Non-Refrigerated Food Storage	II-14
II-15.	Chilled and Frozen Food Storage	II-14
II-16.	Food Preparation Equipment	II-15
II-17.	Food Table Restraints	II-16
II-18.	Fecal/Urine Collector	II-16
II-19.	Foot Restraints at Urine Collector	II-17
II-20.	Fecal Collection Assembly	II-17
II-21.	Fecal Collection	II-17
II-22.	Urine Drawer	II-17
II-23.	Waste Processors	II-18
II-24.	Waste Processing	II-18
II-25.	Trash and Disposal Bags	II-19
II-26.	Trash Disposal Airlock	II-19
II-27.	Sleep Compartment	II-20
II-28.	Sleep Restraint	II-21
II-29.	Handwasher	II-22

LIST OF ILLUSTRATIONS (Continued)

Figure	Title	Page
II-30.	Handwasher Foot Restraints	II-22
II-31.	Shower	II-22
II-32.	Personal Hygiene Kits	II-23
II-33.	Personal Grooming Mirrors	II-23
II-34.	Water Sampler	II-27
II-35.	Iodine Injector	II-27
II-36.	Reagent Container	II-27
II-37.	Iodine Container	II-27
II-38.	Vacuum Cleaner and Accessories	II-30
II-39.	Therapeudic, Dental, Diagnostic and Bandage Kits in Medical Kits Module	II-31
II-40.	Diagnostic Kit	II-32
II-41.	Laboratory Equipment	II-32
II-42.	Therapeudic Kit	II-33
II-43.	Medications Supply Module	II-33
II-44.	Minor Surgery Module	II-34
II-45.	Bandage Kit	II-34
II-46.	Dental Kit	II-35
II-47.	Resupply and Return Container	II-36
II-48.	Medical Accessories Kit	II-36
II-49.	OBS Unsuited Mode	II-37
II-50.	OBS Suited Mode	II-37
II-51.	OWS Refrigeration	II-38
II-52.	Refrigeration System	II-39
II-53.	Heater Locations	II-40
II-54.	Passive Thermal Control	II-41
II-55.	Skylab Power Sources	II-42

LIST OF ILLUSTRATIONS (Continued)

Figure	Title	Page
II-56.	Power Distribution	II-43
II-57.	Cross Section of ATM Solar Cell Module	II-47
II-58.	Cross Section of OWS Solar Cell Module	II-47
II-59.	ATM Solar Wings in Stowed and Partially Deployed Positions	II-48
II-60.	ATM Solar Array Configuration	II-48
II-61.	ATM Electrical Power System	II-49
II-62.	ATM Power Distribution	II-50
II-63.	Skylab Lighting Overview	II-52
II-64.	OWS Lights	II-53
II-65.	CMG Mounting Arrangement	II-56
II-66.	Experiment Pointing Control Subsystem	II-57
II-67.	Control Moment Gyro (CMG)	II-59
II-68.	Star Tracker	II-59
II-69.	Manual Pointing Controller	II-60
II-70.	TACS Components Locations	II-62
II-71.	Intercom Locations	II-64
II-72.	Speaker Intercom Assembly	II-64
II-73.	TV Subsystem	II-66
II-74.	Caution and Warning System	II-69
II-75.	C&W System Elements	II-69
II-76.	C&W Panel 206	II-73
II-77.	C&W Panel 207	II-73
II-78.	C&W Panel 616	II-74
II-79.	Fire Sensor Control Panel	II-74
II-80.	C&W Subsystem	II-75
II-81.	Emergency Subsystem	II-75

LIST OF ILLUSTRATIONS (Continued)

Figure	Title	Page
II-82.	Clothing Module Stowage	II-81
II-83.	Crew Clothing	II-81
II-84.	Crew Underwear	II-82
II-85.	Standard Stowage Locker	II-82
II-86.	Extravehicular Mobility Unit	II-83
II-87.	Skylab ALSA	II-84
II-88.	ODAE Locker	II-85
II-89.	Darts and Board	II-86
II-90.	Exer-Gym	II-86
II-91.	Personal Hygiene Kit	II-86
II-92.	Personal Hygiene Resupply Kit	II-87
II-93.	Skylab Stowage	II-90
II-94.	Stowage Numbering	II-91
II-95.	Command Module Stowage	II-91
II-96.	MDA Stowage	II-92
II-97.	AM Tunnel Stowage	II-93
II-98.	AM STS Stowage	II-94
II-99.	OWS Stowage Ring	II-95
II-100.	OWS Forward Compartment Stowage	II-96
II-101.	OWS Experiment Compartment Stowage	II-97
II-102.	OWS Wardroom Stowage	II-98
II-103.	OWS WMC Stowage	II-99
II-104.	OWS Sleep Compartment Stowage	II-100
III-1.	M071-M073 Experiments	III-6
III-2.	M074 Experiment	III-7
III-3.	M078 Experiment	III-8

LIST OF ILLUSTRATIONS (Continued)

Figure	Title	Page
III-4.	M092 Experiment	III-8
III-5.	M093 Experiment	III-9
III-6.	M131 Experiment	III-12
III-7.	M133 Experiment	III-14
III-8.	M151 Experiment	III-14
III-9.	M171 Experiment	III-15
III-10.	M172 Experiment	III-15
III-11.	Experiment Support System	III-17
III-12.	S020 Experiment	III-22
III-13.	S052 Experiment	III-23
III-14.	S054 Experiment	III-24
III-15.	S055 Experiment	III-25
III-16.	S056 Experiment	III-25
III-17.	S082 Experiment Instrument "A"	III-27
III-18.	S082 Experiment Instrument "B"	III-28
III-19.	H-Alpha Telescope	III-28
III-20.	Advantages of Skylab EREP	III-30
III-21.	EREP Spectral Coverage	III-31
III-22.	EREP Arrangement in MDA	III-32
III-23.	S190A Cameras	III-33
III-24.	S190B Camera	III-33
III-25.	Viewfinder Tracking System (S191)	III-35
III-26.	S191 External Components	III-35
III-27.	S192 Footprint	III-36
III-28.	Experiment S193	III-37
III-29.	Experiment S194	III-37

LIST OF ILLUSTRATIONS (Continued)

Figure	Title	Page
III-30.	Experiment S009	III-40
III-31.	Experiment S019 in SAL	III-41
III-32.	AMS Extension Mechanism	III-41
III-33.	S063 Ozone Photography	III-42
III-34.	Ozone Photography Equipment	III-43
III-35.	S063 Airglow Horizon Photography	III-43
III-36.	Airglow Photography Equipment	III-43
III-37.	S149/T027 Interface	III-44
III-38.	Experiment S150	III-45
III-39.	Experiment S183	III-46
III-40.	Experiment S228	III-47
III-41.	Experiment M518	III-51
III-42.	Experiment D008	III-53
III-43.	Experiment D024	III-54
III-44.	Experiment M415	III-55
III-45.	Experiment M509	III-56
III-46.	Experiment T002 Sextant	III-58
III-47.	Experiment T002 Stadimeter	III-58
III-48.	Experiment T003	III-58
III-49.	Experiment T013	III-59
III-50.	Experiment T020	III-60
III-51.	T027 Sample Array	III-61
III-52.	T027 Photometer System	III-61
III-53.	T027 Photometer Head	III-61
III-54.	MSFC Advisor Steven Hall and Robert L. Staehle	III-66
III-55.	MSFC Advisor Dr. Robert Allen and Todd Meister	III-66

LIST OF ILLUSTRATIONS (Continued)

Figure	Title	Page
III-56.	MSFC Advisor Dr. Robert Allen and Kathy Jackson	III-67
III-57.	MSFC Advisor Dr. Raymond Gause and Judith Miles	III-68
III-58.	MSFC Advisor Loren Gross, Joel Wordekemper and Donald Schlack	III-68
III-59.	MSFC Advisor Charles Cothran and Cheryl Peltz	III-69
III-60.	MSFC Advisor Dr. Raymond Gause and Roger Johnston	III-70
III-61.	Vincent Converse and MSFC Advisor Dr. Robert Head	III-71
III-62.	Terry C. Quist and MSFC Advisor Dr. Raymond Gause	III-71
III-63.	W. Brian Dunlap and MSFC Advisor Dr. Robert Head	III-72
III-64.	Skylab Student Experimenters	III-75
IV-1.	Rescue CM	IV-7
IV-2.	Rescue Response Time	IV-7
V-1.	OWS/S-II Mating	V-8
V-2.	SL-1/IU Mating	V-8
V-3.	AM/MDA in MSOB	V-9
V-4.	CSM/AM/MDA Docking Test	V-10
V-5.	S-IB/S-IVB Mating	V-10
VI-1.	Skylab MOCR Layout	VI-2
VI-2.	MSFC Mission Support Organization	VI-7
VI-3.	HOSC First Floor	VI-10
VI-4.	HOSC Second Floor	VI-11
VI-5.	Mission Support/Voice Distribution	VI-13
VI-6.	Launch Abort Area	VI-16
VI-7.	Merchant Ship Density	VI-17
VI-8.	Skylab Mobile Laboratory	VI-20
VII-1.	Skylab 1-G Trainers	VII-3
VII-2.	Skylab Simulators	VII-4

LIST OF ILLUSTRATIONS (Concluded)

Figure	Title	Page
VIII-1.	Manhour Allocations, SL-2	VIII-1
VIII-2.	Manhour Allocations, SL-3	VIII-3
VIII-3.	Manhour Allocations, SL-4	VIII-3
VIII-4.	EVA Workstations and Lighting	VIII-5
VIII-5.	External Restraints and Mobility Aids	VIII-6
IX-1.	Spaceflight Tracking and Data Network	IX-1
X-1.	OWS at MDAC-WD	X-2
X-2.	MDAC-ED Test Flow	X-3
X-3.	MDA in Vertical Test	X-4
X-4.	ATM Movement	X-6
X-5.	AM/MDA Transporter	X-8
X-6.	USNS Point Barrow	X-9
X-7.	Super Guppy Aircraft	X-10

INTRODUCTION

Three crews, each consisting of three astronauts, will man the experimental Skylab space station in 1973 in what is to be the most ambitious and long-lived manned space research effort yet undertaken. Man will be taking a long, discerning look at the orb that is his home, measuring it a number of ways and deciding how he can improve life thereon. He will also be probing the Sun, his ultimate source of strength, to an unprecedented extent. And he will be looking inwardly at himself, evaluating his own ability to work and live satisfactorily in space over a long period.

Skylab uses, to a large extent, the spacecraft, boosters and knowhow developed in the Gemini and Apollo programs. One Skylab objective was to wring from the Apollo program the greatest possible benefit at the lowest possible cost and to place that legacy in the service of a wide range of scientific and technical disciplines.

Skylab is a three-part program consisting of one 28-day and two 56-day manned visits spanning an 8-month period (Figure 1). One day prior to the launch of the first crew, the unmanned Skylab will be launched and placed in orbit. This constitutes, in effect, a dual launch and marks the start of Skylab flight operations.

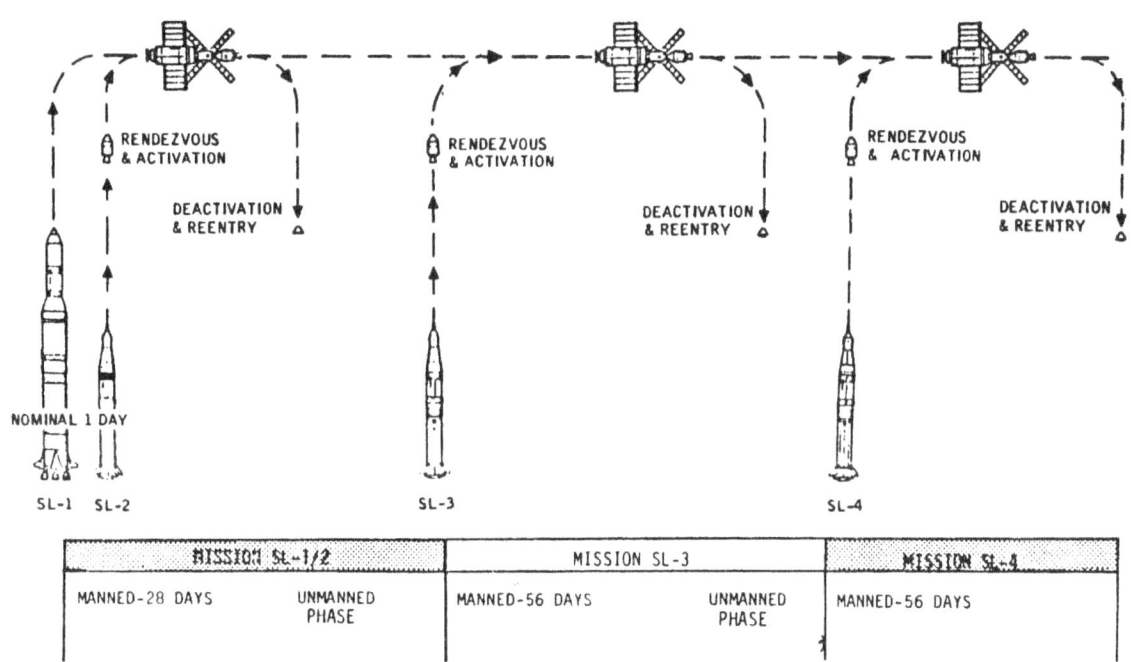

Figure 1 — Skylab Mission Profile

About 270 different scientific and engineering investigations, using 54 different pieces of experimental hardware, will be conducted as part of this first manned orbital research facility. The emphasis will be on the practical human benefits that space can bring to man.

This early space station will be an important step between the exploratory and developmental manned flights of the sixties and early seventies and the more routine space operations seen in the future with the advent of the lower-cost Space Shuttle.

The investigations to be conducted embrace almost every discipline that can take advantage of the unique properties of Skylab's orbital environment — its broad view of Earth, weightlessness and its freedom to measure and monitor the entire solar and celestial electromagnetic spectrum from above the Earth's restrictive atmosphere. Overall, it is expected that more than 1,000 senior scientists and engineers will have a direct function in the analysis and reporting of Skylab data and more than 3,500 astronaut hours will be allocated to the performance of the investigations. That is more than three times the total amount available in all prior U.S. manned Earth orbit missions.

The Skylab experiments fall into these major areas of interest:

Earth's Resources

The NASA Earth Resources Technology Satellite (ERTS) launched last July and the Skylab Earth Resources Experiment Package (EREP) are experimental efforts aimed at demonstrating the feasibility of using space to gather detailed information and apply it to the problems of the environment, diminishing resources and population growth. The data will assist, for example, studies of crop and forest inventories, crop health, mineral and water resources and air and water pollution.

Skylab's EREP will acquire selective data for 146 investigations in 46 task areas. The Skylab instruments generally provide higher spectral and spatial resolution than is available with ERTS. The astronauts will operate these instruments to obtain photographic and infrared images and digital data. About 39,000 Earth resources photographs will be made during the three manned visits as Skylab overflies 75 per cent of the Earth's surface.

Medical

Research in a broad range of medical areas will be an important part of each manned visit to Skylab. Systematic and detailed studies will be made of the effects of prolonged weightlessness on the major body functions. Certain aspects of sleep will be monitored. Blood samples will be collected. Daily food and fluid intake and amounts of uneaten food will all be recorded. Body eliminations also will be recorded, collected and carefully controlled for sampling or return to Earth for post-flight analysis.

The comprehensive Skylab medical studies will provide a base of knowledge so that future decisions on the use of men in space can be made with confidence. Also, it is anticipated that the astronaut's responses to weightlessness will add substantially to the basic understanding of human physiology in man's normal environment and that the technical advances in medical instrumentation developed for Skylab will impact medical diagnosis and treatment in hospitals and doctors' offices.

Studying the Sun

Operating beyond the interference of the Earth's atmosphere, the Skylab solar telescopes will make observations of the Sun not possible from the ground, particularly in the ultraviolet and X-ray regions. The instruments' advanced size and sensitivity make it possible to observe details of form and spectral composition not attainable by the satellite solar observatories orbited to date, and their operation under the direct control of an astronaut allows them to be aimed more quickly and accurately at specific solar details.

Better understanding of solar processes may well lead the way to new means of generating and controlling energy for use on Earth. The Skylab solar investigations address that area of interest and also the problem of explaining mechanisms by which solar events affect the Earth, particularly the streams of high energy particles associated with solar flares that trigger auroras and disrupt ionospheric radio transmissions. Because sunspot activity correlates with temperature and density variations in the Earth's upper atmosphere, it is conceivable that the injection of energy into the atmosphere by solar particles may trigger world-wide weather phenomena.

Skylab missions will provide over 500 hours of astronaut-controlled solar observations, and the five Principal Investigators have agreements for joint investigations with more than 70 astronomers and physicists in the United States and abroad. Concurrent observations from sounding rockets and ballons will provide a wealth of supporting correlative data.

Other Science and Technology

A number of Skylab experiments will be conducted in astrophysics, engineering and technology, materials processing in weightlessness and other areas of interest.

Astrophysics has traditionally sought answers to fundamental questions regarding the nature of the physical universe. The study of matter in previously unknown states and of processes too exotic to occur naturally in our own environment — the generation of thermonuclear energy in the center of stars, for example — have had a great influence on the growth of other physical sciences.

Skylab will provide an opportunity to perform a variety of investigations relating to cosmic rays, measurements of radiation reflected by dust particles in interstellar space, and ultraviolet and X-ray observation of visible and invisible stellar sources. Less stringent weight limitations on Skylab permit larger instruments to be flown, and film and sample return will provide better resolution of data than could be obtained from an unmanned satellite.

The Earth environment determines and limits many materials processes through its ever-present, large gravitational force. The Skylab environment, free of gravitational effect, opens the way to new knowledge of materials properties and processing. It will, for example, permit: small electrostatic and magnetic forces to be applied to materials forming; suppressing of convection and buoyancy in liquids and molten materials; control of voids; and mixing and melting without contamination by containers. Skylab investigations will range from composite structural materials with very specialized physical properties to large, highly perfect crystals with valuable electrical and optical properties, all associated with materials which cannot now be produced on Earth. Examples of the Skylab materials experiments include: behavior of molten metals in free fall and structures formed when they are resolidified; sphere forming to determine sphericity, hardness, surface smoothness and internal microstructure; and growth of single crystals from a solution and by chemical vapor transport. Seventeen investigators will be associated with this research.

Two separate investigations are concerned with techniques for astronaut maneuvering outside future spacecraft.

Skylab History

The concept of using an empty rocket stage as an orbital "house," which is a cornerstone of the Skylab program, is many years old, and efforts to pin down its origin have not met with success. The first documented suggestion concerning the use of a Saturn S-IVB stage in that manner appears to have been in a report published by the Douglas Aircraft Co. in November 1962. By early 1965 NASA was planning the "Apollo Extension Systems" and talking guardedly of a spent stage experiment in which it was envisioned that an S-IVB stage, once emptied of its propellants in orbit, could be outfitted and used as shelter for men in space.

This concept figured prominently in the early planning for the Apollo Applications Program (AAP). The foundation for AAP was the most efficient and economical use of surplus Apollo hardware.

An office to manage this program, which absorbed the Apollo Extension Systems (AES) study program, was formally established by the NASA Office of Manned Space Flight August 6, 1965. For the next year and a half, extensive study was made both within NASA and by NASA-hired firms. The Manned Spacecraft Center and the Marshall Space Flight Center both set up AAP management groups in July 1966.

In early January 1967, the space program had a bright future. There was every optimism that the goal of landing men on the Moon within the decade would be met. The Gemini program had had 10 successful manned flights, providing a bridge between Mercury and the Moon.

Two lunar orbiters had already sent back tremendous — startling — pictures of the lunar terrain, and the first three flights of the Saturn IB had been accomplished and had paved the way for the first manned Apollo launch. The President's budget for space, presented in a press briefing January 26, 1967, called for $454.7 million to bring the AAP from the design definition phase to full development.

At that press briefing, the AAP flights in Earth orbit were announced to begin as soon as the Apollo missions shifted to the Saturn V launch vehicle, making the Saturn IBs available for other use. The Orbital Workshop missions, as the spent stage experiment came to be called, would fly in 1968, concurrently with the preparation for the first manned lunar landing.

According to that early plan, the so-called "wet" workshop would be the Saturn IB second (S-IVB) stage, which would have its emptied hydrogen tank converted in orbit to a habitable volume. A second Saturn IB would launch a modified Apollo Command and Service Module (CSM) to carry the three-man crew and the major part of the mission expendables. Except for some of the major structural elements, all of the furnishings, crew accommodations and experiments were to be carried in the Multiple Docking Adapter during launch and later taken into the workshop by the crew and set up for use.

Following the first mission, that crew would have returned to Earth and, six months later, a second pair of Saturn IB's were to have been launched — one carrying the second crew and the other a solar observatory. The plan was to have those two spacecraft rendezvous and dock with each other then go on to join the Workshop.

Other AAP missions projected at that time included a one year Workshop mission, manned synchronous missions and extended lunar exploration. Follow-on launch vehicle production was planned at four Saturn V's and four Saturn IB's per year, and a six man, land landing, reuseable CSM was to be introduced.

The day following that January 1967 press briefing, plans for going beyond Apollo were swept aside by the Apollo fire and its aftermath. Later that year, Congress cut the AAP funding request by $200 million, beginning a series of budget trimmings that, together with readjustments of the Apollo schedule, led to the present schedule in which Skylab follows Apollo in sequence rather than concurrently. The program was to benefit materially from the delay. Its content and potential came into clearer focus and it became possible to introduce a number of significant improvements. The most important of these, and the key to most of the others, was the shift to the Saturn V and the change from the "wet" to "dry" Workshop in which the S-IVB is not used as an active, fueled stage, hence is dry. When Apollo 11 returned safely from the Moon, it meant that a Saturn V could be diverted from its Apollo assignment, and in August 1969 the decision was made to abandon the "wet" Workshop and shift to "dry," based on studies which had started out as concepts for later AAP missions.

Because the Workshop could now be outfitted on the ground, rather than by the crew in orbit, and because of the larger payload capacity of the Saturn V, additional highly sophisticated equipment and crew provisions could be accommodated. Also, almost all of the expendables for the entire mission could be carried on the initial launch, simplifying the demands on the CSM and reducing the changes needed to adapt it to the Workshop mission, with a corresponding reduction to its cost.

It was also possible to carry the Apollo Telescope Mount (ATM) as an integral part of the Workshop launch, vastly simplifying the operations associated with attaching it to the Workshop in orbit and more than doubling the time available for solar observation. Of equal importance was the ability to operate at a higher inclination for much greater Earth coverage and to add the Earth Resources Experiment Package (EREP).

From the hardware and mission planning viewpoints, the Skylab program has been relatively stable since the July 1969 decision to develop a Saturn V-launched "dry" workshop. Evolution of the cluster of spacecraft had, by then, been completed, and these were (and are) the major elements of the first U.S. space station:

The Orbital Workshop (OWS) provides the crew quarters, the majority of the expendables storage, a major experiment area, structural support for the large solar array, and the cold gas storage and thrusters for the attitude control system.

The Multiple Docking Adapter (MDA) provides the docking port for the Apollo Command/Service Module, houses the control panel for the Apollo Telescope Mount, provides a window for Earth resources viewing, and has other experimental capabilities.

The Apollo Telescope Mount (ATM) houses the solar experiments, the control moment gyros (CMG) installation and a solar array which provides electrical power.

The Airlock Module (AM) has an airlock for extravehicular experiments, main communication/data transmittal links, environmental/thermal system and electrical power control system.

The modified Apollo Command/Service Module (CSM) functions as the manned logistics vehicle for the missions and also provides certain communication functions to the Workshop.

In February 1970 the name Skylab was officially given to the Apollo Applications Program, to recognize and emphasize its role as an exploratory space station in the evolution of man's use of orbital flight for practical benefit.

Mission Sequence

The Skylab mission (Figure 2 shows SL-1/SL-2 missions) will begin with the launch of the unmanned Saturn V vehicle and payload (SL-1) from Launch Complex 39A at the NASA-Kennedy Space Center, Fla. This launch vehicle will consist of the S-IC booster and S-II second stage. Payload elements are the Workshop, ATM, Airlock Module and Multiple Docking Adapter. The Skylab/Saturn V will weigh 2,822,300 kilograms (6,222,000 pounds) at liftoff, and stand 101.7 meters (333.7 feet) tall.

This first launch will place the unmanned cluster in a 435 kilometer (235 nautical-mile) near-circular orbit, with the orbit inclined 50 degrees to the equator.

Payload shroud jettison takes place soon after orbital insertion when the vehicle is pitched through a gravity gradient (nose down) attitude. After insertion and S-II separation, the Saturn second stage is slowed and moved away from the cluster for safety.

ATM deployment is completed while the vehicle is being maneuvered to its normal solar inertial attitude in which the ATM solar cell arrays face the Sun. Solar inertial acquisition having been completed in approximately one-fourth of an orbit, the next sequences are the deployment of the ATM solar arrays, Workshop solar arrays and Workshop meteoroid shield, followed by the control moment gyro spinup.

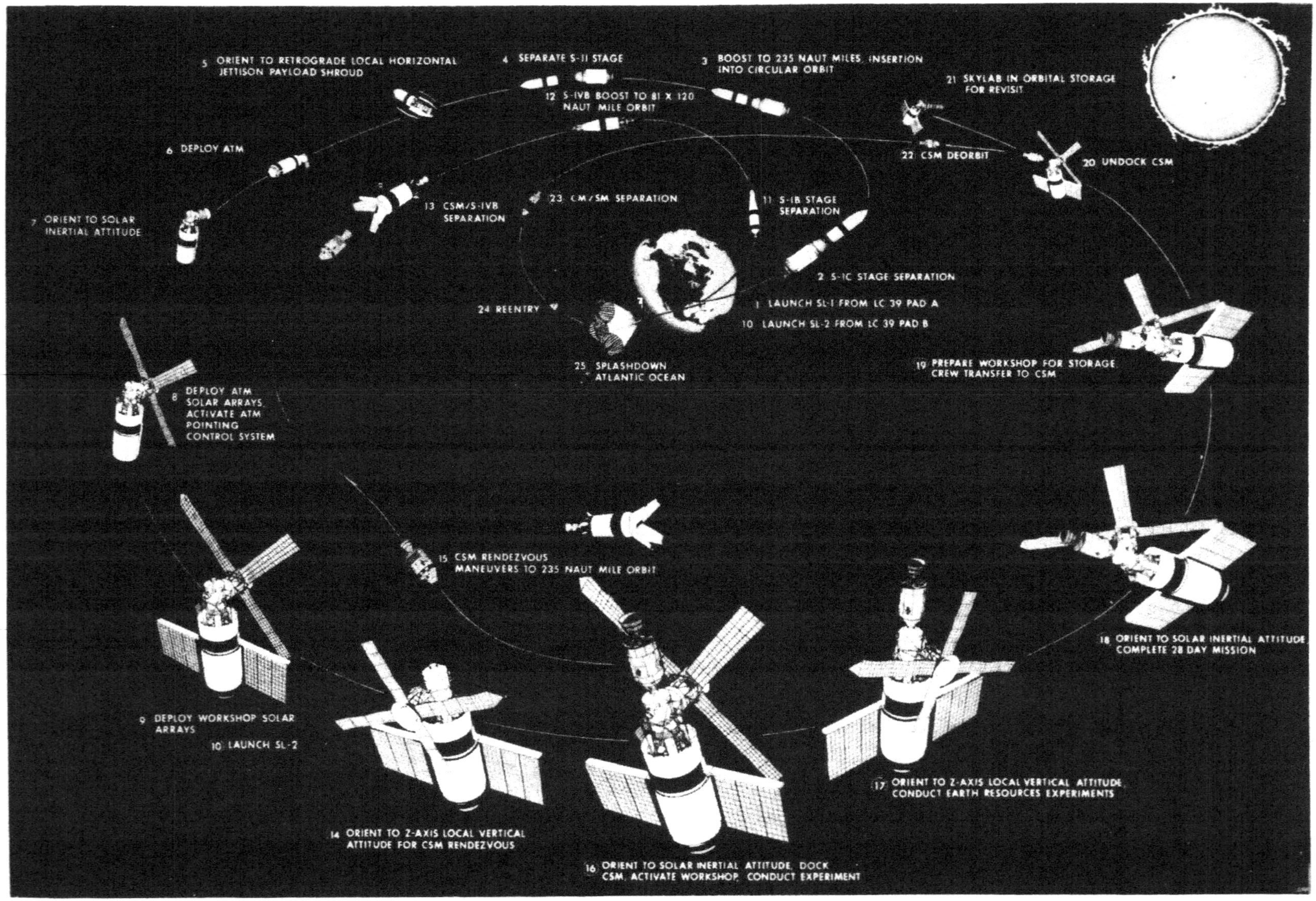

Figure 2 — SL-1/SL-2 Mission Sequence

During this early period, the interior of the Workshop will be pressurized to 3.44 Newtons per square centimeter — five pounds per square inch (psi) — with an oxygen-nitrogen mixture, making it ready to accept docking of the CSM and entry of the flight crew.

Then, about 24 hours after the first launch, the SL-2 launch will take place. The manned CSM will be launched by a Saturn IB vehicle into an intermediate orbit. The crew will rendezvous with the Workshop — using the CSM service propulsion system to attain the required 435 kilometer (235 nautical mile) Skylab orbit — and then dock to the port of the MDA.

The crew will enter and activate the Workshop, which will be their home and work area for the next 28 days. During the remainder of the first manned visit, the experiment program (scientific, biomedical, technological, Earth resources and crew operations) will be conducted. Emphasis will be on the medical research and evaluation of the habitability of the Workshop. ATM experiments will be conducted and satisfactory operation of the hardware will be verified. The Earth Resources Experiment Package also will be operated.

On the 26th day, two of the astronauts will exit the AM, retrieve exposed ATM film, and reload the cameras. Near the end of the 28-day mission, the crew will prepare the Workshop for orbital storage, a dormant period scheduled to last until another crew visits the Skylab.

The CSM separates, reenters and makes a Pacific Ocean landing. Recovery is by normal procedures.

Some 50 days later, SL-3 is launched for a 56-day manned mission using similar hardware and techniques. Greater emphasis will be placed on the solar astronomy and the Earth resources experiments during the second manned mission.

SL-4 will follow the completion of the second manned visit by approximately 30 days and is similar in hardware and mission operations. This final manned visit is scheduled to last 56 days.

A limited rescue capability has been developed for Skylab in the event that a docked CSM becomes unserviceable. A rescue kit to convert Skylab Command Modules to rescue vehicles has been manufactured and qualified. It consists of two additional couches and life support equipment. The rescue mode envisages launch of the next mission spacecraft, with the kit installed, but with only two astronauts on board, thereby providing accommodation for the three astronauts awaiting rescue. Thus SL-3 is potentially the rescue vehicle for SL-2, and SL-4 for SL-3. To provide rescue capability for SL-4, the Skylab back-up spacecraft will be available.

NOTE: The name of the Manned Spacecraft Center was changed to the Lyndon B. Johnson Space Center after the material for this document was prepared for print. Therefore, all references in this document to the center will be to the original name.

SECTION I

HARDWARE

Skylab is a cluster of four major units which will be launched by a two-stage Saturn V rocket into a near-circular Earth orbit with an altitude of 435 kilometers (235 nautical miles) and Command and Service Modules launched by Saturn IB rockets. The orbit will be inclined 50 degrees to the equator.

The Saturn Workshop (SWS) consists of the Orbital Workshop (OWS), Airlock Module (AM), Multiple Docking Adapter (MDA), Apollo Telescope Mount (ATM) and related support structures and thermal and meteoroid shielding. A modified Apollo CSM, in which three astronauts are launched by a smaller Saturn IB rocket, will rendezvous and dock at the MDA. The instrument unit, mounted on the forward end of the OWS, serves the launch vehicle. The entire cluster in orbit, including the CSM, is sometimes called the Orbital Assembly (OA) (Figure I-1).

In orbit, the Skylab cluster will be 36 meters (118.5 feet) long and will weigh 90,607 kilograms (199,750 pounds). The total work space in the OWS, AM, MDA and CSM will be 347 cubic meters (12,398 cubic feet).

Total weight of the payload at liftoff will be about 88,906 kilograms (196,000 pounds). The height of the launch vehicle and payload on the launch pad will be 101.7 meters (333.7 feet). Payload capability for the Saturn V at the proposed Skylab altitude is about 90,720 kilograms (200,000 pounds).

In the launch configuration (Figure I-2), the Skylab elements will be mounted directly above the second stage. An 11,794 kilogram (26,000 pound) Payload Shroud will cover the ATM, MDA and AM during the launch phase. The ATM is held forward of the MDA until the cluster reaches orbit and the shroud is jettisoned. The ATM is then moved by a deployment mechanism 90 degrees to one side. This exposes the docking port on the forward end of the MDA, the port to which the CSM normally will dock. A port on the side of the MDA could be used if necessary.

ORBITAL WORKSHOP

The Orbital Workshop (Figure I-3) uses hardware and techniques developed in the Apollo lunar landing program.

A Saturn IB second (S-IVB) stage was modified and outfitted on the ground as living and working quarters for three astronauts. It contains the majority of the expendables storage, serves as a structural support for a large solar array, and carries the cold gas storage and thrusters for the attitude control system. The stage's liquid hydrogen tank serves as a 292 cubic meter (10,426 cubic foot) space laboratory. The OWS weighs 35,380 kilograms (78,000 pounds).

NASA converted S-IVB-212 as the primary OWS. S-IVB-515 has been outfitted and checked out as the backup hardware. The OWS will have most of its operational equipment in place at launch. Some gear will be stored in the AM and MDA.

The S-IVB converted for Skylab has no engine or propulsive hardware other than the attitude control thrusters. A reusable access hatch replaced an existing manhole in the forward tank dome. (A personnel hatch was also added to the side of the stage to permit workmen and technicians easy access during the checkout and pre-launch phase. This side hatch will be sealed before launch.)

Aluminum open-grid floors and ceilings were installed in the tank to divide it into a two-story "space cabin." An aluminum foil, fire-retardant liner was placed on the inside tank surfaces and a meteoroid shield on the exterior. Two solar arrays are mounted on the outside.

Crew quarters (Figure I-4) are at the aft end of the tank. A ceiling grid separates the quarters from the laboratory area in the forward end. Solid partitions divide the crew quarters into a sleep compartment, wardroom, waste management compartment and an experiment compartment. Lighting fixtures are mounted on the crew quarters ceiling. The waste management compartment is sealed separately with walls and doors to retain odors and loose particles in the weightless environment. Crew quarters contains five radiant heaters. Three radiant heaters are in the forward compartment.

The wardroom has about 9.3 square meters (100 square feet) of area; the waste management compartment has 2.8 square meters (30 square feet) of floor space; the sleep compartment about 6.5 square meters (70 square feet); and the experiment area about 16.7 square meters (180 square feet).

The SWS's thermal control and ventilation system will give the astronauts a habitable environment with a temperature ranging from 15.6 to 32.2°C (60 to 90°F). A two-gas (oxygen and nitrogen) atmosphere will be used with internal pressure kept at 3.45 N/cm² (five psi). Fans will circulate the artificial atmosphere.

Solar arrays on the OWS (Figure I-5) and ATM will provide electrical power for the cluster. The systems are cross-linked for flexibility in handling peak loads and for countering failures. The electrical power distribution system connects OWS areas with power sources in the AM and the solar cell assembly. Light fixtures have individual controls, and portable lights can be used for more illumination as needed.

The meteoroid shields will decrease the probability of hazardous punctures of the OWS. One is a 0.06 centimeter (0.025 inch) aluminum sheet held against the OWS outside surface during launch. Once in orbit, this shield is deployed by swinglinks (powered by torsion bars) and held five inches from the wall. The other shield is a fixed double-wall aluminum alloy cover over the Thruster Attitude Control Subsystem (TACS) cold gas spheres on the aft end of the OWS.

Water and food for Skylab's operational lifetime will be stored inside the OWS, the water in tanks in the forward experiments area and the food in compartments and freezers in that area and in the wardroom.

The wardroom has a window 0.46 meter (18 inches) in diameter in the middle of its wall. The window, double-paned and heated to prevent fogging, will face the sunlit side of the Earth during the mission.

The liquid oxygen tank of the S-IVB stage was converted into a waste container for OWS. An airlock was installed in top of the common bulkhead. The trash disposal section of the tank has 62.5 cubic meters (2,233 cubic feet) of space and the liquid dump area has 7.4 cubic meters (264 cubic feet).

Twenty-three spheres (Figure I-6) containing cold gaseous nitrogen for the TACS and pneumatics are mounted on the aft end of the OWS. These are protected by a thermal shield around the OWS circumference at the aft end and by the aft meteoroid shield. The radiator for life support system (LSS) refrigerators and freezers is mounted aft of the TACS spheres and shield. Two attitude control thrusters of three nozzles each are on the aft end on opposite sides of the OWS.

Two "wings" of solar panels are folded against the OWS on opposite sides for launch. Once in orbit, the arrays are deployed to expose almost 219 square meters (2,355 square feet) of solar cells to the Sun's rays – enough to produce as much as 10,500 watts of power at 55 degrees C (131 degrees F).

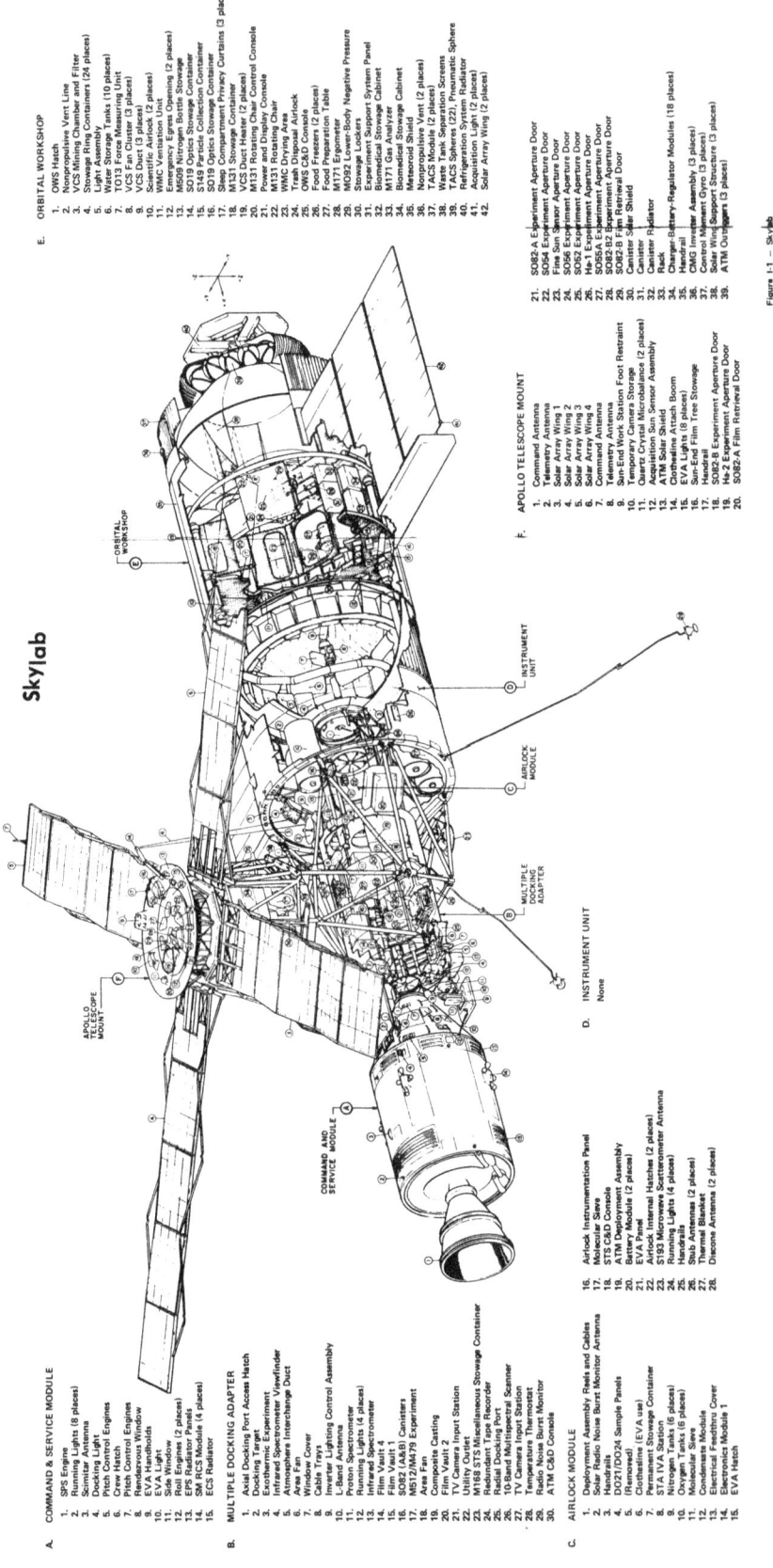

A. COMMAND & SERVICE MODULE
1. SPS Engine
2. Running Lights (8 places)
3. Scimitar Antenna
4. Docking Light
5. Pitch Control Engines
6. Crew Hatch
7. Pitch Control Engines
8. Rendezvous Window
9. EVA Handholds
10. EVA Light
11. Side Window
12. Roll Engines (2 places)
13. EPS Radiator Panels
14. SM RCS Module (4 places)
15. ECS Radiator

B. MULTIPLE DOCKING ADAPTER
1. Axial Docking Port Access Hatch
2. Docking Target
3. Exothermic Experiment
4. Infrared Spectrometer Viewfinder
5. Atmosphere Interchange Duct
6. Area Fan
7. Window Cover
8. Cable Trays
9. Inverter Lighting Control Assembly
10. L-Band Antenna
11. Proton Spectrometer
12. Running Lights (4 places)
13. Infrared Spectrometer
14. Film Vault 4
15. Film Vault 1
16. S082 (A&B) Canisters
17. M512/M479 Experiment
18. Area Fan
19. Composite Casting
20. Film Vault 2
21. TV Camera Input Station
22. Utility Outlet
23. M169 STS Miscellaneous Stowage Container
24. Redundant Tape Recorder
25. Radial Docking Port
26. 10-Band Multispectral Scanner
27. TV Camera Input Station
28. Temperature Thermostat
29. Radio Noise Burst Monitor
30. ATM C&D Console

C. AIRLOCK MODULE
1. Deployment Assembly Reels and Cables
2. Solar Radio Noise Burst Monitor Antenna
3. Handrails
4. DO21/DO24 Sample Panels (Removed)
5. Clothesline (EVA use)
6. Permanent Stowage Container
7. STS/EVA Station
8. S193 Microwave Scatterometer Antenna
9. Handrails
10. Oxygen Tanks (6 places)
11. Molecular Sieve
12. Condensate Module
13. Electrical Feedthru Cover
14. Electrical Feedthru Module 1
15. EVA Hatch
16. Airlock Instrumentation Panel
17. Molecular Sieve
18. STS C&D Console
19. ATM Deployment Assembly
20. Battery Module (2 places)
21. EVA Panel
22. Airlock Internal Hatches (2 places)
23. STS C&D Console
24. Clothesline (EVA use)
25. Handrails
26. Stub Antennas (2 places)
27. Thermal Blanket
28. Discone Antenna (2 places)

D. INSTRUMENT UNIT
None

E. ORBITAL WORKSHOP
1. OWS Hatch
2. Nonpropulsive Vent Line
3. VCS Mixing Chamber and Filter
4. Stowage Ring Containers (24 places)
5. Light Assembly
6. Water Storage Tanks (10 places)
7. T013 Force Measuring Unit
8. VCS Fan Cluster (3 places)
9. VCS Duct (3 places)
10. Scientific Airlock (2 places)
11. WMC Ventilation Unit
12. Emergency Egress Opening (2 places)
13. M509 Nitrogen Bottle Stowage
14. S019 Optics Stowage Container
15. S149 Particle Collection Container
16. S019 Optics Stowage Container
17. Sleep Compartment Privacy Curtains (3 places)
18. M131 Stowage Container
19. VCS Duct Heater (2 places)
20. M131 Rotating Chair Control Console
21. Power and Display Console
22. M131 Rotating Chair
23. WMC Drying Area
24. Trash Disposal Airlock
25. OWS C&D Console
26. Food Freezers (2 places)
27. Food Preparation Table
28. M177 Ergometer
29. M092 Lower-Body Negative Pressure
30. Stowage Lockers
31. Experiment Support System Panel
32. Biomedical Stowage Cabinet
33. M171 Gas Analyzer
34. Biomedical Stowage Cabinet
35. Meteoroid Shield
36. Nonpropulsive Vent (2 places)
37. TACS Module (2 places)
38. Waste Tank Separation Screens
39. TACS Spheres (22), Pneumatic Sphere
40. Refrigeration System Radiator
41. Solar Array Wing (2 places)
42. Solar Array Wing (2 places)

F. APOLLO TELESCOPE MOUNT
1. Command Antenna
2. Telemetry Antenna
3. Solar Array Wing 1
4. Solar Array Wing 2
5. Solar Array Wing 3
6. Solar Array Wing 4
7. Command Antenna
8. Telemetry Antenna
9. Sun-End Work Station Foot Restraint
10. Temporary Camera Storage
11. Quartz Crystal Microbalance (2 places)
12. Acquisition Sun Sensor Assembly
13. ATM Solar Shield
14. Clothesline Attach Boom
15. EVA Lights (8 places)
16. Sun-End Film Tree Stowage
17. Handrail
18. S082-B Experiment Aperture Door
19. He-2 Film Retrieval Door
20. S082-A Film Retrieval Door
21. S082-A Experiment Aperture Door
22. S054 Experiment Aperture Door
23. Fine Sun Sensor Aperture Door
24. S056 Experiment Aperture Door
25. S052 Experiment Aperture Door
26. He-1 Experiment Aperture Door
27. S055A Experiment Aperture Door
28. S082-B Experiment Aperture Door
29. S082-B Film Retrieval Door
30. Canister Solar Shield
31. Canister
32. Canister Radiator
33. Rack
34. Charger-Battery-Regulator Modules (18 places)
35. Handrail
36. CMG Inverter Assembly (3 places)
37. Control Moment Gyro (3 places)
38. Solar Wing Support Structure (3 places)
39. ATM Outrigger (3 places)

Figure 1-1 — Skylab

A. COMMAND & SERVICE MODULE
1. SPS Engine
2. Running Lights (8 places)
3. Scimitar Antenna
4. Docking Light
5. Pitch Control Engines
6. Crew Hatch
7. Pitch Control Engines
8. Rendezvous Window
9. EVA Handholds
10. EVA Light
11. Side Window
12. Roll Engines (2 places)
13. EPS Radiator Panels
14. SM RCS Module (4 places)
15. ECS Radiator

B. MULTIPLE DOCKING ADAPTER
1. Axial Docking Port Access Hatch
2. Docking Target
3. Exothermic Experiment
4. Infrared Spectrometer Viewfinder
5. Atmosphere Interchange Duct
6. Area Fan
7. Window Cover
8. Cable Trays
9. Inverter Lighting Control Assembly
10. L-Band Antenna
11. Proton Spectrometer
12. Running Lights (4 places)
13. Infrared Spectrometer
14. Film Vault 4
15. Film Vault 1
16. SO82 (A&B) Canisters
17. M512/M479 Experiment
18. Area Fan
19. Composite Casting
20. Film Vault 2
21. TV Camera Input Station
22. Utility Outlet
23. M168 STS Miscellaneous Stowage Container
24. Redundant Tape Recorder
25. Radial Docking Port
26. 10-Band Multispectral Scanner
27. TV Camera Input Station
28. Temperature Thermostat
29. Radio Noise Burst Monitor
30. ATM C&D Console

C. AIRLOCK MODULE
1. Deployment Assembly Reels and Cables
2. Solar Radio Noise Burst Monitor Antenna
3. Handrails
4. DO21/DO24 Sample Panels
5. (Removed)
6. Clothesline (EVA use)
7. Permanent Stowage Container
8. STA IVA Station
9. Nitrogen Tanks (6 places)
10. Oxygen Tanks (6 places)
11. Molecular Sieve
12. Condensate Module
13. Electrical Feedthru Cover
14. Electronics Module 1
15. EVA Hatch
16. Airlock Instrumentation Panel
17. Molecular Sieve
18. STS C&D Console
19. ATM Deployment Assembly
20. Battery Module (2 places)
21. EVA Panel
22. Airlock Internal Hatches (2 places)
23. S193 Microwave Scatterometer Antenna
24. Running Lights (4 places)
25. Handrails
26. Stub Antennas (2 places)
27. Thermal Blanket
28. Discone Antenna (2 places)

D.

APOLLO TELESCOPE MOUNT

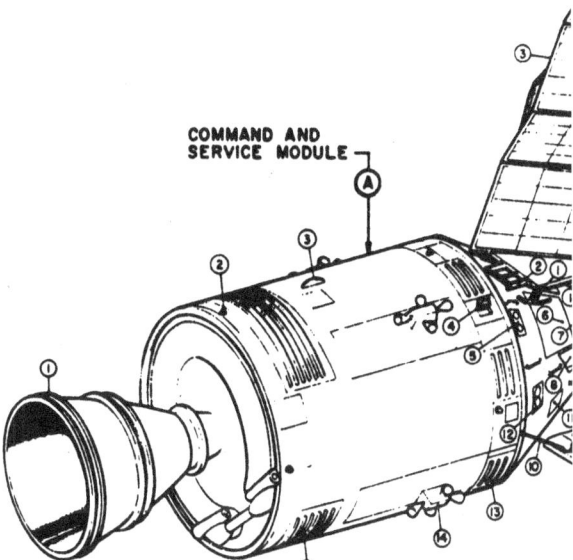

COMMAND AND SERVICE MODULE

Skylab

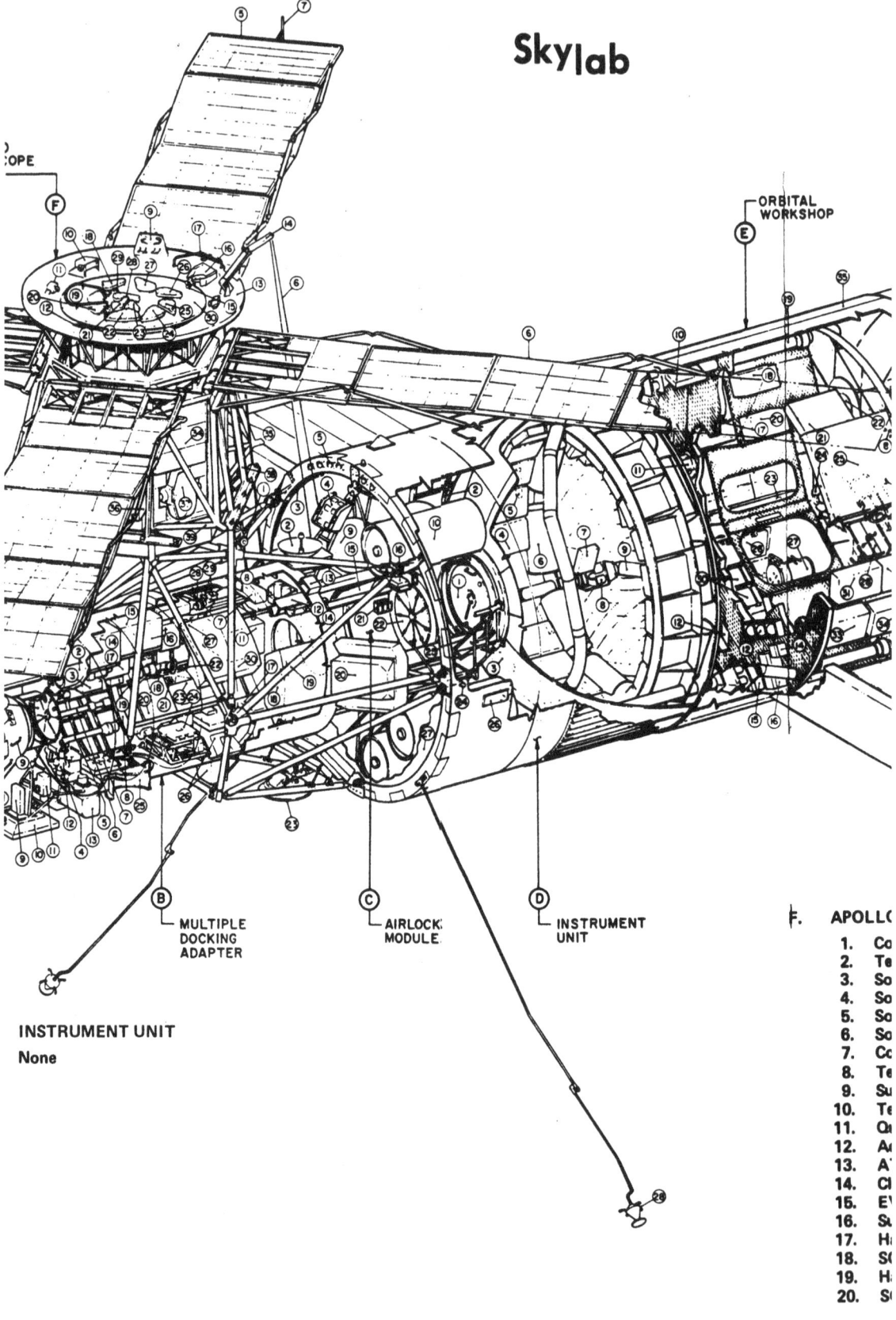

- B — MULTIPLE DOCKING ADAPTER
- C — AIRLOCK MODULE
- D — INSTRUMENT UNIT
- E — ORBITAL WORKSHOP
- F — APOLLO ...

INSTRUMENT UNIT
None

F. APOLLO
1. Co
2. Te
3. So
4. So
5. So
6. So
7. Co
8. Te
9. Su
10. Te
11. Q
12. A
13. A
14. C
15. EV
16. Su
17. H
18. S
19. H
20. S

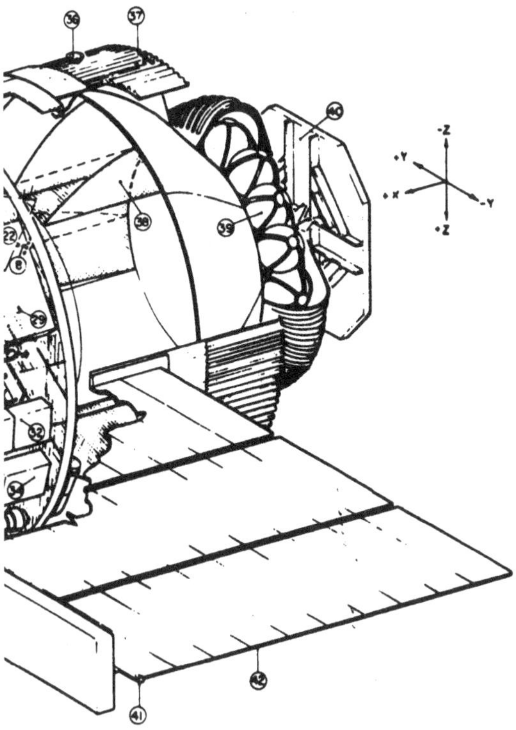

E. ORBITAL WORKSHOP
1. OWS Hatch
2. Nonpropulsive Vent Line
3. VCS Mining Chamber and Filter
4. Stowage Ring Containers (24 places)
5. Light Assembly
6. Water Storage Tanks (10 places)
7. TO13 Force Measuring Unit
8. VCS Fan Cluster (3 places)
9. VCS Duct (3 places)
10. Scientific Airlock (2 places)
11. WMC Ventiation Unit
12. Emergency Egress Opening (2 places)
13. M509 Nitrogen Bottle Stowage
14. SO19 Optics Stowage Container
15. S149 Particle Collection Container
16. SO19 Optics Stowage Container
17. Sleep Compartment Privacy Curtains (3 places)
18. M131 Stowage Container
19. VCS Duct Heater (2 places)
20. M131 Rotating Chair Control Console
21. Power and Display Console
22. M131 Rotating Chair
23. WMC Drying Area
24. Trash Disposal Airlock
25. OWS C&D Console
26. Food Freezers (2 places)
27. Food Preparation Table
28. M171 Ergometer
29. MO92 Lower-Body Negative Pressure
30. Stowage Lockers
31. Experiment Support System Panel
32. Biomedical Stowage Cabinet
33. M171 Gas Analyzer
34. Biomedical Stowage Cabinet
35. Meteoroid Shield
36. Nonpropulsive Vent (2 places)
37. TACS Module (2 places)
38. Waste Tank Separation Screens
39. TACS Spheres (22), Pneumatic Sphere
40. Refrigeration System Radiator
41. Acquisition Light (2 places)
42. Solar Array Wing (2 places)

LO TELESCOPE MOUNT

Command Antenna
Telemetry Antenna
Solar Array Wing 1
Solar Array Wing 2
Solar Array Wing 3
Solar Array Wing 4
Command Antenna
Telemetry Antenna
Sun-End Work Station Foot Restraint
Temporary Camera Storage
Quartz Crystal Microbalance (2 places)
Acquisition Sun Sensor Assembly
ATM Solar Shield
Clothesline Attach Boom
EVA Lights (8 places)
Sun-End Film Tree Stowage
Handrail
SO82-B Experiment Aperture Door
Ha-2 Experiment Aperture Door
SO82-A Film Retrieval Door

21. SO82-A Experiment Aperture Door
22. SO54 Experiment Aperture Door
23. Fine Sun Sensor Aperture Door
24. SO56 Experiment Aperture Door
25. SO52 Experiment Aperture Door
26. Ha-1 Experiment Aperture Door
27. SO55A Experiment Aperture Door
28. SO82-B2 Experiment Aperture Door
29. SO82-B Film Retrieval Door
30. Canister Solar Shield
31. Canister
32. Canister Radiator
33. Rack
34. Charger-Battery-Regulator Modules (18 places)
35. Handrail
36. CMG Inverter Assembly (3 places)
37. Control Moment Gyro (3 places)
38. Solar Wing Support Structure (3 places)
39. ATM Outriggers (3 places)

Figure I-1 — Skylab

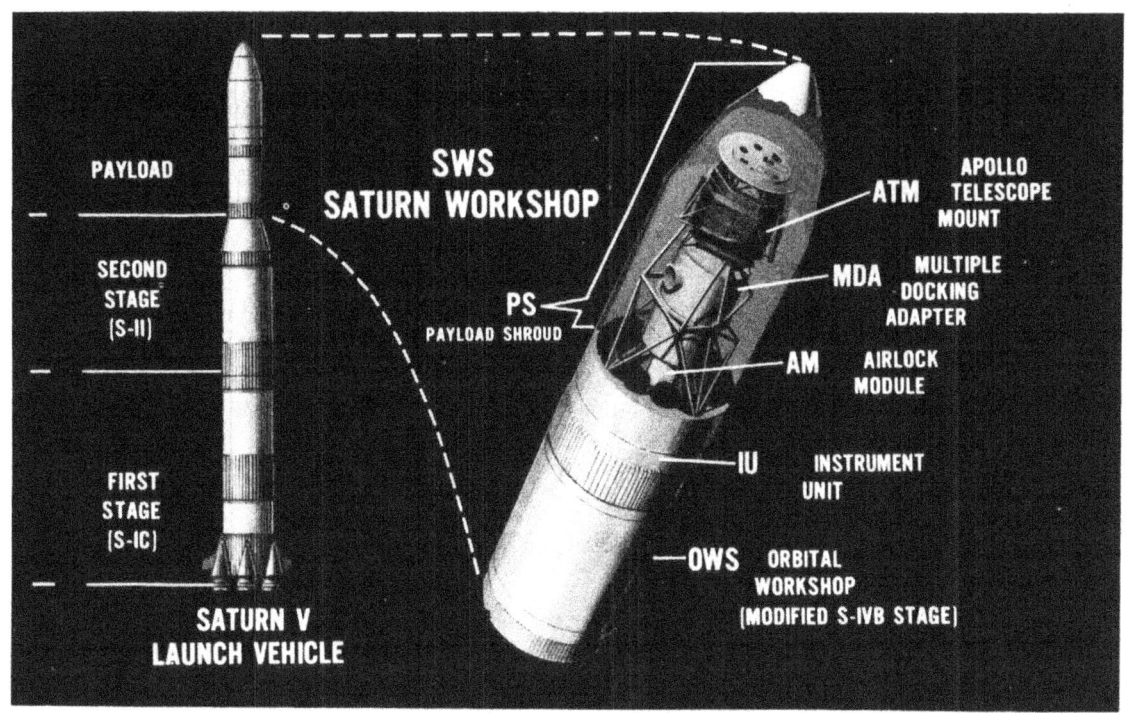

Figure I-2 — Launch Configuration

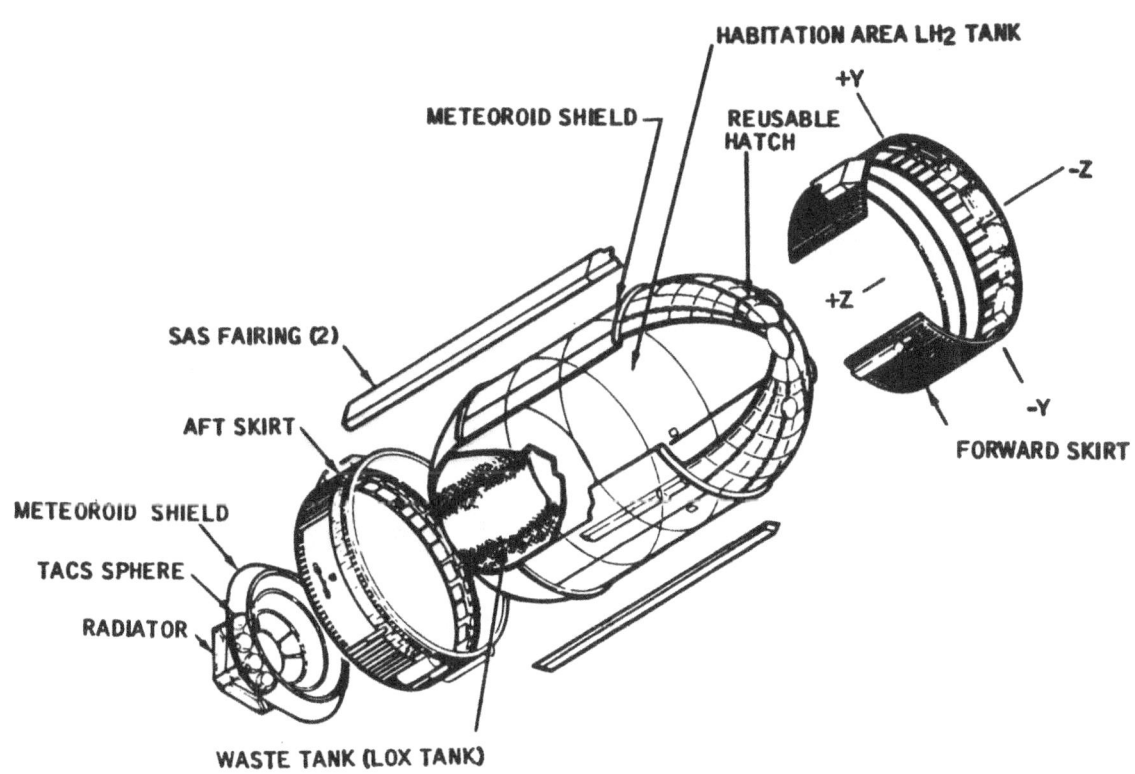

Figure I-3 — OWS Major Assemblies

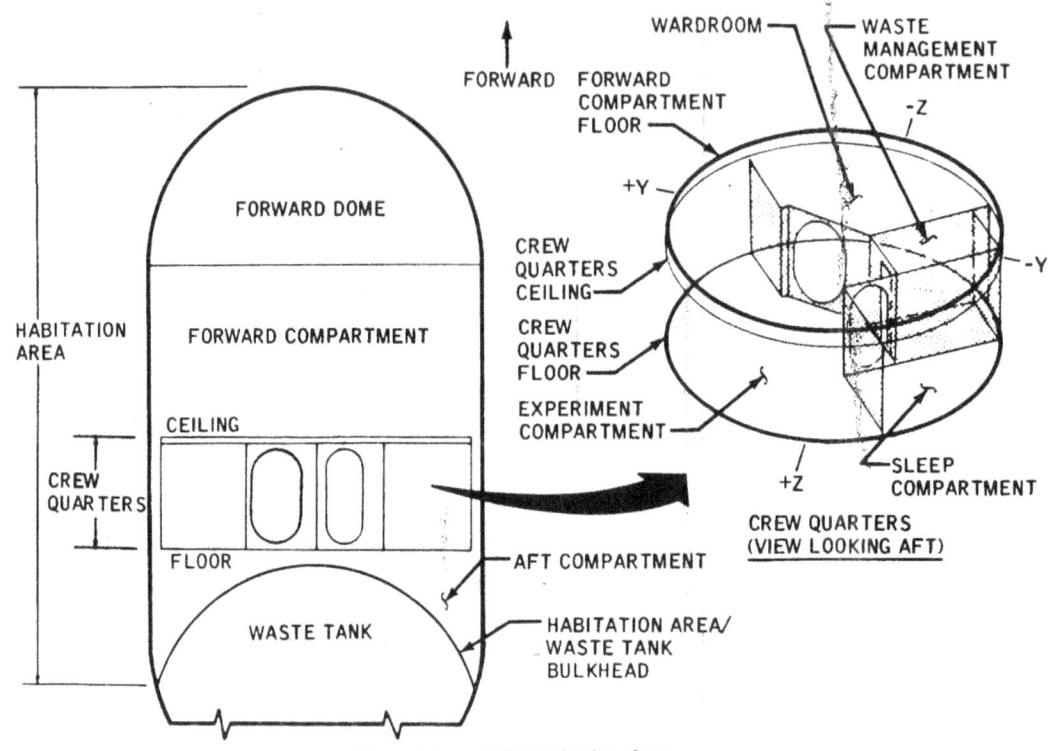

Figure 1-4 — OWS Habitation Area

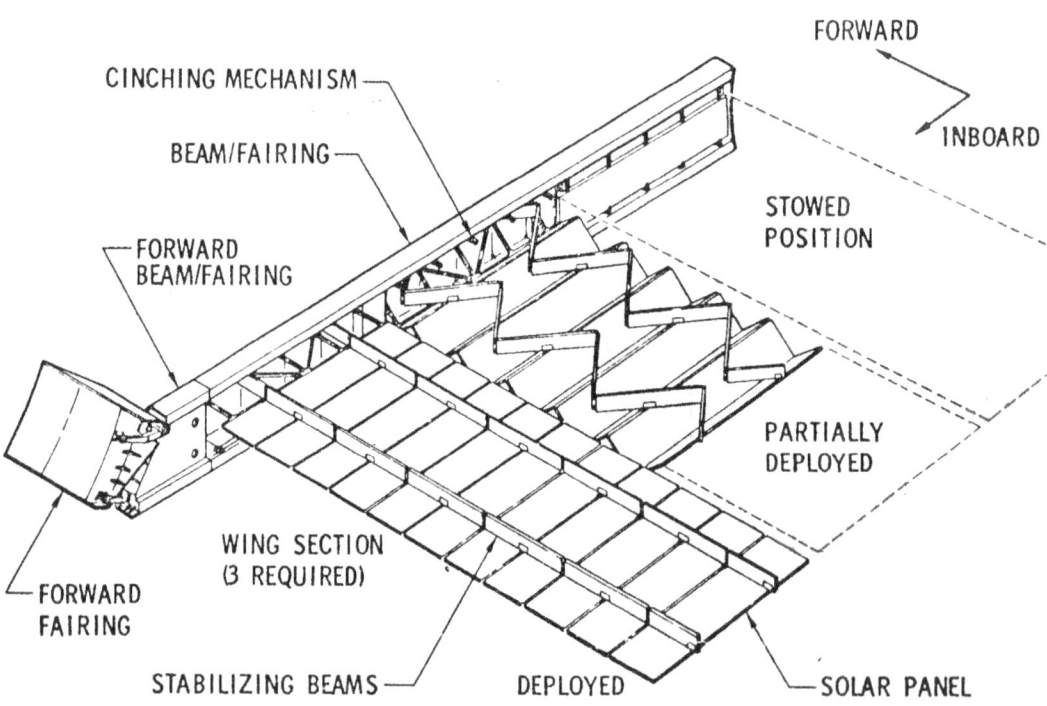

Figure 1-5 — OWS Solar Array Wing Assembly

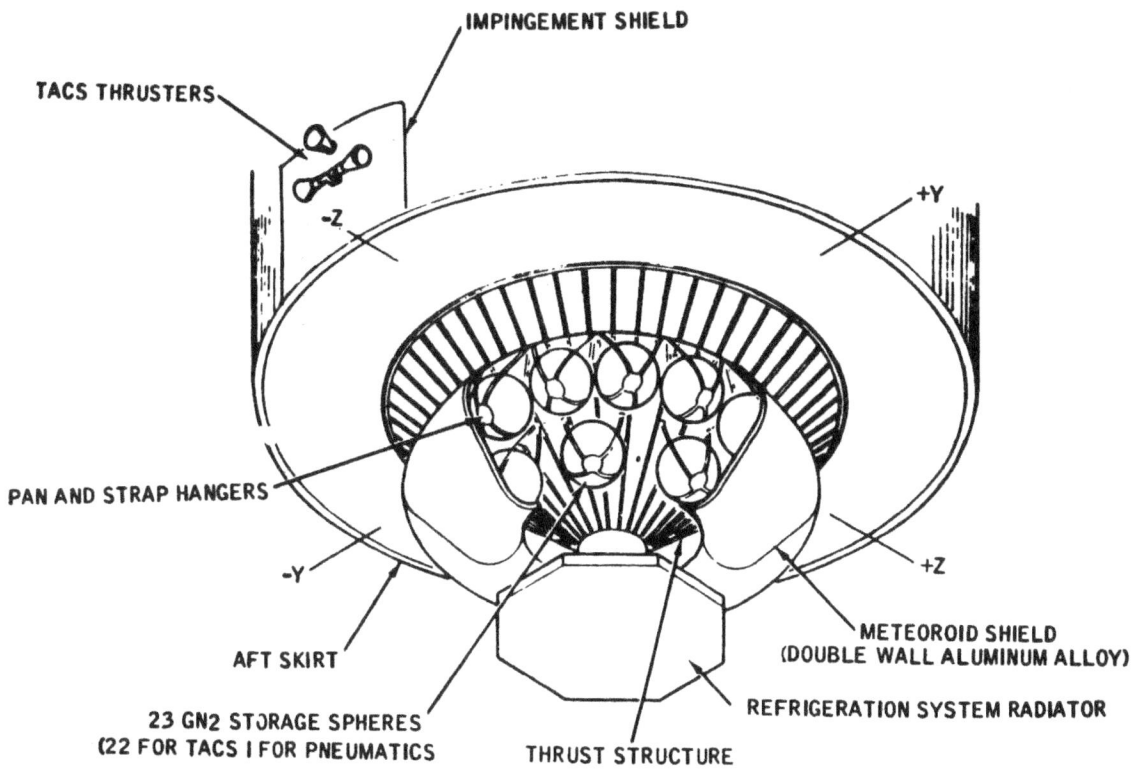

Figure I-6 — TACS Spheres, Meteoroid Shield

Forward Compartment

The forward compartment (Figure I-7 and I-8) occupies the greater part of the habitation area. It is separated from the crew quarters by an eight inch beam structure with an aluminum grid on each side. The forward compartment is divided into three main sections for ready identification of equipment locations, the experiments area, the stowage ring and the dome.

The main items in the dome section include the entry hatch and the ventilation control system mixing chamber and ducts. Other items include light fixtures, a fire extinguisher, electrical cables, handrails, an intercom box, power and instrumentation feed-through provisions, vent sealing devices and power outlets.

The stowage ring (Figure I-9) is at the point where the cylindrical forward experiments area joins the dome. Mounted on the ring are ten water tanks, each having a useable capacity of about 272 kilograms (600 pounds) of water. A portable water tank containing from 11.3 to 12 kilograms (25 to 26.5 pounds) of water is also available in the forward compartment.

Also on the stowage ring are 25 lockers containing supplies needed throughout the OWS – bundles of urine bags, portable lights and electrical cables, hoses, umbilicals, pressure suits, tape recorder, charcoal filters, fans, lamps and intercom boxes. (See section on "Provisions and Stowage" for details on locker numbering and arrangement and lists of items stored in each.)

The forward experiments area contains food lockers and freezers, two Scientific Airlocks, and various items of equipment for performing a number of experiments. Major items include the ultraviolet panorama experiment, the body mass measurement device, contamination measurement equipment, photographic equipment, astronaut maneuvering equipment, EVA suits, film vault and scientific instruments.

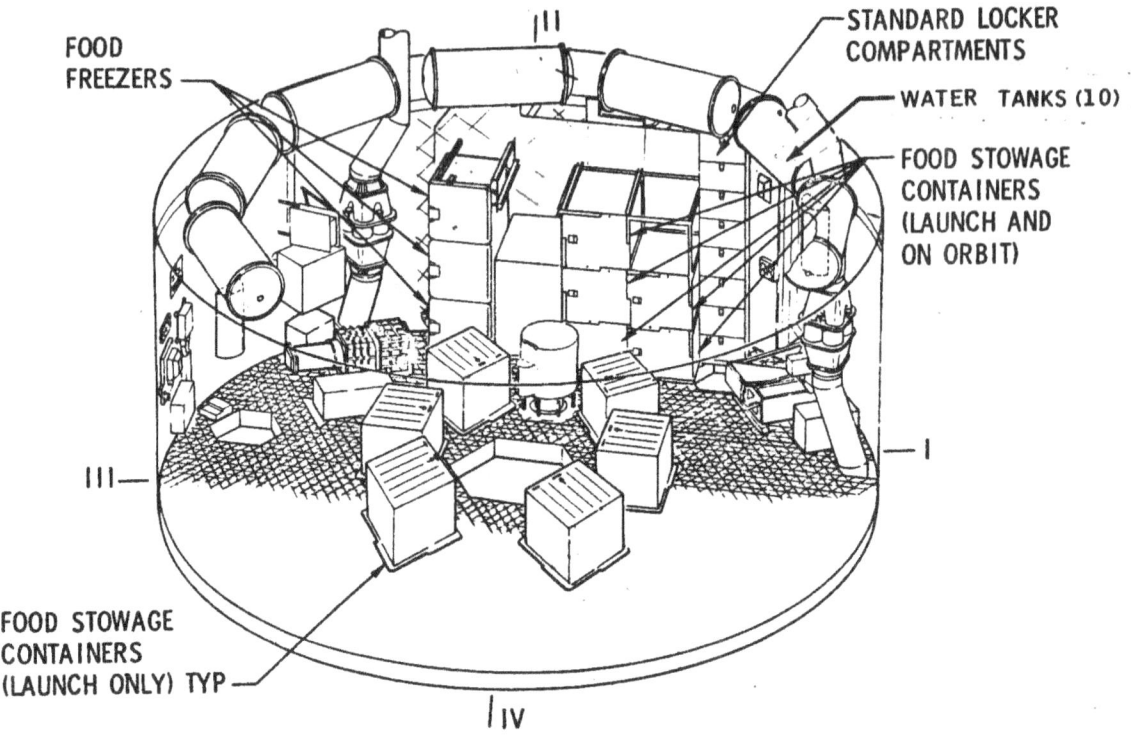

Figure I-7 — OWS Forward Compartment, Launch Configuration

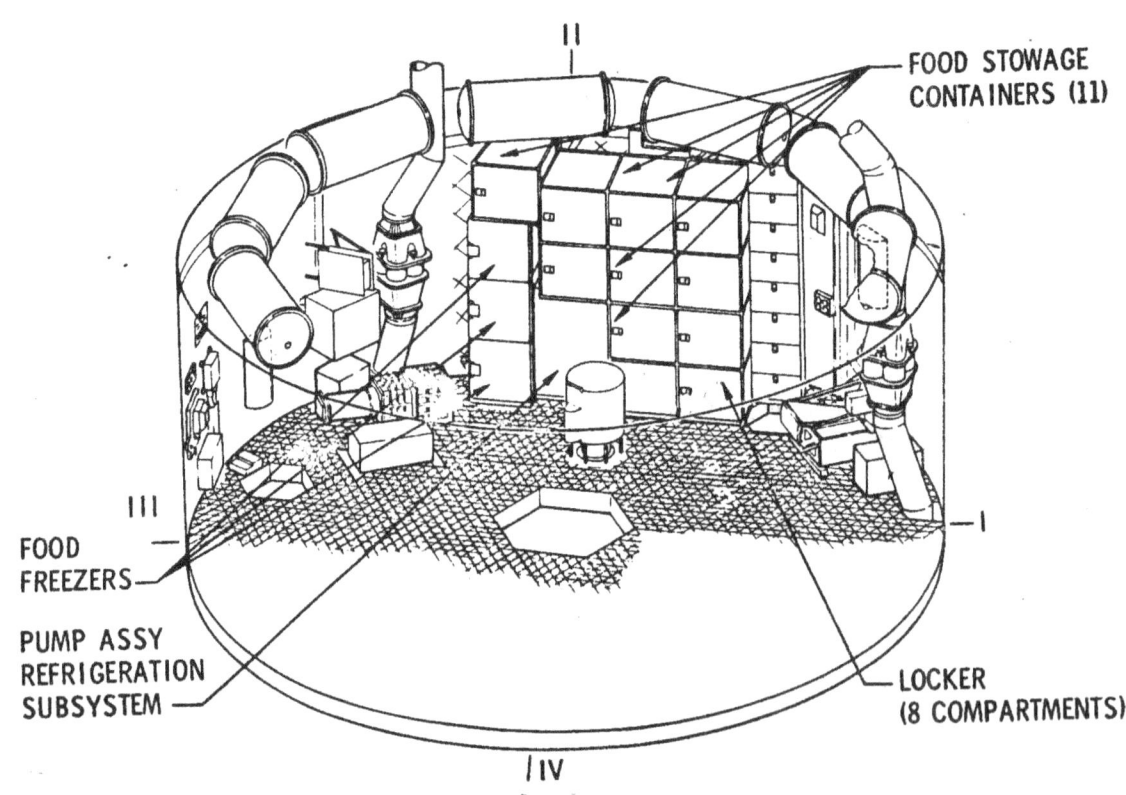

Figure I-8 — OWS Forward Compartment, On-Orbit Configuration

Figure I-9 — OWS Stowage Ring

Two Scientific Airlocks (SAL) are located in the forward compartment (Figure I-10). They are bolted directly on a flange on the OWS wall over a hole 0.21 meter (8.25 inches) square in the tank wall. These SALs provide a method of deploying experiments through the wall of the OWS with depressurization.

Control and display units in the dome area include: OWS hatch and pressure equalization valve, intercom box, utility power outlets, solenoid vent port, pneumatic vent port and TV input station.

Control units in the forward area other than the dome include: utility power outlets; TV input station; two intercom boxes; two fire sensor control panels; experiment recorder control panel; water system pressure panel; and experiments T020, T027, T013, S019, S020, M509 and M172.

Water purification equipment is mounted on the wall of the experiments section. It consists of iodine containers, waste sample container, water samplers, reagent container, iodine injector, a color comparator and control devices.

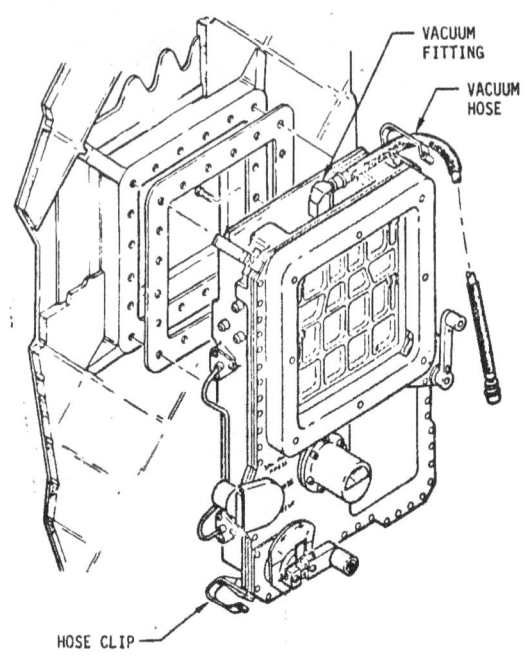

Figure I-10 — Scientific Airlock

Crew Compartment

The crew compartment is separated into four basic areas: a wardroom, waste management compartment, sleep compartment, and an experiment compartment.

The <u>Wardroom</u> (Figure I-11) with about 9.3 square meters (100 square feet) of floor area occupies most of the space in the -Z area of the crew compartment. It provides facilities for food preparation and serving, earth observations, and crew relaxation. (See Figure I-1 for spacecraft coordinates.)

The Wardroom has a window slightly left of the -Y axis. It is double-paned and heated to prevent fogging. An intercom station is located above and to the left of the window, and the flight data file is in a compartment below the window. The room has four general illumination light fixtures and emergency egress openings in the floor and ceiling.

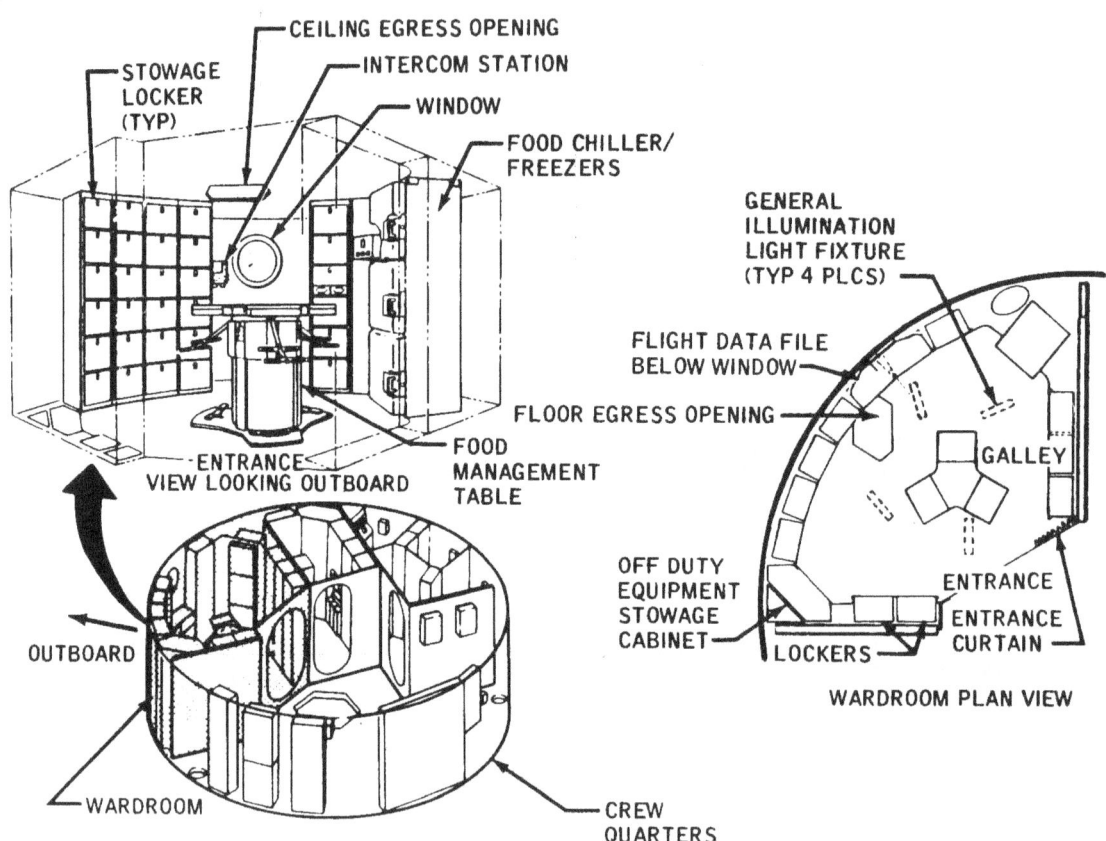

Figure I-11 — Wardroom Arrangement

The Wardroom has 58 stowage compartments (lockers), a food chiller and two food freezers. In these containers are food, tissues and wipes, medical kits, off-duty equipment, clothing modules, towels, flight data files, trash bags and scientific equipment. (See section on "Provisions and Stowage" for details.) The food management table is located near the center of the room.

The astronauts have three storage lockers available for temporary storage of cans of food, one compartment for storage of snacks and beverages, and a six-well empty food can disposal unit. Also in the wardroom is the compartment designed specifically for storing food trays during launch and the tray lids during eating periods.

Off-duty equipment is stowed in a corner cabinet. Mounted on the door are a tape player and two racks holding 48 tape cassettes. Items in the cabinet include headsets, microphones, batteries, playing cards, 36 paperback books, a dartboard and darts, balls, exercising equipment and binoculars.

Control panels in the wardroom include: control panel for water dump and window heater; control panels for S063 and M074 experiments; intercom box; food preparation table; three food tray outlets; a power outlet; and the water dump valve.

The Waste Management Compartment (WMC) (Figure I-12) is a rectangular room with 2.8 square meters (30 square feet) of floor space between the Wardroom and the Sleep Compartment. It contains fecal and urine collection equipment, waste processing and urine management facilities, personal hygiene facilities, crewman restraint provisions and privacy and contamination control accommodations. Waste management and personal hygiene equipment and supplies are stored in the WMC. (See section on "Provisions and Stowage" for details.)

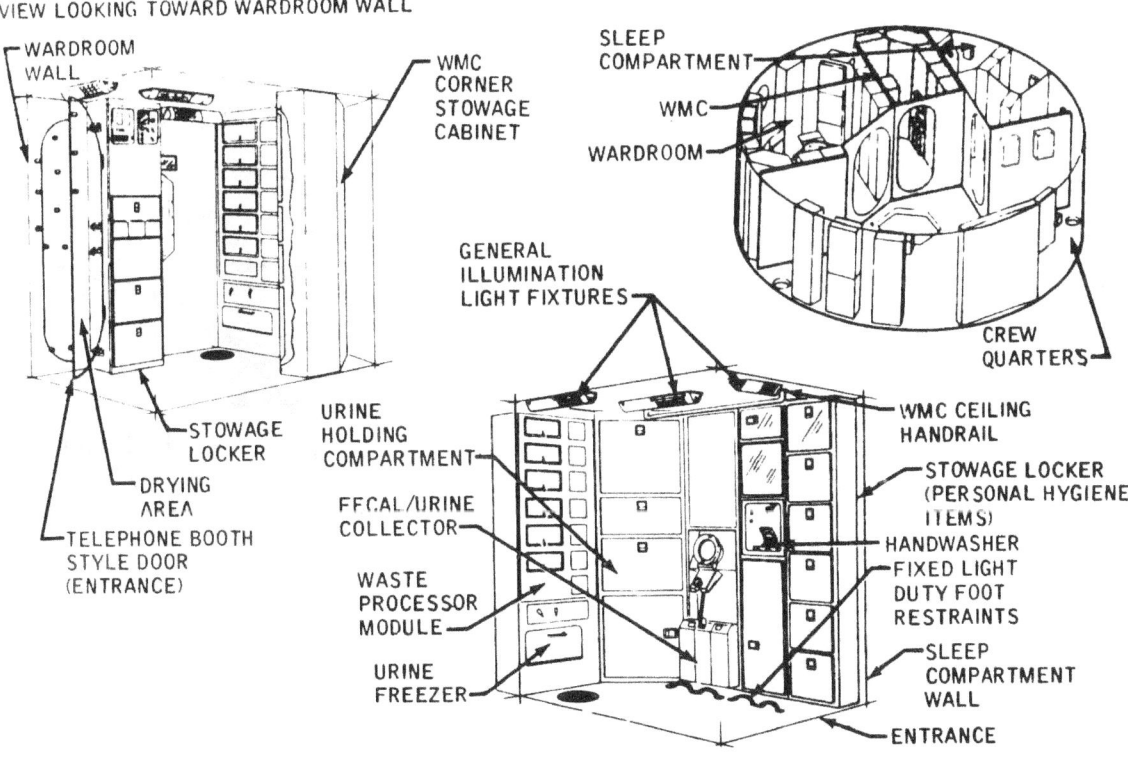

Figure I-12 — WMC Arrangement

The WMC has three general illumination light fixtures and a breakaway door for emergency egress. It has 15 stowage compartments, a urine holding compartment, a fecal/urine collector (toilet), a waste processor module, a urine freezer, a handwasher, foot restraints and overhead handrails, sample return containers, a vacuum cleaner and four mirrors, one fixed on the wall next to the wardroom, an articulating mirror mounted below the handwasher, and one each on the inside of the doors to compartments above and to the right of the handwasher.

Three individual personal hygiene modules (IPHM) are stored in the WMC, one for each astronaut. A typical hygiene kit contains shaving and dental equipment and supplies, soap, emollient, swabs, hair groom brush and cream, nail clippers, deodorant and expectorant collectors. Each kit occupies a separate locker. In the same row of lockers is one with washcloths and towels. Kits and towel/washcloth dispensers are color coded with a "Snoopy" emblem (Figure I-13) — red background for the commander, white for the scientist pilot and blue for the pilot. Towels and washcloths for the astronauts are also color coded.

On the wall opposite the handwasher are other containers, one containing a supply of soap, tissues and disinfectant pads. To the left of this row of lockers is a towel/washcloth drying panel.

The fecal/urine collector is mounted on the WMC wall. The collector, analogous to a toilet seat, is mounted in such a position that the weightless user appears to be sitting on the wall facing the floor.

Waste processing and urine management facilities include a waste processing chamber, urine holding compartment and urine freezer. Three insulated urine sample return containers are stowed in the OWS forward compartment until transferred into the CM for return to Earth. Specimen containers are bag assemblies that will be strapped to CM lockers for the trip to Earth.

Also located in the WMC are a vacuum cleaner with three different nozzles and a handwashing unit which has magnetic soap-holders, (note: bars of soap containing steel bars are held in place by the magnetic holders), a pushbutton water dispenser, a washcloth squeezer, a waste water bag and a handrail.

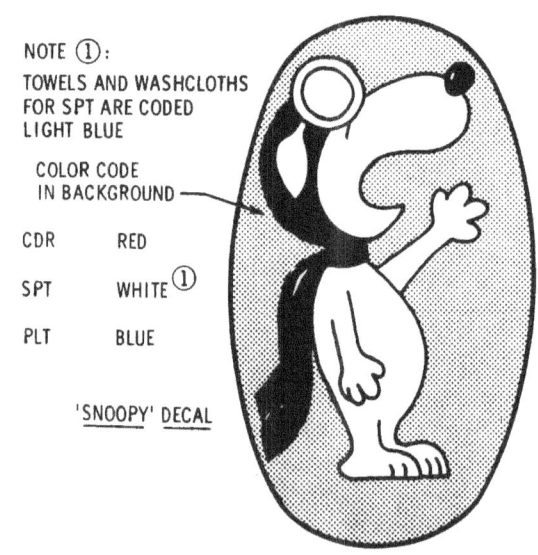

Figure I-13 — Crewman Color Coding

Controls in the WMC include: WMC control panel; intercom box switches and displays; power outlets; circuit breaker panel for the waste processor; M074 experiment panel; fecal/urine collector controls; and water dump valve.

The Sleep Compartment (Figure I-14) is roughly triangular and is subdivided into three individual private rooms, each with supplies and equipment needed by an astronaut. The compartment provides noise abatement and light baffling provisions, sleep restraints, personal and mission equipment and supplies, individually controlled lighting and emergency egress provisions.

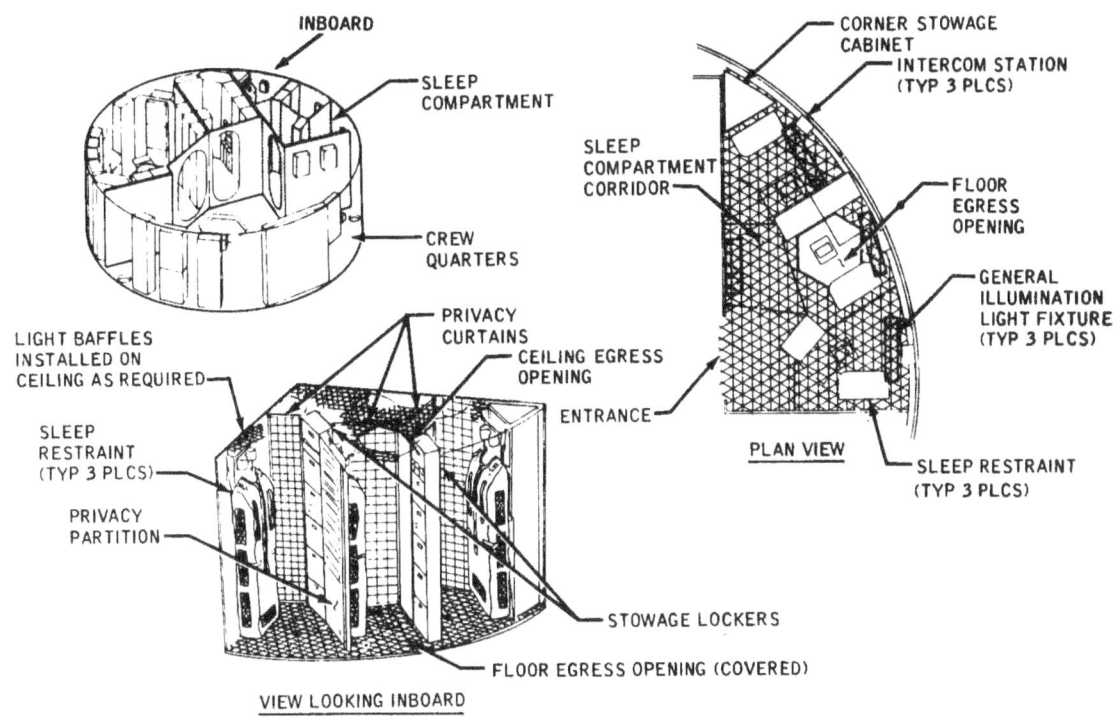

Figure I-14 — Sleep Compartment

Each of the areas in the Sleep Compartment holds a sleep restraint, a sleeping-bag type arrangement into which the astronaut encloses himself to keep him from floating about the area while sleeping. The sleep restraints "hang" from the ceiling with the bases attached to the grid floor.

The Sleep Compartment has light baffles on the ceiling and overhead privacy curtains. The areas are separated by hard walls and fabric doors. A total of 22 lockers are in the sleep compartment. The top locker in each area has a personal hygiene mirror mounted on the inside of the door. (See section on "Provisions and Stowage" for details.)

Facilities are provided for the performance of experiment M133, sleep monitoring. Control panels in this section are the three intercom boxes.

The Experiment Compartment (Figure I-15) occupies approximately half of the crew living and working section (crew compartment). This section is the experiments area and OWS control center. It has trash disposal facilities and mission and experiment equipment stowage facilities. It is lighted by 14 general illumination fixtures which can be supplemented by portable lights as needed.

The compartment has two fire extinguishers, one mounted on the wall separating the compartment from the sleep compartment and the other on the wardroom partition. Emergency egress openings are also provided. A remote Caution and Warning display panel is located in the area.

Major items of equipment in the experiments compartment include the trash disposal airlock, a rotating litter chair, lower body negative pressure device, an ergometer, metabolic analyzer and the experiment support system (ESS).

Six stowage lockers containing trash bags and other equipment are in the experiments area. Also in this section are the remote control WMC and Wardroom lighting panel, portable restraints and experiments compartment.

I-13

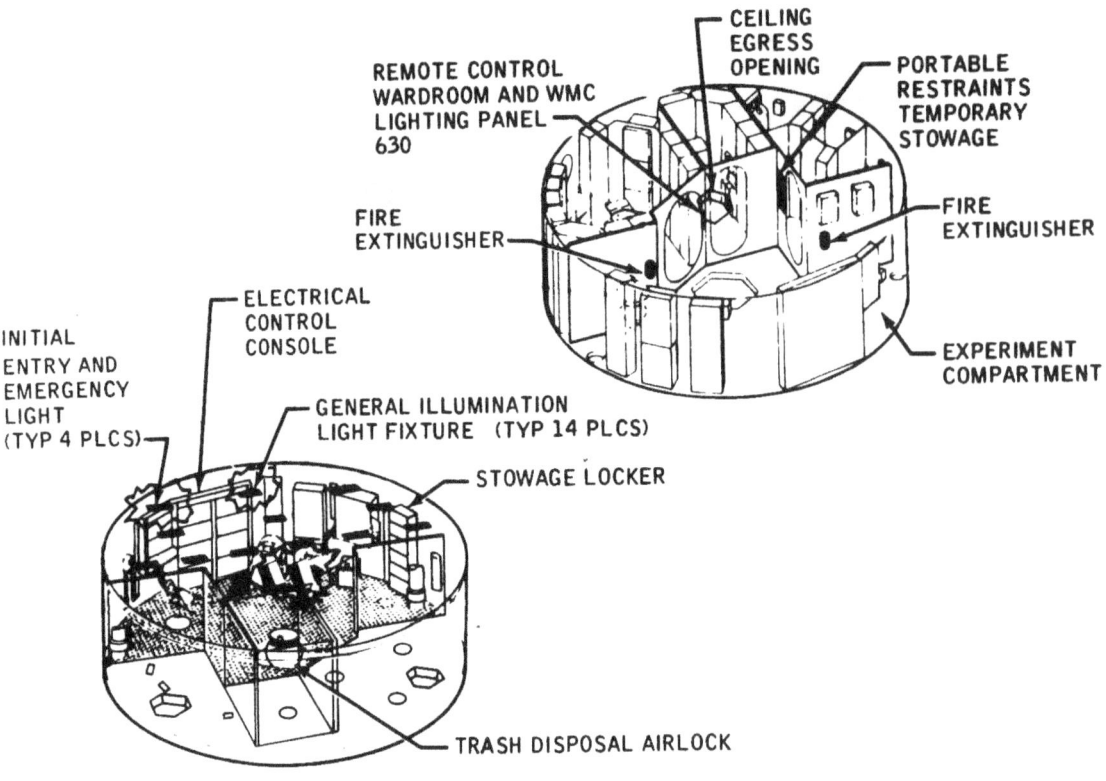

Figure I-15 – Experiment Compartment

Facilities are provided to perform the following experiments: M071, Mineral Balance; M092, Lower Body Negative Pressure; M093, Vectorcardiogram; M131, Human Vestibular Function; M151, Time and Motion Study; M171, Metabolic Activity; and T003, Inflight Aerosol Analysis. The rotating chair will be used in M131, the ergometer in M093, M151 and M171, and the LBNP device for M092.

A replaceable Van Allen Belt dosimeter is located on the experiment compartment wall to permit ground monitoring of the radiation dosage rate to which the crewmen are exposed. (A spare unit is stowed in the OWS forward dome).

The experiments section has 30 control panels, six of which are directly related to experiments. Seven circuit breaker panels are for: refrigeration; high power and utility outlets; TV; lighting, experiments, caution and warning; habitability support system; thermal control system; electrical power system; and instrumentation system. Two panels are intercom boxes, five are fire sensor control panels, three are power outlets and one is the solar flare alert panel.

Astronauts will have tool kits, repair kits, a tool caddy, and spare parts for orbital maintenance (Figure I-16).

The experiment compartment contains the tool kits, repair kits, tool caddies and selected spare parts for orbital maintenance. Many of the tools contained therein are duplicated within the cluster in strategic locations to facilitate maintenance and Skylab initial activation and deactivation tasks. Additionally, spare tools are carried in the event of inadvertent damage or loss.

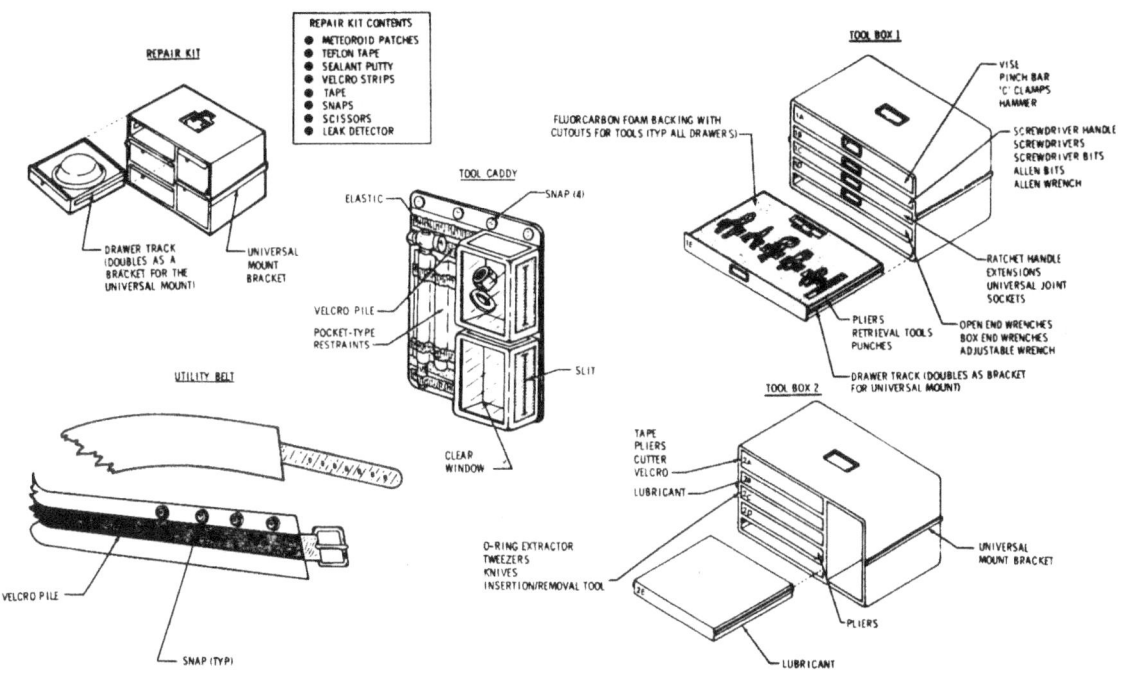

Figure I-16 — OWS Maintenance Equipment

Waste Tank

Engineers created a "trash dump" to be carried on Skylab because there is no such depository in space. The S-IVB liquid oxygen tank has been modified to serve as a storage container for solid trash and a dumping facility for waste liquids. It has a total volume of 80 cubic meters (2,826 cubic feet).

The waste tank (Figure I-17) is divided into compartments by screen enclosures, the largest of which is for trash disposal. This volume of 62.5 cubic meters (2,233 cubic feet) is used to contain solid waste and bagged liquid.

Uncontained waste liquids enter the liquid dump compartment at three points. This compartment has a volume of 7.4 cubic meters (264 cubic feet). The unconfined liquid rapidly evaporates or solidifies and then sublimates so that it can be vented overboard as a gas. The tank is vented to space through two non-propulsive vents which are enclosed by screens (about 10 micron) to minimize the particulates vented overboard. This same type screen material is used to form a compartment around the liquid waste dump probes.

MULTIPLE DOCKING ADAPTER

The Multiple Docking Adapter, or MDA, provides a permanent interface with the Airlock Module and a docking interface with the Command and Service Modules (CSM). The MDA permits the transfer of personnel, equipment, power and electrical signals between the docked module, the AM and the Workshop.

The MDA general configuration consists of a forward conical bulkhead and a cylindrical structure 5.2 meters (17.3 feet) long and 3 meters (10 feet) in diameter. It weighs 6,260 kilograms (13,800 pounds) and contains about 32 cubic meters (1,140 cubic feet) of volume. The MDA has a primary axial docking port at the forward end and a backup or rescue port at the +Z axis.

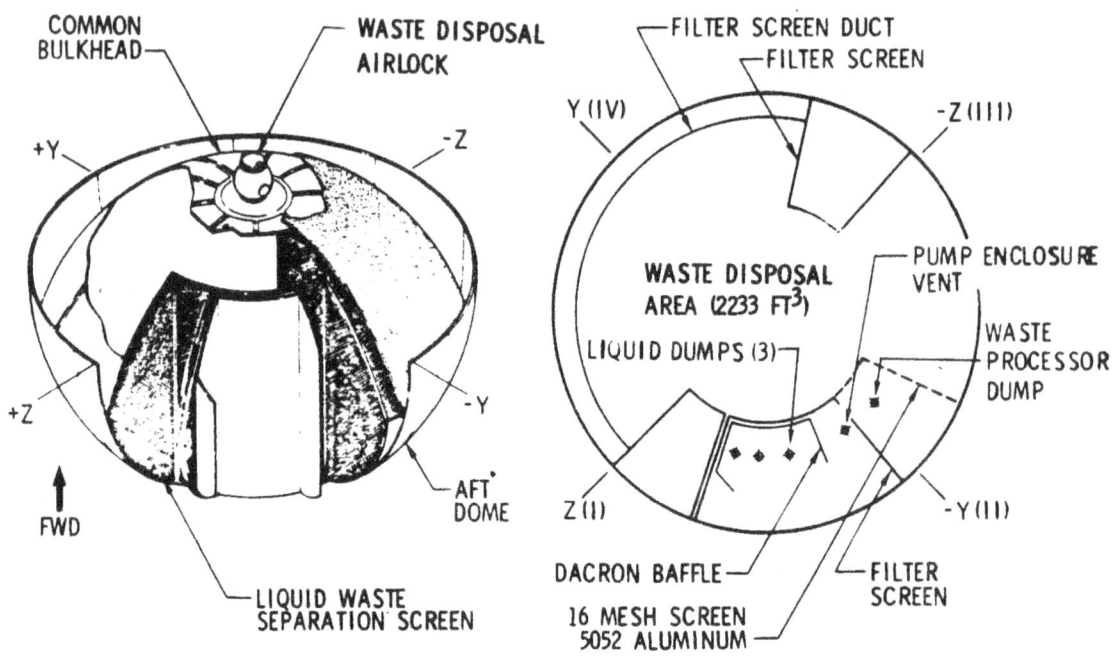

Figure I-17 — OWS Waste Tank

In orbit the MDA functions as a major experiment control center for solar observations; metals and materials processing; and the Earth Resources experiments. Once on station, the MDA will be positioned with the +Z axis pointing earthward to provide an orientation for the Earth Resources Experiment Package (EREP) or positioned with the -Z axis pointed toward the Sun for solar observations. The EREP equipment includes a multispectral photographic facility, an infrared spectrometer, a 13-band multispectral scanner, an L-Band radiometer and a microwave radiometer/scatterometer altimeter.

The Apollo Telescope Mount will be operated by the astronauts from the Control and Display (C&D) console in the MDA. (Figure I-18). There the crew will actively control the telescopes that will greatly increase understanding of solar physics. Repeated coronal observations will provide the ability to develop a three-dimensional structure of coronal forms by obtaining high resolution photographs of the corona. Three other telescopes will be operated from the C&D console to gather scientific data on photographic film mounted in cameras and stored in the film vaults occupying space in the MDA.

The "Materials Processing in Space" facility (Figure I-19) mounted in the module provides a furnace or vacuum work chamber with an electron beam generating device. A smaller yet important experiment controlled and operated from the MDA is the Nuclear Emulsion (S009).

The external surface of the MDA (Figure I-20) is covered by a radiator/meteoroid shield structure that stands 7.6 centimeters (3 inches) from the pressure skin. Items mounted on the outside of the MDA include: MDA vacuum vent; MDA/AM electrical feed-through cover assembly; ATM feed-through power distributor assembly (4); instrumentation signal conditioner (all covered by raised portion of the meteoroid shield); S192 10-band multispectral scanner; S191 infrared spectrometer; cover for the S190 experiment window; S194 L-band antenna; proton spectrometer; inverter lighting control assembly; and orientation lights. Associated docking lights and docking targets are mounted on the outside.

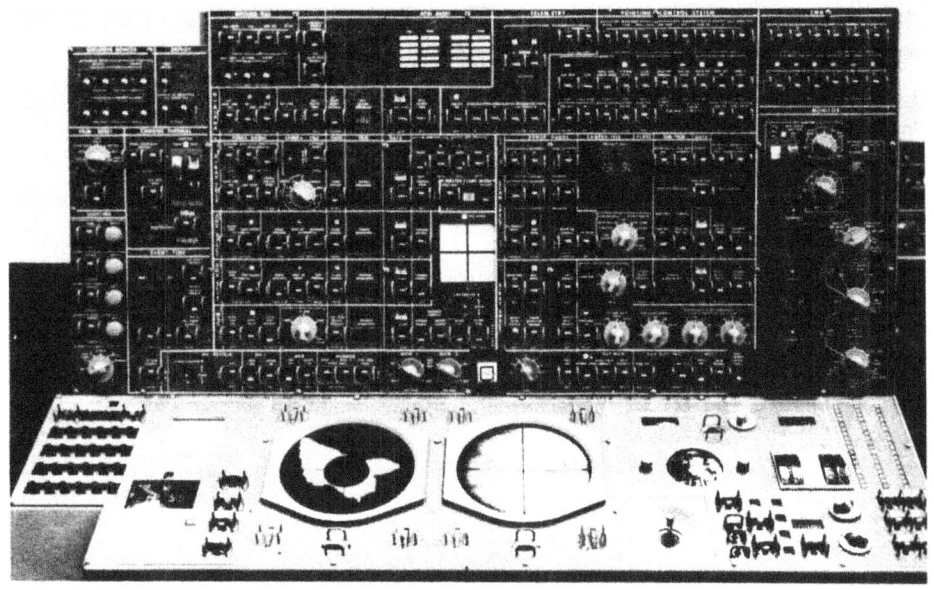

Figure I-18 — ATM Control and Display Console

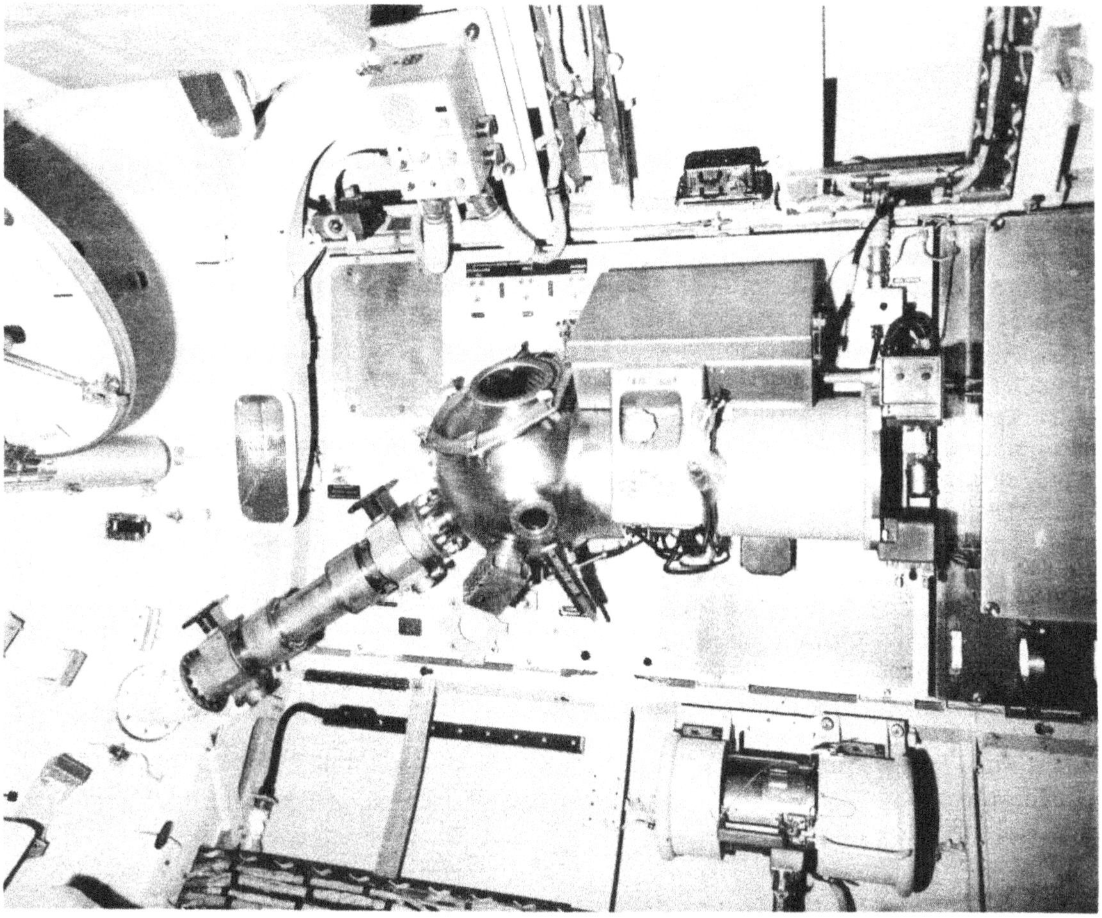

Figure I-19 — M512 Facility

I-17

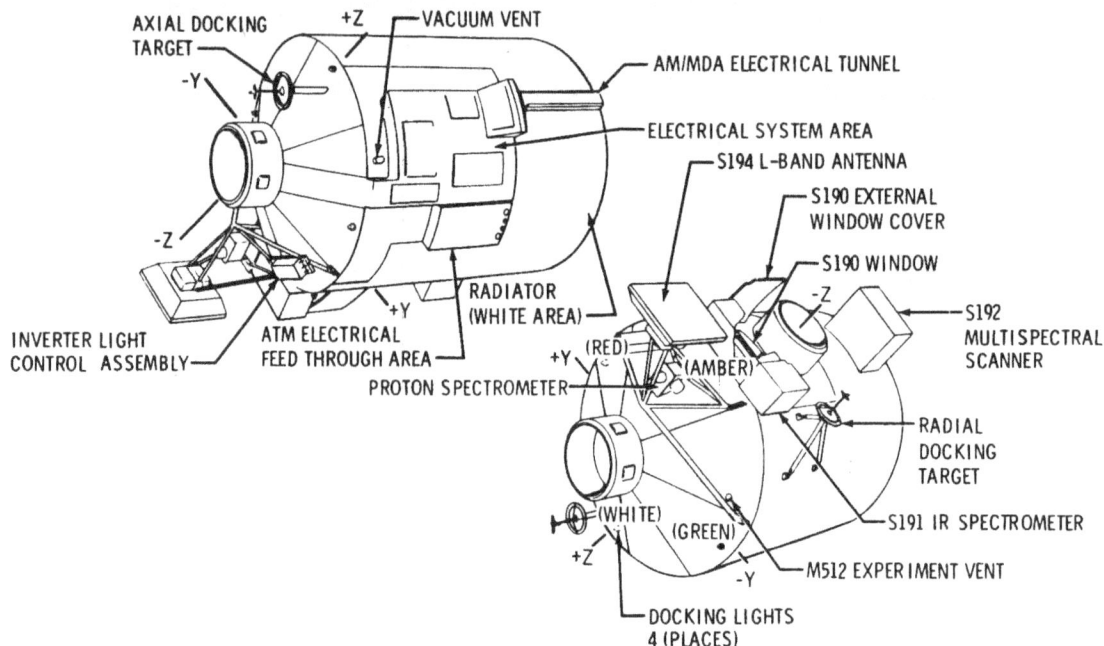

Figure I-20 – Multiple Docking Adapter

The MDA is structurally cantilevered from the AM and can withstand launch loads, docking loads, in-orbit stabilization maneuvering loads and internal pressure loads.

Although surrounded by ATM and Deployment Assembly members, the MDA does not interface structurally with either the ATM or the deployment assembly.

The major hardware and experiments contained in the MDA are identified in Figures I-21 and I-22.

The MDA access hatch (Figure I-23) is at the aft end of the axial docking tunnel. A like hatch is in a similar position in the radial port. On the CSM side (outside) of the hatch are a pressure equalization valve and handle, a pressure gauge, a hatch opening handle, six latch connecting rods; and contingency tool kit for opening an inoperative hatch; like items, except for the connecting rods, are on the inside of the hatch.

The docking system is a means of connecting and disconnecting the CSM/MDA during a mission and of providing for intravehicular transfer between the two modules.

Docking is achieved by maneuvering the CSM close enough to the MDA so that the extended probe engages the drogue on the MDA (Figure I-24). When the probe engages the drogue through the capture latches, the probe retract system is activated to pull the MDA and CSM together. Upon retraction, the MDA tunnel ring activates the 12 automatic latches and effects a pressure seal between the modules through the two seals in the CM docking ring face.

Following hard mate of the two modules and when the MDA pressure hatch is opened, providing a passageway between the two modules, the probe and drogue are transferred for stowage in the MDA.

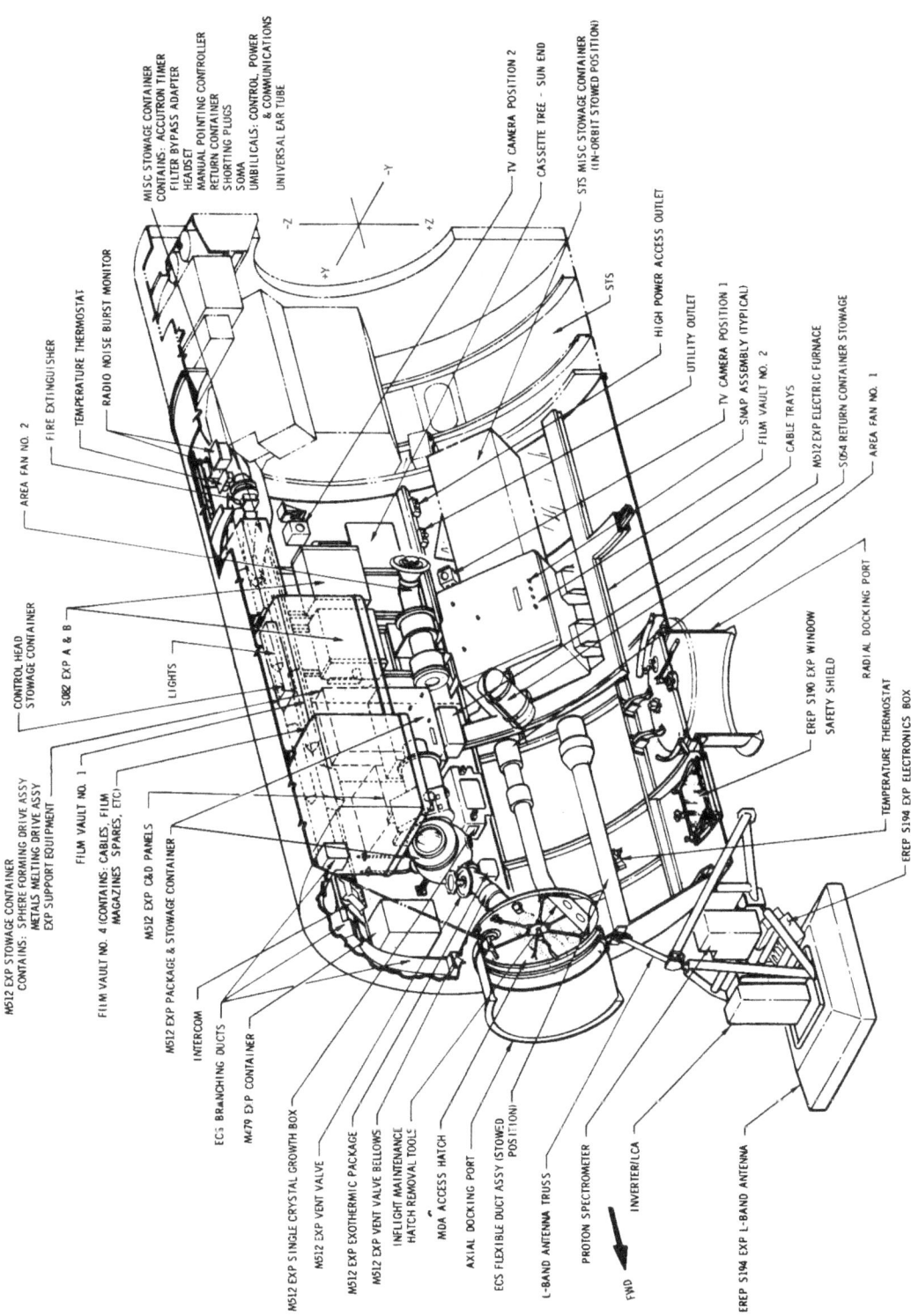

Figure I-21 — MDA Internal View, +Y

I-19

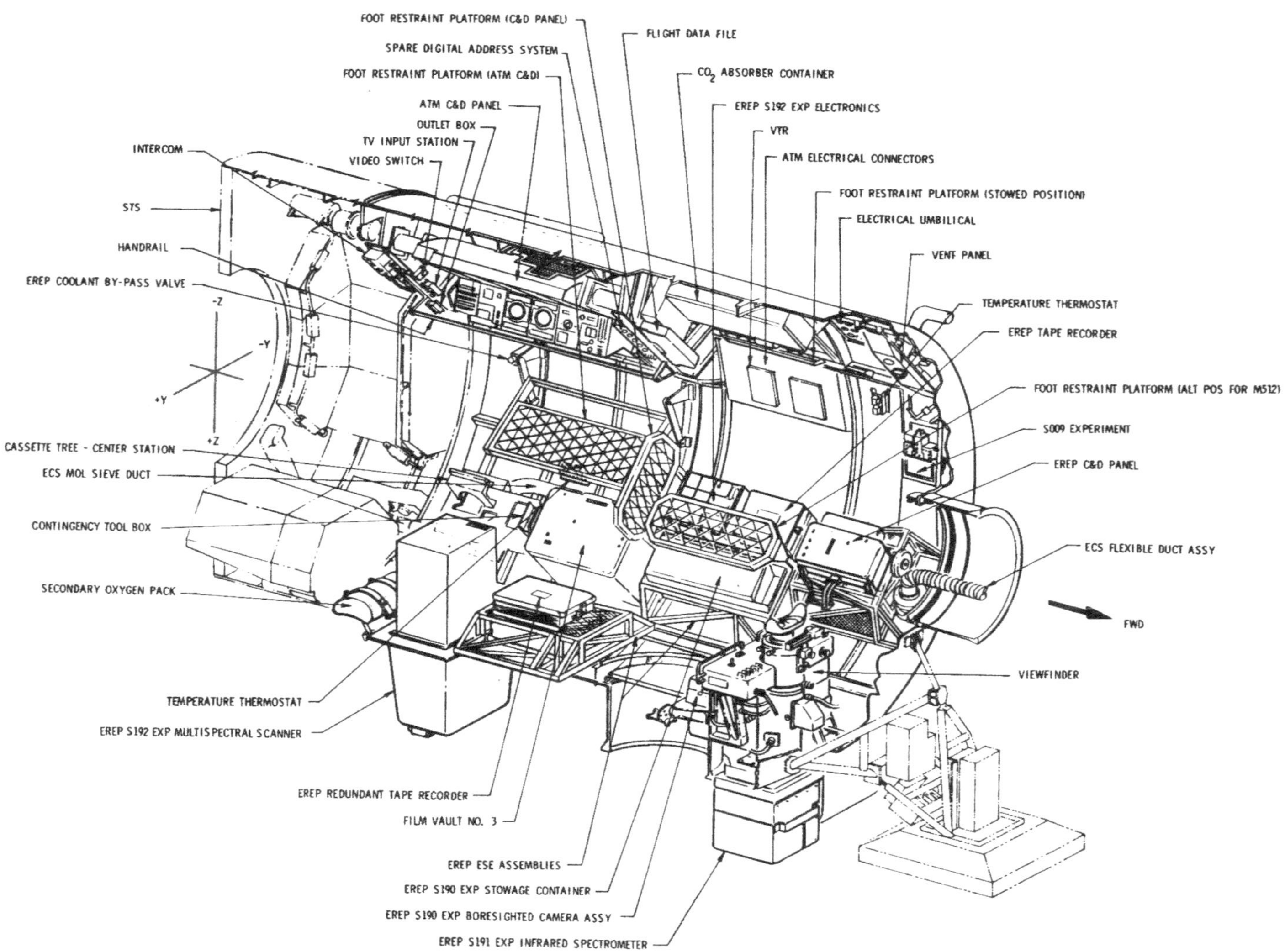

Figure I-22 – MDA Internal View, -Y

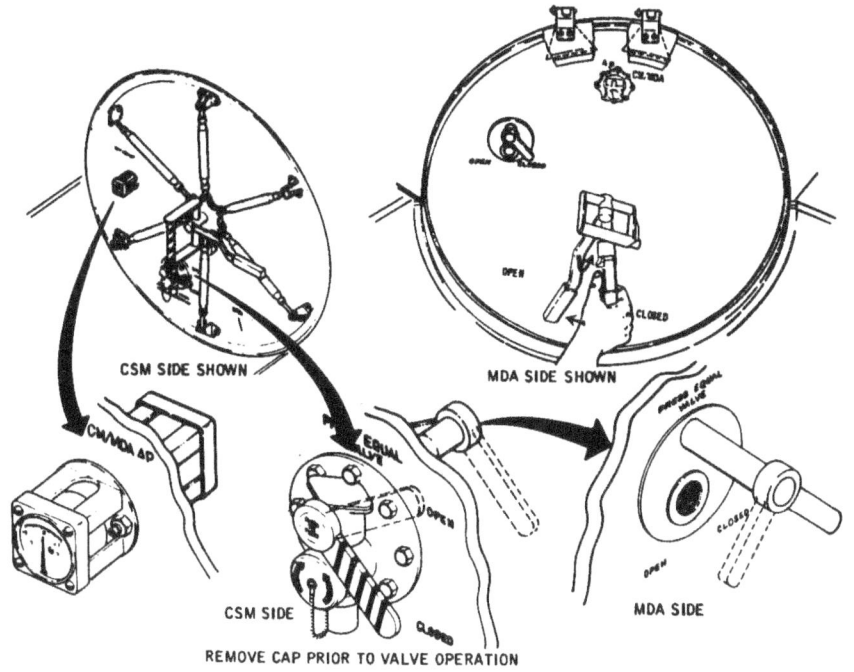

Figure I-23 — MDA Hatch

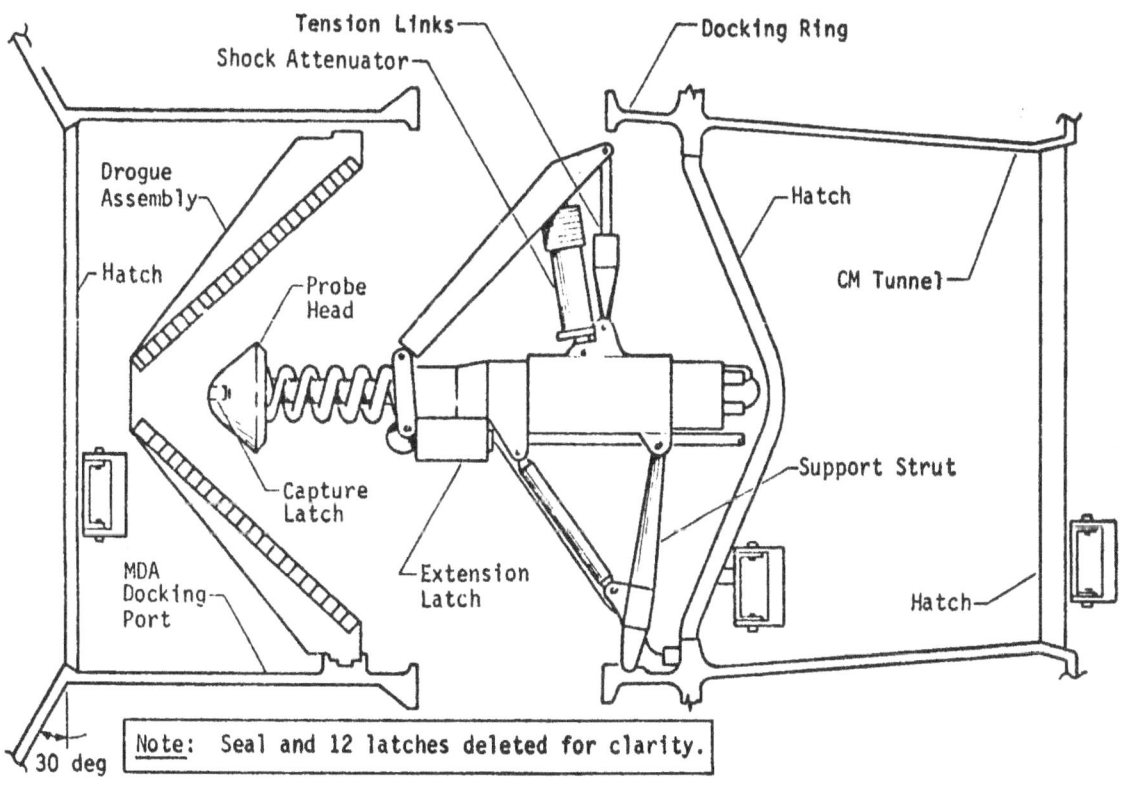

Figure I-24 — MDA Docking Mechanism

I-21

Special tool kits and spare parts are contained in the MDA for selected types of orbital maintenance and activation/deactivation sequences to be performed by the astronauts.

Contingency hatch opening tools provide a combination of specially designed tools to be utilized in the event the MDA axial entry hatch is inoperable in the conventional manner. These tools are stowed on the MDA hatch (CM side) affixed to the hatch in a specially designed tool box.

These tools include: ring removal tool, pin removal tool, seal loosening tool and mallet assembly.

Activation/Deactivation tools provide a combination of tools to be utilized in the activation/deactivation sequences performed on initial entry and final exit of the Skylab. Also contained are tools necessary to open the OWS hatch in the event it is inoperable in the conventional manner. This tool box is located on a film vault in the MDA. These tools include: speeder handles, ratchet extension, socket wrenches, screwdriver bits, pinch bar, open box wrench, tool caddy, utility belts, leak repair materials, phillips screwdriver and tape.

S190 Tool Kit provides a combination of tools specifically required for the operation/maintenance of Experiment S190. These tools are used frequently during the experiment operation and therefore are stowed within the S190 experiment envelope for ease of astronaut operation. These tools include; large spanner wrench, small spanner wrench, screwdriver, scissors and tape.

Spare parts are contained in the MDA for orbital maintenance. These spares are stowed on a special spares pallet and include such items as: TV input station, video switch, manual pointing controller, crewman communication umbilical, lightweight crewman communication umbilical and control head.

AIRLOCK MODULE

The Airlock Module (AM) (Figure I-25) is the structural assembly between the OWS and the MDA. It is 5.3 meters (17.5 feet) long, weighs 22,226 kilograms (49,000 pounds) and has 17.4 cubic meters (622 cubic feet) of habitable volume. It consists of a Structural Transition Section (STS), tunnel assembly, four truss assemblies, the lower truss of the Deployment Assembly, a flexible tunnel extension and a Fixed Airlock Shroud (FAS) (Figures I-26) - I-29).

The STS connects the tunnel assembly to the MDA structure. The tunnel, a passageway for the astronauts, has an airlock and hatch to permit the astronauts to perform extravehicular activities without depressurizing the complete spacecraft. A flexible extension connects the tunnel assembly to the OWS to continue the passageway while isolating structural loads from the OWS forward dome.

The FAS which is a continuation of the OWS cylinder, provides a shroud around the aft portion of the AM and structural mounting for the AM and MDA modules, the ATM Deployment Assembly and the Skylab oxygen supply tanks. It supports the Payload Shroud, the ATM, AM and MDA during boost.

The truss assemblies attach the AM to the FAS and provide exterior mounting structures for battery, electronic, thermal and experiment equipment.

Some of the basic requirements provided to Skylab by the AM are:

Oxygen and nitrogen storage for atmosphere supply.
Thermal control for Skylab atmosphere.
Purification of Skylab atmosphere.
OWS/AM electrical power control and distribution.
Lock, hatch and support for extravehicular activity.

Instrumentation for real time and delayed data transmission.
Caution and Warning displays and tones.
Command link with ground network.
Ranging link for CSM rendezvous.
Tracking lights.
Teleprinter.
Experiment support.
Equipment stowage.

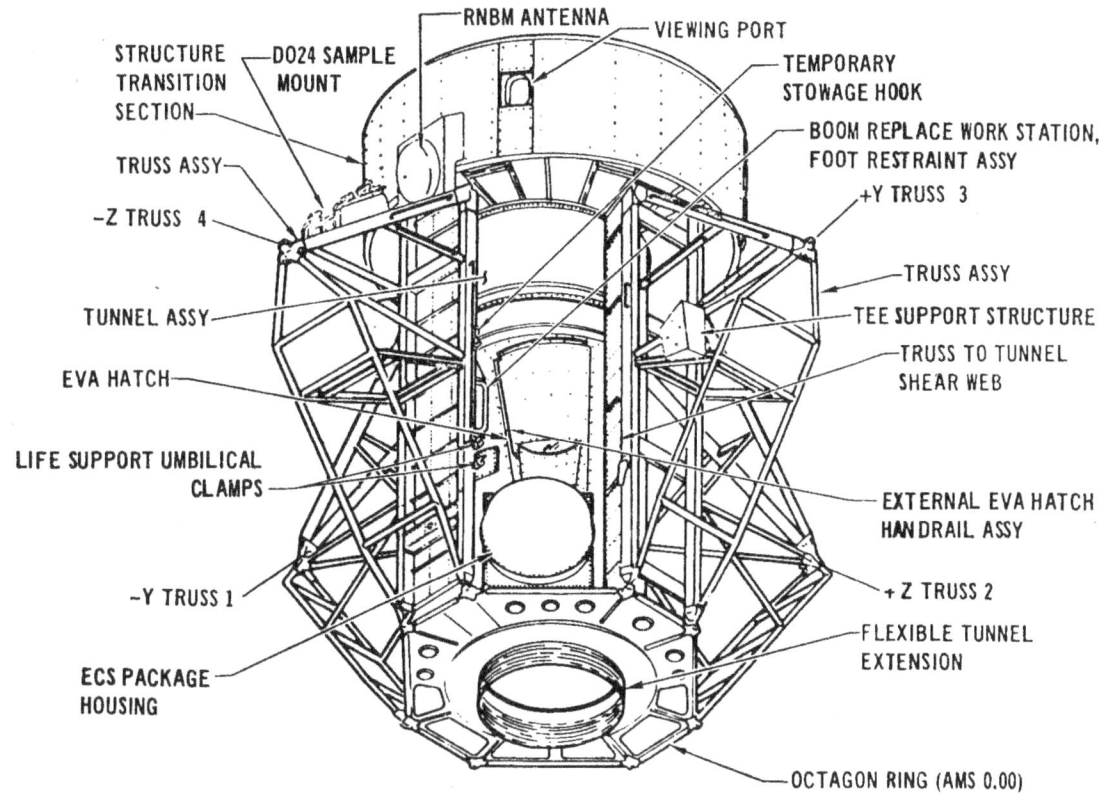

Figure I-25 — Airlock Module

Structural Transition Section

The STS is at the forward end of the airlock tunnel and is physically secured to the MDA. It provides the structural transition from the MDA to the airlock tunnel and its trusses. It is constructed as a welded aluminum cylinder of stressed skin in a semimonocoque configuration. Four double pane glass viewing ports, one in each quadrant, are provided for visibility.

Contents of the STS include: AM data file; STS control panel; six 10-watt instrument panel area lights; eight 10-watt STS area lights; a circuit breaker panel; the ECS supply duct to the MDA; the teleprinter paper stowage container; a spare stowage container (removable); molecular (mol) sieve; cabin heat exchanger module; ATM tank module; oxygen/nitrogen control panel; water tank module; condensate module; STS intravehicular activity (IVA) station; and carbon dioxide sensor module (Figures I-27 and I-28).

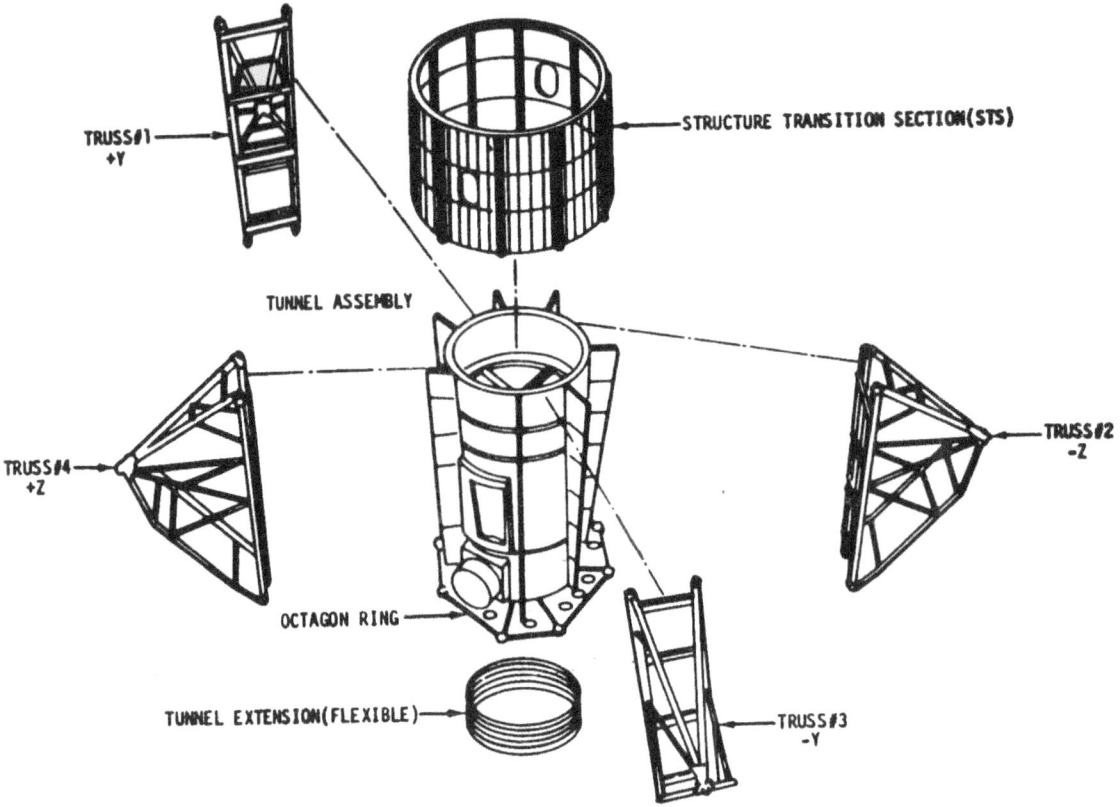

Figure I-26 — AM Structural Elements

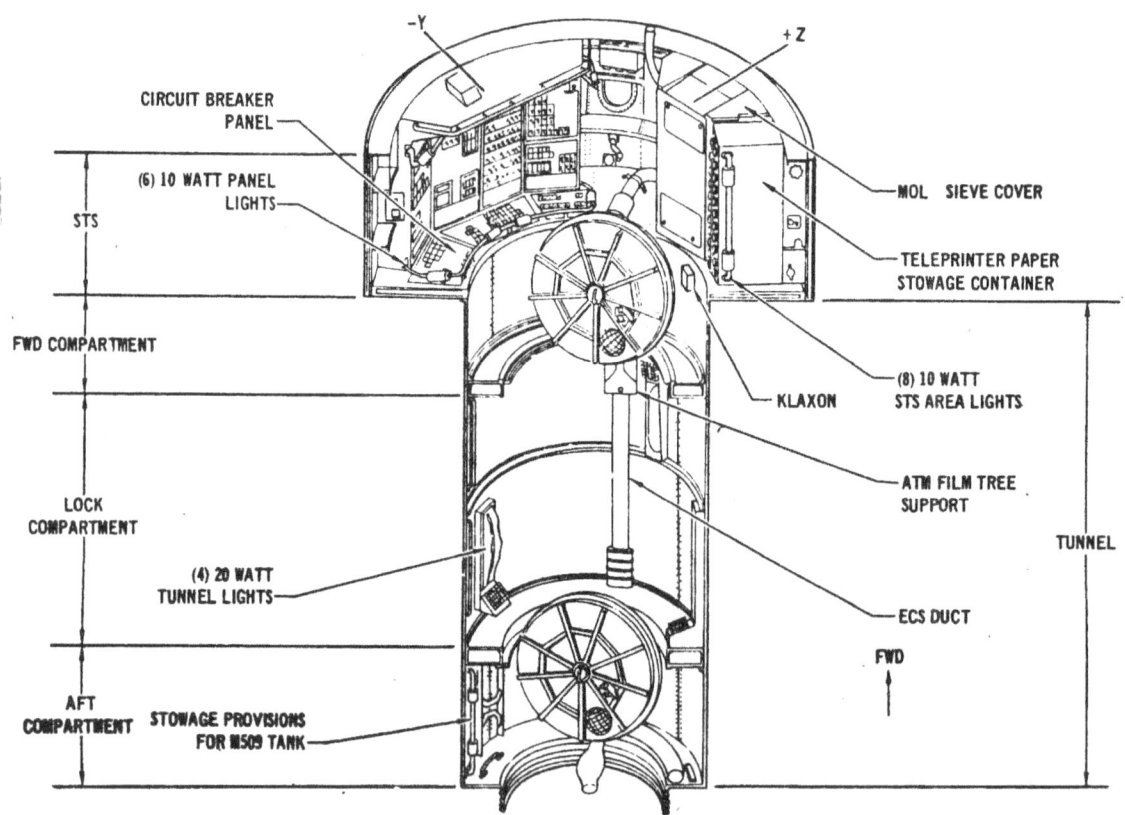

Figure I-27 — AM Interior, -Y+Z

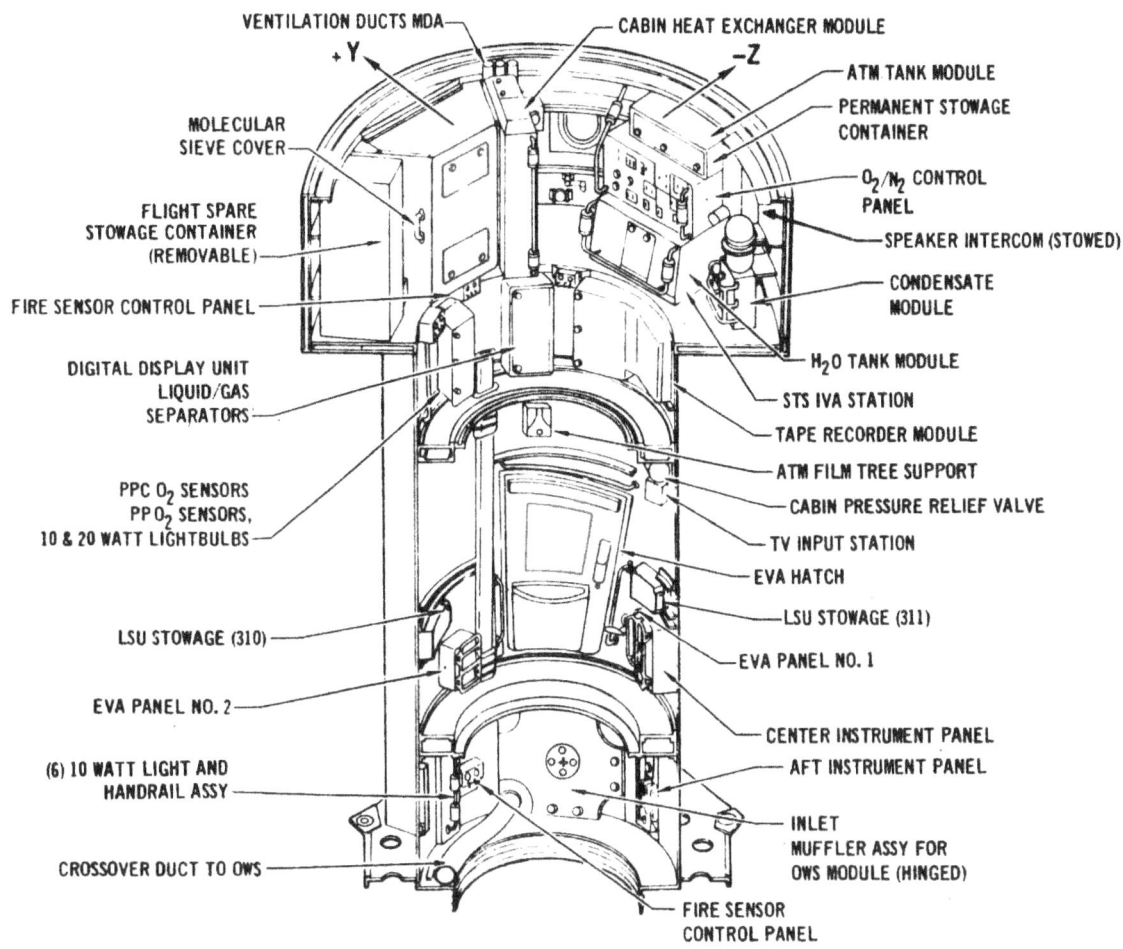

Figure I-28 — AM Interior, +Y-Z

Airlock Tunnel Assembly

The AM tunnel assembly provides the passageway from the MDA/STS to the OWS. It is constructed of aluminum and is cylindrical in shape. The tunnel is divided into three compartments by two internal bulkheads equipped with hatches. The forward hatch leads to the STS via the forward tunnel, and the aft hatch to the OWS through the tunnel extensions. The center (or lock) compartment includes a crew hatch for EVA. It is essentially the same as the hatch used on Gemini spacecraft, the first from which astronauts emerged to "walk in space." The hatch is roughly trapezoidal in shape and curved to match the wall of the AM. It has internal and external hatch handles and a window and is kept closed and sealed by 12 latches. When unlatched, the hatch swings outward on a hinge somewhat like a piano hinge. A rod on the inside of the hatch holds it open.

The two AM internal hatches are quite similar in appearance and function. Both are circular and both swing outward from the lock compartment. Each has a hatch opening of 120.1 centimeters (47.3 inches) in diameter. Each hatch has a window 21.6 centimeters (8.5 inches) in diameter with a stainless steel grid shield, pressure equalization valve, stiffeners and nine latches. The latches are the same type used on the EVA hatch.

Each hatch swings toward the lock compartment to close. A seal bead around the hatch rests in a molded silicone rubber seal when the latches are closed. Drawing a vacuum inside the lock compartment, as when depressurized for EVA, tends to reinforce the seal against leaks. The hatches are held in place by Velcro straps when open.

Although relatively small, the AM tunnel contains dozens of items of equipment and supplies. The small aft compartment contains stowage provisions for the Experiment M509 tank, a 10-watt light and handrail assembly, a crossover duct to the OWS, the inlet muffler assembly for the OWS, and the aft instrument panel.

Between the two tunnel hatches, the lock compartment contains four 20-watt tunnel lights, ATM film tree support, environmental control systems (ECS) duct, aft compartment vent valve, life support umbilical (LSU) stowage, EVA panels 1 and 2, center instrument panel, umbilical end stowage, ATM film tree support, EVA hatch and cabin pressure relief valve.

The forward tunnel contains the spare molecular (mol) sieve fan and replacement liquid/gas separators, tape recorder module, portable timers and spare batteries, spare 10-watt and 20-watt lightbulbs, and spare teleprinter head.

Fixed Airlock Shroud

The Fixed Airlock Shroud (FAS) (Figure I-29) is a circular structure that joins with the IU and extends forward to surround the AM aft compartment and about two-thirds of the lock compartment. It serves as a structural support for the ATM, AM, MDA and Payload Shroud (PS) and supports mounts for oxygen tanks.

Also carried on the FAS are two discone antennas which are deployed away from the OA once in orbit. The antennas are stowed for launch in the +Y+Z and -Y+Z quadrants halfway between the trusses. Pivot/attach points are on the inside of the FAS wall. Each antenna is in two sections. The sections are folded together at a rotary joint which rests against the AM. In orbit the antennas are deployed; one is 45 degrees from -Z toward -Y and the other is 45 degrees from -Z toward +Y.

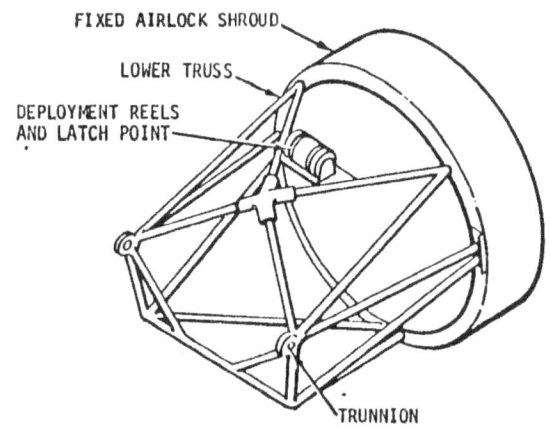

Figure I-29 – Fixed Airlock Shroud, Deployment Assembly

Truss Assemblies

The four AM truss assemblies located outside provide the longitudinal support for the AM between the FAS and STS. They provide mounting support for experiments, consumable containers and other hardware. The four trusses are located symmetrically around the tunnel assembly. A single point on each truss attaches the assembly to the FAS. Truss 1 is at +Y, truss 2 is at -Z, truss 3 at -Y, and truss 4 at +Z. Two nitrogen spheres are mounted in each of trusses 2, 3, and 4. Two umbilical stowage spheres are between trusses 1 and 4. Battery modules are mounted on the forward ends of trusses 1 and 2. Sample panels for experiments D021 and D024 are mounted atop truss 4. Electronic modules and equipment and thermal control equipment are also mounted on the truss assemblies.

Thermal curtains cover the trusses and surround the AM from the octagon ring at the aft end to the STS. Meteoroid curtains cover the battery modules and extend from the FAS to the aft end of the AM. The EVA hatch is not covered because it must be opened by astronauts for EVA activities.

APOLLO TELESCOPE MOUNT

The Apollo Telescope Mount (ATM) is the first manned astronomical observatory for performing solar research from Earth orbit. In addition to accommodating telescopic equipment for the solar investigations, the ATM has a solar array which generates about half of Skylab's electrical power, and the ATM structure houses primary stabilization and attitude control components for the total spacecraft.

The ATM weighs 11,092 kilograms (24,656 pounds), is 4.4 meters (14.7 feet) tall, and measures nearly 6 meters (20 feet) across with the solar arrays folded (prelaunch and launch position) (Figure I-2) and 31 meters (102 feet) across with the solar arrays extended (orbital configuration) (Figure I-1). The operational ATM (Figure I-30) system consists of five major hardware elements: the cylindrical experiment canister (Figure I-31) which houses the solar astronomy experiments; the Attitude Pointing and Control System; the solar array wings; the Control and Display (C&D) console (Figure I-18) which is in the MDA and which provides the capability for astronaut operational control and monitoring; and the rack assembly, a large octagonally-shaped structural frame which surrounds the canister and provides structural attachment points for the solar and thermal shields, outrigger assemblies, solar arrays, Deployment Assembly, experiment pointing control-roll positioning mechanism and numerous subsystem components.

The aluminum rack is made up of two large octagonally-shaped rings separated by eight vertical beams attached to the eight corners of the rings. Equipment-mounting panels are provided in seven of the bays between the vertical beams. One bay is left open for the astronaut work station (Figure I-32) to provide access to the canister's MDA-end film doors. The opposite end of the rack includes attachment points for mounting the solar shield assembly and the acquisition Sun sensor. The solar shield assembly protects the ATM rack

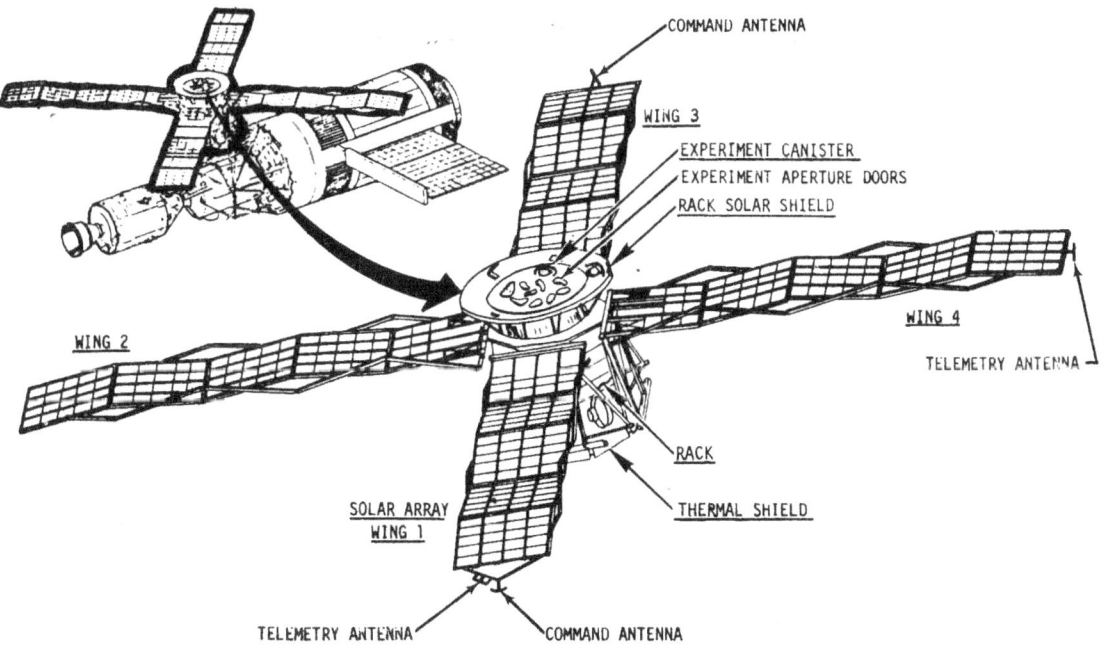

Figure I-30 — Operational ATM

I-27

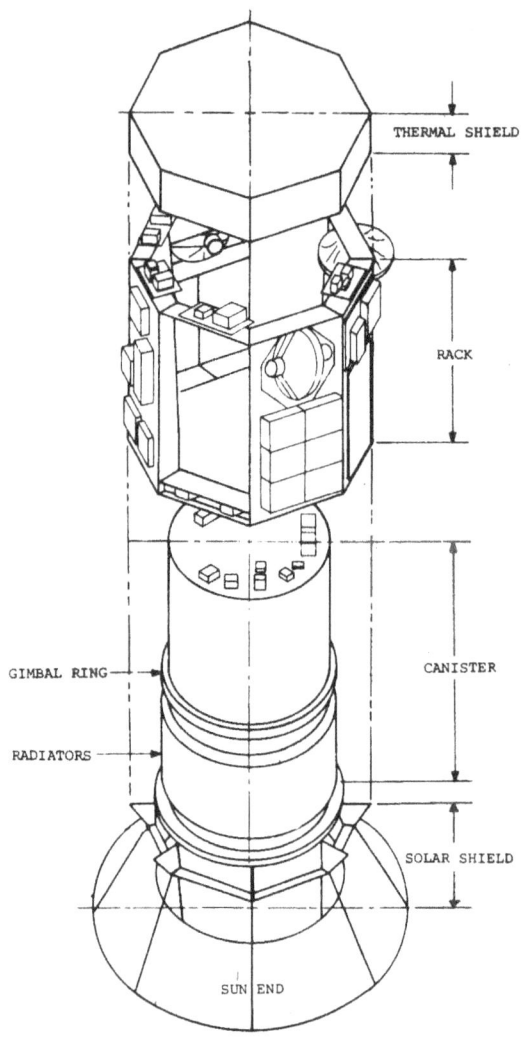

Figure I-31 — ATM Major Structural Components

mounted equipment from direct solar radiation and includes a work station for access to the experiment package's Sun-end film retrieval doors (Figure I-33). The MDA end of the rack provides the attachment points for the thermal shield. Four outrigger assemblies on the sides of the rack provide structural support of the ATM by the Payload Shroud during launch. Major spar-mounted components are shown in Figure I-34.

The experiments canister consists of the Spar, the MDA and Sun-end canister halves and the canister girth ring.

The spar (Figure I-35) is a cruciform structure constructed of three 1-inch thick insulation covered aluminum plates which provides structural support of the experiments and experiment pointing control components (fine Sun sensor, rate gyros, etc.). Girdling the center of the spar is a girth ring which provides the structural interface between the experiments canister and the rack mounted experiment pointing control-roll positioning mechanism. The girth ring provides attach points for the canister halves which enclose the spar mounted experiments to provide a contamination-free environment for the experiments.

The MDA end canister includes four film retrieval doors which are used for in-orbit experiment film retrieval and replacement. The Sun-end canister half contains two film retrieval doors and ten aperture doors on the Sun end bulkhead. These aperture doors cover the fine Sun sensor and experiment apertures during non-operating periods to prevent optical contamination. The experiments canister includes the EPCS and an active thermal control system to provide a stable thermal environment for the experiments.

Mounted on the ATM are major elements of Skylab's Attitude and Pointing Control System (APCS) that provides three-axis attitude stabilization and maneuvering capability for the orbiting vehicle. It also provides the capability of pointing experiments at desired locations, such as the Sun, the Earth and other targets of interest.

The APCS is comprised of the Instrument Unit/Thruster Attitude Control Subsystem (IU/TACS), Control Moment Gyros Subsystem/Thruster Attitude Control Subsystem (CMGS/TACS) and Experiment Pointing Control Subsystem (EPCS). (For details, see Attitude and Pointing Control in Section II-4).

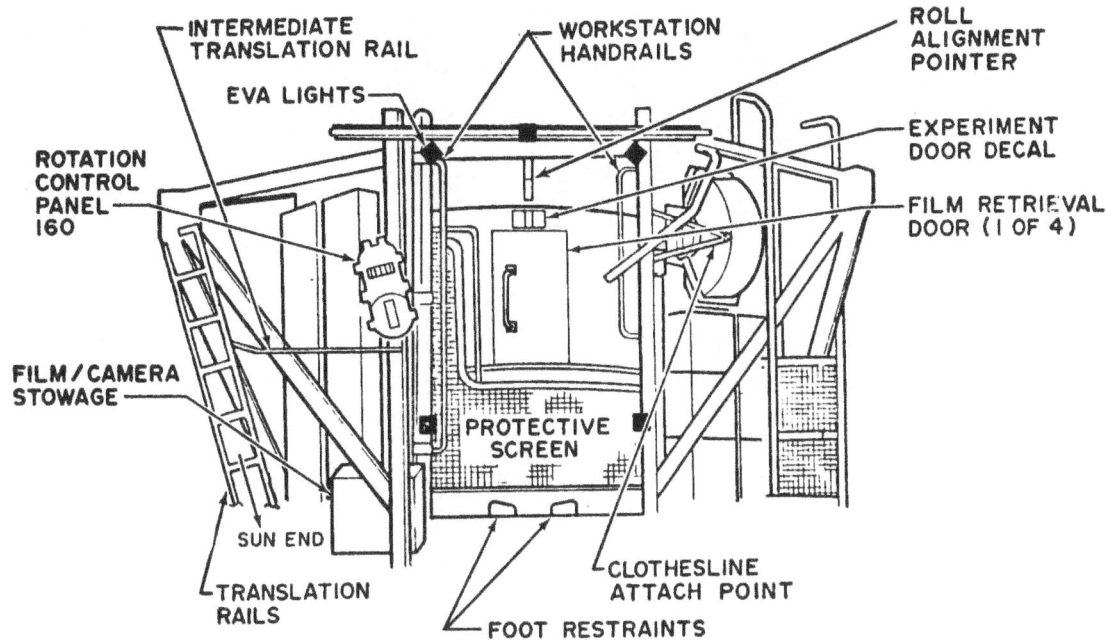

Figure I-32 — ATM Center Work Station

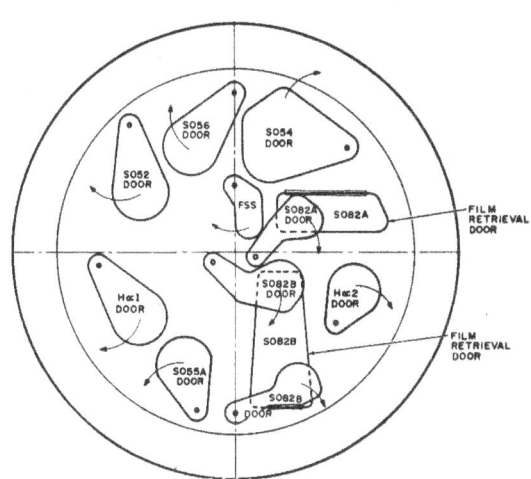

Figure I-33 — ATM Sun End

The EPCS maintains fine attitude pointing and control about two axes for the ATM instrument package. The third axis (roll) is controlled from the C&D Console. The ATM instrument package can be manually pointed to any desired location on the solar disk or its outer perimeter.

Vehicle attitude information is derived from strapdown reference computations in the ATM Digital Computer (ATMDC) using rate gyro information, and, during orbital daytime only, Sun sensors. The ATMDC processes the sensor signals to generate the CMG gimbal rate commands. The astronaut has the capability of manually controlling the CMGS through his keyboard on the ATM C&D Console.

The ATM CMG consists of an induction-motor-drive constant-momentum rotor, gimbal supported to provide two degrees of freedom. Associated with each CMG is an Electronics Assembly (CMGEA) for positioning the gimbals and controlling the gimbal rates, and an Inverter Assembly (CMGIA) for providing power. Each CMG has an angular momentum storage capability of 2,300 foot-pound-seconds.

Three double-gimbaled CMGS hardmounted at 90 degree angles to the ATM actuate the system. A CMG is basically a spinning wheel that provides the forces required for vehicle control.

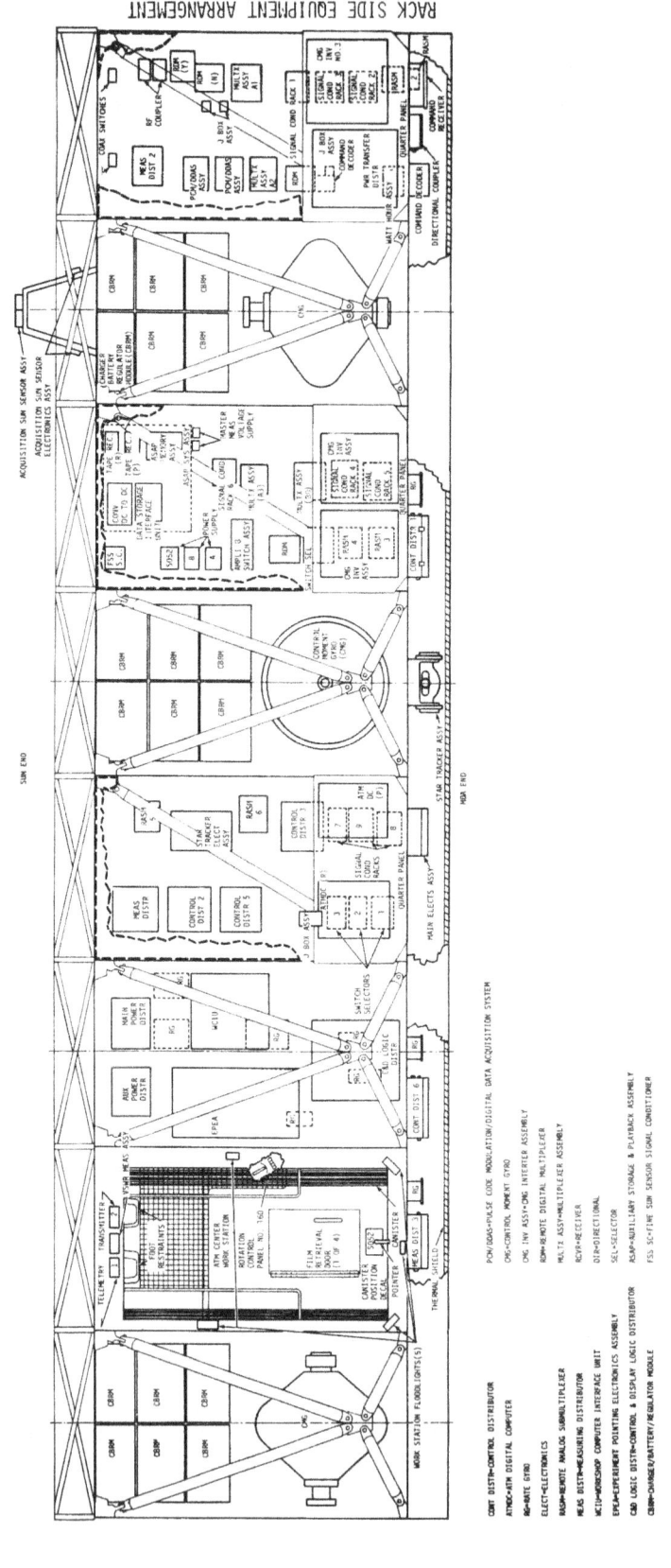

Figure I-34 — ATM Rack Mounted Equipment

I-30

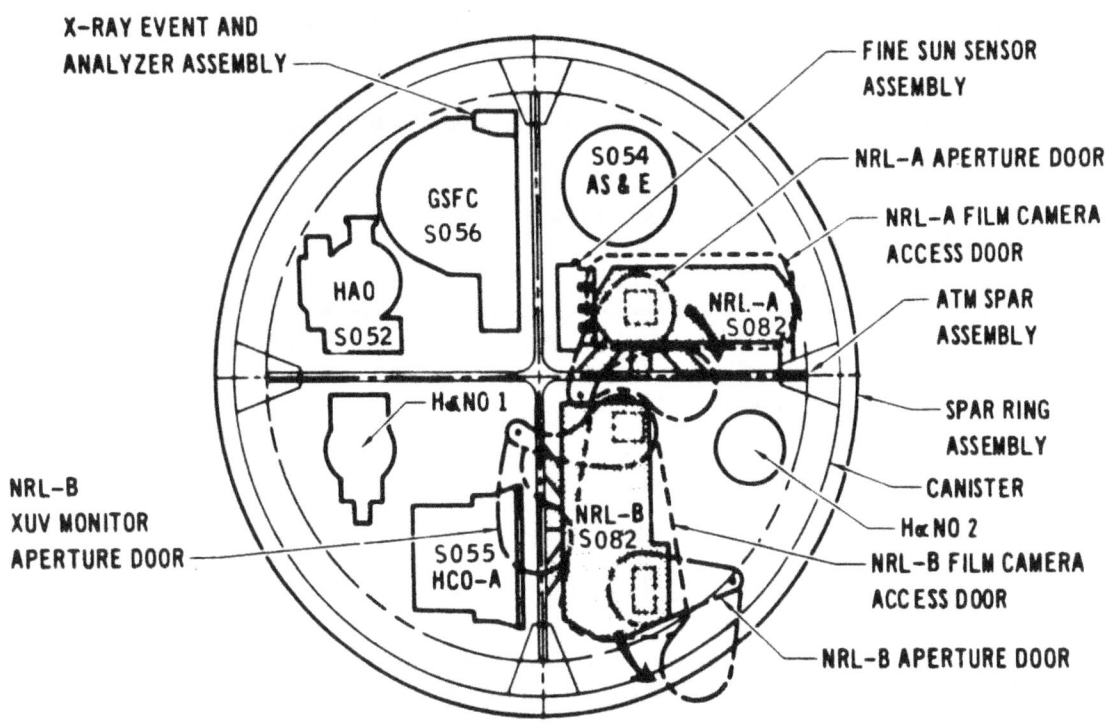

Figure I-35 — ATM Spar, Sun End View

The EPCS provides fine pointing control and stability for the ATM experiments, further isolating them from any disturbance torques from the Skylab assembly. This system provides control to within 2.5 arc seconds for periods up to 15 minutes, utilizing fine pointing Sun sensors for attitude reference. The experiment package can be off-set pointed within a ±24 arc min square centered on the solar disc. It can also be rotated to any desired roll orientation throughout ±120 degrees.

To help maintain stability and alignment for the scientific instruments, a thermal control loop is incorporated within the skin of the experiment canister to circulate liquid coolant. This active thermal control system is self contained within the canister. The water/methanol cooling fluid transfers heat absorbed from cold plates to radiators on the exterior side of the experiment canister, where it radiates into space. This active coolant systems maintains an average temperature within the canister of approximately 12°C (53°F). Each experiment also has its own thermal control heaters, designed to maintain its temperature within about ±0.6°C (1°F) throughout the length and width of the instruments. Precautionary measures have been taken to avoid any fluid leakage which could contaminate the optical elements of the scientific instruments. All fluid lines and components are on the outside to avoid leakage inside. In addition to the active thermal controls for the experiments, a passive system regulates the temperature or the ATM supporting rack structure and the components mounted on it. The thermal shield attached to the Sun end of the canister and rack minimizes solar heating of system components when the ATM points directly at the Sun for data acquisition.

As a part of the ATM control and displays (Figure I-18), two selectable video presentations will be available to the astronaut conducting the experiments. Pictures of the Sun can be displayed in various wavelengths from several of the solar instruments; thus a crewman onboard the spacecraft will be able to assure proper identification and tracking of solar events of interest and point the instruments with a high degree of accuracy or spatial resolution (1 arc second corresponds to about 700 kilometers (434 statute miles) on the solar disk). (See Section III-2.) The presence of the astronaut at the C&D console will permit rapid response to transient events on the Sun, such as the initial phase of an explosive solar flare.

The ATM solar array (Figure I-30) consists of four individual wing assemblies which are stowed in a folded configuration during launch and deployed in orbit at 90° to each other to expose a total of 11.2 square meters (1,200 square feet) of solar cell surface to the Sun. Each individual wing assembly weighs 486 kilograms (1,071 pounds) and consists of four and one half solar cell panels (total of 18 panels for all wings), electrical cabling and the wing support structure, which includes the ordnance-activated decinching mechanism, and the motor driven deployment mechanism. Three solar arrays include telemetry and/or command antennas at their tips.

The solar cell panels are made up of a framework of rectangular aluminum tubing mounting solar cell modules. The four outboard panels of each wing mount 20 modules. The inboard panel mounts 10 modules on the outboard portion of the panel.

Each solar cell module consists of an aluminum honeycomb sandwich covered by a fiberglass insulation and mounting silicone solar cells with quartz cover slides. Cover slides are provided to protect the solar cells from micrometeoroid and radiation damage.

The ATM is mounted on the ATM Deployment Assembly (DA) (Figure I-36) which provides in-orbit structural support between the ATM and the Fixed Airlock Shroud (FAS) and deploys the ATM upon reaching orbit. The DA also provides a mounting fixture for some experiments, acquisition lights, VHF ranging antenna and wire routing from the ATM to OWS. The Deployment Assembly consists of upper and lower tubular truss assemblies, a release ordnance system, a rotation system and a means to latch the DA in the deployed position. The upper truss assembly, attached to the ATM, includes a rigidizing assembly which prevents ATM launch loads from being imposed on the DA. It provides structural rigidity between the ATM and DA following Payload Shroud (PS) separation (ATM launch loads are supported by the PS during launch).

I-5

PAYLOAD SHROUD

The Payload Shroud (PS) is a smooth structure which surrounds and protects the ATM, MDA, AM and associated hardware during the launch and climb-out phase of the mission. The principal requirements are to provide an aerodynamic and environmental protection cover for Skylab elements and structural support to the ATM during the prelaunch and launch phases. Once in orbit, the PS is split into four sections or quadrants by pyrotechnic devices and jettisoned. Each quadrant has a cylinder section, aft cone, and forward cone. One quadrant also has a nose cap at the end of the forward cone. The four shroud sections are joined with "shear" rivets and latching link mechanisms.

Two ordnance systems are used for separation and jettisoning, the discrete latch system and the longitudinal separation system. With the SWS in a nose down attitude, signals from the IU cause detonation of ordnance. This creates a pressure that retracts pins in the latch actuators and circumferentially unlocks the quarter sections. Seconds later an IU signal is sent causing longitudinal separation into four segments by expandable bellows with a linear explosive inside (Figure I-37). When longitudinal separation is initiated, the bellows are expanded by the internal pressure buildup from the linear explosive causing the longitudinal separation joint rivets to shear. The bellows continue to expand, imparting a velocity to the PS quarter sections causing them to move away from the payload at a rate of about 6 meters (20 feet) per second (Figure I-38).

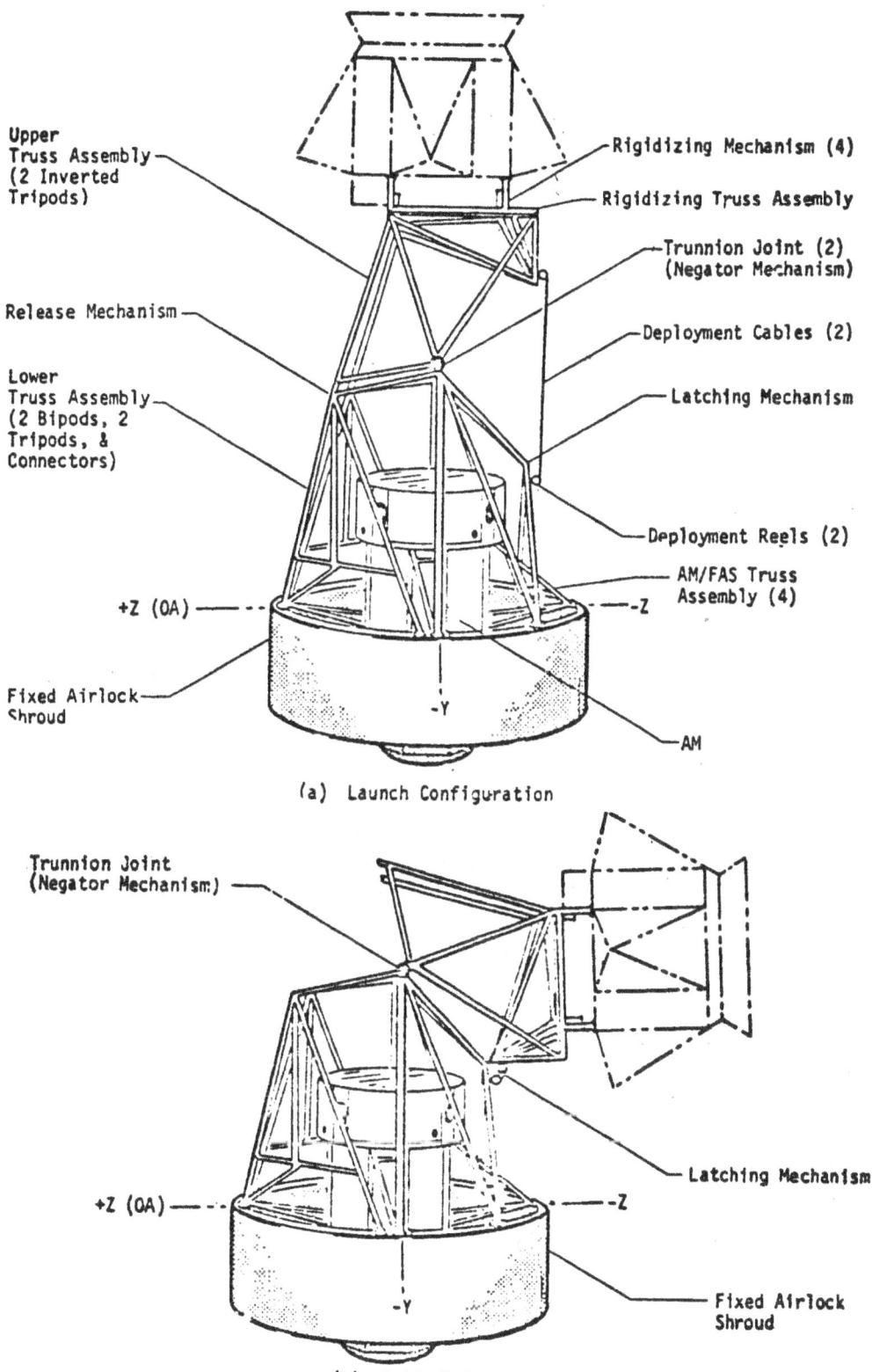

Figure I-36 — ATM Deployment Assembly

The total PS is 6.5 meters (21.7 feet) in diameter at the aft end (interface locking ring), 16.8 meters (56 feet) long and weighs 11,794 kilograms (26,000 pounds). The forward cone tapers at a 25-degree angle and the aft cone at 12.5 degrees. The nose cap is a structural shell made of aluminum and stiffened with rings and intercostals. The aft cone and cylinder section are structural shells stiffened by rings.

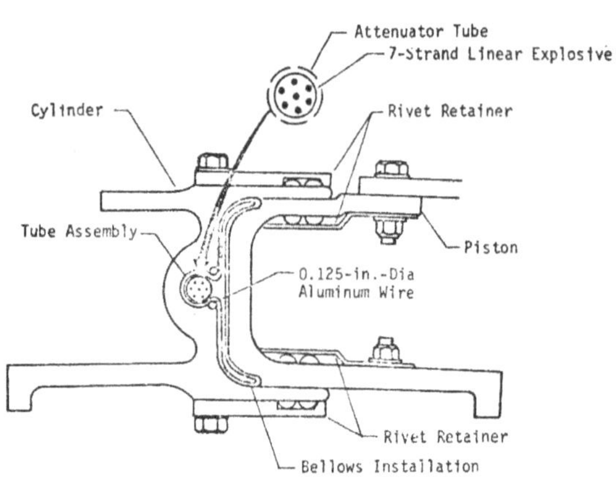

Figure I-37 — PS Separation Bellows

COMMAND AND SERVICE MODULE

The Skylab Command and Service Module (Figure I-39) is basically the same CSM (J-type) used during the Apollo program. (See "Apollo Spacecraft News Reference.") Numerous modifications and certain deletions have been made to accommodate the unique mission requirements of the Skylab missions.

Figure I-38 — PS Separation

CSM 116, 117 and 118 will be flown on SL-2, SL-3 and SL-4 respectively. CSM 119 will be used as backup or rescue vehicle if required.

Major modifications include addition of a 12-tank reaction control system (RCS) propellant storage module, with a total of 680 kilograms (1,500 pounds) of propellants to more than double the former RCS propellant capacity, expansion of the spacecraft's thermal control system, addition of a 50-gallon water tank to eliminate water dumps, addition of three descent 500-ampere-hour batteries, deletion of one of the vehicle's three fuel cells, and deletion of two of the four service propulsion system (SPS) propellant tanks and one of the two helium tanks.

CSM system changes have been:

Deletion of high gain antenna which is not needed for Earth orbit missions.

Addition of a capability to switch all S-band omni antennas from the ground.

Pyrotechnic batteries (0.75 ampere-hour) deleted and replaced with entry/postlanding-type 40

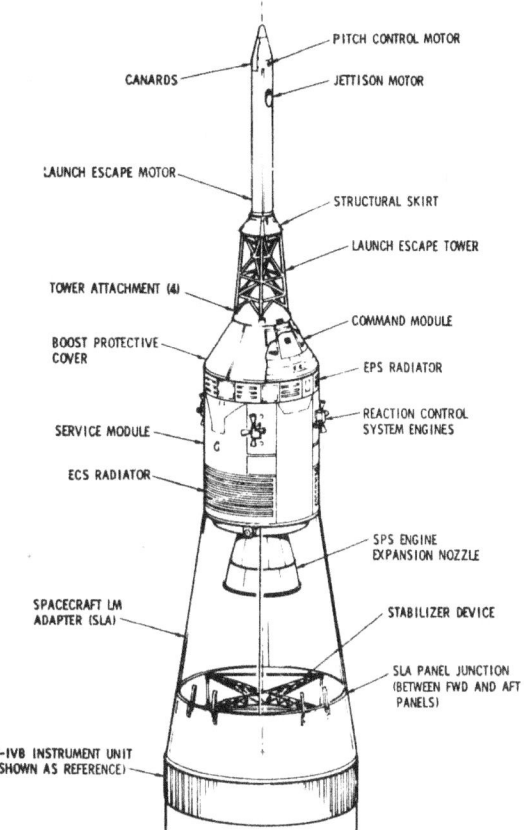

Figure I-39 – CSM With LES

ampere-hour batteries giving a total of five entry-type batteries. Proven life of the entry batteries is 140 days, as opposed to the 36-day life of the pyro batteries.

Several blanket heaters added to various systems for thermal control during the prolonged solar-inertial attitudes of the OA.

Addition of electrical power transfer umbilicals to supply OWS power to the CSM during the powered-down phase, after fuel cell shut-down.

Addition of oxygen and hydrogen nonpropulsive vent valve to prevent propulsive torques.

Additional stowage capacity has been provided by increasing locker size and configuration on the aft bulkhead.

Command Module

The Command Module (Figure I-40) will transport 3 crewmen and between 453-680 kilograms (1,000-1,500 pounds) of stowed equipment to and from Skylab, serve as the primary communications vehicle and command station for the SWS, provide backup attitude control, and have the capability of being reactivated after 56 days of semi-dormancy in space.

Communications and Instrumentation – The CSM high bit rate telemetry will be the normal mode of operation. Data for the Circadian Rhythm experiments (S071/S072) will be dumped by ground control via the CSM real-time TM link.

I-35

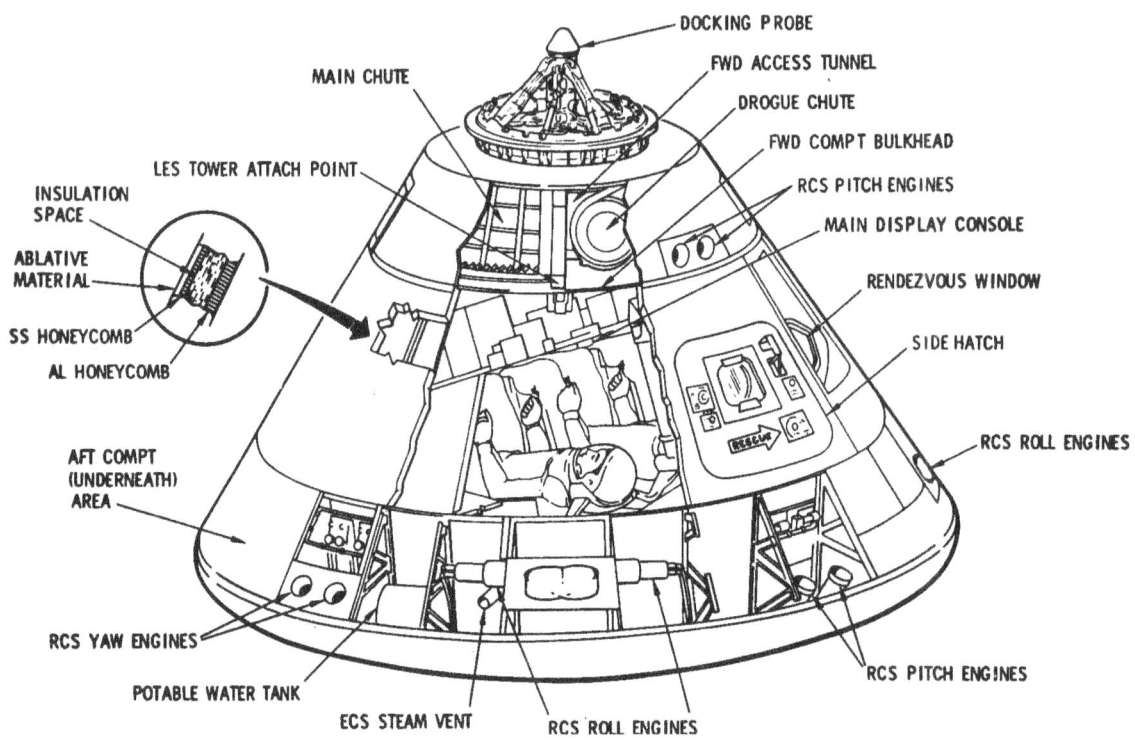

Figure I-40 — Command Module

The CSM tape recorder will not normally be used to record routine systems data, with the exception of the periods from launch through activation and deorbit through reentry. The recorder is required for radiation experiment (D008) operations during the initial 2 weeks at approximately 6 to 7 hours per day.

All voice transmission will be through the CSM communication system. The CSM is an integral part of the intercom system and will also be used for downlinking real-time voice on S-band, with VHF backup.

Guidance, Navigation, and Control — The CSM guidance and navigation system will be powered down and the command module computer will be maintained on standby during orbit activities. Unless operationally required, the Inertial Measuring Unit (IMU) will not be aligned periodically during the CSM quiescent mode.

Environmental Control — At launch the CSM cabin atmosphere will contain 60 percent oxygen and 40 percent nitrogen. Lithium hydroxide (LiOH) cartridges are changed after 12 hours of use or if CM carbon dioxide pressure exceeds 5.5 mm mercury pressure.

LiOH cartridge replacements are stored in the MDA and deployed in the CM. The LiOH system is operative only during CSM use and during molecular sieve bakeout deactivation.

The CSM fuel cells remain on hydrogen cryogenic supply for approximately 13 days or until the supply is depleted. Residual hydrogen will be vented to space and excess oxygen will be vented through the cluster or vented overboard.

Electrical Power — Entry and post landing batteries (A and B) will be turned on five minutes prior to an SPS burn and turned off immediately after completion of the burn. The batteries will be charged immediately after docking.

I-36

Spacecraft LM Adapter

The Spacecraft Lunar Module Adapter, or SLA (Figure I-39), is a large truncated cone which connects the CSM and second stage of the launch vehicle. For Skylab flights SL-2, SL-3 and SL-4 the adapter contains only a stabilizing device for providing structural support to the outer shell.

Service Module

The Service Module (Figure I-39) is a cylindrical structure which serves as a storehouse of critical subsystems and supplies. It remains attached to the Command Module from launch until just before Earth atmosphere entry. Its main propulsion engine is used to perform deorbit burns for return to Earth.

Launch Escape System

The launch escape system (Figure I-41) will take the CM containing the astronauts away from the launch vehicle in case of an emergency on the pad or shortly after launch. The subsystem carries the CM to a sufficient height and to the side, away from the launch vehicle, so that the drogue chutes and three main parachutes can operate to lower the spacecraft and crew safely.

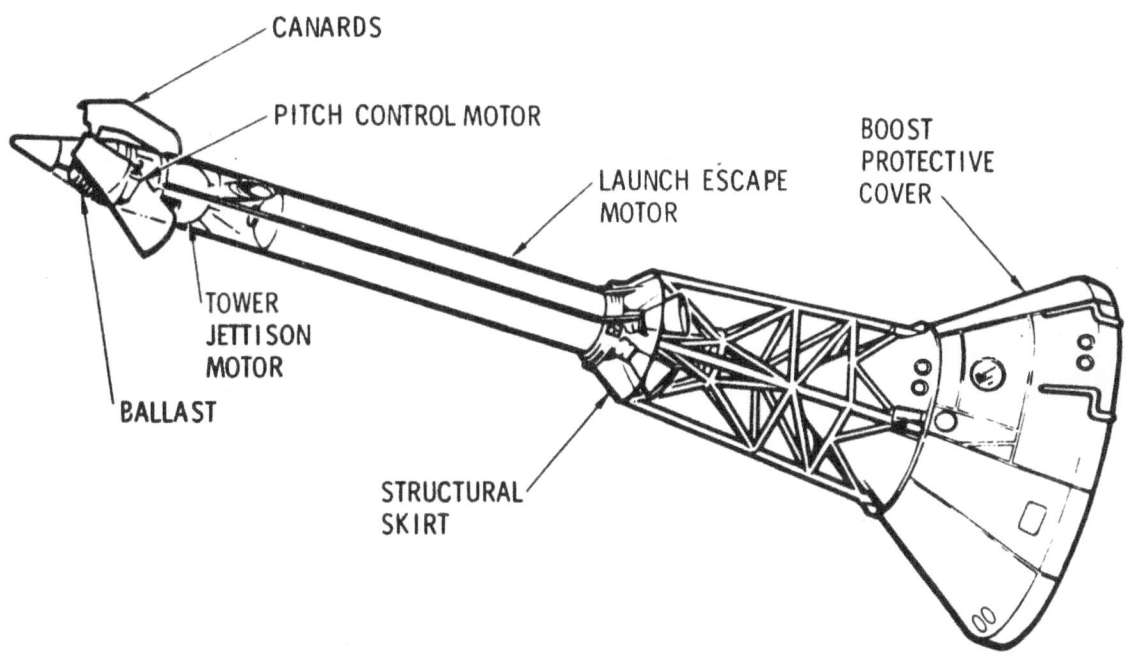

Figure I-41 — Launch Escape System

SATURN V

Saturn V vehicle AS-513 (Figure I-42) has been designated SL-1 for the first Skylab launch. The S-IC and S-II stages of the launch vehicle are essentially the same as those used in the lunar landing missions. From the S-II stage forward, the vehicle is considerably different. The Apollo launch vehicle had an S-IVB stage, IU, SLA, LM, CSM, and LES. On

PS   Payload Shroud
Diameter 6.6 meters (21.7 feet)
Length 16.8 meters (56 feet)
Weight 11,794 kilograms (26,000 pounds)

ATM   Apollo Telescope Mount
Wi   Width 3.3 meters (11 feet)
Height 4.4 meters (14.7 feet)
Weight 11,181 kilograms (24,650 pounds)

MDA   Multiple Docking Adapter
Diameter 3 meters (10 feet)
Length 5.2 meters (17.3 feet)
Weight 6,260 kilograms (13,800 pounds)

AM   Airlock Module
Diameter STS 3 meters (10 feet)
Diameter FAS 6.6 meters (21.7 feet)
Length 5.3 meters (17.5 feet)
Weight 22,226 kilograms (49,000 pounds)

IU   Instrument Unit
Diameter 6.6 meters (21.7 feet)
Length 0.9 meter (3 feet)
Weight 2,064 kilograms (4,550 pounds)

OWS   Orbital Workshop
Diameter 6.6 meters (21.7 feet)
Length 14.6 meters (48.5 feet)
Weight 35,380 kilograms (78,000 pounds)

S-II   Second Stage
Diameter 10 meters (33 feet)
Length 24.8 meters (81.5 feet)
Weight 488,074 kilograms (1,076,000 pounds) fueled
       35,403 kilograms (78,050 pounds) dry
Engines J-2 (5)
Propellants: Liquid Oxygen 333,837 liters (88,200 gallons)
                 Liquid Hydrogen 1,030,655 liters
                   (272,300 gallons)
Thrust 5,150,000 Newtons (1,150,000 pounds)
Interstage Approx. 5,171 kilograms (11,400 pounds)

S-IC   First Stage
Diameter 10 meters (33 feet)
Length 42 meters (138 feet)
Weight 2,245,320 kilograms (4,950,000 pounds) fueled
       130,410 kilograms (287,500 pounds) dry
Engines F-1 (5)
Propellants: Liquid Oxygen 1,318,315 liters
                  (348,300 gallons)
                RP-1 (Kerosene) 814,910 liters
                  (215,300 gallons)
Thrust 31,356,856 Newtons (7,723,726 pounds)

Figure I-42 — SL-1 Vehicle

the SL-1, the S-IVB is replaced by the Orbital Workshop (OWS). Forward of the OWS is the IU, AM, MDA, and ATM; all except the IU are enclosed in the PS. The LES is not required since SL-1 is not manned. Total weight of the SL-1 launch vehicle and payload is approximately 2,822,300 kilograms (6,222,000 pounds). Height of the total assembly is 101.71 meters (333.7 feet). Payload capability is near 90,720 kilograms (200,000 pounds). The S-IC and S-II propulsive stages of the launch vehicle are each 10.06 meters (33 feet) in diameter. The S-IC is 42.06 meters (138 feet) long and the S-II is 24.84 meters (81.5 feet) long.

Several major changes were to meet requirements of the Skylab mission. The boost acceleration limit was increased to 4.7 g for Skylab. On Apollo missions the limit was set at 4 g because those vehicles were manned. Since the SL-1 mission is unmanned, the restriction on g-force is not as great.

The S-IC engine cutoff sequence for Apollo missions was center engine first and then the outboard engines. On SL-1 the center engine will cut off first, followed by cutoff of two opposing outboard engines and then the remaining two outboard engines. The 1-2-2 cutoff sequence was programmed to gradually reduce slowdown instead of initiating it suddenly. Cutting off all four outboard engines at once would subject the ATM to a dynamic load that could cause problems.

On Apollo missions, the terminal stage (last to provide propulsion) was the S-IVB; on SL-1 it will be the S-II. Therefore, a number of changes had to be made to the S-II. First, provisions were made for S-II engines to be cutoff by guidance signal instead of by propellant depletion. This enables cutoff precisely at the desired velocity. Also, provisions were made to separate the S-II from the Skylab payload in such a manner as to preclude recontact. (Analyses have verified that the nearest the S-II will approach the Skylab after initial separation is about 1.42 kilometers (4,659 feet) nominal. Provisions were also made for making the S-II stage inert almost immediately after separation to assure that it remains intact. The safing sequence will be initiated by a timer 3.5 minutes after separation.

The IU will provide the initial payload attitude control signals and events sequencing to enable systems activation and checkout functions. These include: jettison payload shroud; initiate ATM deployment; acquire solar inertial attitude; activate the Attitude and Pointing Control System (APCS); and transfer attitude control to the ATM (switchback capability will exist only through IU lifetime).

The IU also will disable its functional interfaces at the end of its active lifetime (about 7.5 hours). These include: OWS switch selector inhibit; electrical power depletion through normal loads; fourth battery loads depletion through the Command and Communications System (CCS) transponder and an added heater; and gas and liquid (nitrogen and water) normal depletion. The structural lifetime of the IU has been increased to eight months instead of the few hours required of it during Apollo missions. (Adhesives have been verified by thermal and vacuum testing and structural integrity verified by analysis and test).

Changing the launch azimuth from 72-100 degrees on Apollo missions to 40.88 degrees for Skylab made it mandatory to revise vehicle maneuvers at liftoff. On Apollo launches, the launch azimuth was slightly north of due east. A yaw maneuver caused the vehicle's boattail to move toward the launch tower in a direction slightly west of due north. This aimed the vehicle slightly east of South, causing it to move away from the tower as it rose. On the SL-1 launch, a yaw maneuver alone would move the boattail northwest and aim the vehicle southeast, which would not result in sufficient clearance of the tower for safety. Therefore, a pitch maneuver will be combined with the yaw to cause the boattail to move directly northward toward the tower, aiming the rising vehicle southward away from the tower.

The Emergency Detection System (EDS) on Apollo launches was on a closed loop, providing abort capability and retention of telemetry of critical functions. On SL-1 the EDS will be on an open loop which will eliminate the abort feature but retain critical functions on telemetry. On SL-1 the only abort command possible must be initiated by the range safety officer. He can destroy the unmanned vehicle if it presents a hazard to inhabited areas.

A preliminary analysis of control, stability and dynamics for SL-1 shows that the control wind limit is greater than 90 meters-per-second at all altitudes, probability for a successful launch is 98 percent, and Skylab engine out capability is less than that for Apollo. SL-1 capability to withstand winds has been verified at 56.5 knots with damper attached and free-standing and at 38 knots during vehicle launch release. This is slightly less than the 64 and 39 knots, respectively, on Apollo.

The safing sequence for the second stage is as follows: separation from OWS at 9.5 minutes GET; safing sequence initiated at 13 minutes GET; start tank and helium control tanks safe at 23 minutes; liquid hydrogen tank safe at 58 minutes; and liquid oxygen tank safe at 1 hour and 13 minutes.

I-8

SATURN IB

The SA-206, -207, and -208 launch vehicles which will carry the three Skylab crews into orbit and rendezvous with the Skylab cluster are basic Saturn IB vehicles with modifications to:

Update them to the SA-205 configuration which was manned. (SA-206, -207, and -208 were originally configured for unmanned flight.)

To incorporate changes to adapt the vehicles for the Skylab mission.

Each of the three Saturn IB vehicles consist of the S-IB and S-IVB (first and second propulsive stages, respectively) the Instrument Unit (IU) and the payload.

The SL-2 launch vehicle, SA-206, is made up of the S-IB-6, S-IVB-206, S-IU-206, and CSM-116. The payload, derived from a modified Apollo Block II Command and Service Module (CSM), includes a Command Module (CM), Service Module (SM), Spacecraft and Lunar Module Adapter (SLA), and the Launch Escape System (LES). The SL-3 launch vehicle, SA-207, consists of the S-IB-7, S-IVB-207, S-IU-207, and CSM-117. The SL-4 launch vehicle, SA-208, consists of the S-IB-8, S-IVB-208, S-IU-208, and CSM-118. (Refer to Section IV-5 for details on rescue vehicle hardware.)

Each launch vehicle with payload stacked stands 68.3 meters (224 feet) tall and weighs about 589,680 kilograms (1,300,000 pounds). See Figure I-43 for dimensions and weights. The payload capability for the launch vehicle is 16,012 kilograms (35,300 pounds) for SA-206, 16,057 kilograms (35,400 pounds) for SA-207, and 16,239 kilograms (35,800 pounds) for SA-208.

Each Saturn IB will be launched from Launch Complex (LC) 39 at KSC. LC-39 is normally used only for launching the larger Saturn V vehicle. The Saturn IB will rest on a 39 meter (127 foot) tall pedestal on the launch platform. The pedestal will hold the vehicle at the proper height above the platform to use existing tower swingarms with fueling, power and instrumentation facilities (Figure I-44).

Most of the changes made to the Saturn IB were a part of the continuing effort to improve the vehicle. Some changes, however, were necessary to support the Skylab Program. A number of changes eliminated single-point electrical failures and provided redundancy. Engine thrust was uprated to improve the payload capability. Other changes improved the

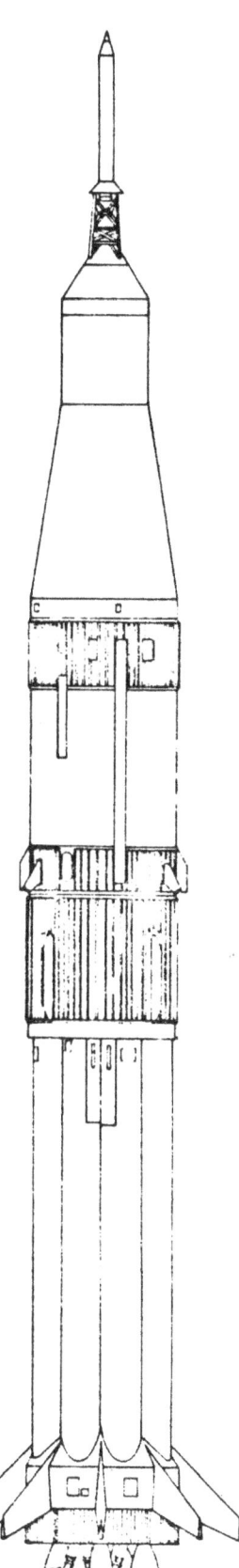

LES	**Launch Escape System** Length 10 meters (33 feet) Weight 3,629 kilograms (8,000 pounds) Thrust 653,856 Newtons (147,000 pounds)
CM	**Command Module** Diameter 3.8 meters (12.8 feet) Height 3.5 meters (11.5 feet) Weight 6,033 kilograms (13,300 pounds)
SM	**Service Module** Diameter 3.8 meters (12.8 feet) Length 7.4 meters (24.8 feet) Weight 7,938 kilograms (17,500 pounds) at Launch Weight 7,462 kilograms (16,450 pounds) at Docking
SLA	**Spacecraft/LM Adapter** Diameter forward end 3.8 meters (12.8 feet) Diameter aft end 6.6 meters (21.7 feet) Length 8.5 meters (28 feet) Weight 1,952 kilograms (4,300 pounds)
IU	**Instrument Unit** Diameter 6.6 meters (21.7 feet) Height 0.9 meter (3 feet) Weight 1,996 kilograms (4,400 pounds)

S-IVB	**Second Stage**
	Diameter 6.6 meters (21.7 feet)
	Length 17.8 meters (58.4 feet)
	Weight 114,760 kilograms (253,000 pounds) fueled
	       10,433 kilograms (23,000 pounds) dry
	Engine J-2 (1)
	Propellants: Liquid Oxygen 75,700 liters (20,000 gallons)
	              Liquid Hydrogen 242,240 liters (64,000 gallons)
	Thrust 1,000,800 Newtons (225,000 pounds)
	Interstage Approx. 2,948 kilograms (6,500 pounds)

S-IB	**First Stage***
	Diameter 6.5 meters (21.4 feet)
	Length 24.4 meters (80.2 feet)
	Weight 452,240 kilograms (997,000 pounds) fueled
	       38,347 kilograms (84,540 pounds) dry
	Engines H-1 (8)
	Propellants: Liquid Oxygen 249,810 liters (66,000 gallons)
	              RP-1 (Kerosene) 157,077 liters (41,500 gallons)
	Thrust 7,294,700 Newtons (1,640,000 pounds)

* Approximations for AS-206 vehicle.

Figure I-43 — SL-2 Vehicle

vehicle's reliability by eliminating possible problems, such as leaks and corrosion. One change provided a means of dumping residual fuel through the J-2 engine of the second stage. Other changes added sensors to collect needed data.

Engine cutoff circuits in both propulsive stages were redesigned to eliminate inadvertent engine cutoff. Two vibration sensors and one pressure measurement device were added for evaluation of possible longitudinal oscillations ("Pogo") and load responses if such should occur in flight.

Each of the eight H-1 engines was uprated from 889,600 Newtons (200,000 pounds) to 911,840 Newtons (205,000 pounds) thrust to increase the payload capability. Total stage thrust was increased from 7,116,800 Newtons (1,600,000 pounds) to 7.294,720 Newtons (1,640,000 pounds).

In the hydraulics system, a number of standard "O" rings were replaced with Quad-Bond seals because the latter resist spiraling. Spiraling of "O" rings in the past has resulted in leaks that caused failures in some components.

Additional instrumentation was added to provide measurements to identify low frequency vibration levels experiences during flight.

In the pneumatic system, the Pneumatic Power Control Module was replaced with a dual regulator and Actuation Control Module to eliminate marginal system performance and provide specified actuation pressure for the S-IVB J-2 engine start tank vent and relief valve.

The LH_2 tank relief and latching bypass valves in the S-IVB propellant system were replaced with a latching vent and relief valve. The open latch feature provides tank safing beyond the life of stage batteries. The "LOX Depletion Timer" was removed to eliminate a single-point failure. Safe engine shutdown using LOX depletion was demonstrated which made protection by the timer unnecessary.

Redesign of LOX and LH_2 prevalves improved reliability, and redesign of the directional propellant control valve eliminated materials subject to stress corrosion. An orbital safing system was installed to provide the capability of dumping residual LOX and LH_2 and engine pneumatics through the J-2 engine. The tanks will be safed through their respective vent systems.

Figure I-44 — SL-2 On Launch Pedestal

Redesign of the LOX and LH$_2$ low pressure feed incorporates multi-ply bellows and features which increase resistance to flow induced vibrations. An alternate propellant loading capability was added to the S-IVB for improved loading accuracy and reliability in alternate loading mode.

Structural changes include the addition of a drag-in cable door in the S-IVB aft interstage and another such door in the forward skirt so that LC-39 can be used for launching Saturn IB vehicles.

In the Instrument Unit, redesign of the inverter detector will reduce its susceptibility to negative transients, and modification of the flight control computer will minimize cracking of solder joints over an extended period.

Redesign of circuitry in the power distributor minimizes voltage transients of short duration momentarily interrupting coolant pump No. 1 operation. The flight control computer was isolated from coolant pump noise on the +6D41 bus to avoid erroneous engine movement signals. Redundant power has been provided to the IU switch selector and S-IB stage to insure operation in case of +6D10 or +6D30 battery failure.

In the Environmental Control System (ECS), a hydraulic snubber assembly was installed between the ECS coolant pump outlet and pressure transducer to prevent the transducer from responding to dynamic pressure fluctuations from the coolant pump.

Four relays were added to the IU Emergency Detection System (EDS) circuitry to allow the dumping of propellant through the J-2 engine by inhibiting EDS cutoff from turning off engine control power. In conjunction with this change, a third EDS cutoff signal was added to the S-IB for a 2 out of 3 voting logic in the S-IB.

Redundant power was provided for the ST-124 platform stabilized system, and the command decoder was redesigned to improve aging characteristics and overcome a possible problem of cracked solder joints.

Two sensor panels were added to the exterior of the IU to provide data to evaluate short-term stability of the optical properties of the thermal control coatings (AS-206 only).

Other telemetry changes made will reduce electrical noise of the IU and vehicle affecting the telemetry system, improve launch range coverage for Skylab missions, and increase the capability to detect a battery failure.

Increasing the size of GHe storage spheres provided more storage volume for longer missions and reduced total weight.

The methanol/water coolant in the thermal conditioning system was changed to Oronite Flo-Cool 100 because the latter is less susceptible to corrosive galvanic currents and will reduce the alkaline content which reacts with the cooling system as compared to methanol/water.

The capability has been added to deorbit the S-IVB/IU stages of SL-2, SL-3 and SL-4 by dumping residual propellants through the J-2 engine to alter the vehicle's trajectory. By controlling the vehicle's attitude and the time and duration of the dump, the S-IVB/IU can be impacted in an ocean. The maneuver will be commanded in real time after considering vehicle trajectory, condition and capability. Sending the spent stage into the ocean eliminates the hazard of space debris impacting a populated land area.

I-9

RESTRAINTS AND MOBILITY AIDS

Placing a spacecraft into orbit results in a condition known as weightlessness which presents problems both for the design engineer and the astronaut.

Astronauts have another "law" within the bounds of which they are held – the principle of action and reaction. An attempt to turn a handle in one direction would cause a crewman to turn in the opposite direction unless he was held in place.

These two conditions made it necessary to design into the Skylab a complete system of restraints and mobility aids — internal for normal operations and external for EVA (Figures I-45 through I-49). One type of device is for crew restraint and mobility and the other is for equipment.

Crew restraints and mobility aids include open grids, foot restraints, handrails and handholds, tethering devices, and thigh, sleep and waste management restraints. Equipment restraints include straps, bungees, towel holders, universal mounts, liquid cooled garment hangers, Velcro and snaps.

The grid structure for walls and ceilings is predominant throughout the OWS. Some sections have covers on one side that keep articles from passing through and other sections are uncovered or "open grid" structures. The open grids are handy for restraints. The triangular openings are uniform and sized to fit triangular plates fastened to the soles of shoes. A crewman wearing the shoes can insert the plate or "cleat" into a grid opening and turn his foot to lock himself into place. These grids are placed at points in the OWS and MDA where crewmen will be working with their hands but will need to be held in place. The stowage ring in the forward section of the OWS also has cleat receptable holes.

Astronauts will have other devices for holding their feet in place. One is a pliable arrangement similar to the straps on shower "clogs" placed at various points, such as three pairs on the floor of the wardroom at the base of the food table and a pair in the WMC on the floor beneath the handwasher. A third type of foot restraint consists of toe bars and heel fittings fastened to the floors or walls. These can be moved wherever needed.

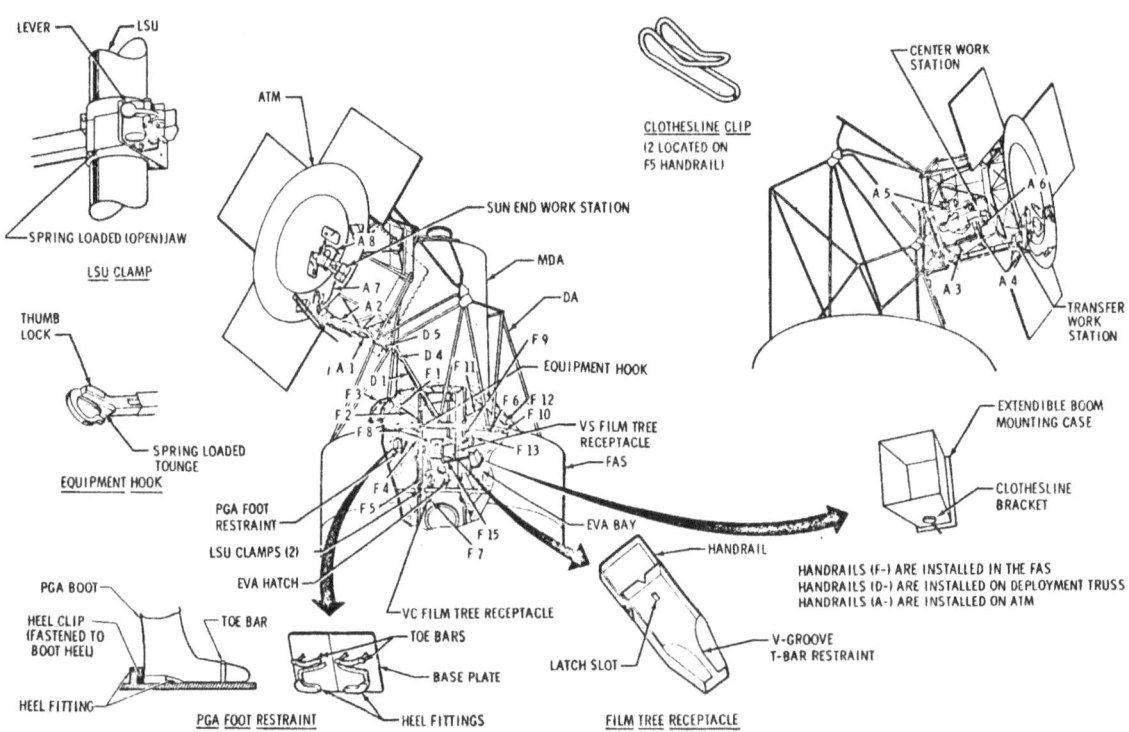

Figure I-45 — External Restraints And Mobility Aids

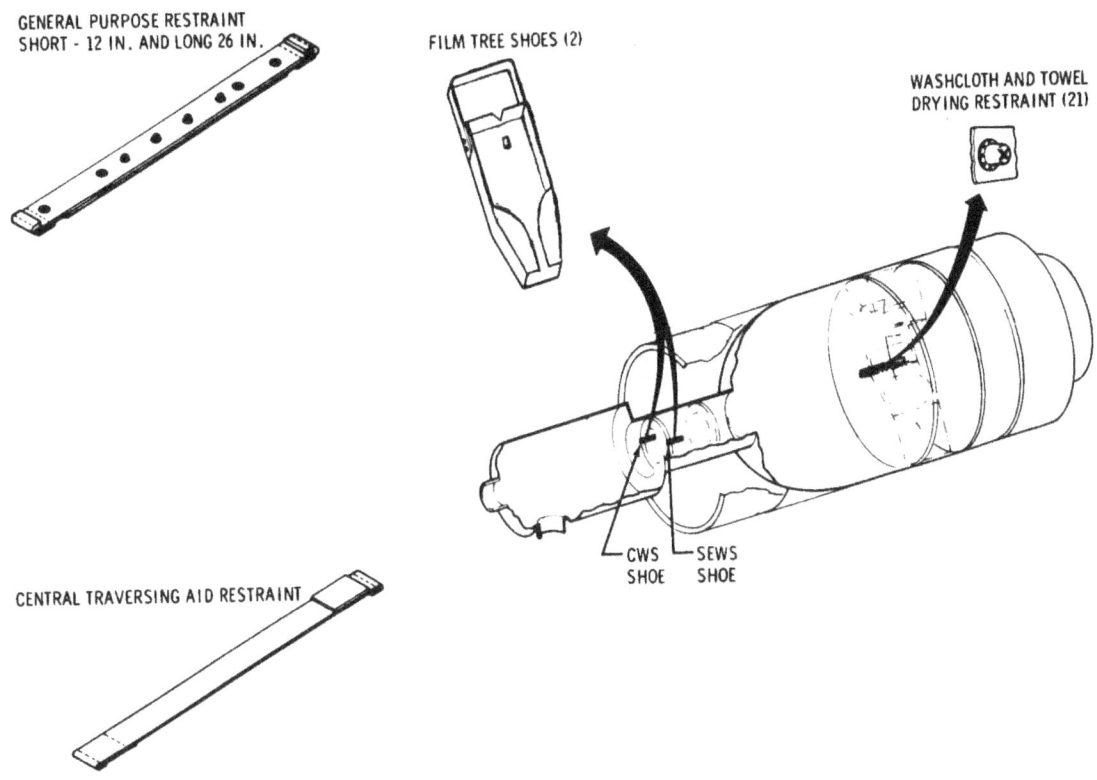

Figure I-46 — Equipment Restraints

Handrails and handholds are used extensively in all components of the OA, both inside and outside. Most of these are fixed but crewmen will have a number of detachable handholds they can move to various locations. They will use the handrails and handholds to propel themselves about the station and hold themselves in position temporarily. They will use an extensive system of rails in performing EVA.

Tethering devices are simply straps with hooks which crewmen can attach to themselves and fixed points to keep them from drifting too far. Thigh restraints are bars which extend from the food table. Each has two other bars that form a double "T" at the end. Astronauts put one leg on each side of the main bar and between the other two bars. Sleep restraints are similar to sleeping bags. The astronaut gets into the bag and zips up the front. Attached at the floor and ceiling, the bag holds the crewman in place while he is sleeping. Waste management control restraints consist of handholds and foot restraints. Astronauts will also have a handy device for moving from one section of the OWS to another — the "fireman's pole," which is removable and collapsible. All handrails and fixed handholds are painted blue for ease of identification.

Straps of varying lengths and stretchable bungees are stored on the OWS for use in holding and putting pressure on equipment to keep it in place. Rubber cylinders with a cross cut in the end are used to hold washcloths and towels. Universal mounts can be moved to sites needed and snapped into place to anchor equipment. Temporary restrainers include Velcro material and a uniform snap pattern.

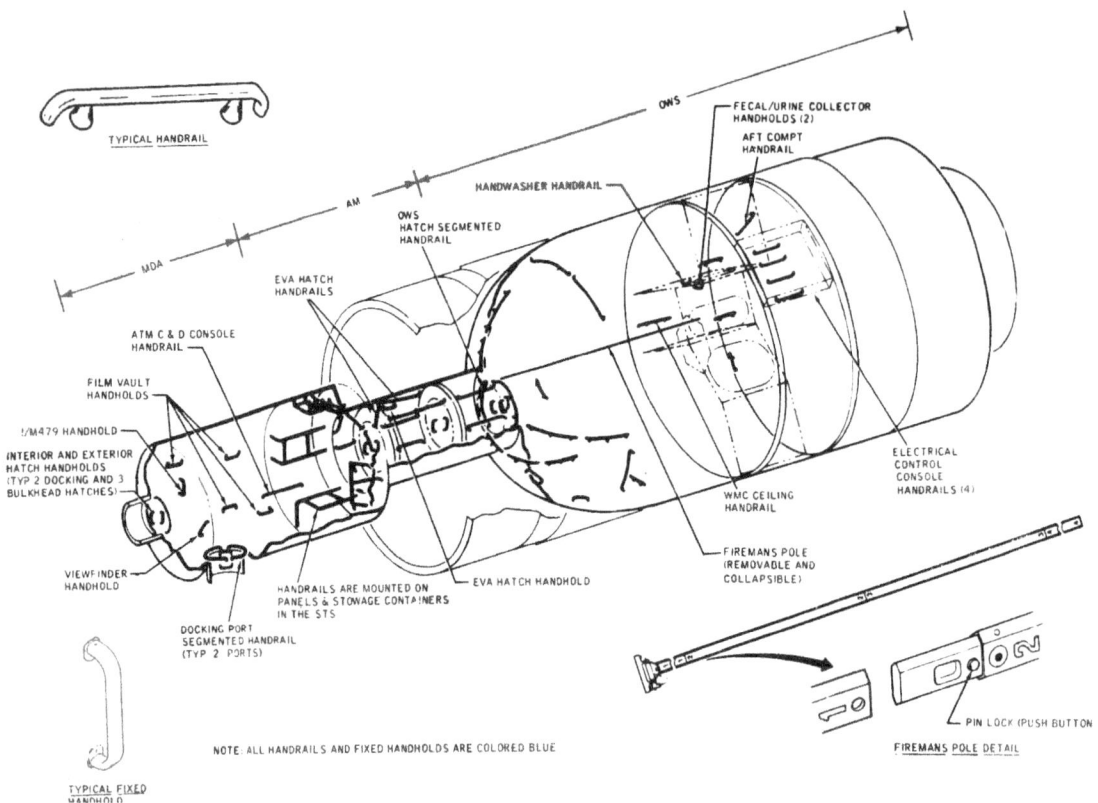

Figure I-47 — Hand Restraints

BACKUP HARDWARE

An identical set of hardware will be available the day the flight hardware for Skylab is launched. The backup hardware has two prime purposes: It can be used in place of the flight hardware in case of a failure of the latter; and it can be used for troubleshooting, confidence testing and systems engineering during the Skylab mission. Backup hardware will be kept available at least through completion of the second manned visit.

NASA will have the capability to launch the backup Skylab within 12 to 15 months after a decision is made to do so. The launch vehicle for a backup launch would be Saturn V vehicle AS-515 composed of S-IC-15, S-II-15 and S-IU-515, now in storage at KSC.

The backup Workshop will not be given post-manufacturing test and checkout unless required for flight, and experiments will not be installed. It will be stored at the McDonnell Douglas plant in Huntington Beach, Calif.

Manufacturing test and checkout was completed on the MDA before it was shipped to the McDonnell Douglas plant at St. Louis for mating with the AM. The mated AM/MDA was tested through simulated flight test, a confidence test program in support of SL-1, instead of post-manufacturing and pre-delivery checkout. The unit is at the St. Louis plant ready for use in duplicating conditions of possible problems encountered in the flight hardware during the mission.

Refurbishment of the ATM prototype structural and electronic hardware is now in work at MSFC and at the many ATM experimenter and vendor facilities. This is scheduled for completion in 1973. Limited subsystems checkout of this hardware will be performed at MSFC following assembly and the unit will be maintained as required for provide a flight backup unit.

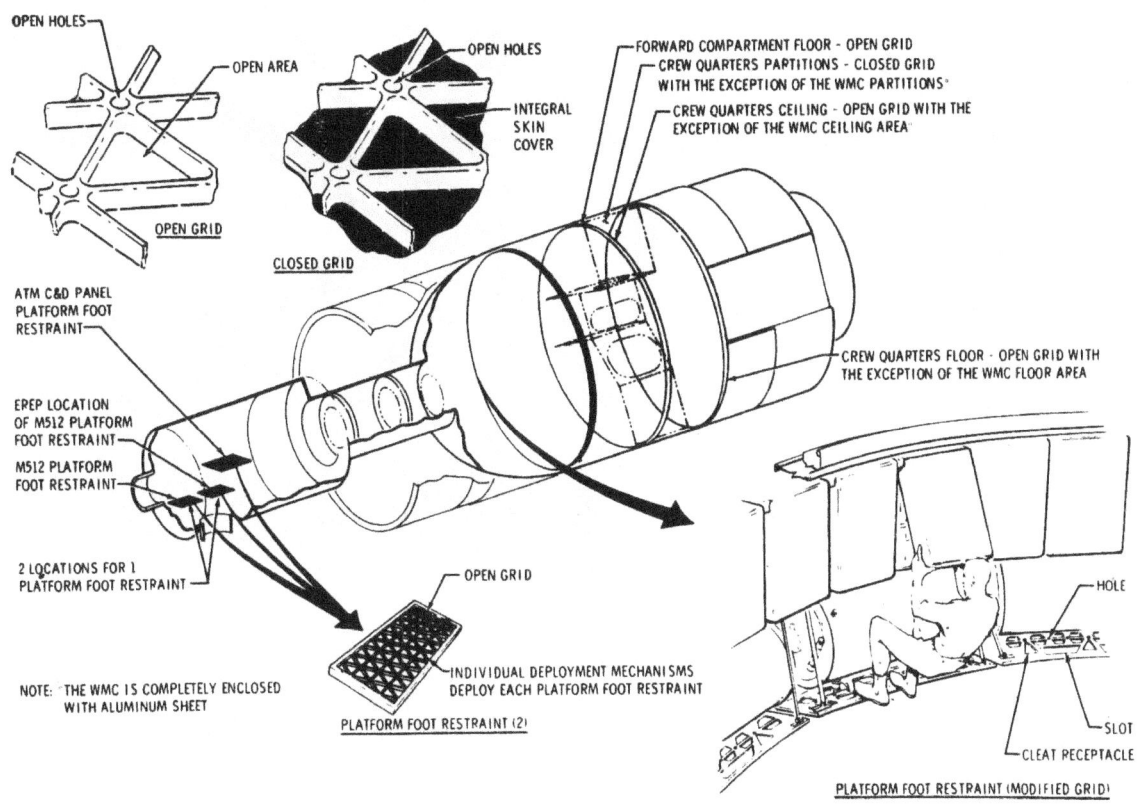

Figure I-48 — Grid Type Restraints

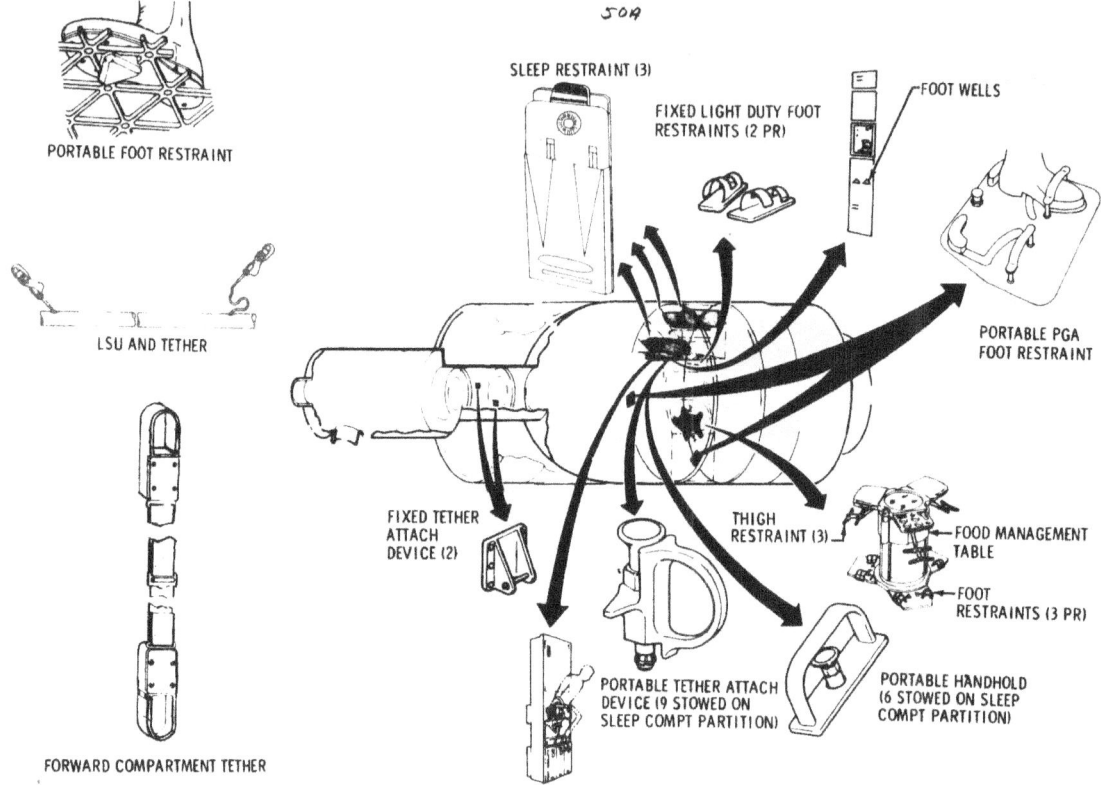

Figure I-49 — Crewman Restraints

SECTION II

SYSTEMS

LIFE SUPPORT SYSTEM

The Skylab Life Support System encompasses:

Environmental control provisions to decontaminate, deodorize, dry, heat and circulate a two gas atmosphere.

Food and water management equipment to store, preserve and prepare palatable food and drinking water.

Waste management equipment to collect, process and dispose of body wastes and mission trash.

Sleep provisions to provide restraint and privacy during sleep periods.

Personal hygiene arrangements to maintain personal health and comfort.

Microbial control precautions to assist in maintaining crew health and comfort.

Inflight Medical Support and Medical Accessory Kit equipment to enable remedial activities in the event of illness or injury.

Operational biomedical equipment to monitor physiological parameters during critical tasks such as EVA.

Environmental Control

Skylab is the first U.S. manned spacecraft pressurized with a two gas atmosphere. The characteristics of this atmosphere are compared to the normal Earth environment in Table II-1.

The Environmental Control System (ECS) provides for distribution and purification of the atmosphere in Skylab, control of pressure, ventilation, cooling and heating of the atmosphere, and cooling and heating for the crew during EVA. The ECS is comprised of the Gas Distribution and Pressure Control Subsystem, the Ventilation and Atmosphere Control Subsystem, and the EVA/IVA Support Subsystem.

Characteristic	Skylab	Earth
Oxygen	72%	21%
Nitrogen	24 to 28%	79%
Pressure	3.45 N/cm² (5 psia)	10.13 N/cm² (14.7 psia)
Temperature	15.6°-32°C (60-90°F)	Variable
Dew Point	8°-15.6°C (46°-60°F)	Variable

Skylab uses a gas distribution and pressure control subsystem (Figure II-1), with a total gas pressure of 3.45 N/cm² (5 psia), an oxygen partial pressure of 2.48 N/cm² (3.6 psia) nominal, a nitrogen nominal partial pressure of 0.6 N/cm² (1.4 psia) and a circulation velocity of 4.6 to 30.5 m (15 to 100 ft) per min. The oxygen is stored in six 114 cm (45-inch) diameter 228.6 cm (90-inch) long cylindrical steel lined fiberglass tanks at 2067 N/cm² (3000 psia), mounted on the internal portion of the Fixed Airlock Shroud. The nitrogen is stored in six 102 cm (40-inch) diameter titanium spherical tanks at 2067 N/cm² (3000 psia), mounted on the external AM trusses.

The two-gas control system supplies either oxygen or nitrogen to the Skylab cluster. Oxygen is supplied until the partial pressure of the oxygen reaches 2.55 N/cm² (3.7 psia) and the nitrogen is supplied until the partial pressure of oxygen drops to 2.41 N/cm² (3.5 psia). Potential over-pressurization of the space vehicle is prevented by three pressure relief valves located in the AM.

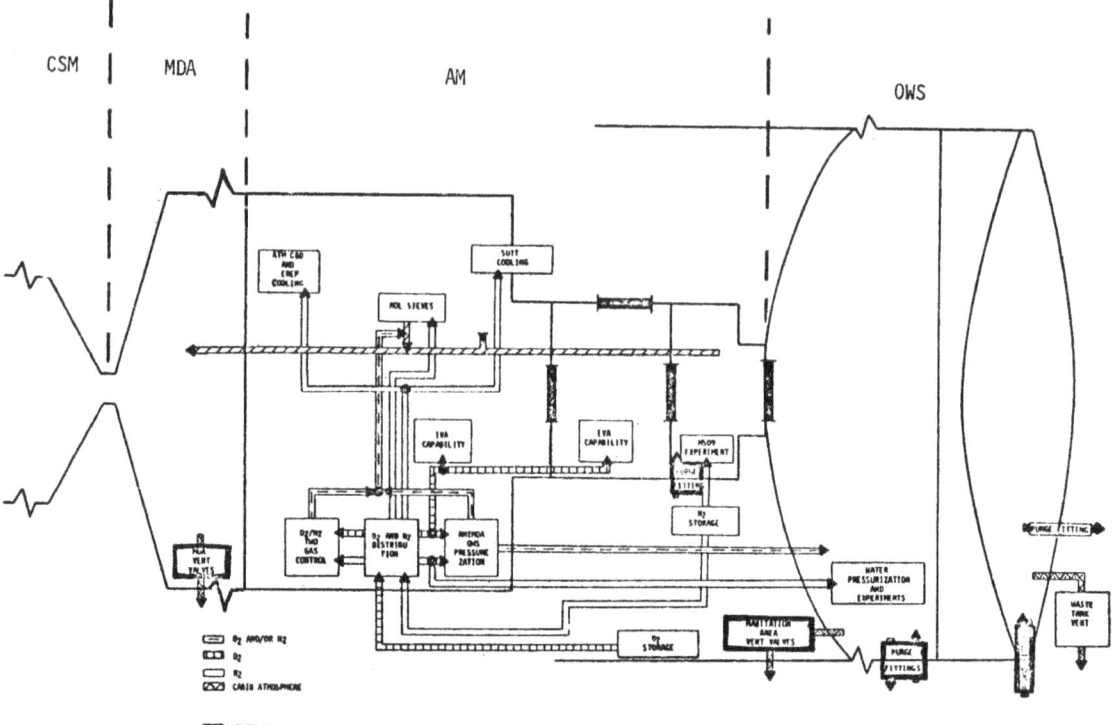

Figure II-1 — Pressurization and Gas Distribution

Prior to launch, the AM/MDA/OWS is pressurized with nitrogen. During ascent the nitrogen from the MDA/AM is allowed to bleed down through two MDA vent valves until a timed command closes the valves. Nitrogen from the OWS is vented through its own vent valve system during and following ascent. Nitrogen from the waste tank is vented through its own system also, after orbit is attained. It has been estimated that 0.9 N/cm² (1.3 psia) of nitrogen will be left in the MDA/AM/OWS when the valves are closed. The waste tank is allowed to vent to vacuum. A ground command is then used to open oxygen system solenoid valves, allowing oxygen to flow into the habitable volume until a total pressure of 3.45 N/cm² (5 psia) is reached. After the solenoid valves are closed, the oxygen/nitrogen makeup system reverts to a demand basis.

During the unmanned storage periods the orbital assembly pressure is vented to 1.38 N/cm² (2 psia) and then allowed to decay by normal leakage to a minimum pressure of 0.34 N/cm² (0.5 psia), after which it is repressurized by command as required.

Skylab uses the ventilation and atmosphere control subsystem (Figures II-2 and II-3), to limit contamination and to maintain safe and comfortable levels of CO_2, humidity and temperature. Fans in the MDA, AM and OWS are used to circulate the atmosphere for crew comfort. Three cabin heat exchangers in the AM Structural Transition Section (STS) provide cooling to the MDA/STS region and four heat exchangers in the aft AM (Figure II-4) provide cooling to the OWS. Fans direct the flow of cool air through OWS, MDA and AM areas.

Humidity, heat, carbon dioxide and odors are removed from the atmosphere by one of two redundant molecular sieve assemblies, A or B (Figure II-5). Each assembly contains two screen filters, two compressors, two heat exchangers, two molecular sieve beds and one activated charcoal canister. During operation of the molecular sieve assembly, atmosphere flows from the AM through the screens which filter large particle contaminants. The

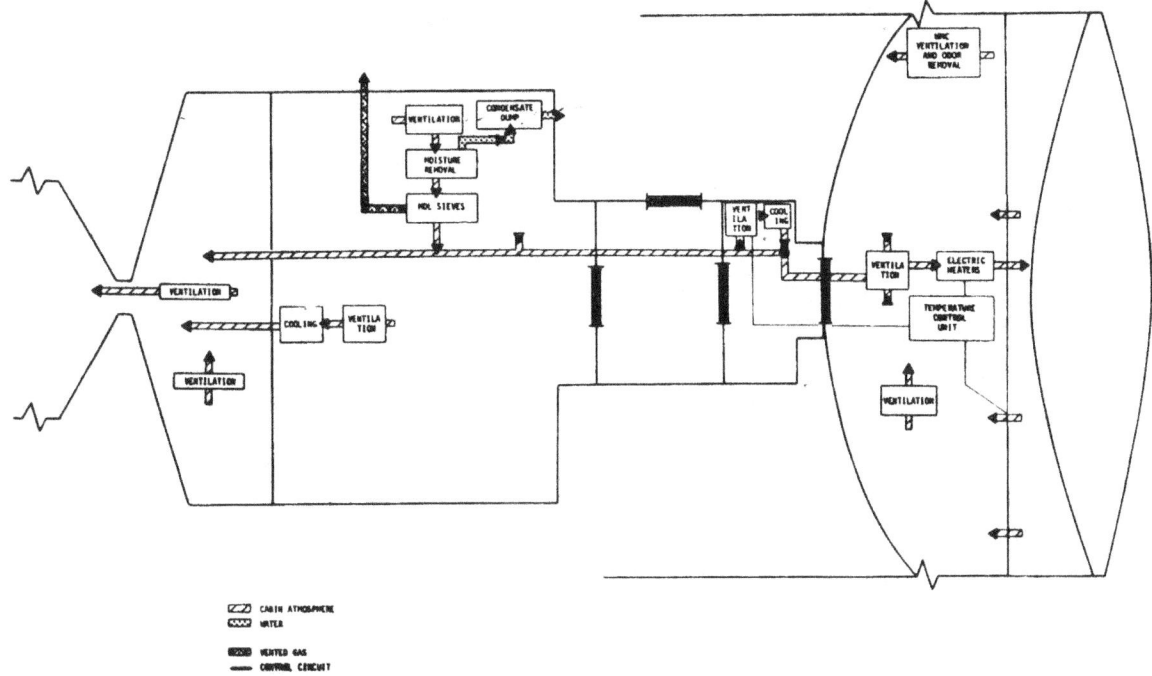

Figure II-2 — Atmosphere Control Diagram

compressors force the atmosphere through one of the condensing heat exchangers which cools the atmosphere and removes some water. Part of the atmosphere then flows into one of the two molecular sieve beds where water and carbon dioxide are removed. The rest of the flow goes through either the activated charcoal canister where odors are removed, or through a bypass. Under the control of valves in each molecular sieve bed, atmosphere flows through one molecular sieve bed while the other is exposed to space vacuum to remove most of the water and CO_2 which has been adsorbed by the molecular sieve bed. The beds are cycled every 15 minutes, automatically, so that one bed is active while the other is exposed to space vacuum. The space vacuum water removal method does not, however, remove all the water, necessitating the use of a 12-hour 204°C (400°F) bake-out cycle after approximately 28 days of sieve operations.

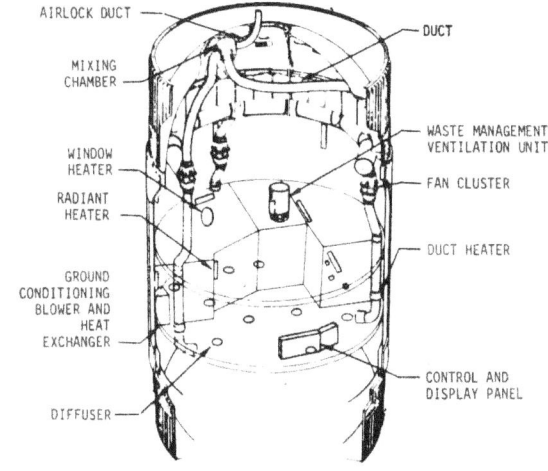

Figure II-3 — OWS Atmosphere Control Equipment

The Environmental Control System provides thermal control and a supply of dry oxygen for suit pressurization, ventilation and breathing to the crewman performing EVA or suited IVA. Suit pressure is maintained for EVA at 2.48 N/cm² (3.6 psia) with a variable controlled rate. Used oxygen is rejected to space through a controller in the suit. A pressure relief valve in the space suit prevents overpressurization. One IVA O_2 supply station is in the forward section of the AM and two EVA/IVA O_2 supply stations are in the lock portion of

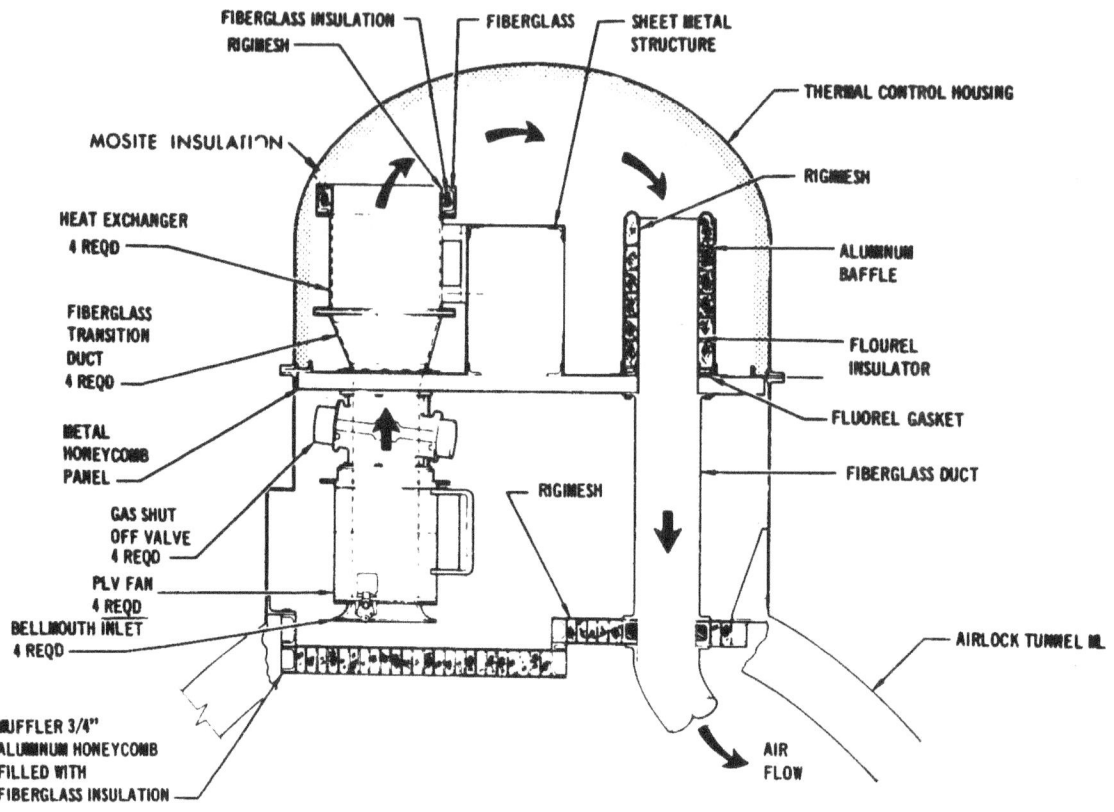

Figure II-4 — OWS Fan-Muffler

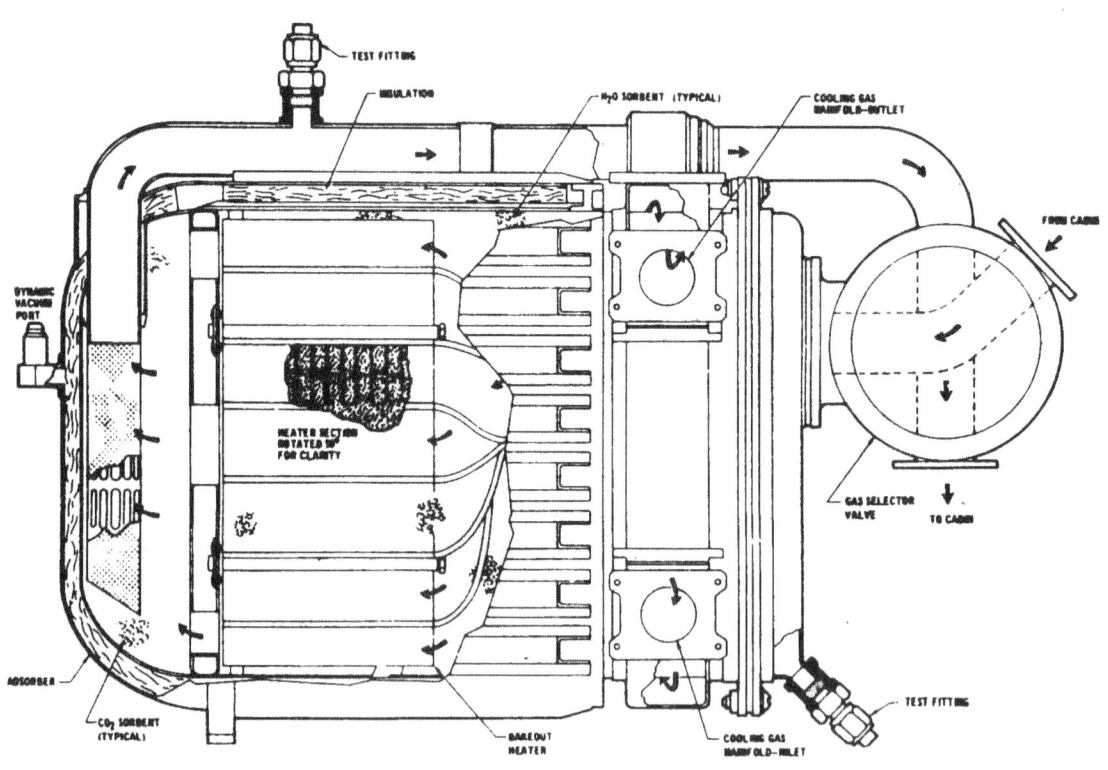

Figure II-5 — Molecular Sieve Bed, Adsorb Mode

the AM. During EVA operations the airlock is depressurized and O_2 is provided to the EVA crewman. After the EVA the AM airlock is repressurized to 3.45 N/cm^2 (5 psia) by gas from the atmosphere.

Active cooling is provided to the crewmen through a water cooled undergarment. The water cooled system is a closed loop system, the water being re-cooled in the EVA heat exchanger in the AM.

Water Management

Unlike the Apollo spacecraft, which generated sufficient potable water as a by-product of fuel cell activity, Skylab contains enough stowed water at launch to serve all of the needs of three men for 140 days of operation. This water is stowed and managed by the OWS water subsystem. The OWS water subsystem provides for storage pressurization, distribution, purification, thermal control and conditioning, and dispensing of the water. The water is provided principally for food reconstitution, drinking, crew hygiene, housekeeping and urine separator flushing, if needed (Table II-2). Water is also provided for the Life Support Umbilical/Pressure Control Unit (LSU/PCU), the AM EVA/IVA cooling loop, the ATM C&D Panel/EREP cooling loop, the M512 facility experiment and the OWS shower.

Water is stored in ten stainless steel storage tanks in the forward compartment area of the OWS (Figure II-6). The water tanks (WT) each having a useable capacity of about 272 kilograms (600 pounds) are identified as 1 through 10 and are assigned to particular water networks: WT 1, WT 2, WT 3, WT 4, WT 5, and WT 10 are assigned for Wardroom use; WT 7 and WT 8 are assigned to the WMC water network; and WT 6 and WT 9 are provided as contingency water tanks in case of excessive water usage or failure in the water supply, in either the Wardroom or the WMC. The capacity of one of the ten water tanks could be lost without decreasing the water supply below mission requirements.

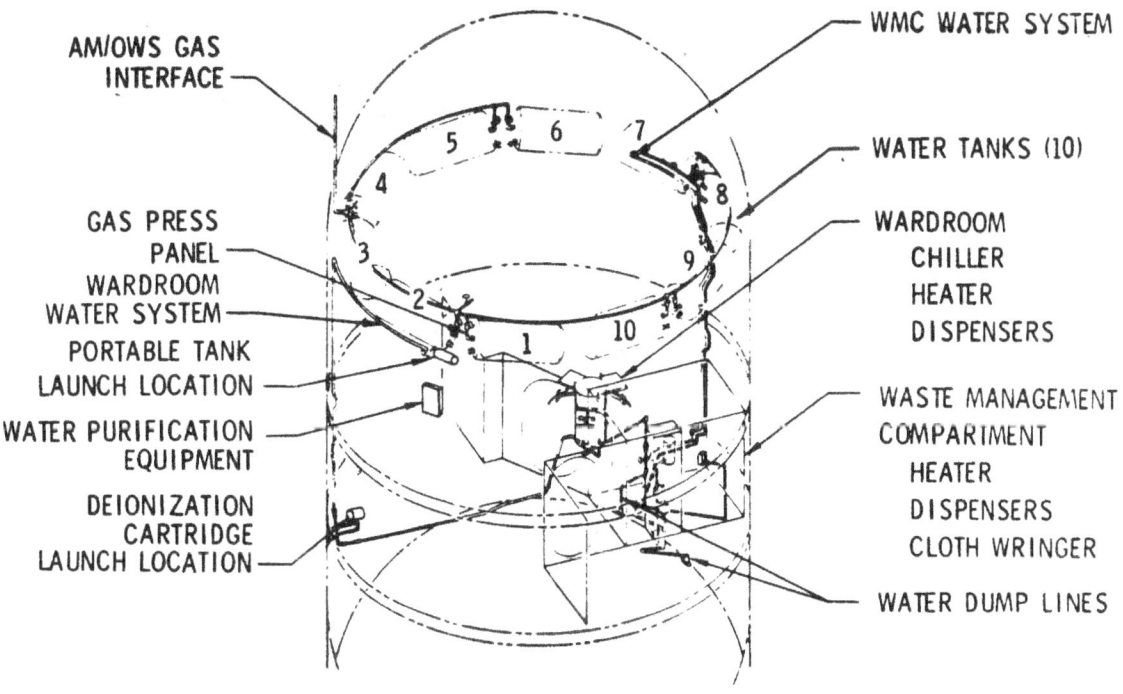

Figure II-6 — OWS Water Systems

Table II-2 – Water Tank Budget

Function	Maximum 3 Crewmen Use Rate	Total Requirement, kg (lb)	Containers Required*	Remaining Usable, kg (lb)
CM return	10.9 kg (24 lb)/mission	32.7 (72)	[1, 10, 2, 3, 4, 5]	(266.79) 121
Metabolic	10.2 kg (22.5 lb)/day	1373.9 (3,028.9)	6	
WR system bleed (end of mission)	6.4 kg (14 lb)/mission	20.4 (45.00)	(3,549.6 lb) 1610 kg	
WR system microbiological flush	9.5 kg (21 lb) start 1st mission 26.3 kg (58 lb) start 2 and 3 missions	62.1 (137.00)		
WM system bleed (end of mission)	4.5 kg (10 lb)/mission			
Housekeeping	1.8 kg (4 lb)/day	243.4 (536.52)	[7, 8, 9] 3	(453.39) 205.7
Personal Hygiene	1.4 kg (3 lb)/day	182.5 (402.39)	(1,774.8 lb) 805 kg	
OWS shower	2.7 kg (6 lb)/shower (1 shower/man/week)	163.3 (360.00)		
Urine flush	600 ml/day	82.1 (180.99)	[6] 1	
EVA & C&D panel		29.5 (65.06)		

*Based on actual usable quantity of 266.2 kg (591.6 pounds) per container.

Each water tank is a stainless steel cylinder (Figure II-7) with a sealed metal bellows inside (Figure II-8). Each tank is an independent unit, supplied with a nitrogen gas pressurant to maintain water supply pressure. The sealed bellows assembly forms a nitrogen gas chamber and provides a constant pressure during use. The bellows extends as water is withdrawn from the tank. Two water tank heater blankets are used on each tank to keep water temperature at approximately 15.6°C (60°F) during all mission phases. Each water tank contains water tank servicing equipment to facilitate ground filling and to permit water purification. Purity of the water is maintained by using iodine as a biocide. The water is periodically sampled on-orbit by use of the sample port valve. If the on-orbit sample analysis reveals a need to purify the water, iodine will be injected into the tank through the iodine injection port, with mixing from operation of the agitator pump.

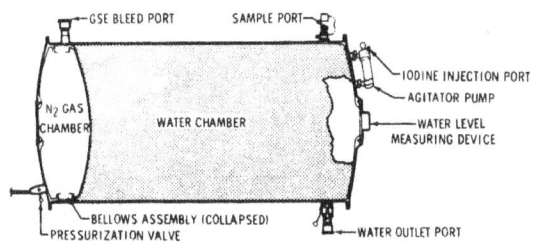

Figure II-8 — Water Tank Cutaway

Figure II-7 — Water Tank

The three water distribution networks are the Wardroom network, the WMC network and the urine flush network, if needed. Each distribution network is designed as a totally independent system to prevent cross contamination.

The Wardroom network distributes water to the food table where the water is chilled or heated for food reconstitution and drinking. The network consists of two water hoses, a water supply line, a filter, relief valves, a water heater which heats the water to 67°C (152°F), and a water chiller which chills the water to 7°C (45°F).

The Wardroom has two water dispenser valves, one cold and one hot, for reconstituting dehydrated foods and beverages. The water chiller, in the table pedestal, provides chilled water to the cold Wardroom water dispenser valve; the internal table-mounted water heater supplies hot water to the Wardroom hot water dispenser valve.

Water is distributed to three color coded water guns for drinking purposes. They are located on the food table pedestal (Figure II-9), adjacent to each eating station. The water guns, fitted with replaceable drinking mouth-pieces, discharge chilled water in small, drinkable quantities.

The WMC water network distributes water to the water heater in the WMC, where the water is heated to 53°C (127°F), and routed to a dispenser for body cleansing and housekeeping purposes. The network is composed of two hoses, a supply line, relief valves and a water heater.

The network supplies water to a WMC water dispenser valve in the hand washer. The dispenser is provided hot water from the WMC heater. The dispenser contains a plunger which, when depressed, expels three jets of water in a continuous stream into the crewman's hand-held washcloth.

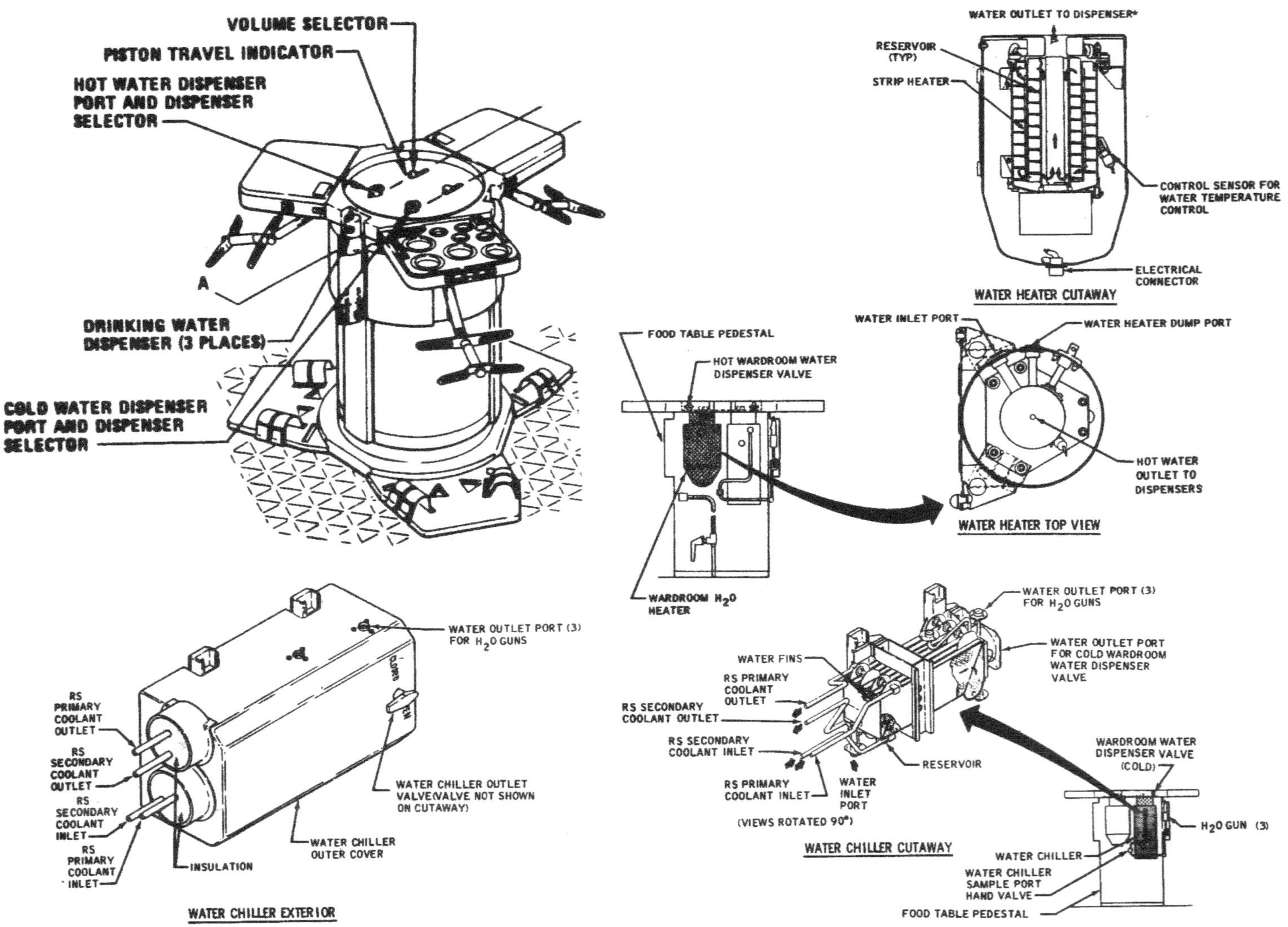

Figure II-9 — Food Table Pedestal

II-8

The urine system flush water network distributes water from the tank to the flush equipment in the WMC corner stowage compartment. This equipment is used only if needed to flush residual urine from the urine separators in the fecal/urine collector. The network is composed of a water hose, a water supply line and a filter.

The wardroom water dump provides for wardroom water network evacuation into the waste tank. Dumping is through quick-disconnects, flexible hoses, tubing, a hand valve, and a heated waste tank discharge nozzle (dump heater probe).

The WMC water dump provides for evacuation of the WMC water network, urine system flush water network (if used), washcloth squeezer bag and the condensate control system into the waste tank.

The portable water tank (Figure II-10) is independent and completely portable. It accommodates a self-contained pressurization unit and a 11.8 kilogram (26 pound) capacity water supply and is mounted in the OWS forward compartment on a wall bracket below WT 1 and WT 2. It may be taken to or near the area needed if one of the water networks fails. The quick-release fastener on the tank permits retention on any grid surface.

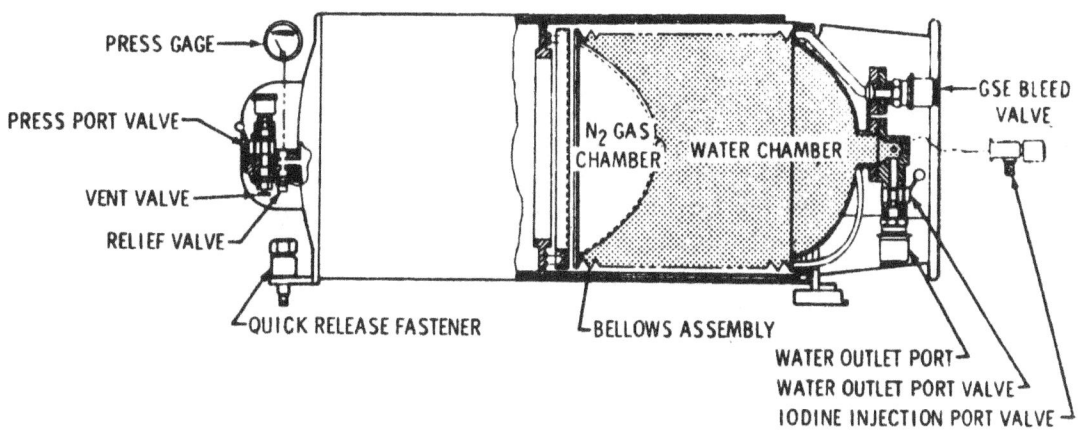

Figure II-10 — Portable Water Tank Cutaway

Food Management

The meals on Skylab are more palatable than those of previous manned programs, due primarily to advancements in technology which allow the inclusion of frozen and canned foods.

The Skylab food system must meet the rigid requirements and objectives of medical experiments which demand precise knowledge of nutrient intake and must fulfill the experimental objectives of the M487 Habitability experiment.

The food system provides the energy requirements of each individual astronaut based on his body weight and age. The ration will insure a daily intake of between 750 and 850 mg of calcium, 1500 to 1700 mg of phosphorous, 3,000 to 6,000 mg of sodium, 300 to 400 mg of magnesium and 90 to 125 g of protein. Each astronaut will maintain a constant level of intake of these controlled nutrients within two percent. The diet is baselined to provide at least the dietary allowances of carbohydrates, minerals, vitamins, and fats recommended by the National Academy of Science.

About 950 kilograms (2,100 pounds) of food and accessories for all three manned visits is stowed aboard the workshop prior to launch. The frozen food items will be secured in the five food freezers which have a combined stowage space of 0.3 cubic meter (10.6

cubic feet) and the other food items are stowed in lockers on the floor of the workshop at launch and later placed in assigned locations during the activation period by the SL-2 crew. The 11 food lockers have a combined storage area of 2.5 cubic meters (88.3 cubic feet).

The more than 70 food items provided in Skylab have been taste tested by crew members. The items aboard the OWS will meet the following criteria: The food is a familiar kind; the food portions are processed to be prepared, served and eaten in a familiar manner; and the prepared food is satisfactory with regard to taste, aroma, shape, color, texture and temperature.

The Skylab menu includes the following food types:

<u>Dehydrated</u> — ready to eat rehydratable foods with a moisture content reduced to less than 3 percent.

<u>Intermediate Moisture</u> — precooked, thermally stabilized, or fresh food with the moisture content partially reduced so that the final moisture content is approximately 10 to 20 percent.

<u>Thermostabilized</u> — precooked, thermally stabilized or fresh food with the temperature reduced below -23°C (-10°F) prior to launch to retard spoilage.

<u>Frozen</u> — precooked fresh food with temperature reduced below -40°C (-40°F) before launch to retard spoilage and maintained in freezers in the OWS.

<u>Beverages</u> — rehydratable drinks including: black coffee, tea, cocoa, cocoa flavored instant breakfast, grape drink, limeade, lemonade, orange drink, grapefruit drink, cherry drink and apple drink.

An example of a typical three day menu for one crewman is given in Table II-3.

The crewman assigned to duty of chef for the day consults the day's menu to determine what food and beverages are required for the next meal. He removes the items from the food lockers and carries them to the OWS food table (Figure II-11) for preparation. One portable food tray is provided for each crewman. The food trays (Figure II-12) have eight food cavities — four for large food cans and four for small cans. Three of the large cavities are heated. Foods which require heating are placed in those cavities. The astronaut sets the timer and turns on the warmer. A removable tray lid is used when the food is heating and is stowed in the food tray stowage area when not in use. Each of the trays and lids are color coded. Magnets hold the utensils to the tray until used. At the end of each meal, the food trays are filled for the next meal and the heater timers set for the appropriate interval.

Foods which require rehydrating are prepared at mealtime by each astronaut through the use of the cold or hot water gun in the center of the food table (Figure II-9). The food table pedestal houses the water chiller and the wardroom water heater.

A plastic membrane is installed inside each food can to prevent spillage in the zero-g environment. The membrane is fitted with a one-way spring loaded valve for rehydratable food which permits the addition of water without leakage and prevents the escape of contents during and after lid removal and mixing. The crewman slits the membrane with his knife and the membrane retains the balance of the contents until consumed. Figures II-13 through II-16 illustrate food storage and preparation equipment.

Rehydratable beverages are packaged individually in collapsed accordion-shaped beverage packs which vary in length according to the type of beverage. These packs expand in length as the cold or hot water is added to the beverage powder.

The astronauts sit at the food table in special thigh restraints with their feet in portable foot restraints (Figure II-17).

Table II-3 — Sample Menu

Commander Charles Conrad

Meal	Day 1	Day 2	Day 3
A	Scrambled Eggs Sausage Patties Strawberries Bread Jam Orange Juice Coffee	Cornflakes Chocolate Instant Breakfast Grape Drink Coffee	Scrambled Eggs Sausage Biscuit Cocoa Coffee
B	Chicken and Gravy Asparagus Peaches Biscuit Cocoa Lemonade	Cream of Potato Soup Chicken and Rice Pre-Buttered Roll Peaches Lemonade	Cream of Potato Soup Pork & Scalloped Potatoes Green Beans with Cheese Sauce Pears Grape Drink
C	Veal & Barbecue Sauce Mashed Potatoes Green Beans with Cheese Sauce Peach Ambrosia with Pecans Grapefruit Juice Snacks: Coffee x 2 Butterscotch Pudding	Filet Mignon German Potato Salad Pears Biscuit Cocoa Grape Drink Snacks: Coffee x 2	Lobster Newberg Mashed Potatoes Asparagus Vanilla Wafers Vanilla Ice Cream Lemonade Snacks: Dried Apricots Coffee x 2

Waste Management

The management of body wastes for three men and the mission trash from 140 days of manned operation posed a challenging problem for NASA design teams. To complicate the situation, body wastes are required as samples for a group of experiments conceived to assess the muscle and bone changes in crewmen exposed to weightlessness over a long period (M071, Mineral Balance, and M073, Bioassay of Body Fluids).

A collector module (Figure II-18) in the Waste Management Compartment (WMC) of the OWS performs the same function as an Earth toilet, but in a different manner. The collector module is used by a crewman in a seated position facing the compartment floor. This collector permits one crewman to defecate and urinate simultaneously while seated, using a lap belt and two handholds to hold himself in place. The urine receptacle, which is connected through the centrifugal separator to the urine inlet valve of the urine bag, is also a convenient height for standing urination. A pair of light-duty foot restraints in front of the fecal/urine collector (Figure II-19) permits standing urination and allows the crewman to conduct maintenance on the fecal/urine collector.

Figure II-11 — OWS Food Table

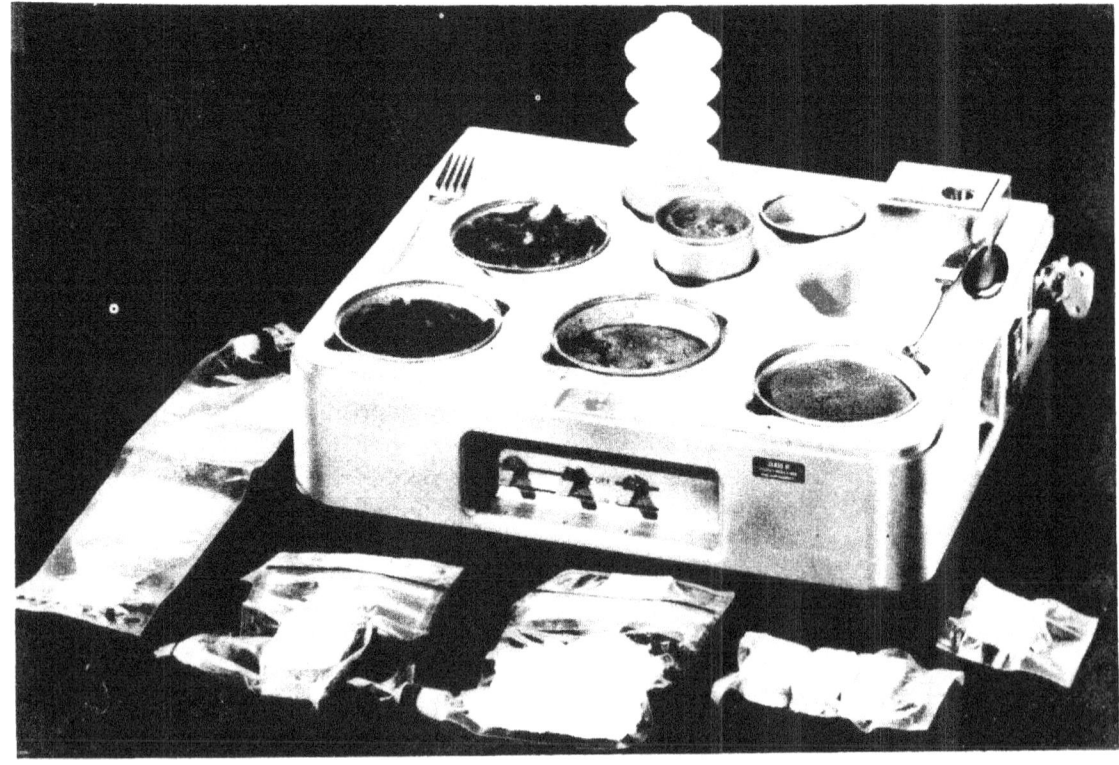

Figure II-12 — Food Tray

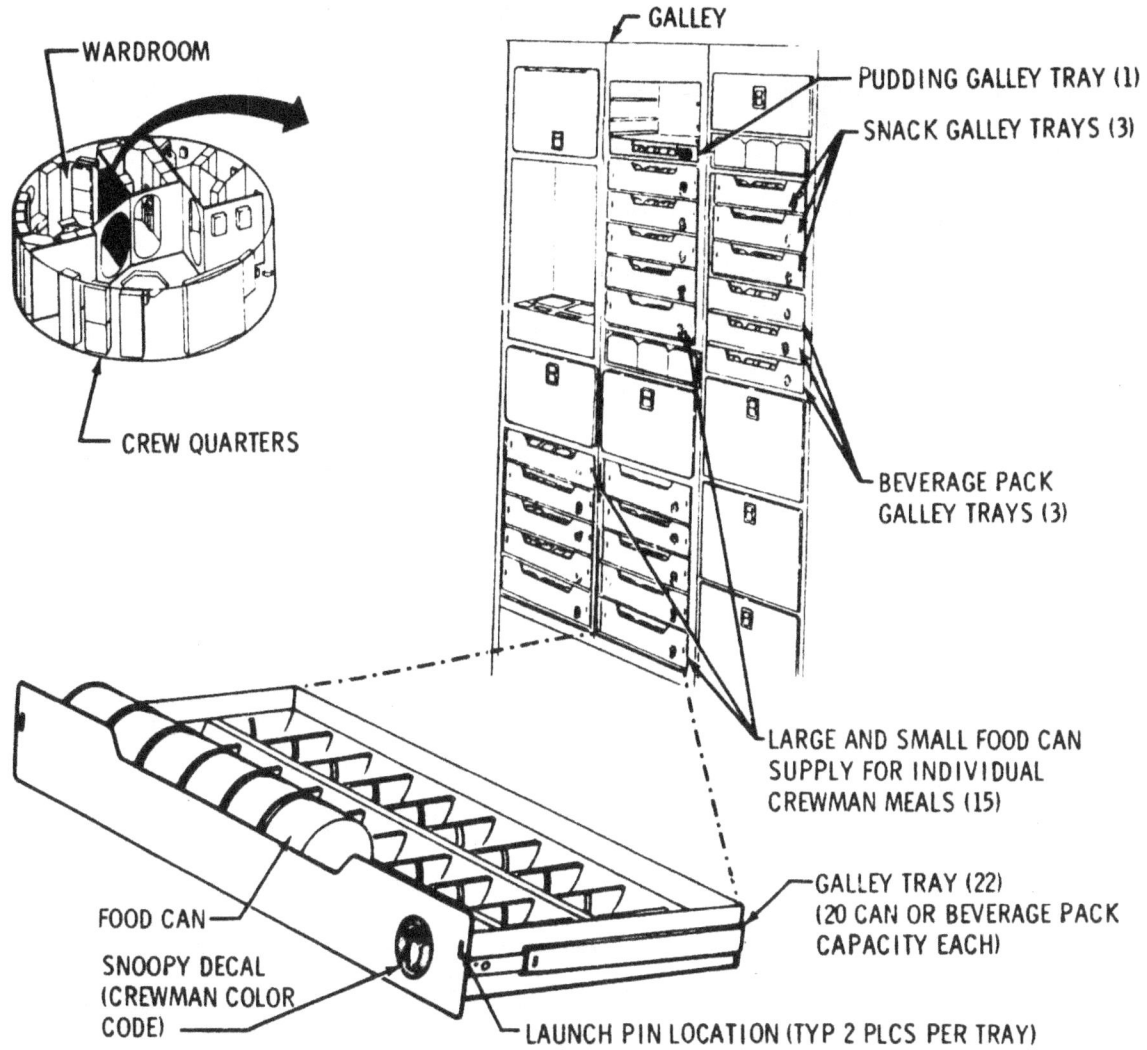

Figure II-13 — Non-Refrigerated Food Storage — Daily

The fecal/urine collector is a rigid, wall-mounted unit. It contains one fecal bag (Figure II-20) for collection of a single defecation and three urine drawers, one per crewman, to collect each urination and to store the urine in a chilled state for 24 hours. A fecal/urine collector blower unit provides a gravity substitute airflow (suction) to draw and retain the waste material into the fecal collector (Figure II-21) and into the urine drawer (Figure II-22) during waste collection. The airflow is filtered to remove odors prior to its recirculation back into the cabin by the blower unit. The fecal bag is removed from the fecal collector after each defecation and replaced immediately with a new bag. The fecal bag with its contents is weighed on a mass measuring device and then vacuum dried in a waste processor to facilitate on-orbit storage.

The waste processors (Figure II-23) preserve, in collection bags, those organic and inorganic constituents of vomit and feces required to support the medical experiments. Six independent waste processors are wall-mounted in the WMC. Each processor uses mechanical pressure, an electric heating element and venting to the waste tank vacuum to dry the waste material. Each processor will accommodate one fecal bag or one contingency

AMBIENT FOOD STORAGE

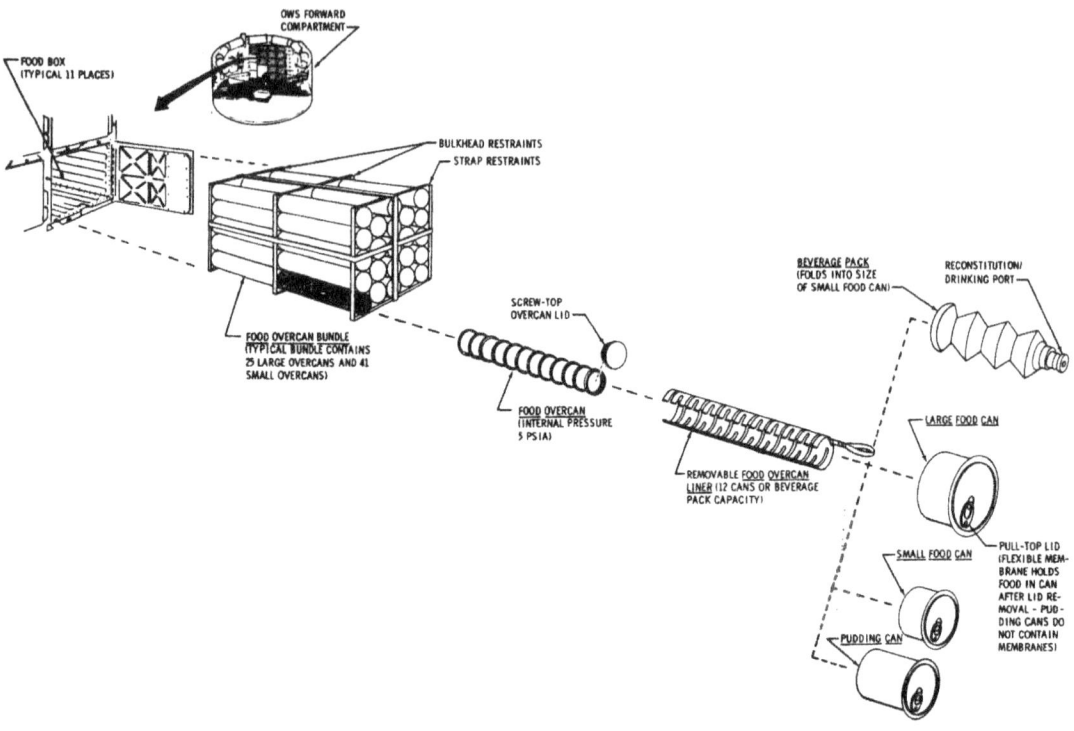

Figure II-14 — Non-Refrigerated Food Storage

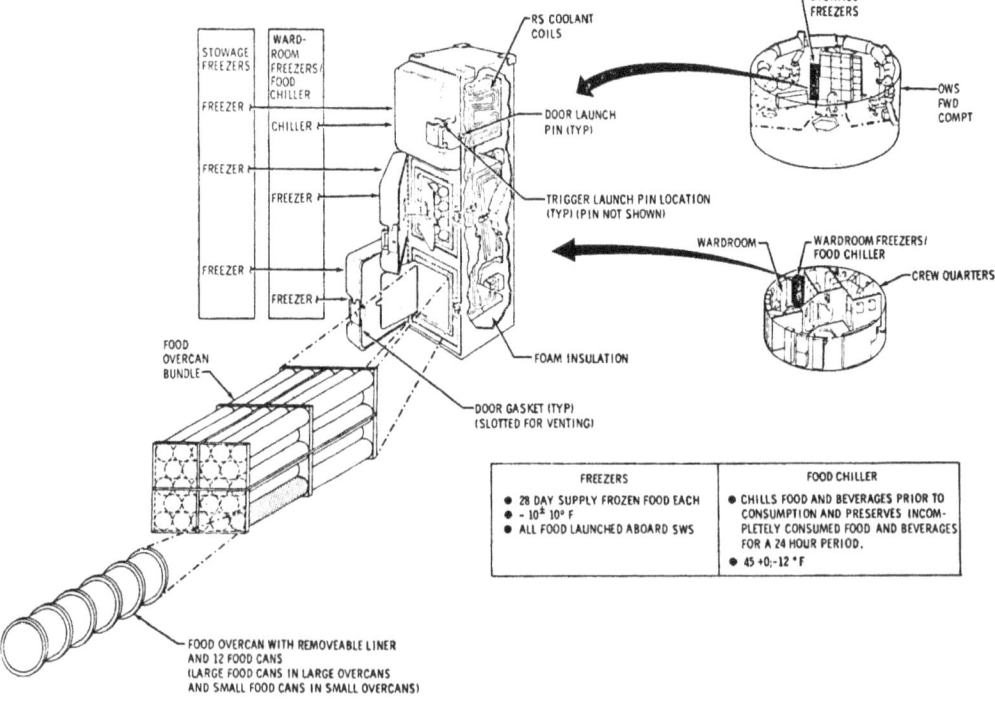

Figure II-15 — Chilled and Frozen Food Storage

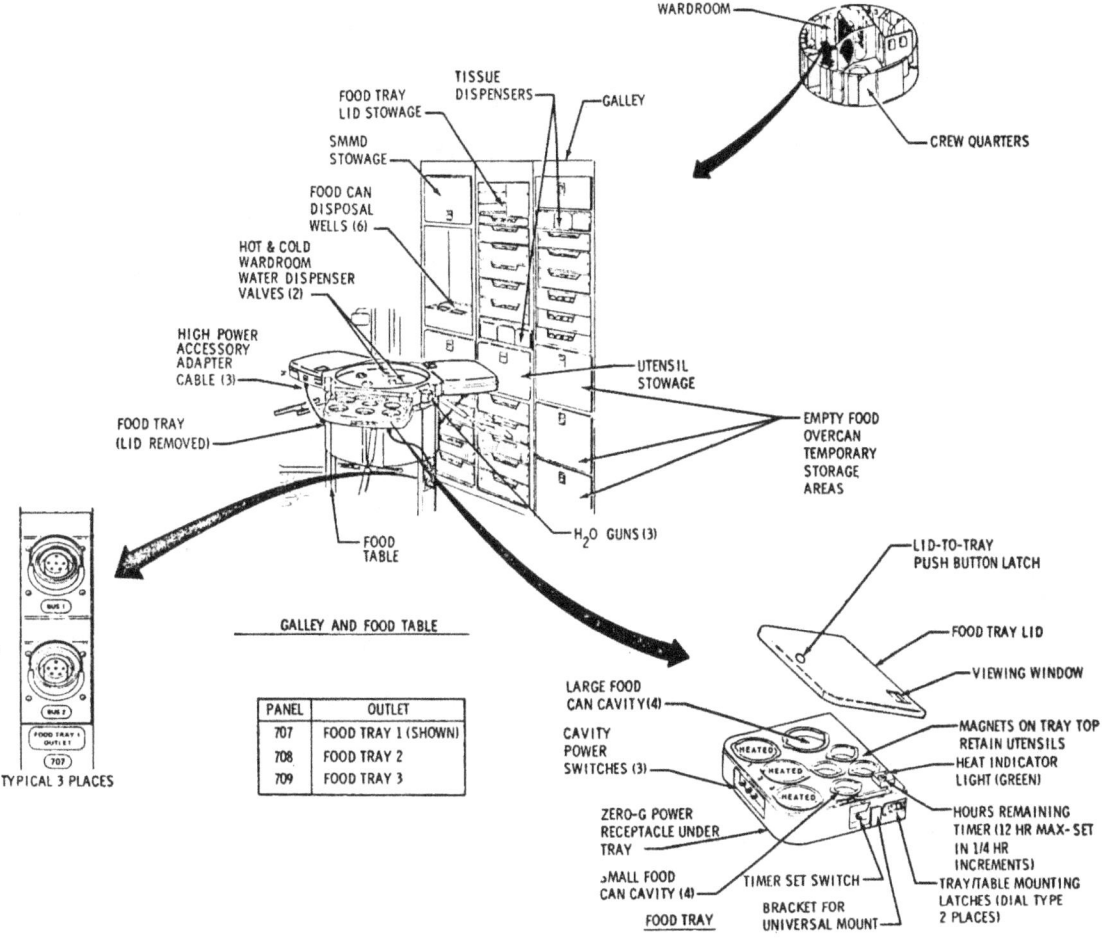

Figure II-16 — Food Preparation Equipment

fecal bag. Each is controlled by individual control and display panels which include a timer, manually set by the crewman to automatically initiate and terminate the drying cycle. The drying time is selected as a function of the waste material's mass. After processing, the collection bags are removed and transferred to a storage area in the WMC for eventual return to Earth.

Three urine drawers are at the base of the fecal/urine collector, one assigned to each crewman. The drawers are used for collecting, storing for 24-hours, and measuring and sampling the urine. Each drawer contains the facilities to accept the urine, to separate the air, to collect and store the urine, and to withdraw a urine sample once daily. Each drawer is serviced with refrigeration subsystem coolant to refrigerate the urine bag and cool the urine separator. Each drawer is provided with a gravity substitute airflow from the collector's blower unit. Cabin air, used to collect the urine, is removed by the separation device through centrifugal action prior to freezing the samples.

The urine freezer, immediately below the waste processors, provides interim low-temperature storage of urine samples for return to Earth at the end of the mission. The 120 ml urine samples in sample containers are retained in urine trays which hold 42 sample containers (2 weeks accumulation) in partitioned segments (Figure II-24). Two urine trays, with an integral thermal capacitor of dodecane wax, are stacked in the freezer at all times. The wax, after being thermally conditioned in the freezer, keeps the sample containers below -10 degrees C (14 degrees F) during the return to Earth in a urine sample container.

Any item that is biologically active after its use (clothing, filters, food cans, urine bags, sleep restraints, hygiene kits, tissues, wipes, towels, washcloths, etc.) is considered trash and disposed of in the waste tank in trash collection bags. Controlled venting via the trash bags prevents excessive waste tank pressures and minimizes the formation of large ice crystals, which may collect on and clog the screens in the waste tank. Two types of trash collection bags are provided: trash bags that serve as trash receiving stations within the OWS, and disposal bags for use in bagging large items. Eight trash containers are located in the OWS; one in the experiment compartment, two in the wardroom, one in the WMC, three in the sleep compartment (one in each sleep area), and one in the OWS forward compartment.

The trash bags (Figure II-25) in the wardroom and the WMC are replaced daily, others weekly. When a trash bag is removed from the trash container, a bag-mounted, adhesive-backed diaphragm cover is sealed into place over the diaphragm to seal-off the opening. Filled trash bags are put into the waste tank through the trash disposal airlock.

Figure II-17 — Food Table Restraints

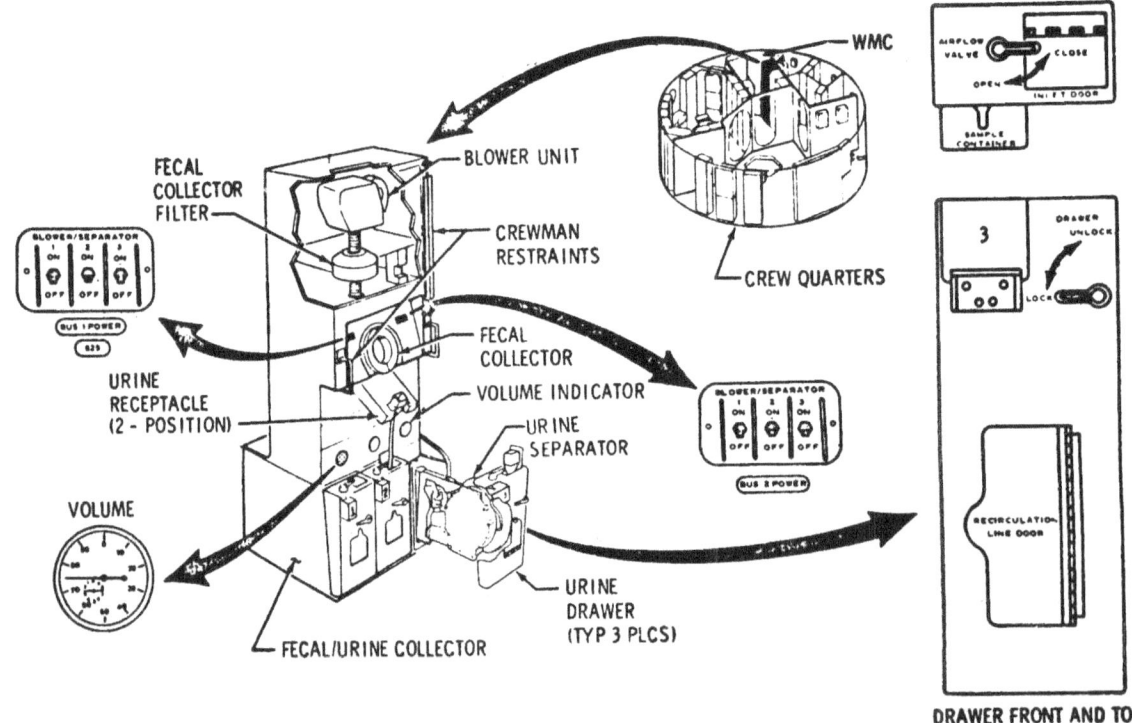

Figure II-18 — Fecal/Urine Collector

II-16

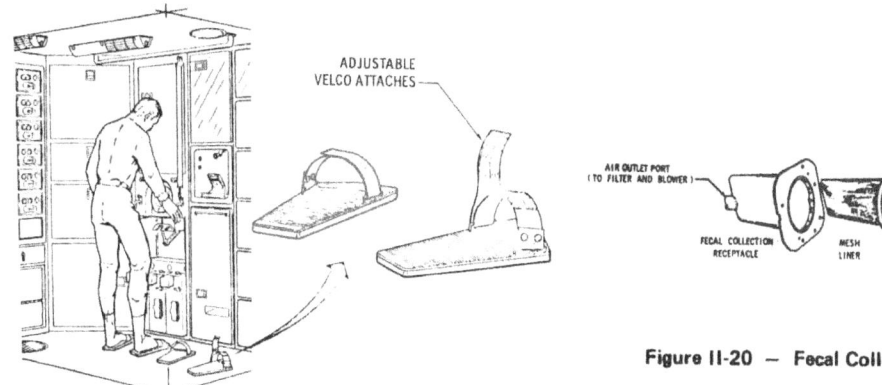

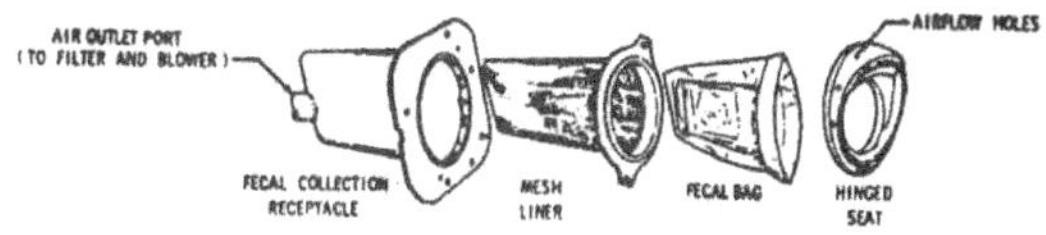

Figure II-20 — Fecal Collection Assembly

Figure II-19 — Foot Restraints at Urine Collector

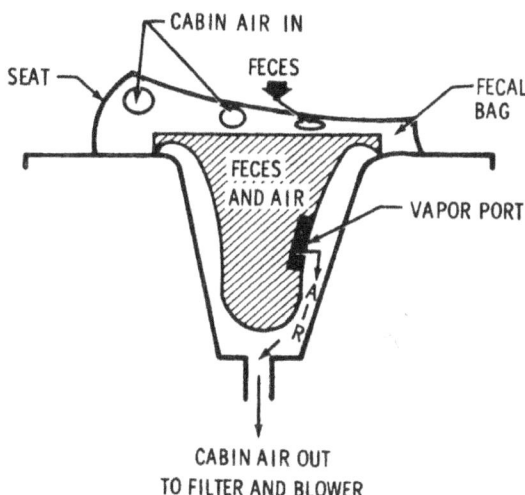

Figure II-21 — Fecal Collection

Figure II-22 — Urine Drawer

II-17

Disposal bags are used for large items (urine bags, sleep restraints, food overcans, charcoal filters, etc.) which do not fit into the trash bags. When a large item is to be disposed of, a disposal bag is obtained from one of the stowage compartments and transferred to the work area. The bag is secured near the work area by the bag's Velcro lining or its snaps. "Stays" in the bag opening maintain the bag open or shut. After use, the bag is sealed shut by a tab on the bag and put into the waste tank.

The trash disposal airlock (Figure II-26) is located in the center of the crew compartment floor. The airlock will be used about five times a day.

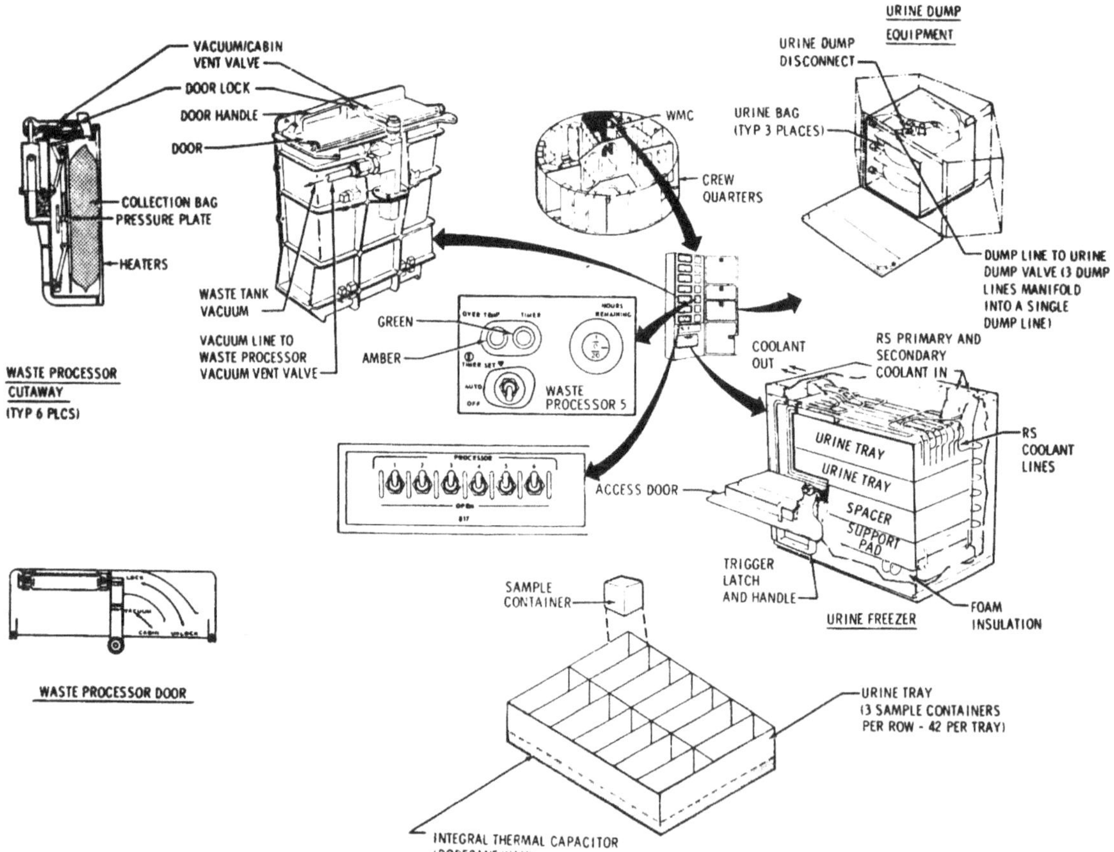

Figure II-23 — Waste Processors

Figure II-24 — Waste Processing

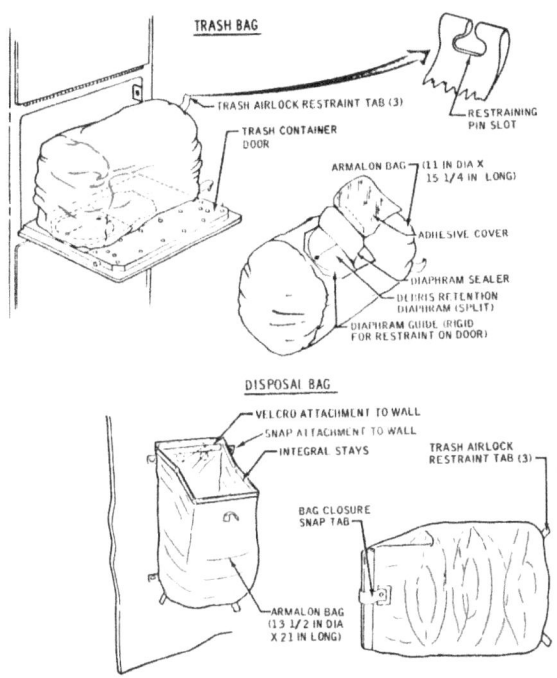

Figure II-25 — **Trash and Disposal Bags**

Sleep Provisions

Sound sleep periods of sufficient length are vital to enable crewmen to continue functioning under the stresses of long duration missions in weightlessness. The provisions for Skylab are consequently more elaborate than those of earlier, shorter duration flights. During sleep periods, each crewman is provided with an individual compartment for privacy and isolation from illumination and noise (Figure II-27). Each compartment contains a removable sleep restraint, storage compartment for personal items and a removable light baffle for the ceiling.

The sleep restraints provide for a variety of sleep positions, including the fetal position, and also thermal comfort during sleep periods. The restraints are vertically mounted on a frame attached to one of the sleep area partitions (Figure II-28). This frame is strapped to the floor and ceiling grid to provide rigidity. The neck opening is used for normal ingress and egress and for emergency egress. The restraint also has openings for crewmen's arms. The total sleep restraint is replaced every 14 days but individual components such as the headrest and blankets may be replaced at the discretion of the crewmen. Replacement restraints, headrests and blankets are stowed in sleep area stowage compartment.

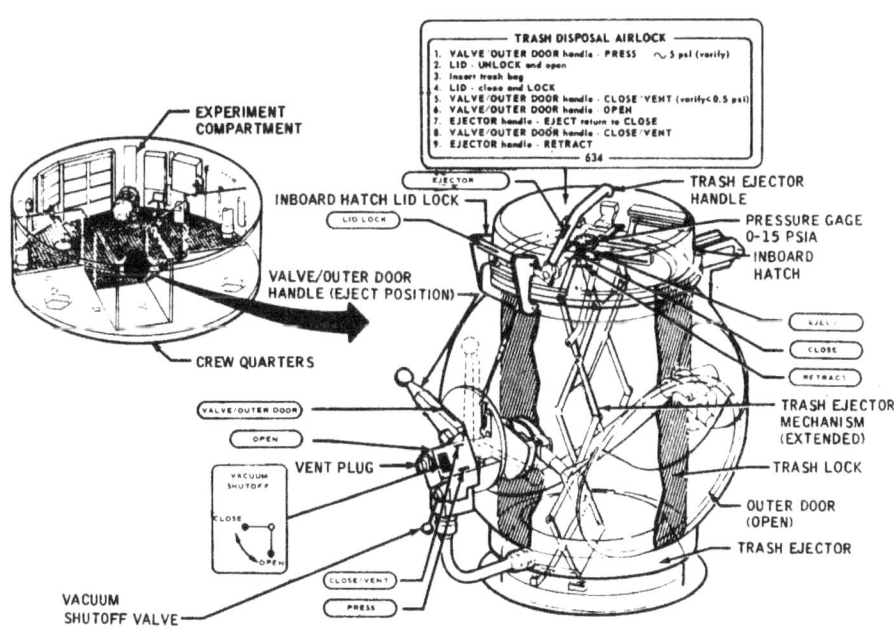

Figure II-26 — **Trash Disposal Airlock**

II-19

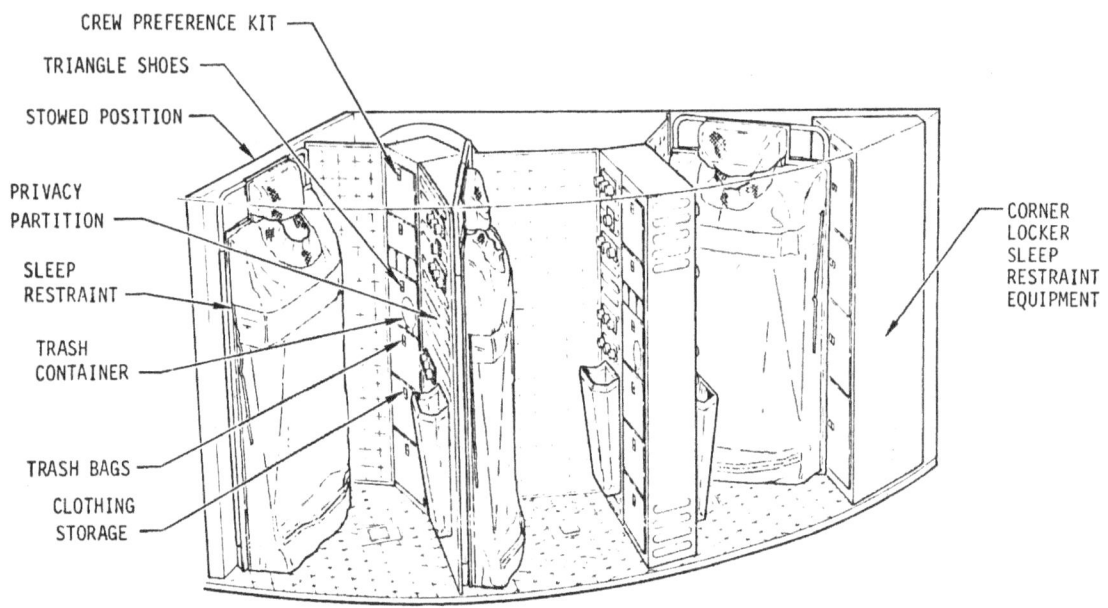

Figure II-27 — Sleep Compartment

Crewmen place four light baffles on the ceiling grid of the sleep areas for light abatement during sleep periods.

Personal Hygiene

On relatively short flights, the lack of personal cleanliness constitutes an acceptable annoyance. On longer missions personal hygiene becomes necessary to support good health and comfort. Consequently, Skylab has been equipped with a partial body cleanser, a whole body shower and personal hygiene items.

A partial body cleansing facility is provided in the form of a handwasher in the WMC (Figure II-29). Crewmen will dispense water to a wash cloth and use a squeezer to wring excess water from cloths. The handwasher installation includes a waste water bag which collects the water from the squeezer. Four magnetized soap holders in the handwasher retain bars of soap, each of which contain a metallic insert. In using the handwasher, crewmen will position themselves in front of the module in the foot restraints (Figure II-30).

For the first time in space, crewmen are able to take a shower using a "wet-soap-rinse" ritual (Figure II-31). Once a week they shower using an allotment of three quarts of water. Crewmen enter the shower, raise the enclosure and attach it to the ceiling grid. Water is delivered by flexible hose from a portable water bottle to a pushbutton shower nozzle. A vacuum head on a flexible hose draws the used water into a disposable bag, which is later deposited in the waste tank.

Personal hygiene items are shown in Figures II-32 and II-33.

Microbial Control

The longer duration of Skylab missions makes the control of microorganisms in space an important consideration. The microbial control program includes: the collection and disposal of organic trash; personal hygiene measures with periodic changes of clothing; air drying of Liquid Cooled Garments (LCG) and Pressure Garment Assemblies (PGA) after use; periodic replacement of components highly susceptible to the collection of microorganisms; scheduled wipe-downs of susceptible areas with disinfectant; sampling and purification of potable water; decontamination of water distribution lines; and clean up of organic matter and water spills.

Certain micro-organisms are shed from the crewmen's skin surfaces into their clothing. Skin cells and moisture provide an ideal environment for their growth. Consequently, all articles of personal clothing are disposed of after use.

The PGAs are dried after each EVA by inserting the boots into portable foot restraints near one wall of the OWS, securing the neck to the water ring grid in the forward compartment by straps, and blowing warm air through the PGA oxygen inlet quick-disconnects. The LCGs are passively dried by placing them on hangers and restraining the legs to the floor grid.

Equipment which readily collects micro-organisms but is difficult to decontaminate by wiping with disinfectant or by cleaning is replaced periodically (Table II-4). Components and areas of Skylab which are particularly conducive to microbial growth (in regions with optimum temperature and availability of oxygen, moisture, and nutrients) are wiped down periodically with disinfectant wipes (Table II-5).

The potable water contains iodine in solution as a biocide at launch. Iodine, maintained above two parts per million, is an acceptable method of preventing microbial growth. The crews use the equipment (Figures II-34 through II-37) to periodically draw samples from the tanks, to add a reagent, to determine the iodine concentration by comparison to a color chart, and to add iodine when necessary.

To prevent the growth of micro-organisms in the Wardroom water distribution network, the lines are filled with an iodine solution, allowed to soak for one hour, and flushed clear. The iodine

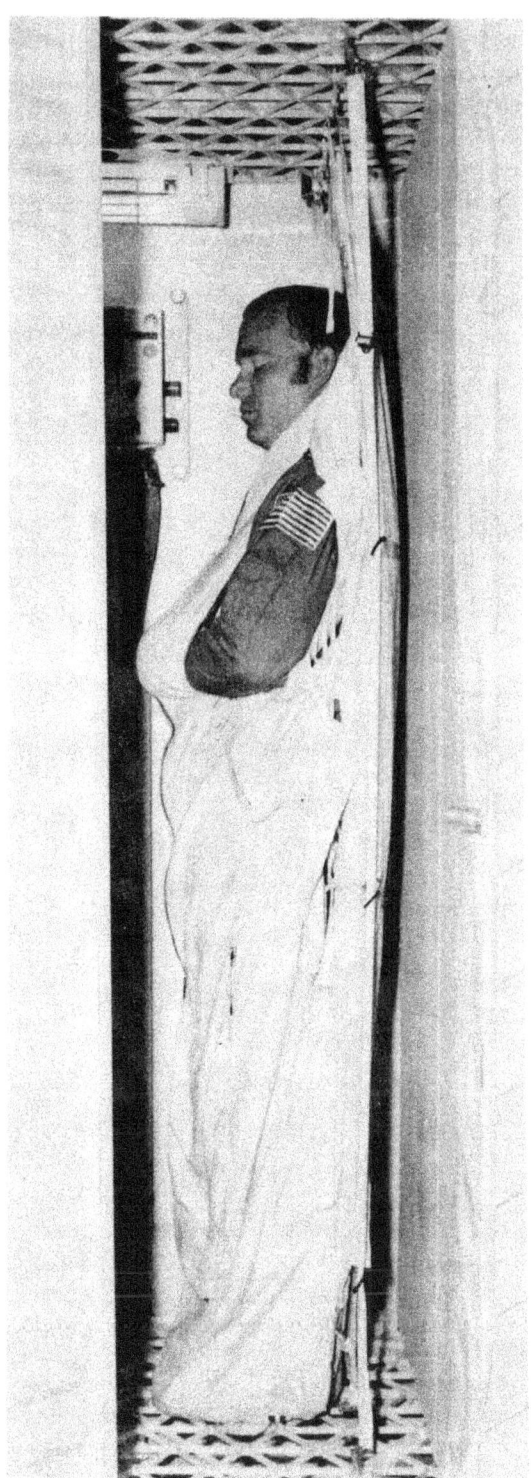

Figure II-28 — Sleep Restraint

injector is used to inject 100 parts per million of iodine into the portable water tank for transfer to the wardroom water lines. This decontamination is performed during deactivation of SL-3 and SL-4.

To limit the nutrients and moisture necessary for microbial growth, organic debris and moisture are cleaned up as required. Some equipment (Table II-6) is cleaned periodically. Any spills of organic material or water are immediately cleaned up by the use of the vacuum cleaner (Figure II-38).

Inflight Medical Support System

The Skylab Inflight Medical Support System (IMSS) provides an extensive inflight diagnostic and treatment center for emergency medical care of an "out patient" nature. The IMSS consists of three basic groups of equipment: diagnostic, laboratory and therapeutic which crew members have been trained to use. The equipment supplied is normally standard off-the-shelf medical items. The IMSS provisions are made up of kits, modules and a work table stowed in locker compartments W706 through W709. The kits are diagnostic, therapeutic, dental and

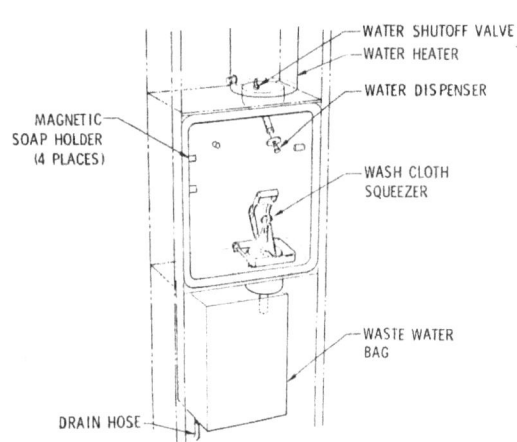

Figure II-29 — Handwasher

Figure II-30 — Handwasher Foot Restraints

Figure II-31 — Shower

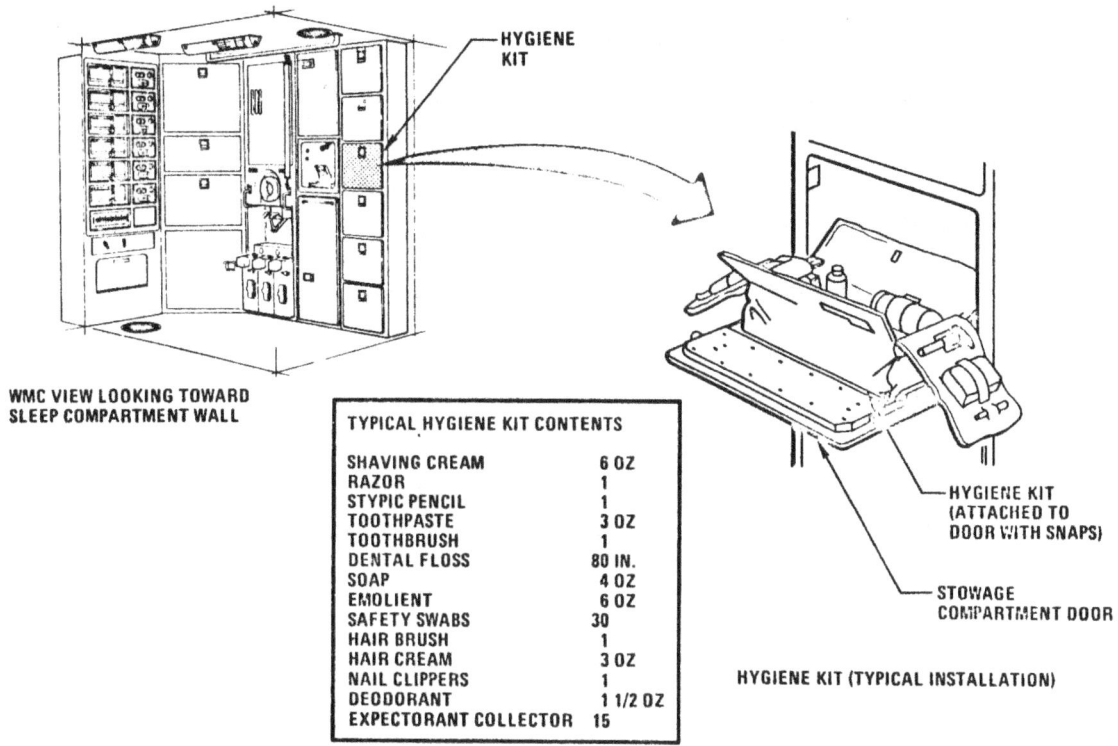

TYPICAL HYGIENE KIT CONTENTS	
SHAVING CREAM	6 OZ
RAZOR	1
STYPIC PENCIL	1
TOOTHPASTE	3 OZ
TOOTHBRUSH	1
DENTAL FLOSS	80 IN.
SOAP	4 OZ
EMOLIENT	6 OZ
SAFETY SWABS	30
HAIR BRUSH	1
HAIR CREAM	3 OZ
NAIL CLIPPERS	1
DEODORANT	1 1/2 OZ
EXPECTORANT COLLECTOR	15

Figure II-32 — Personal Hygiene Kits

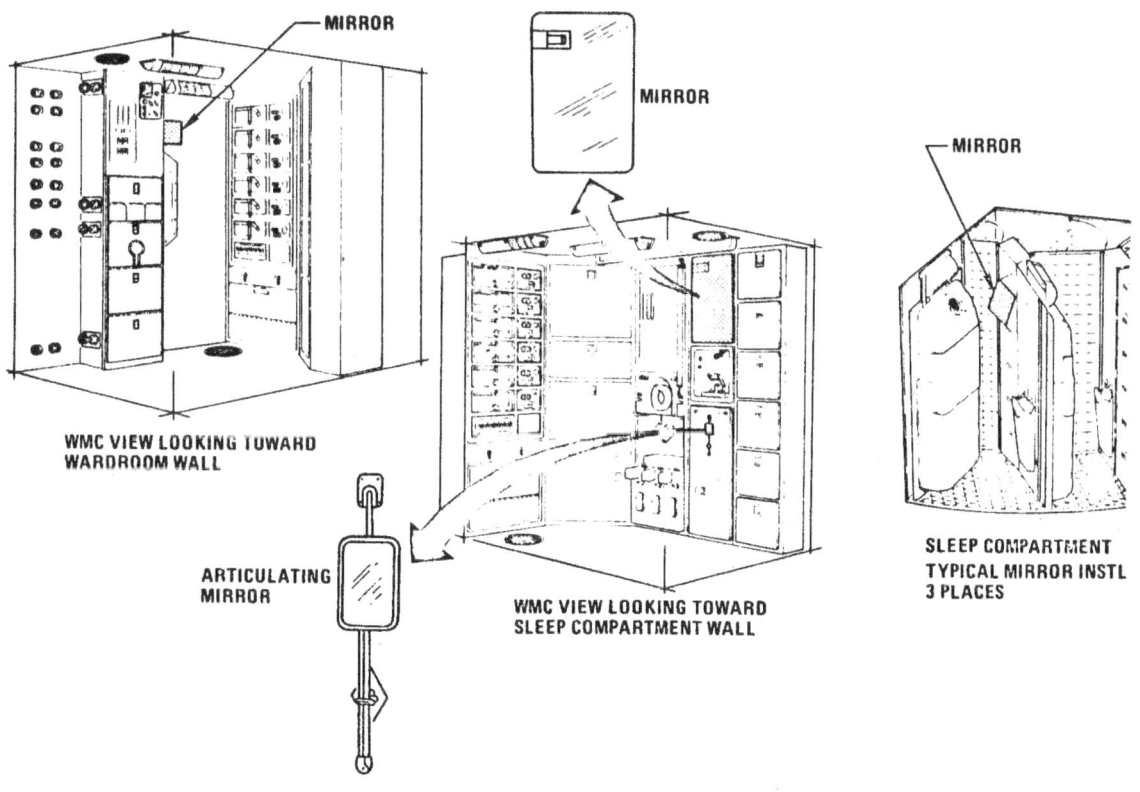

Figure II-33 — Personal Grooming Mirrors

II-23

Table II-4 — Microbial Control Replacements

Component	Replacement Schedule
Absorption Cartridge, Metabolic Analyzer	After Each Use of Analyzer
Protective Assembly, Ear Plug	After Each Use
Urine Collection and Sample Bags	Daily
Wash Cloth Squeezer Bag Assembly	During Activation of SL-3 and SL-4
Top Blanket of Sleep Restraint	As Necessary
Stabilizer Assembly, CO_2 Stabilizer	Every 28 Days
Urine Inlet Hose Assembly	Daily
Trousers	Weekly
Socks, Shirts, Underwear	Every 2 Days

bandage, stowed together in the medical kits module (Figure II-39). All of the kits are portable and are fitted with snaps which attach to a portable work platform or to other snap locations in the OWS. Medications and supplies with limited shelf life will be resupplied on each mission.

For diagnostic purposes the IMSS is supplied with the standard clinical tools the physician uses to perform a physical examination. The kit (Figure II-40) includes: A stethoscope, a blood pressure measuring device, neurological examination instruments and eye, ear and oral examination equipment. The kit also contains an ophthalmoscope and a headmount light. Altogether the diagnostic kit contains more than 20 items.

The laboratory equipment module (Figure II-41) provides the equipment to perform inflight medical analysis through the use of an incubator, a microscope kit, a slide stainer, a microbiology kit and a hematology and urinalysis kit.

The incubator is permanently mounted in locker compartment W708 and is used to grow cultures for analysis in isolating microorganisms to aid in diagnosing the probable cause of crewmen's illnesses. A slide drying compartment is provided as part of the incubator for rapid fixing of the stains onto tempered glass slides for microscopic analysis.

Table II-5 — Disinfectant Wipe Schedule

Component/Area	Disinfectant	Schedule
Vectorcardiogram Electrodes	Zepherin	Prior to Each Use
Eating Utensils	Zepherin	After Each Use
Food Trays	Zepherin	After Each Use
Urine Receiver Support	Betadine	Once Each Week
Fecal Collector Support Hardware	Betadine	Once Each Week
Trash Airlock	Betadine	Once Each Week
Urine Drawer	Betadine	Once Each Week
Urine Receiver Storage Area	Betadine	Once Each Week
Fecal Collector Seat	Betadine	Once Each Week
Food Chiller	Betadine	Every Two Weeks
Trash Locker Diaphragms	Betadine	Every Two Weeks
Urine Dump & Holding Area	Betadine	As Required
Vacuum Cleaner Accessories	Betadine	As Required
LBNPD Back Rest, Waist Seal, Inner Walls, Saddle, & Foot Restraints	Betadine	Every Two Weeks
Tool Kit	Betadine	Contingency

Table II-5 (Continued)

Component/Area	Disinfectant	Schedule
Food Management Area	Betadine	Contingency
Fecal Bags	Betadine	Contingency
SMMD	Betadine	Contingency
Blood Pressure Cuff	Betadine	Contingency
Metabolic Analyzer Chest Board, Exhalation Hose, & Seat	Betadine	Once Every Two Weeks
Biomedical Harness Electrodes	Zepherin	Prior to Each Use
Urine Chiller	Betadine	Once Every Two Weeks
Urine Inlet Support and Track Assembly	Betadine	Once Every Two Weeks
Urine Centrifugal Separator to Urine Storage Bag & Chiller	Betadine	Contingency
Fecal Collector	Betadine	In Event of Bag Rupture
Waste Processor	Betadine	In Event of Bag Rupture
Waste Process	Betadine	Contingency
Urine Collector or Other Areas	Betadine	In Event of Bag Rupture
Towel & Washcloth Drying Holder Assembly	Betadine	Once Each Week

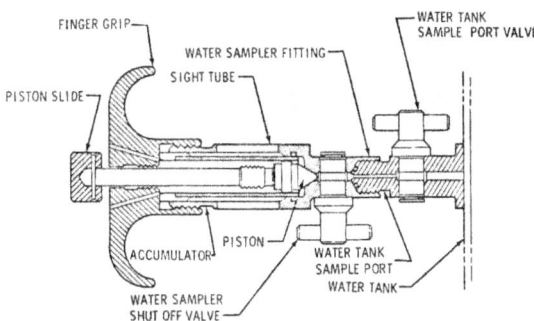
Figure II-34 — Water Sampler

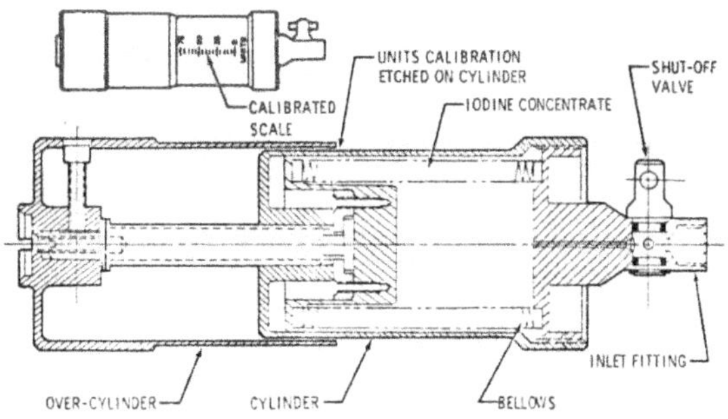

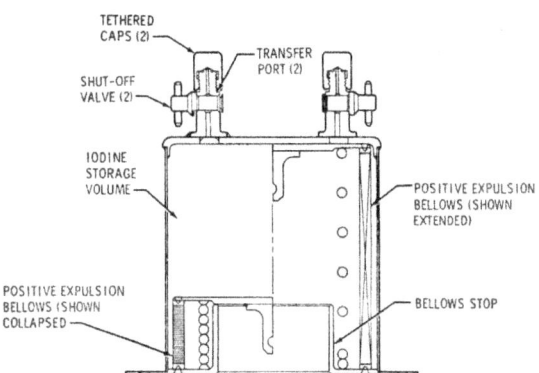
Figure II-35 — Iodine Injector

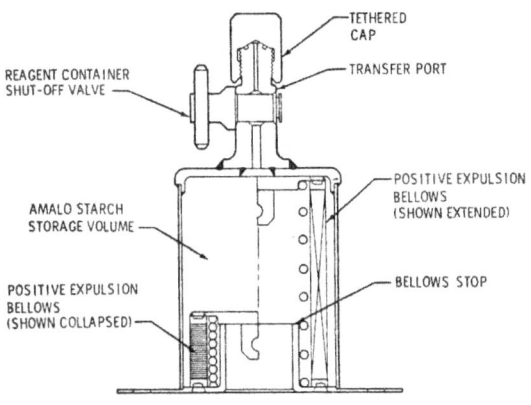
Figure II-36 — Reagent Container

Figure II-37 — Iodine Container

The microscope kit includes a portable microscope, lens tissue, tempered slides and slide treatment equipment. The slide staining equipment functions as a self-contained and totally enclosed fluid system for staining slides for microbial analysis and for blood smears to perform white blood cell identification.

The microbiology kit provides equipment: to perform antibiotic sensitivity testing as a diagnostic aid; to collect microbial samples from crewmen at the time of illness for inflight analysis; and to obtain environmental and crew microbial samples throughout the mission for post-flight analysis at MSC.

Supplies and equipment for performing urinalysis (the urinalysis kit) include a specific gravity refractometer. The hematology kit supplies the equipment, including a hemoglobin meter, to obtain white blood cell and hemoglobin information.

The locker module containing the laboratory equipment has a hinged, sheet metal surface for use as a laboratory workstation platform. Operational areas are provided on the workstation for mounting the microscope and for securing slides and associated equipment for processing cultures while analysis is being conducted. When not in use, the workstation platform folds into the module.

The therapeutic kit (Figure II-42) contains items such as injectable medications, syringes and a laryngoscope which operates on batteries. For therapeutic purposes the kit contains a wide variety of medications, both oral and injectable, for the treatment and prevention of infection, disease and allergies. Minor surgery tools for the care of injuries are also included.

Table II-6 — Dry Wipe Schedule

Equipment	Method of Cleaning	Frequency of Cleaning
LBNPD Waist Seal, Inner Walls, & Saddle	Utility Wipes	After Each Use
Vestibular Function Otolith Test Goggles & Bite Board	Utility Wipe or Tissue	After Each Use
Metabolic Analyzer Mouthpiece	Utility Wipe	After Each Use
Metabolic Analyzer BTMS Probe	Utility Wipe or Tissue	Weekly
Food Reconstitution H_2O Dispenser	Utility Wipe or Tissue	After Each Use
Drinking H_2O Dispenser	Utility Wipe or Tissue	After Each Use
Galley	Utility Wipes	After Each Use
Toothbrush	Utility Wipe or Tissue	After Each Use
Fecal Collector Seat	Utility Wipes	After Each Use
WMC Hand Washer	Utility Wipes	After Each Use
Urine Drawer	Cleaner Wipe & Utility Wipe	Daily
Food Chiller	Utility Wipe	Twice Each Week
Window Frames & Girth Ring	Utility Wipe	Weekly

Table II-6 (Continued)

Equipment	Method of Cleaning	Frequency of Cleaning
Fecal Bags	Utility Wipe or Tissue	As Required
Washcloth Squeezer	Utility Wipe	Contingency
OWS, MDA, AM Walls	Utility Wipe	Contingency
Personal Hygiene Equipment	Utility Wipe or Tissue	Contingency
Urine Chiller Cold Plate	Cleaner Wipe & Utility Wipe	Daily
Waste Processor Chamber	Utility Wipe	Daily
Urine Dump Compartment	Cleaner Wipe & Utility Wipe	Contingency
Urine Flush Dispenser	Utility Wipe	After Each Use
Urine Centrifugal Separator Compartment	Cleaner Wipe & Utility Wipe	Daily
Waste Processor Exterior	Utility Wipes	Weekly
WMC H_2O Dispenser Valve	Utility Wipes	After Each Use
Food Management Area	Vacuum Cleaner	Contingency
WMC Charcoal Filter Screen	Vacuum Cleaner	Once a Week

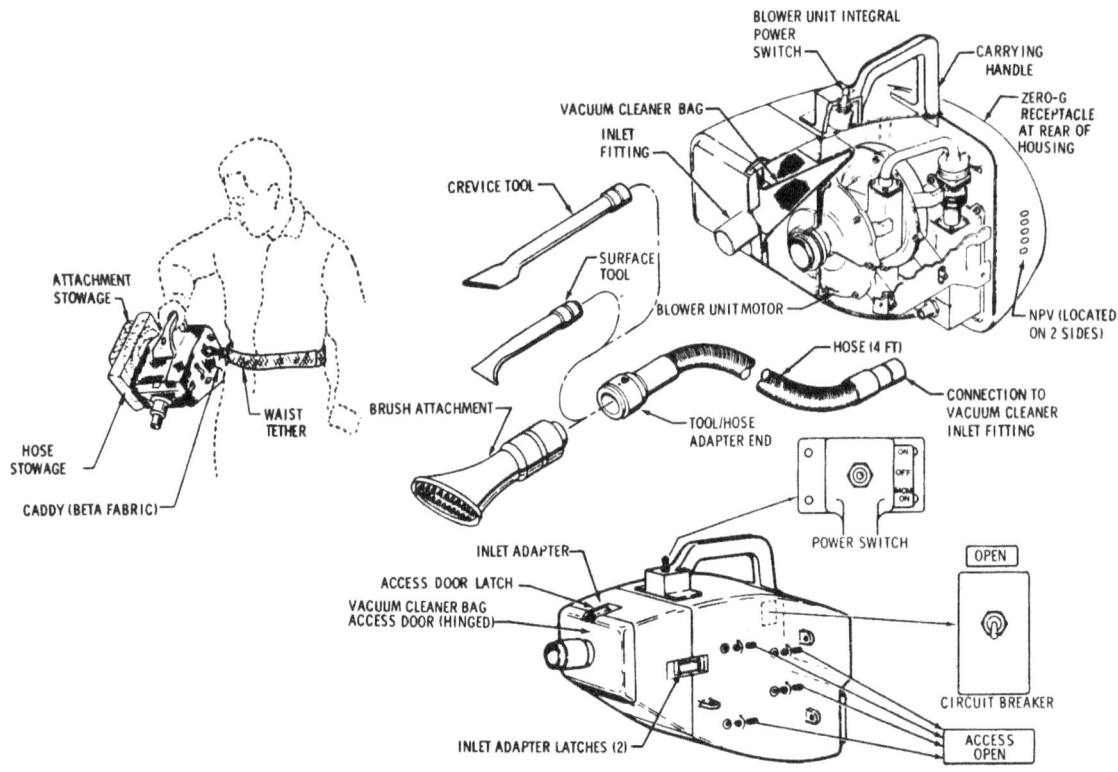

Figure II-38 — Vacuum Cleaner and Accessories

Several dozen types of medicines (Figure II-43) are in separate medication cans (two supplied per mission) stowed in W706 in the OWS. Prior to launch, the pills for SL-2, SL-3 and SL-4 are individually packaged and placed in the cans which are then sealed and opened when required. About 190 separate packages of pills, ointments, injectors and other medicines are aboard Skylab. Cans which have been opened and their unused contents will be disposed of at the end of each mission.

A catheterization kit is in a pocket inside the therapeutic kit. The equipment in the kit serves to relieve urinary blockage and to aspirate a crewmember's stomach if necessary.

Two treatment kits for minor surgery, containing more than 15 items each, are provided in the minor surgery module. The sterile surgical kits, in cylindrical cans, provide the necessary equipment to permit extraction of small foreign items from the skin or for creating and or closing small lacerations and puncture openings. The module also contains such surgical items as scissors, forceps, hemostats, scalpels and retractors (Figure II-44).

The bandage kit (Figure II-45) contains more than 20 various configurations of bandages, swabs, dressings, gauze and associated equipment for treating minor injuries and fractures.

In the event any crewmember has a toothache or for some reason requires repair or extraction of a tooth, a 20-item dental kit provides the necessary equipment to cope with inflight dental emergencies (Figure II-46).

The resupply and return container (Figure II-47) is used to carry medical supplies to Skylab and return samples to Earth.

Figure II-39 — Therapeudic, Dental, Diagnostic and Bandage Kits in Medical Kits Module

Medical Accessories Kit

The Medical Accessories Kit (MAK) provides the crew with all medical supplies and bioinstrumentation accessories necessary for the Command Module (CM) portions of the Skylab mission (Figure II-48). The MAK is stowed in the CM and is for use only in the CM. Designed as a rucksack for easy accessibility, it contains a variety of medications that can be generally categorized as antibiotics, anti-motion sickness medicines, pain relievers and anti-histamines. Also, normal first aid equipment is provided, such as bandages and ointments. Space bioelectrodes and bioelectrode paste are also in the MAK.

Operational Bioinstrumentation

The purpose of the Operational Bioinstrumentation System (OBS) is to provide a means of transmitting to ground in real time certain critical physiological information during launch and recovery, EVA and on demand, as necessary. With such data at its disposal, the medical staff will be able to maintain a constant surveillance of health and physiological condition throughout selected phases of the Skylab mission.

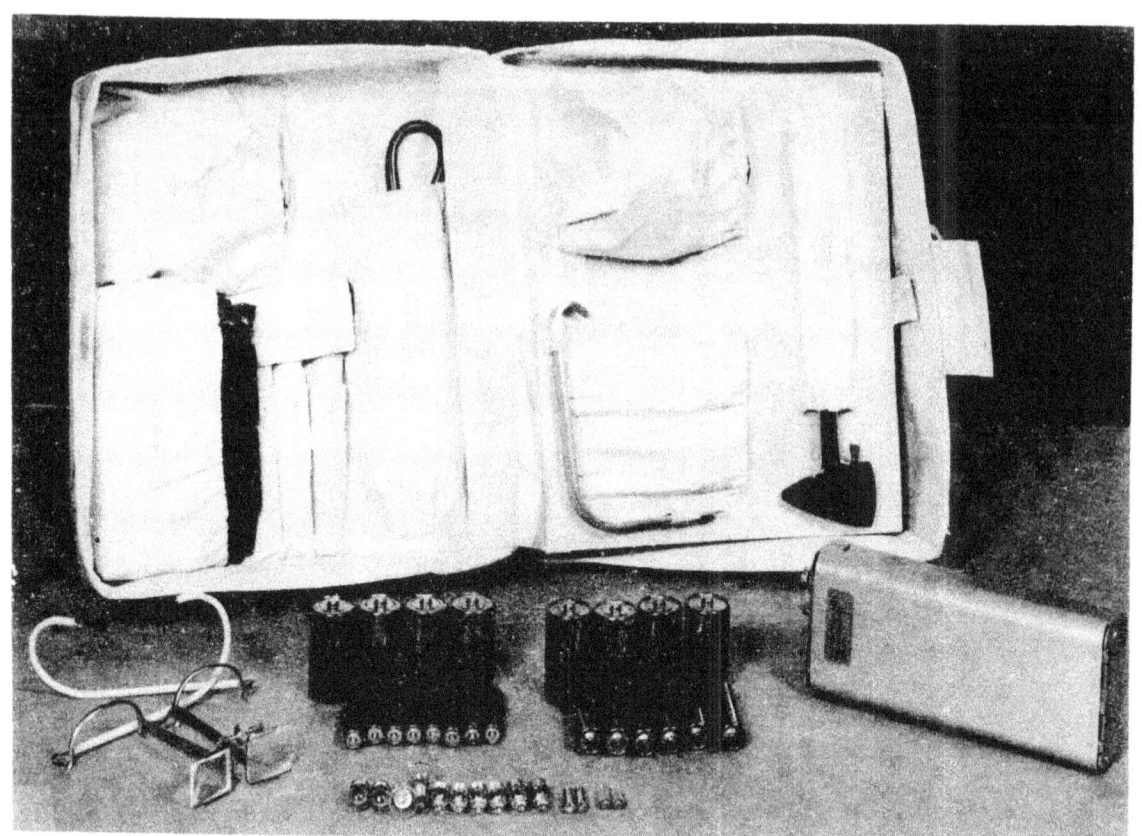

Figure II-40 — Diagnostic Kit

Figure II-41 — Laboratory Equipment

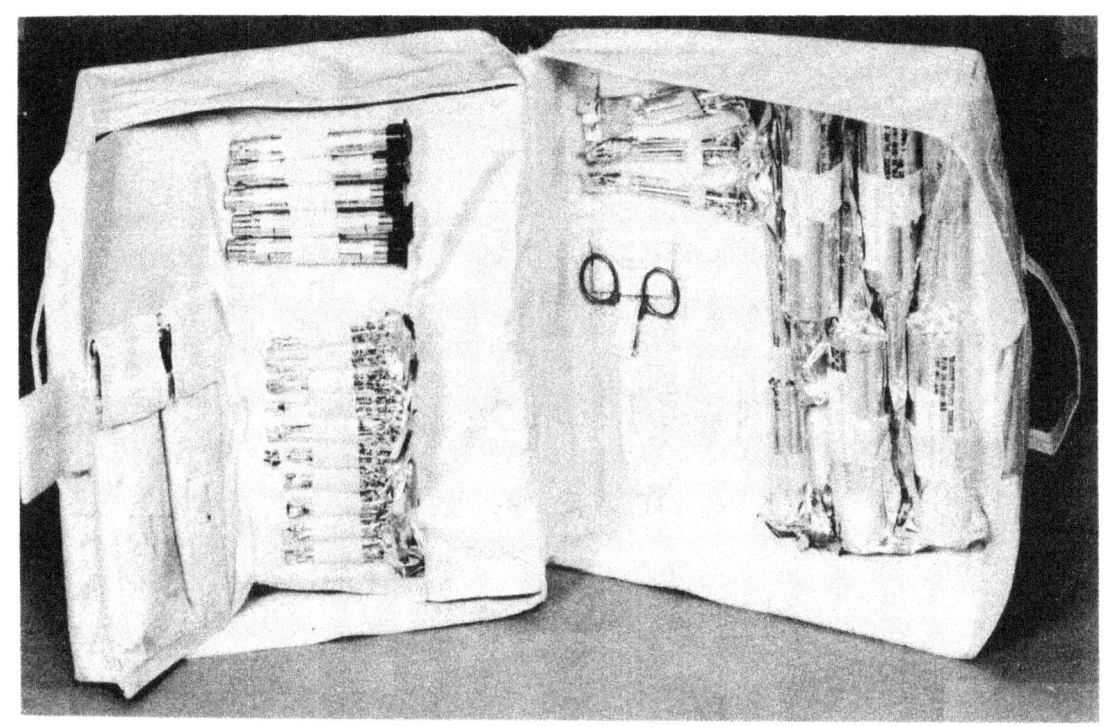

Figure II-42 — Therapeudic Kit

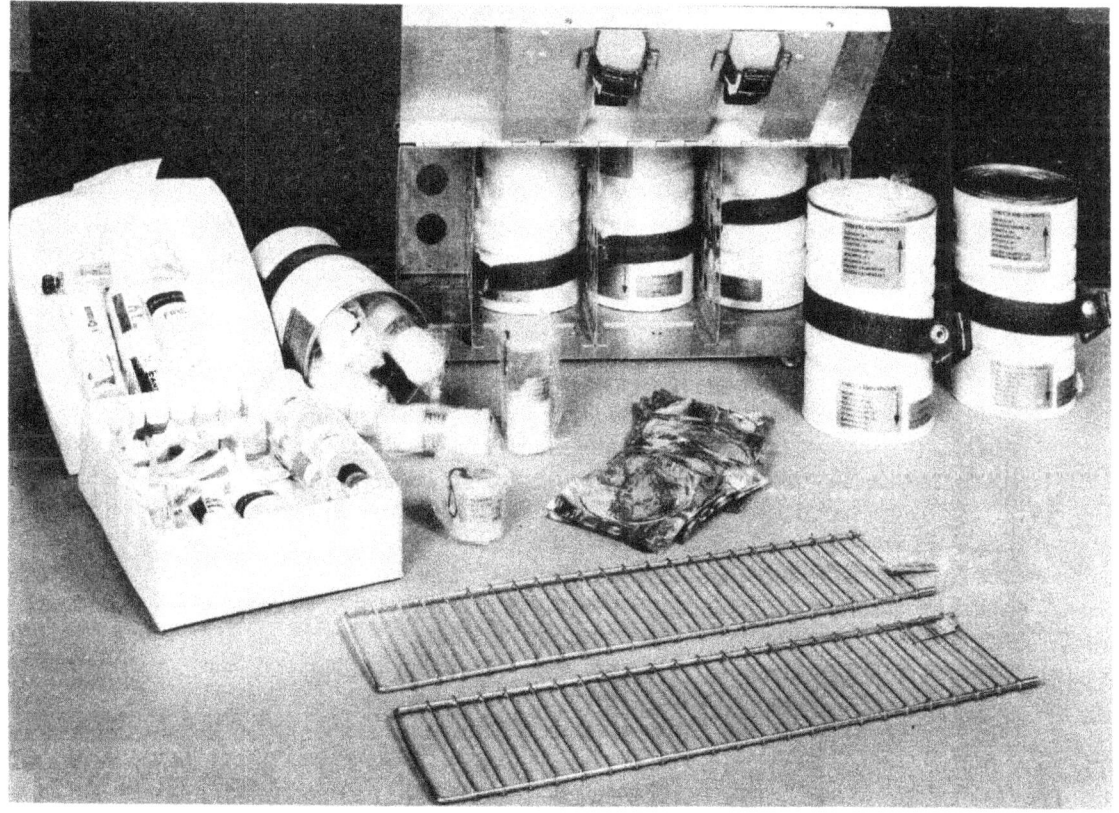

Figure II-43 — Medications Supply Module

II-33

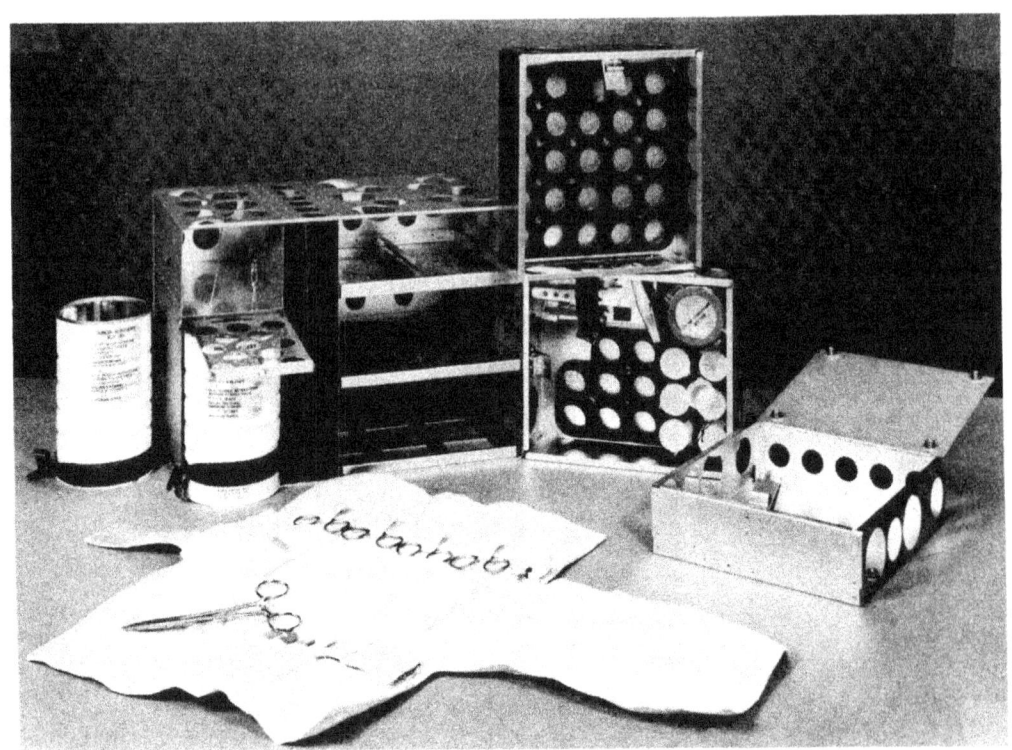

Figure II-44 — Minor Surgery Module

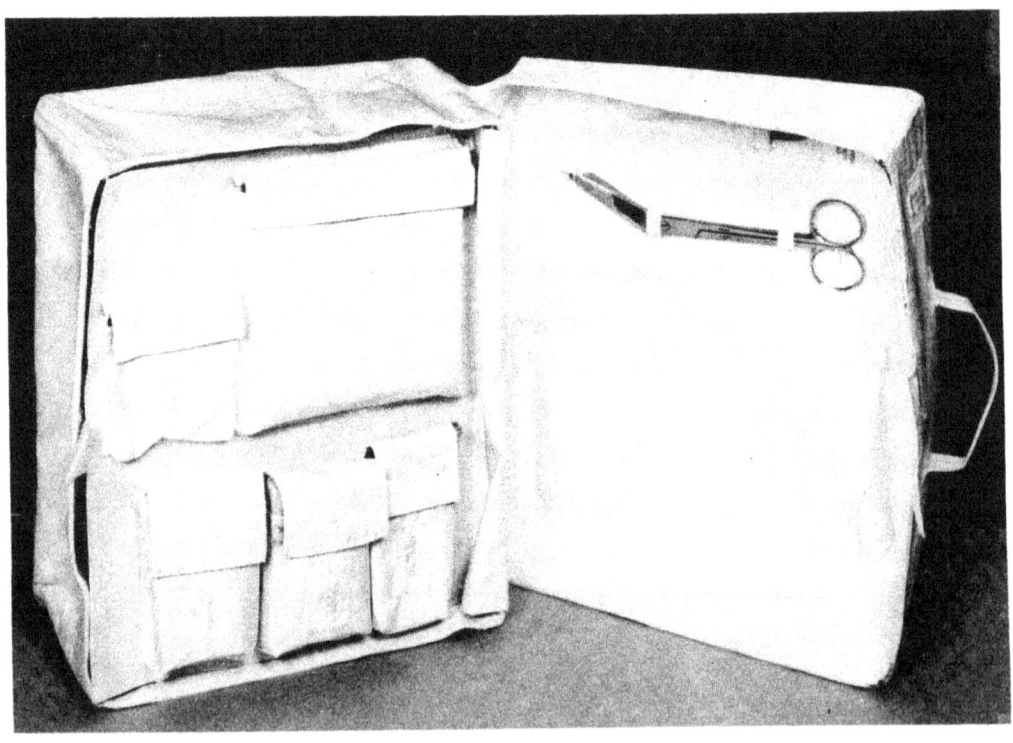

Figure II-45 — Bandage Kit

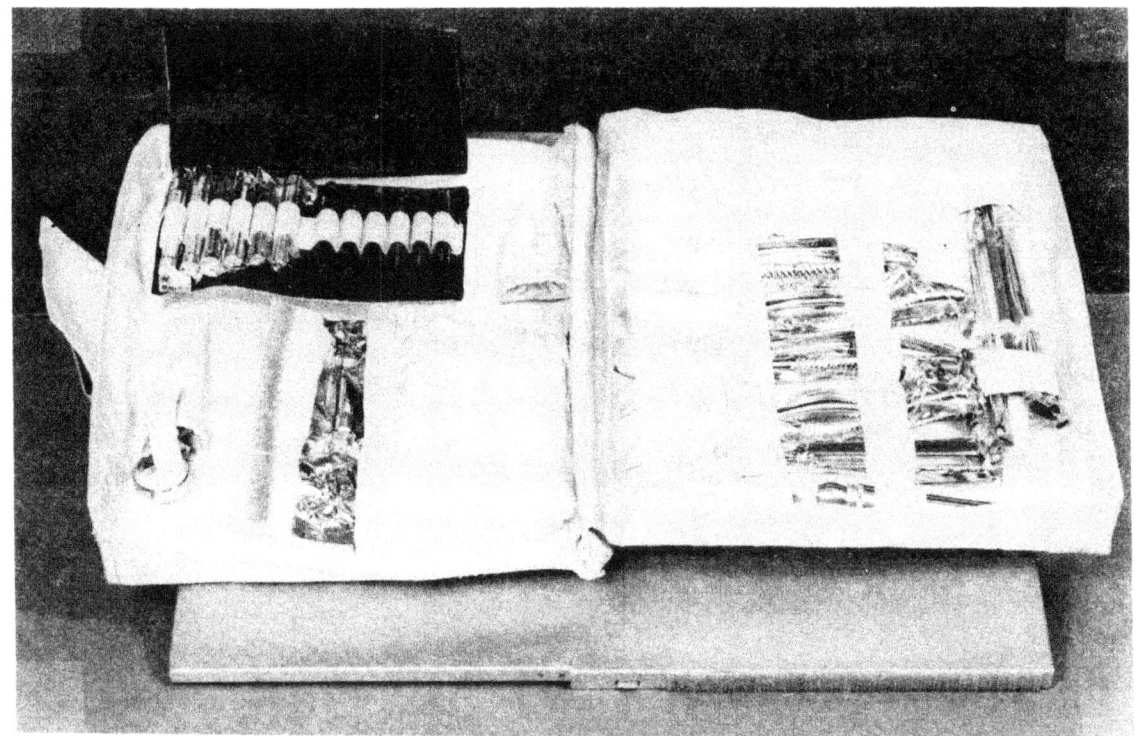

Figure II-46 — Dental Kit

The OBS (Figures II-49 and II-50) is an electronic system including sensors, signal conditioners and telemetry interfaces. Also included as a part of the system is the Electrical Harness Assembly into which the signal conditioners are placed and which each crewman wears.

The system returns the following data via telemetry:
Electrocardiogram (ECG)
Heart Rate (Cardiotachometer [CTM])
Respiration Rate (Impedance Pneumogram [ZPN])
Subject Identification (SID)
The following is a description of the various hardware elements of the system:

Signal Conditioners — There are three signal conditioners, one each for ECG, CTM and ZPN. These are small electronic devices, the purpose of which is to electronically prepare (or condition) the incoming signals from the various sensors to interface with the telemetry system and on-board readout devices.

DC to DC Power Converter (DDC) — The DDC accepts the current-limited spacecraft power and converts it to an isolated balanced power supply for each signal conditioner.

Biosensors — The biosensors are electrodes that are affixed to the crewman's body. They sense the bioelectric potentials associated with heart action (ECG and CTM) and chest cavity expansion and contraction (ZPN) and send them through their attached leads to the appropriate signal conditioner. These electrodes are similar to those used in association with Medical Experiment M093.

Subject Identification Module (SID) — The SID is an electrical device that emits a unique voltage output for each crewman.

Figure II-47 — Resupply and Return Container

Figure II-48 — Medical Accessories Kit

Electrical Harness Assembly (EHS) — In addition to providing a means of affixing the signal conditioners to the body as already described, the EHS includes the wiring between the sensors and their conditioners and between the DDC and the conditioners and SID. It also provides the electrical interface connectors to spacecraft power and telemetry system.

Portable CO_2/Dew Point Sensor

The Portable CO_2/Dew Point Sensor provides the crew with a reliable means of measuring CO_2 concentration, humidity and ambient gas temperature anywhere within the Skylab cluster. This device will serve as supplementary instrumentation to the fixed Skylab instrumentation which is supplied as a part of the environmental control system. Such adjunct instrumentation is required because it is anticipated that there will be areas of static

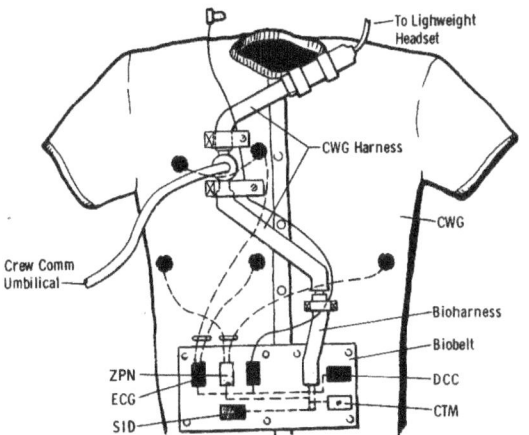
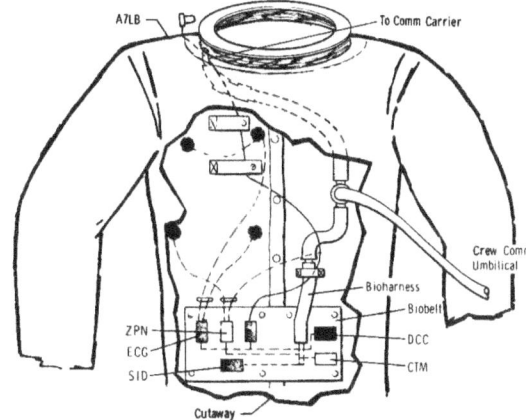

Figure II-49 — OBS Unsuited Mode   Figure II-50 — OBS Suited Mode

or nonconvective gas masses due to the circulatory pattern of the Skylab fans. The Portable CO_2/Dew Point Sensor is designed as a portable hand-held device utilizing components of the Apollo Dew Point Hygrometer System and the Apollo Portable Life Support System CO_2 Sensor for its measurement functions. It measures gas and dew point temperatures from 4.4°C to 38°C (40° to 100°F) and carbon dioxide partial pressures from 0.1 to 30 mm Hg. The portable sensor is completely self-contained, requiring no spacecraft support. It is designed to physically lock into the floor grid so that it can be placed in a fixed position when desired.

II-2

THERMAL CONTROL

Thermal control within Skylab is maintained by the combined influences of the passive system, which limits heat fluxes through the compartment sidewall, and the active systems which dissipate atmospheric and cold-plate heat loads to the radiator heat sink, provide refrigeration and suit cooling, and heat components during storage.

Active Thermal Control Systems

The AM active thermal coolant loop removes and dissipates waste heat which is due to cluster equipment operation and metabolic heat loss. Active cooling is provided to the EVA/IVA suit cooling module, condensing heat exchangers, three tape recorder cold plates, oxygen heat exchanger, ATM Control and Display (C&D) panel heat exchanger, battery modules and six electronic modules. A bifilar (each loop split into two parallel passes) radiator and a thermal capacitor (charged with tridecane wax) immediately downstream of the radiator are used for heat rejection. The capacitor absorbs heat from the coolant while the vehicle is on the sunny side of the orbit, and adds heat to coolant on the dark side.

The ATM C&D/EREP cooling system provides single-loop water cooling by circulating water through a heat exchanger that interfaces with both loops of the Airlock Module coolant system. The fluids of the C&D panel cooling system (water) and the AM coolant system are isolated by heat exchanger walls.

The suit cooling system provides astronaut cooling for EVA and IVA by circulating water through umbilicals and liquid-cooled garments (LCG). The system consists of two identical subsystems (one subsystem per AM coolant loop). Each subsystem can deliver a minimum of 90.7 kg (200 lb) per hr of water. The water flowrate through each LCG is controlled by an adjustable flow divider valve.

During the 7.5 hours that the IU equipment is required, the IU-mounted equipment is thermally controlled by an active thermal control system. Flow Cool 100 coolant is circulated by a centrifugal pump and heat rejection capacity is provided by a sublimator, using stored-water supply.

The active Thermal Control System for the ATM is a closed-loop fluid circulation system designed to maintain the methanol-water at the ATM canister inlet at a temperature of 10 ± 2 degrees C (50 ± 3 degrees F). Two parallel flow paths contain eight cold plates each. The cold plates are on the internal circumference of the canister; two cold plates are connected in series in each quadrant on the MDA end, and two cold plates, also connected in series, are in each quadrant on the Sun end.

The heat picked up by the fluid while in the cold plates is rejected to space by radiators on the exterior of the ATM canister. Temperature control of the cold plates is obtained by using a modulating flow control valve, which proportions the fluid flow between the radiator and the heater flow passages.

A refrigeration subsystem is provided aboard the OWS for food refrigeration, potable water chilling, food freezing and urine-sample freezing. Also, this subsystem provides temperature control during the initial storage period through a range of 3.9 to -25 degrees C (39 to -14 degrees F). A single-phase liquid refrigerant, Coolanol-15, circulates through the low-temperature storage units and accepts heat, which in turn is delivered to an external radiator and rejected to space (Figures II-51 and II-52).

The Wardroom food freezer provides low-temperature storage -16.7 degrees C (0 degrees F) maximum for 45.4 kg (100 lb) (a 56-day supply) of frozen food, 4.4 ± 3 degrees C (40 ± 5 degrees F) storage for up to 22.7 kg (50 lb) of perishable food, and the required food-chilling capability (from room temperature) for beverages, desserts, etc.

The food storage freezer provides low-temperature storage -16.7 degrees C (0 degrees F) maximum for 68 kg (150 lb) (an 84-day supply) of frozen food at a pressure of 0.7 N/cm^2 (1 psia) minimum. The low temperature is maintained by transferring heat from the freezer to the refrigeration subsystem coolant loop.

Heat is rejected from the food compartments to the refrigeration subsystem coolant loop. Valving, pressure-tight covers, seals and transducers are provided to maintain the required food pressure and to provide pressure readout. Transducers are provided for temperature readout.

A water chiller is provided to satisfy the requirement for chilled drinking water at 4.4 + 3 degrees C (40 + 5 degrees F).

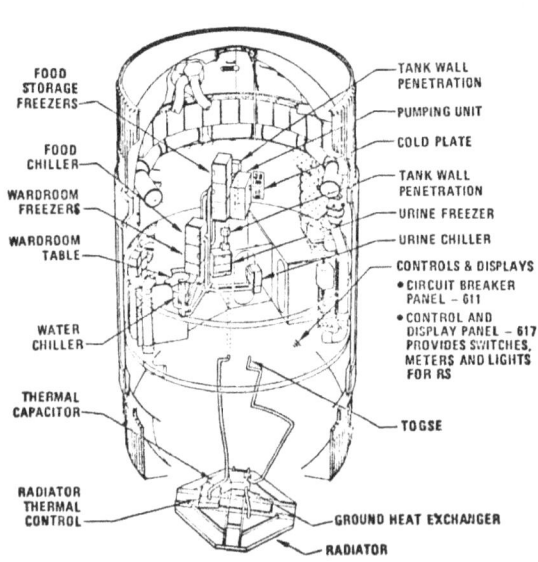

Figure II-51 — OWS Refrigeration

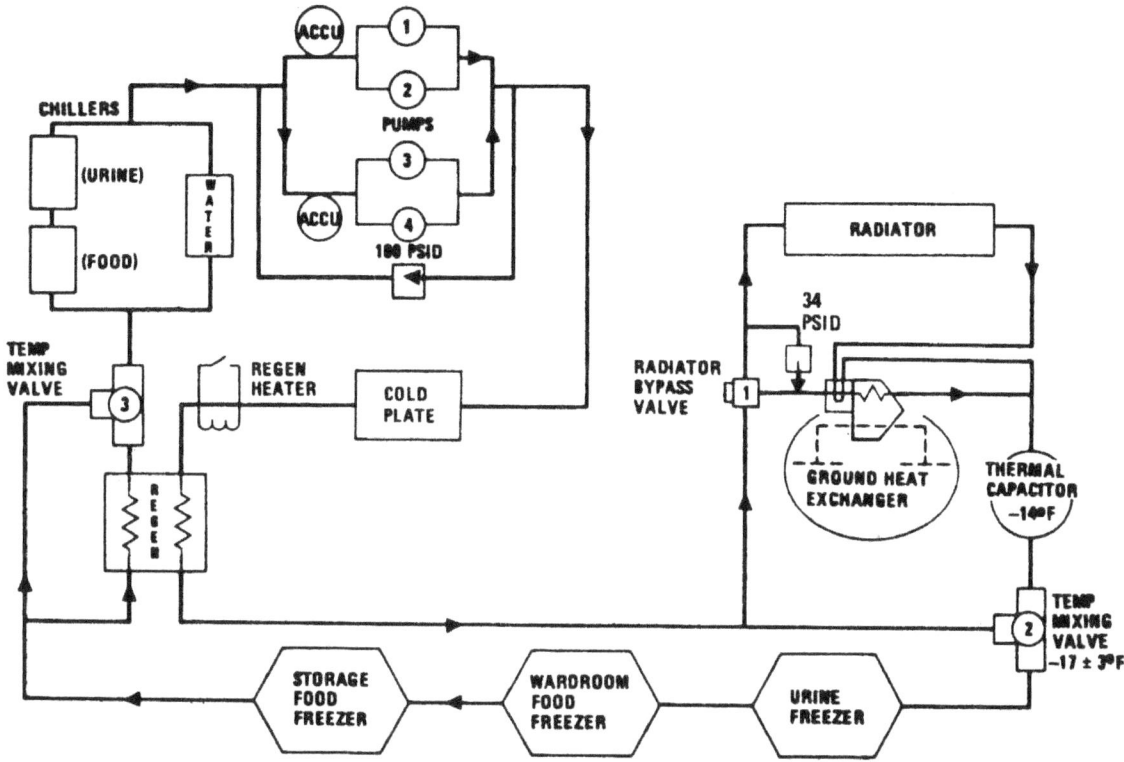

Figure II-52 — Refrigeration System

The chiller can hold 4.5 kg (10 lb) of chilled drinking water. Only 2.3 kg (5 lb) of chilled water can be extracted at one time. Once this amount has been extracted, it takes approximately 1 hour to rechill the incoming water. The refrigerant is controlled to an inlet temperature of 4.6 ± 2 degrees C (39 ± 3 degrees F).

Electrical heaters are provided to maintain SWS wall temperatures at an acceptable level during storage-mode operation and to prevent freezing in the water storage container plumbing, suit-cooling, and ATM C&D panel cooling systems during the unmanned phase of the mission (Figure II-53).

Passive Thermal Control System

The passive thermal control system employs appropriate surface coatings and thermal insulation (Figure II-54) to maintain the internal walls and truss-mounted equipment within acceptable temperature limits. Due to the dependence of passive system performance on the external radiative exchange, temperatures are influenced by the vehicle orientation.

The external-surface thermal coatings are black and white paints. The radiator on the MDA and STS uses a white paint (zinc oxide) with a low ratio of solar absorptivity to surface emissivity to provide low effective sink temperatures and higher heat-rejection rates. The design value used for the radiator surface is slightly higher than values measured for a clean surface to account for degradation during the mission caused by exposure to ultraviolet radiation, meteoroids, exhaust plume impingement, etc. Black paint is used for the forward skirt, IU, FAS and MDA.

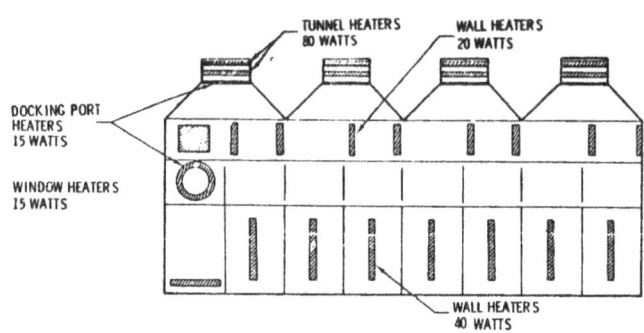
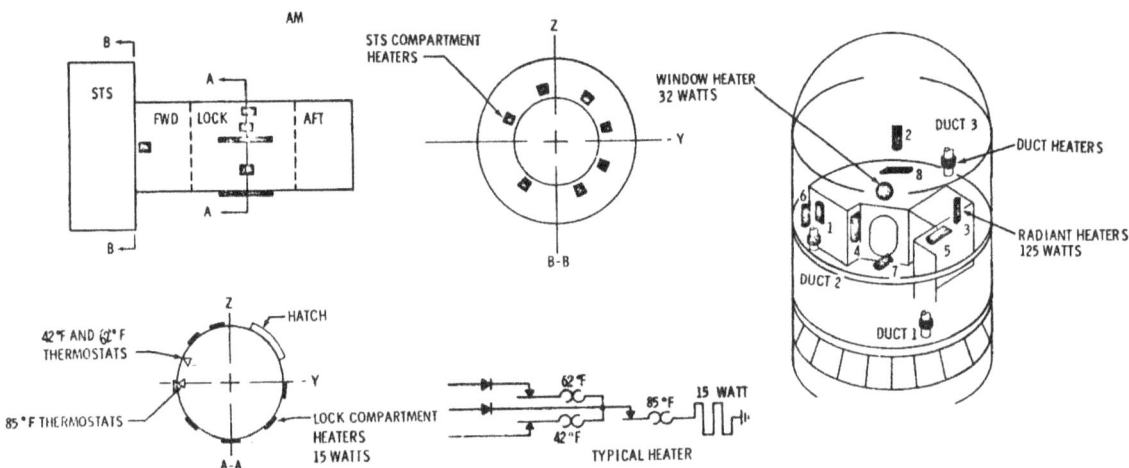

Figure II-53 — Heater Locations

Rack-mounted equipment has surface thermal coatings to maintain operable temperatures by radiation heat transfer. The canister lip and the Sun shield prevent direct solar radiation from striking the rack-mounted equipment in the nominal mission attitude. Reradiation from the Earth and the surrounding portions of the cluster is balanced by radiation from the rack-mounted equipment to space, thereby maintaining the components within the upper operable, low-power and orbital-night periods of the mission.

Passive protection against solar radiation or cold space for crewmen during EVA is provided by a thermal garment consisting of seven layers of aluminized mylar separated by layers of spacers.

II-3

ELECTRICAL POWER SYSTEM

The human body is kept alive and well by the proper flow of vital force throughout the complex circulatory system. Likewise Skylab is kept alive and functioning by the flow of electric current throughout the complex power distribution systems.

Skylab's electric supply is provided by three electrical power systems: IU EPS; SWS EPS; and CSM EPS (SWS refers to the mated OWS, ATM, MDA, AM and IU). Each system has a specific function and operating profile for which it was designed.

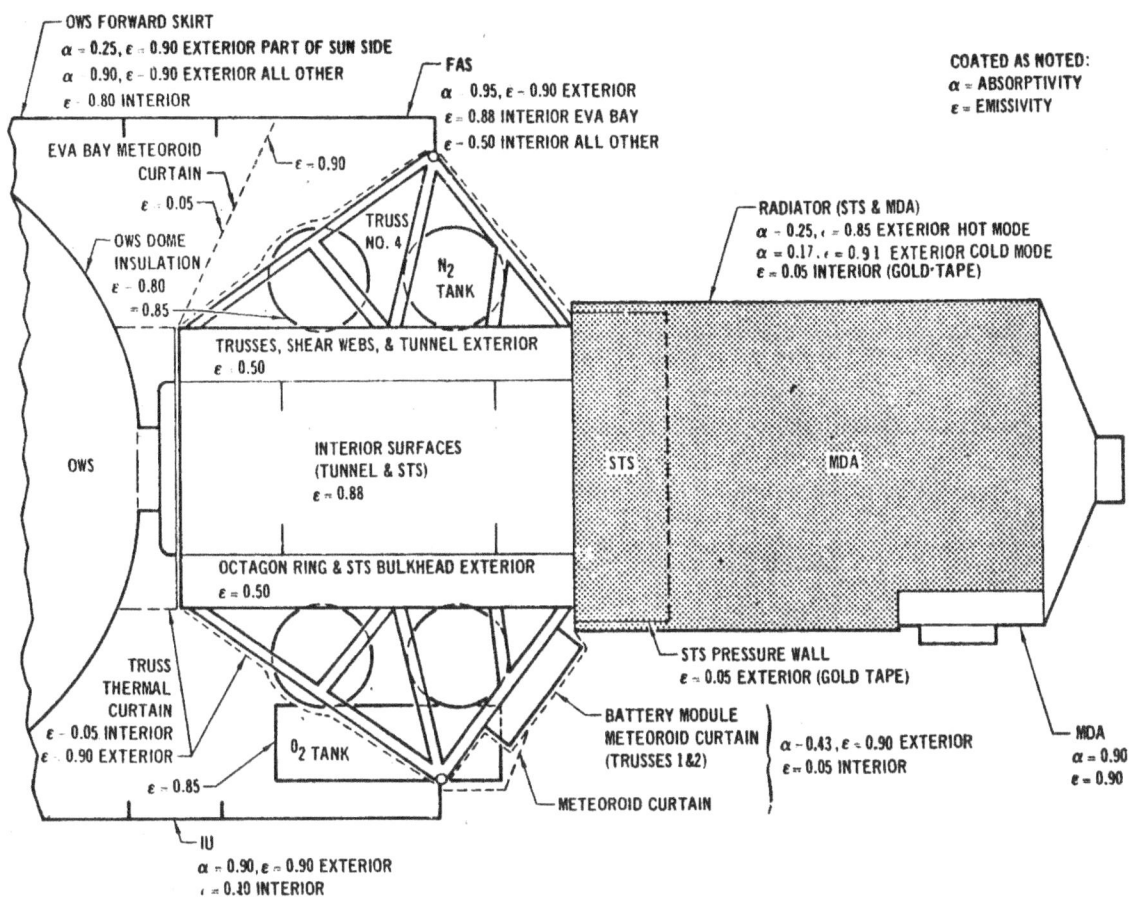

Figure II-54 — Passive Thermal Control

The IU EPS is required only during countdown, launch and orbital insertion. The IU initiates automatic, programmed signals which control critical Skylab functions during that period. The maximum required operating life of the IU EPS is 7.5 hours. This electrical power system is independent of the other electrical power systems.

The IU EPS consists of the following major components:

Batteries	4
Platform AC Power Supply	1
56 Vdc Power Supply	1
5 Vdc Power Supply	1
Distributors	7
Power Transfer Switch	1

The four batteries are silver-oxide-zinc 20-cell dry charge type with a capacity of 350 ampere-hours each. Activation of each battery before launch is through addition of potassium hydroxide electrolyte. A connector with jumpers to connect 18, 19 or 20 cells is provided on the battery case to match the battery output to its load.

The Platform AC Power Supply converts unregulated 28 Vdc power to three-phase, 400 Hertz, 26.6 Vac. This power is fed to the Control Distributor for use by the Platform electronics.

Both the 56-volt power supply and the 5-volt power supply are DC to DC converters whose outputs are fed to the Control Distributor and the Telemetry System, respectively.

The seven distributors include a Power Distributor, two Auxiliary Power Distributors, one Control Distributor, one Emergency Detection System Distributor and two Measuring Distributors. The pressurized distributors serve as junction and switching boxes distributing 28 Vdc power to IU components and housing the various necessary electrical hardware.

The sources of primary electrical power for the SWS cluster are solar cell arrays on the ATM and the OWS. These arrays collect and convert incident visible sunlight into electrical power. This power is then made available to recharge batteries and is power conditioned before distribution to designated loads (Figures II-55 and II-56).

The ATM EPS and AM/OWS EPS are operated in parallel and are capable of delivering power to, or receiving power from each other, as well as delivering power to the CSM EPS.

The end-of-mission average per orbit power requirements of the two EPSs are:

	Solar Inertial	Z-LV-Rendezvous	Z-LV-EREP
ATM	3716 W	1300 W	3000 W
AM/OWS	3814 W	1300 W	3000 W
Total	7530 W	2600 W	6000 W

Table II-7 is the load distribution summary.

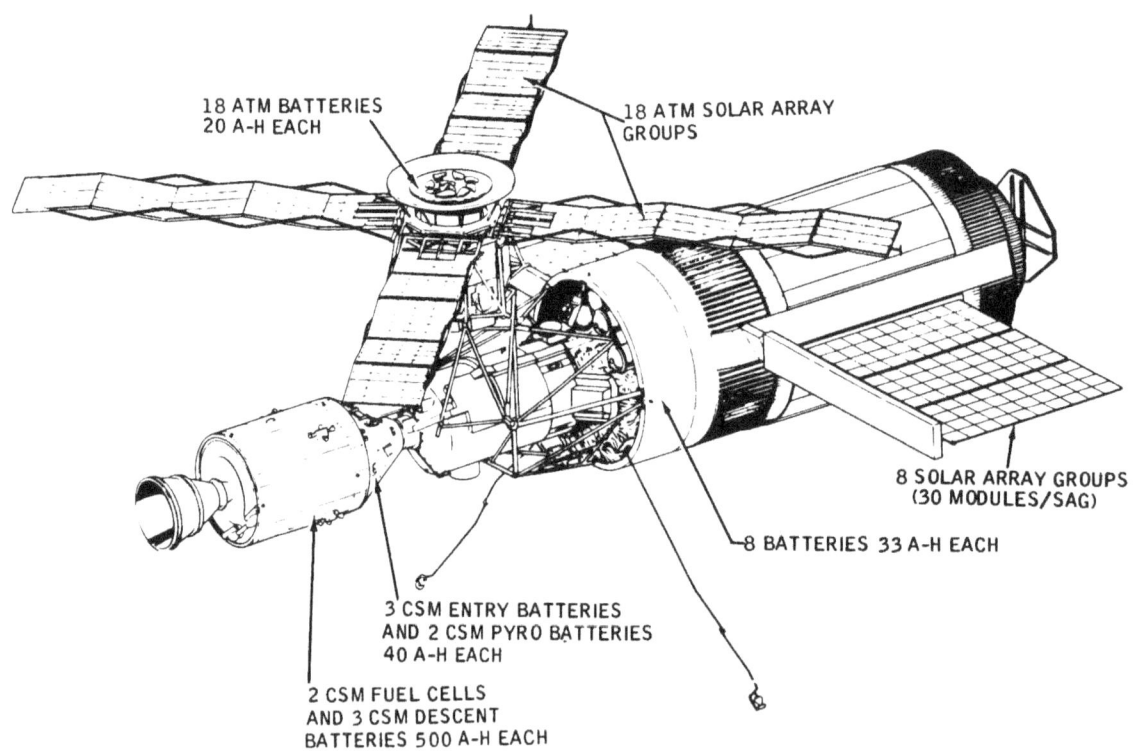

Figure II-55 — Skylab Power Sources

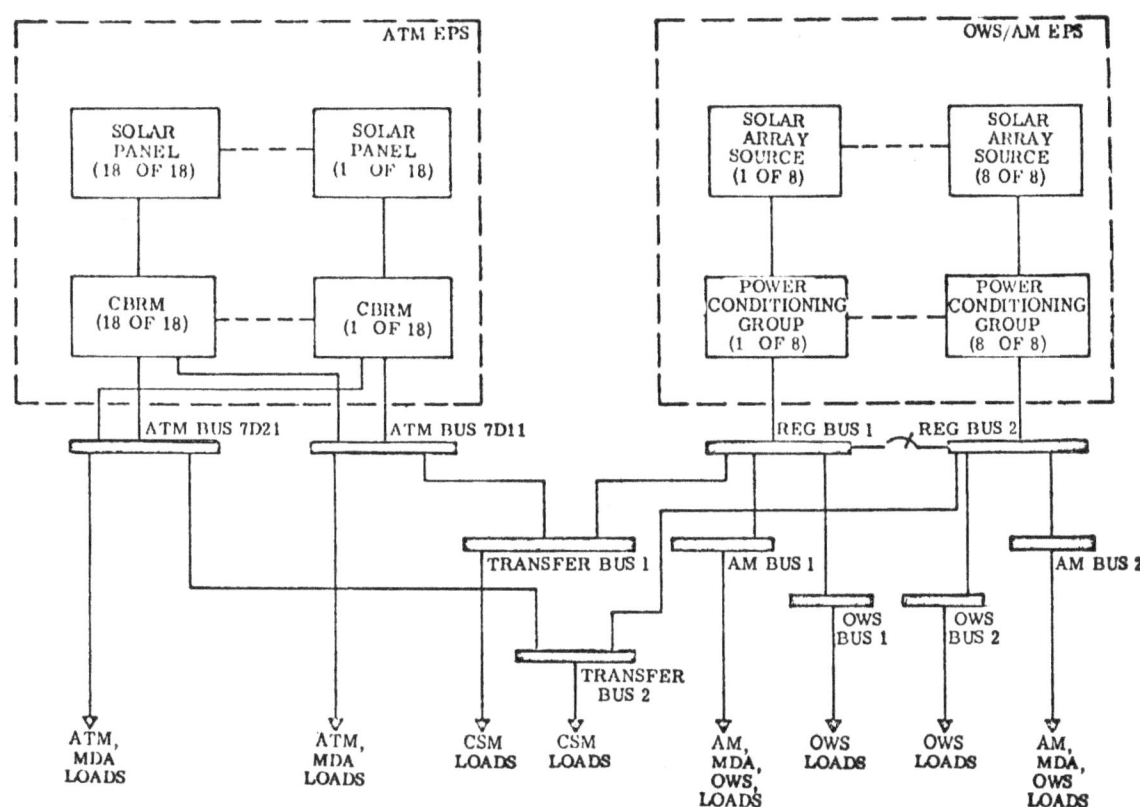

Figure II-56 — Power Distribution

Table II-7 — Load Distribution Summary (Percent)

	Maximum SI Orbit	Maximum 120° Z-LV Orbit	Maximum 60° Z-LV Orbit
Life Support	36	34	35
Housekeeping	56	53	55
Experiments	8	13	10

The CSM EPS consists of the following major components:
   2 fuel cell power plants (Bacon type)
   3 solid state inverters
   3 entry/post-landing batteries (40 amp-hr)
   2 pyrotechnic batteries (40 amp-hr)
   3 descent batteries (500 amp-hr)

The fuel cells are capable of supplying up to 1,420 W. When CSM power is required after fuel cell life, it is supplied by the batteries or by the SWS EPS (during the docked mode). CSM power is provided by the fuel cells up to and after docking for a maximum of 13 days. Just prior to fuel cell depletion, the CSM EPS is connected to the SWS EPS and power for CSM loads is provided by the SWS until undocking.

During deactivation, prior to undocking, the descent batteries are operating. The SWS EPS is paralleled with the CSM EPS during switch-over. The parallel operation periods are about four minutes.

When docked, the CSM main buses are connected to the AM transfer buses through circuit breakers and motor driven switches with power connections in the MDA docking ring umbilical.

Power Generation

The physical characteristics of the solar cell power generation systems are listed in Tables II-8 and II-9. Figures II-57 and II-58 show cross section cuts through typical solar cell modules in each subsystem. Differences exist due to individual supplier design and assembly preferences, and each is considered capable of satisfying all mission requirements. Figures II-59, II-60 and I-5 show the solar array systems in various stages of deployment and detail.

The raw power capability, prior to power conditioning at the beginning of the mission, of the as-launched solar cell arrays is:

$$ATM - 10{,}480 \text{ watts at } 55°C, \beta = 0°$$
$$\text{Illumination} = 1400 \text{ W/m}^2$$
$$AM/OWS - 12{,}400 \text{ watts at } 77°C, \beta = 0°$$
$$\text{Illumination} = 1400 \text{ W/m}^2$$

The difference between the total generated power capability of about 23 kilowatts and that available to spacecraft loads is due to power conditioning losses and cable resistances. Further reduction in solar array output is caused by the dependent influences on the solar cells such as UV exposure, solar flare charged particles, and temperature cycling fatigue of materials due to the more than 4000 day-night periods imposed by the mission. Other factors include the seasonal Sun distance effect on illumination intensity, orientation angle with the Sun and shadow patterns from vehicle protuberances.

The above factors and many others have been considered in assuring that Skylab load requirements will be satisfied at all times.

Power Conditioning and Energy Storage

The ATM uses 18 Charger Battery Regulator Modules (CBRMs) while the AM/OWS uses 8 Power Conditioning Groups (PCGs). A CBRM and a PCG perform essentially the same functions.

Each CBRM is designed to operate at various power levels, which are determined partly by the amount of energy supplied by its associated solar cell panel. The CBRM conditions the power by converting the higher solar array or battery voltage to the lower 28 Vdc level required by Skylab power buses. The maximum capability of each CBRM is 415 watts at an efficiency of 92 percent. The output of the CBRM is fed to two buses (Figure II-61).

Table II-8 — Physical Characteristics of ATM SAS

Array	
Size	13.2 m (521 inches long), 2.7 m (104.5 inches wide) (per wing)
Weight	1723 kg (3,800 pounds) (including deployment structure)
Panels Per Wing	5 (inboard panels are half covered with modules)
Total Panels	20
Total Solar Cells	2 × 2 cm — 123,120 and
	2 × 6 cm — 41,040 Total = 164,160
Total Modules	360
Solar Panel	
Size	2.7 m (104.3 inches long), 2.7 m (104.5 inches wide)
Weight	66.2 kg (146 pounds) (including panel frame)
Modules Per Panel	20 (inboard panels contain 10 modules each)
Total Cells Per Full Panel	2 × 2 cm — 13,680 or
	2 × 6 cm — 4,560
Solar Cell Module (Both Types)	
Size	.5 m (20.0 inches long), .63 m (24.625 inches wide)
Weight	2.2 kg (4.93 pounds)
Series Cells	114
Parallel Cells	2 × 2 cm — 6 or
	2 × 6 cm — 2
Total Cells	2 × 2 cm — 684 or
	2 × 6 cm — 228
Cell Interconnector	2 × 2 cm — Expanded Silver Mesh
	2 × 6 cm — Solder Plated Copper
Cell to Substrate Adhesive	0.127 mm (0.005 inch) Silastic 140
Substrate	Aluminum Facesheet/Aluminum Honeycomb
Dielectric Insulation	0.127 mm (0.005 inch) Micaply
Solar Cell	
Type	N/P
Size	2 × 2 cm and
	2 × 6 cm (both 0.014 inch thick)
Base Resistivity	7 to 14 ohm-cm
Cell Contact	AgTi, fully solder covered contacts

Table II-9 — Physical Characteristics of OWS SAS

Array	
Size	9.5 by 8.3 m (372 by 328 inches) (per wing)
Weight	1840 kg (4,056 pounds) (including deployment and stowage structures)
Panels Per Wing	30
Panels Per SAS	60
Total Solar Cells Per SAS	147,840
Modules Per SAS	240
Solar Panel	
Size	0.7 m (27.13 inches wide), 3 m (120.7 inches long)
Weight	12.7 kg (28 pounds)
Modules Per Panel	4
Series Cells Per Module	154
Parallel Series Strings Per Module	4
Total Cells Per Panel	2,464
Substrate	Aluminum Facesheet/Aluminum Honeycomb
Dielectric Insulation	Perforated 0.002-inch Kapton
Solar Cell Group	
Type	Overlapped
Cells Per Group	11 series
Cell Interconnector	0.025 mm (0.001 inch Kovar) (Solder Plated)
Group Interconnector	0.075 mm (0.003 inch Kovar) (Solder Plated)
Cell to Substrate Adhesive	RTV 3145
Solar Cell	
Type	N/P
Size	2 x 4 cm (0.014 inch thick)
Efficiency	AMO, 28°C–11.1 percent minimum, Average Bare, New
Base Resistivity	2 ohm-cm
Cell Contact	AgTi Machine-Pressed Fully Solder Covered Contacts

A portion of the solar array power is required to replace the energy removed from the batteries during the dark part of the orbit. The function of the charger is to optimize the power available from the solar panel. It distributes power to the load regulator and the batteries when required. The charger output is connected to the battery only when charging power is available and is disconnected when either the battery is fully charged or when the available solar array power is too low to charge the batteries.

The regulator provides regulated 28 Vdc power to appropriate buses within the ATM. The regulator can receive power from either the battery or the solar panel directly. The average and maximum steady state power levels delivered by the regulator are 235 W and 415 W respectively.

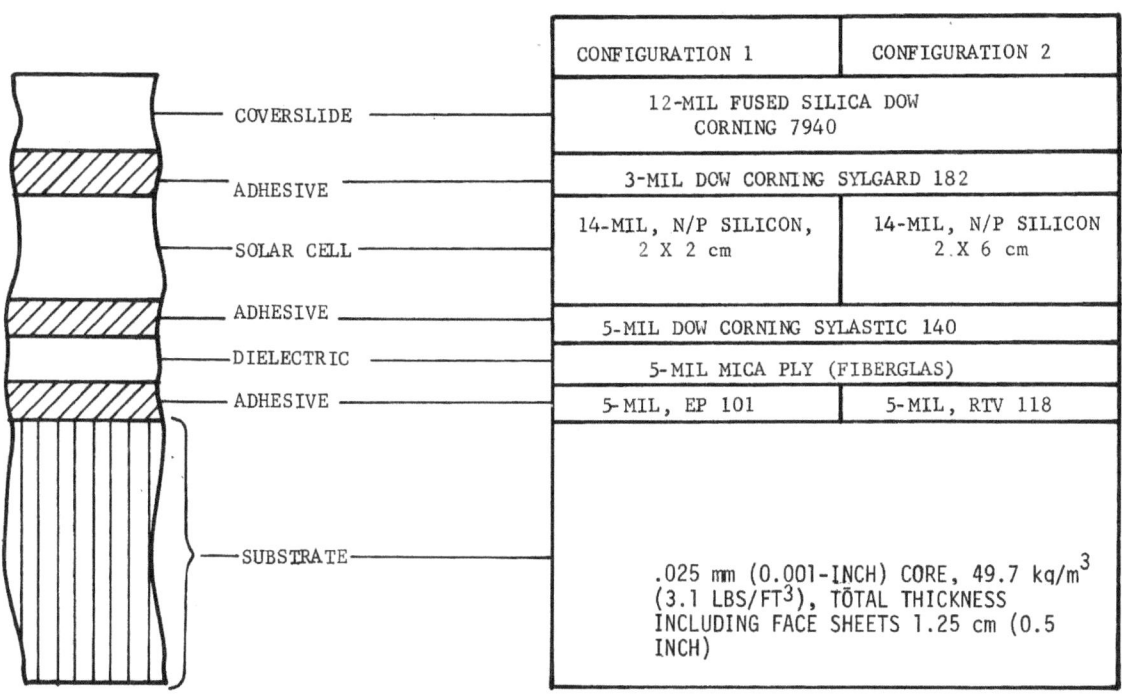

Figure II-57 — Cross Section of ATM Solar Cell Module

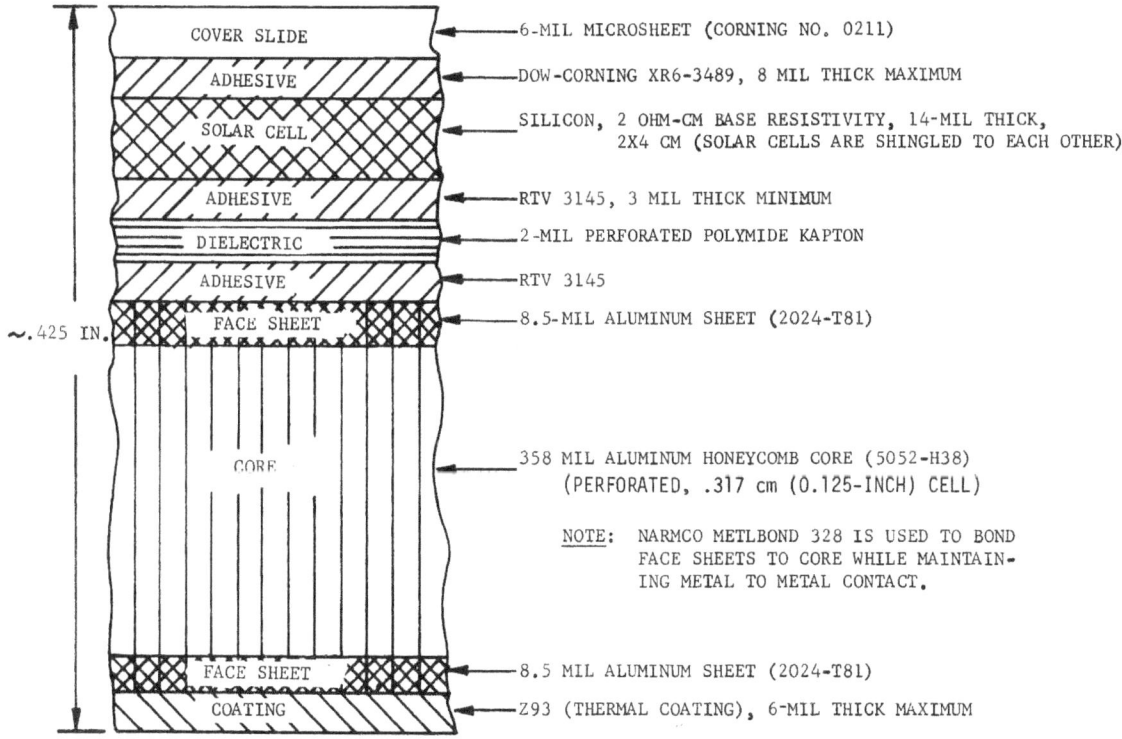

Figure II-58 — Cross Section of OWS Solar Cell Module

II-47

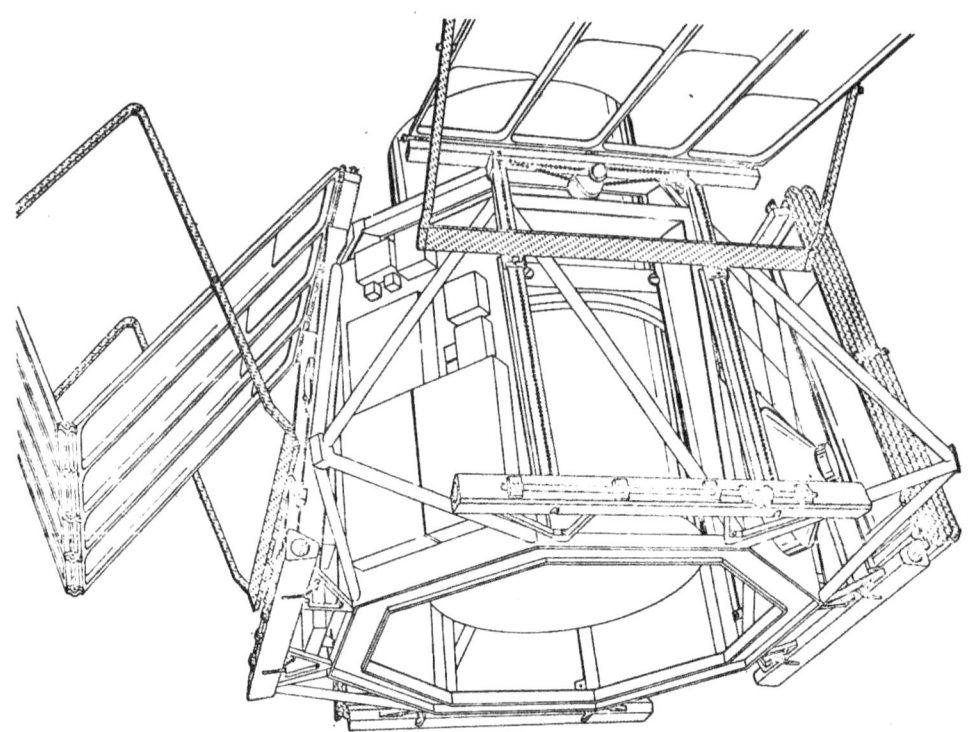

Figure II-59 — ATM Solar Wings in Stowed and Partially Deployed Positions

NOTES:
- 684 2X2 cm - or 228 2X6 cm SOLAR CELLS MAKE UP A MODULE
- 20 MODULES MAKE UP A PANEL (POWER SOURCE)
- 5 PANELS MAKE UP A WING
- 4 WINGS MAKE UP THE ARRAY
- THE NUMBER ON EACH PANEL INDICATES THE CBRM TO WHICH IT IS WIRED
- VIEWED FACING ACTIVE CELL SIDE

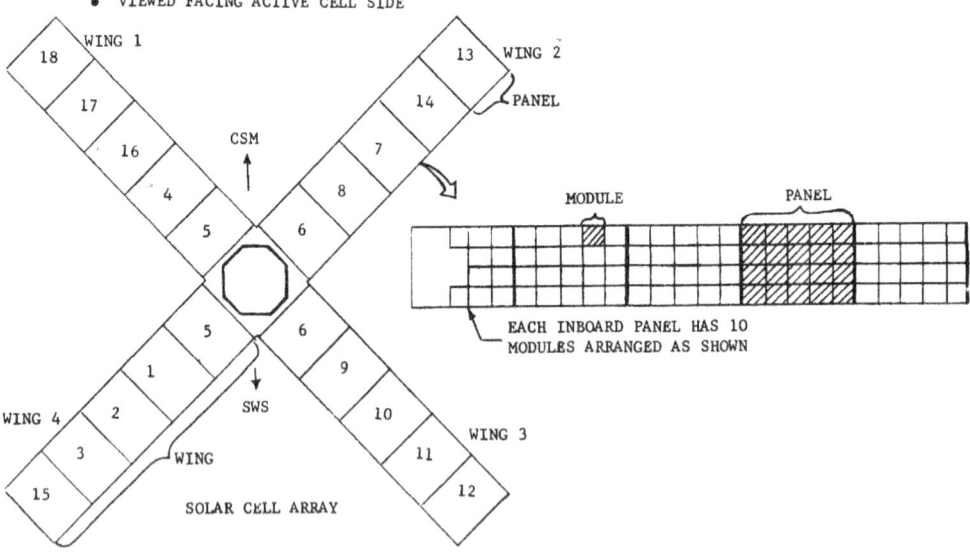

Figure II-60 — ATM Solar Array Configuration

II-48

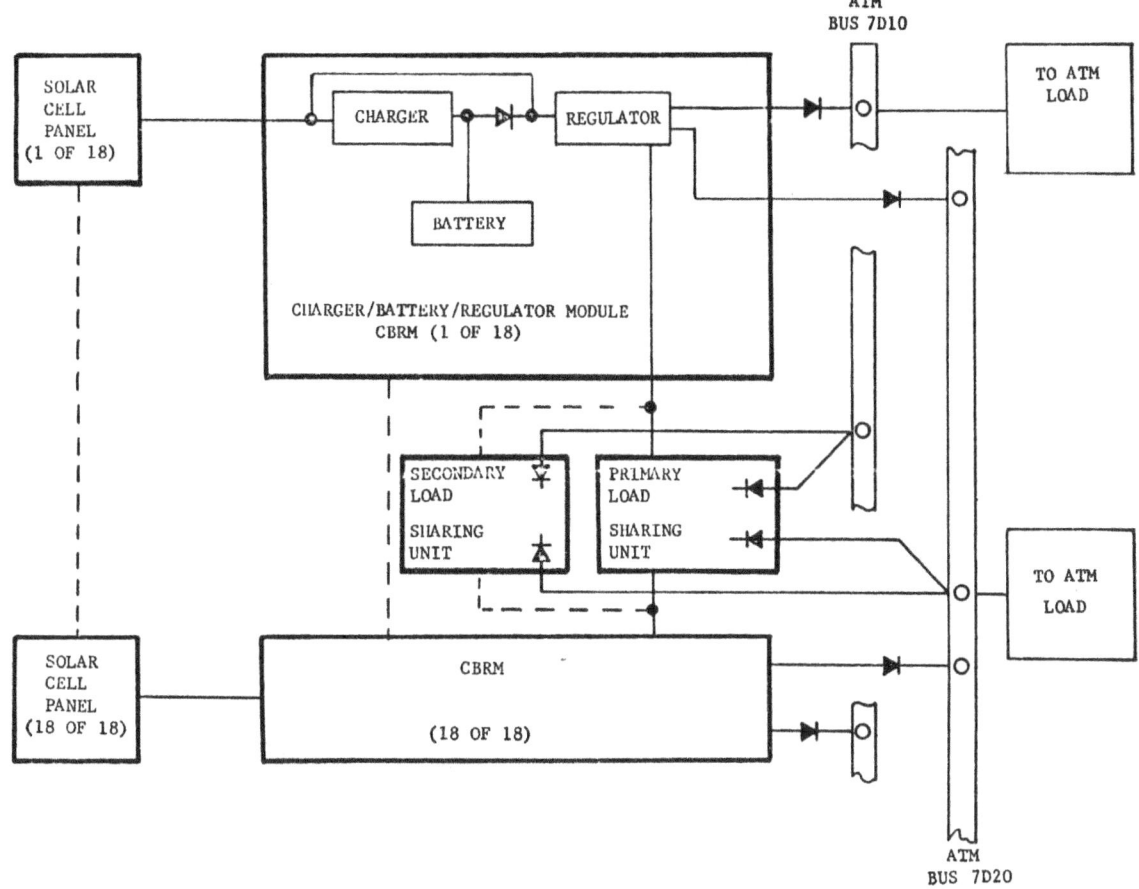

Figure II-61 — ATM Electrical Power System

The operational modes for the AM are similar to those stated for the ATM CBRM except that the control limits are different. Charge termination results when the battery thermal switch opens at 51.7 degrees C (125 degrees F), and/or when the array cannot supply sufficient power.

The voltage regulator regulates power to both the EPS control and regulated buses. This power is supplied in one of four ways:
    From the battery charger only
    From the battery charger and battery in parallel
    From battery only
    From SAG only

Fuses are provided for each regulator to isolate malfunctioning systems from the remaining normal regulators on the same bus.

The OWS/AM EPS is provided with a Shunt Regulator on each EPS Control bus to restrict over-voltage transients resulting from voltage regulator malfunctions. This is accomplished by providing an additional current path until circuit interruption occurs in the malfunctioning circuit.

Power Distribution Systems

Figure II-62 shows the ATM Power Distribution block diagram indicating the quantity for each distributor type.

Both power distribution systems have their negative returns connected to a single point ground in the AM when the CSM is not docked and to the CM vehicle ground when CSM is mated with the SWS.

Power Control

Control of the SWS EPS is normally automatic but manual or ground control is also possible. Various ground command systems can be used for remote control of the SWS EPS.

ATM power-up/power-down may be accomplished by activating or deactivating each CBRM independently or by activating all simultaneously.

Both the ATM EPS and the AM/OWS EPS are designed to operate while interconnected. This provides for power sharing between the two EPSs. The SWS EPS in turn is parallel with the CSM EPS. This paralleling provides power to the CSM when required. In this paralleled condition the EPS with the highest bus voltage will supply the major portion of the total power requirement. Balanced power sharing is an important capability of this configuration.

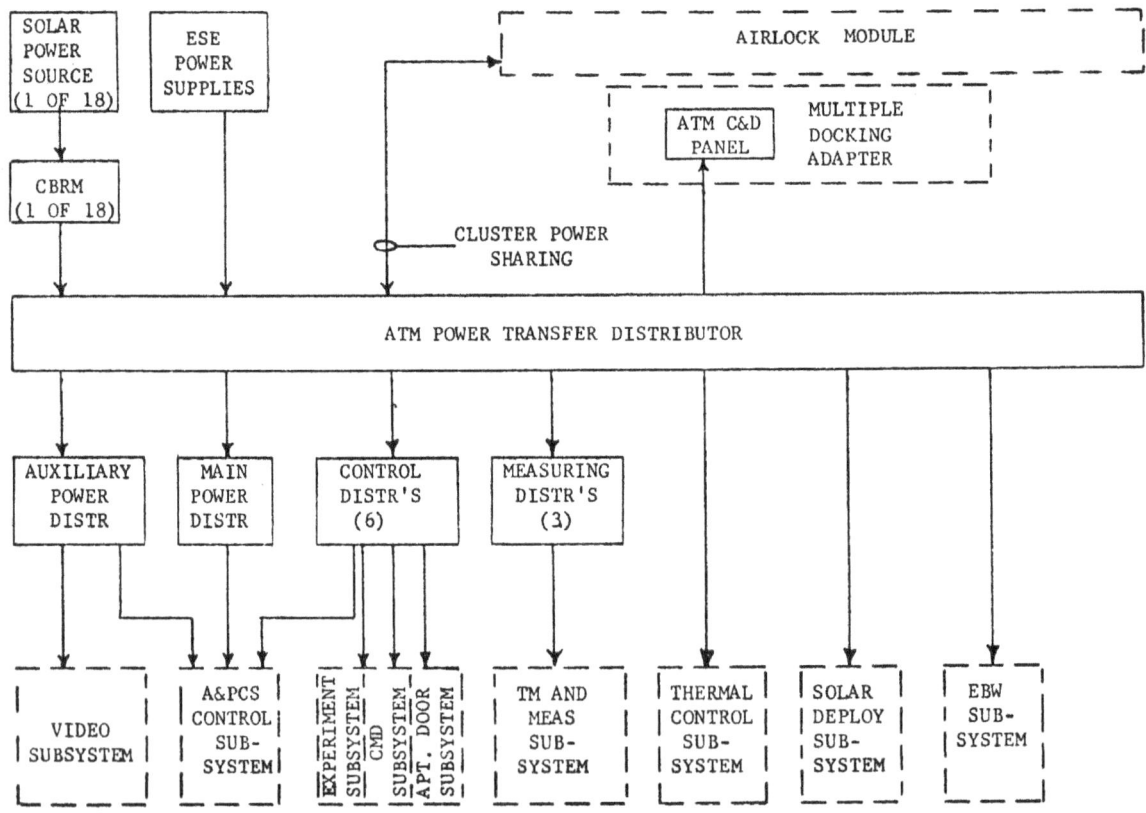

Figure II-62 — ATM Power Distribution

Monitoring Systems

On-board monitoring capabilities for the SWS EPS are contained in various types of readout devices. Also, the telemetering system permits status information to be displayed on the ground.

System Circuit Protection

Protective devices for the cluster's electrical power system consist of fuses and circuit breakers. Their function is mainly to protect system wiring, being placed ahead of associated components and as near as practical to associated feeder buses. The design of the protection system is such that individual failures will not affect remaining normal loads.

Lighting

Lighting of Skylab was somewhat of a challenge to engineers because they had to be sure "every nook and cranny" had sufficient light, but at the same time, not place too great a drain on the power supply. Another consideration was the weight factor — too many lights would raise the OWS total weight significantly. Still another factor was types of lighting for various purposes. Also, special provisions had to be made for mobility of some light fixtures, shading and protection against breakage.

Skylab lighting requirements are internal and external. The internal lighting system is for normal operations of the spacecraft and for conducting experiments. The external lights are for docking, rendezvous tracking and EVA.

External Lights — Four strobe lights are mounted on the ATM Deployment Assembly near the pivot points, two each at -Y and +Y. These serve as tracking lights as they flash regularly. These lights cast a cone of brilliant light 45 degrees from the +X axis of the Skylab, giving a 90 degree spread of light. The light would appear as a third magnitude star if viewed from an Apollo Command Module window at a distance of 267 kilometers (144 nautical miles) or through a sextant at a distance of 1,165 kilometers (629 nautical miles).

All eight docking lights are 20-watt incandescent bulbs protected by grid enclosures. Two are red, two white, two green and two amber. Four are mounted on the forward end of the MDA, the white at +Z, red at +Y, amber at -Z and green at -Y. The four mounted on the forward end of the FAS are placed with colors matching those coordinates of the MDA lights, resulting in a cross of lights when viewed from ahead of the OWS. As the Command Module approaches the forward docking port, the two white lights will be "up," the two amber lights "down," the red lights to the right and green lights to the left. Two other lights will be visible during the docking maneuver. Four 0.7 watt white bulbs protected by a transparent dome will be shining at the tip of each of the two discone antennas.

Internal Lights — Interior lighting of Skylab uses 78 fixed lights and five portable spares. General illumination light fixtures include 24 in the crew quarters, 18 in the OWS forward compartment and 8 in the MDA. In the crew quarters, the Wardroom has four general illumination lights (numbered 1 through 4 in the illustration, Figures II-63 and II-64), the WMC has three (1-3), the sleep compartment three (1-3) and the experiments compartment 14 (1-14). In the forward compartment, 10 lights are on the upper wall (1-10) and eight on the dome (1-8). Four general illumination lights are in the forward end of the MDA and four in the aft section.

Handrail lights and tunnel light fixtures are used in the smaller spaces. Four handrail lights are in the forward part of the Airlock Module's Structural Transition Section (STS) and four in the aft section of the STS. Six handrail lights illuminate the instrument panel in the STS, and six of the lights are in the AM aft compartment. Four tunnel light fixtures are installed in the AM lock compartment. Of these interior lights, certain ones have been

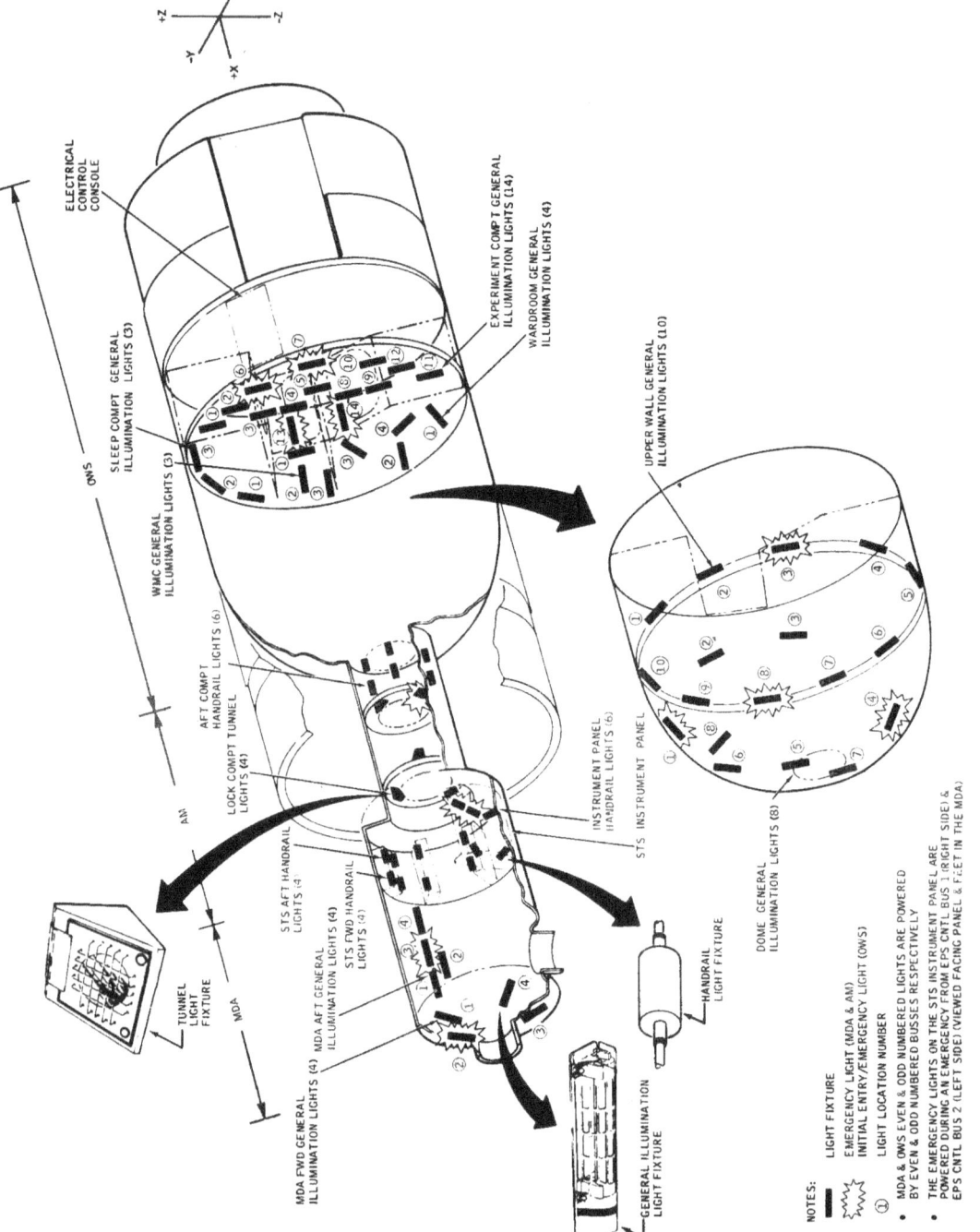

Figure II-63 — Skylab Lighting Overview

OWS LIGHTS

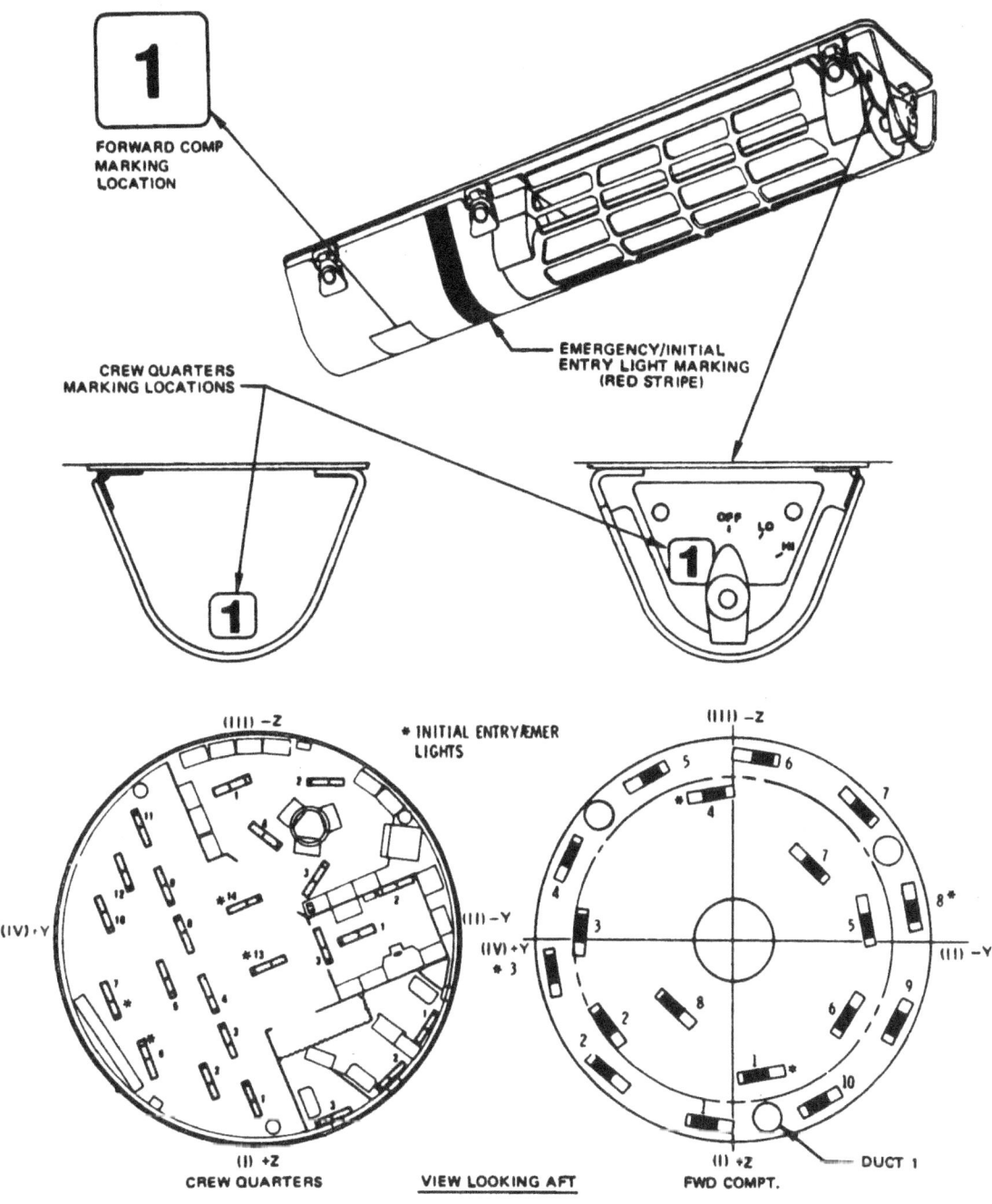

Figure II-64 — OWS Lights

designated as emergency and initial entry lights. One general illumination light in each end of the MDA, two in the OWS dome, two on the OWS forward compartment upper wall and four in the OWS experiment compartment fall into this category, as do two handrail lights in the STS and one tunnel fixture in the AM lock compartment.

High intensity lights are used in conducting many of the experiments. Four of these lights are in the OWS forward compartment, one at the +Y-Z quadrant near the M509 experiment bottles, one at the -Z-Y quadrant for the T013, M151 and S149 experiments, one at the +Z-Y quadrant next to the M172 experiment, and one at the +Y+Z quadrant next to the M509 experiment.

The four high intensity lights in the OWS forward compartment are fixed. Two spare high intensity lights are stowed for use if needed. They can be plugged into any high power outlet. Three portable hand lights, stowed for use as needed, are powered by any available utility outlet.

ATTITUDE AND POINTING CONTROL SYSTEM

The Saturn Workshop (SWS) Attitude and Pointing Control System (APCS) provides three-axis attitude stabilization and maneuvering capability of the orbiting vehicle. It also provides the capability of pointing experiments at desired locations such as the Sun, the Earth and other targets of interest. The APCS is comprised of the Instrument Unit/Thruster Attitude Control Subsystem (IU/TACS), Control Moment Gyros Subsystem/Thruster Attitude Control Subsystem (CMGS/TACS) and Experiment Pointing Control Subsystem (EPCS).

The IU/TACS provides control and maneuvering of the SWS from insertion to control transfer (switch-over) to the CMGS/TACS. Switchback to IU/TACS control capability is provided; however, the IU/TACS lifetime is limited to launch plus approximately 10.5 hours.

The CMGS/TACS provides control, maneuvering and offset pointing of the SWS and OA commencing with control transfer from IU/TACS. This subsystem utilizes three primary control modes (solar inertial, Z local vertical and experiment pointing) with the control forces provided by the CMGS augmented by TACS firings as required.

The EPCS provides highly accurate pointing and stabilization of the ATM experiment package along with limited offset pointing from Sun center in the pitch and yaw axes and roll axis positioning over a wide range. The EPCS is activated as required to meet experiment objectives.

The TACS APCS is activated at orbital insertion (S-II cutoff). The OWS thrusters accept commands from the IU Launch Vehicle Digital Computer (LVDC) after orbit insertion to maneuver the SWS for Payload Shroud jettison and then to the solar inertial (SI) attitude. The SI attitude points the vehicle -Z axis to the center of the solar disk and places the X-principal axis near the orbit plane. CMG spinup is initiated and vehicle control is switched to the CMG control subsystem (TACS augmented). Shortly thereafter, the complete CMG control program is enabled, including automatic management of the Skylab momentum.

The APCS can be commanded by manual inputs from the astronauts using the ATM C&D Console in the MDA or from the ground via the vehicle Digital Command System. System redundancy is provided by using component backup units or by operating the system in an alternate configuration, e.g., although three CMGs are used in maintaining vehicle control, two can provide this function if one unit fails. The APCS can receive commands and control input data from ground operations through the Command System. The Telemetry System is used to transmit APCS information to the ground for operational support and evaluation.

Six control modes are addressable by C&D Console switches and also by the DAS and DCS system for APCS operation. Two of the six modes, Solar Inertial (SI) and Z Local Vertical (Z-LV), are the attitudes used most of the time during the mission. All vehicle attitudes other than SI or Z-LV are attained by maneuvering or offsetting from SI or Z-LV. The attitude used most of the time for Skylab is SI.

The six modes are:

<u>Solar Inertial Mode</u> — During orbit daytime, this mode is used for maintaining the vehicle's minimum moment of inertia axis (X-principal axis) near the orbital plane and the Z-axis parallel to the sunline. During orbit "nighttime," it is used to perform gravity gradient momentum dump maneuvers for desaturating the CMGs.

<u>Z-Local Vertical (Z-LV) Mode</u> — This mode is entered during the rendezvous phases of the mission or when Earth pointing experimentation periods are required.

<u>Experiment Pointing Mode</u> — This mode is identical to the SI mode except that the EPCS is automatically activated each orbital sunrise and deactivated each orbital sunset.

<u>Attitude Hold (CMG) Mode</u> — In this mode, the vehicle can be maneuvered to any inertial-oriented attitude and held using the CMGs only.

<u>Attitude Hold (TACS) Mode</u> — This mode is used to maneuver the vehicle to any inertial-oriented attitude and held using the TACS only.

<u>Standby Mode</u> — Used when vehicle control is not required of the APCS, e.g., during CSM control of the OA.

The pointing accuracy and stability requirements are as follows:

CMG System	Accuracy	Stability
Solar Inertial X, Y	±6 arc-min	±9 arc-min/15 min
Z	±10 arc-min	±7.5 arc-min/15 min
Z-LV (EREP) X, Y, Z	±2°	
Z-LV (rendezvous) X, Z	±6°	
Y	±12°	
EPC System		
X, Y	±2.5 arc-sec	±2.5 arc-sec/15 min
Z	±10 arc-min	±7.5 arc-min/15 min

Thruster Attitude Control Subsystem

The IU/TACS provides for attitude control of the Skylab about all axes following separation from the boost vehicle. Upon activation of the CMGS, the TACS augments the CMG Control Subsystem. When both the CMGS and TACS are enabled, TACS is used when the vehicle's total system momentum exceeds a defined limit. The vehicle is controlled by TACS only when redundancy management disables CMGS control, CMG control is inhibited via DAS/DCS, or the Attitude Hold (TACS) Mode is selected when the attitude error exceeds 20 degrees during CMG operations.

Control Moment Gyro Subsystem (CMGS)

Vehicle attitude information is derived from strapdown reference computations in the ATM Digital Computer (ATMDC) utilizing Rate Gyro information. The Acquisition Sun Sensors provide data for updating of vehicle attitude information for the X and Y control axes. Switching from the orbital daytime to the orbital nighttime configuration and vice versa is performed automatically upon command from the ATMDC. Orbit nighttime commences either at sunset or at the start of momentum desaturation maneuvers, whichever begins first. The minimum period of the commanded nighttime configuration is 35 percent of the orbit. Orbit daytime commences either at sunrise or at the end of the momentum maneuvers, whichever ends later. The ATMDC processes the sensor signals with the CMG control law to generate the CMG gimbal rate command (control forces). The astronaut has the capability of manually controlling the CMGs by means of the Digital Address System (DAS) keyboard on the ATM C&D Console.

Three double gimbaled CMGs hardmounted at 90 degree angles to each other on the ATM provide the control forces (Figure II-65). A CMG is basically a large spinning wheel that provides the forces required for vehicle control by changing the orientation of the wheel's spin axis. The momentum is monitored and managed by the ATMDC. The CMGs also have the ability to store a certain amount of force or momentum. When they have absorbed the maximum amount they are saturated and the accumulated momentum must be reduced to desaturate the CMGs. The momentum stored in the CMGs is reduced by special maneuvers which reorient the CMG wheel spin axes. These are performed automatically during orbit nighttime.

Experiment Pointing Control Subsystem (EPCS)

The EPCS (Figure II-66) is used to maintain pointing and stabilization of the ATM experiments package. The package is provided with an independent control system to isolate it from perturbations due to disturbances on the vehicle.

The EPCS provides automatic control of the experiment package X and Y axes. Manual positioning of the two axes and offset pointing of the experiment package are provided by the Manual Pointing Controller in conjunction with the Fine Sun Sensor (FSS). The FSS is used for sensing ATM experiment package pointing errors, with rate gyro sensing rates. The Experiment Pointing Electronics Assembly (EPEA) conditions the sensor signals to provide command signals to the actuators.

The experiment package can be offset pointed using the FSS in the X and Y axes over a range of ±24 arc-minutes, with the center of the solar disk being the zero position. The actuators allow approximately ±2 degrees of experiment package rotation about the X and Y axes. The Roll Positioning Mechanism (RPM) provides ±120 degrees rotation about the Z axis. The solar north pole is the experiment package zero roll reference position.

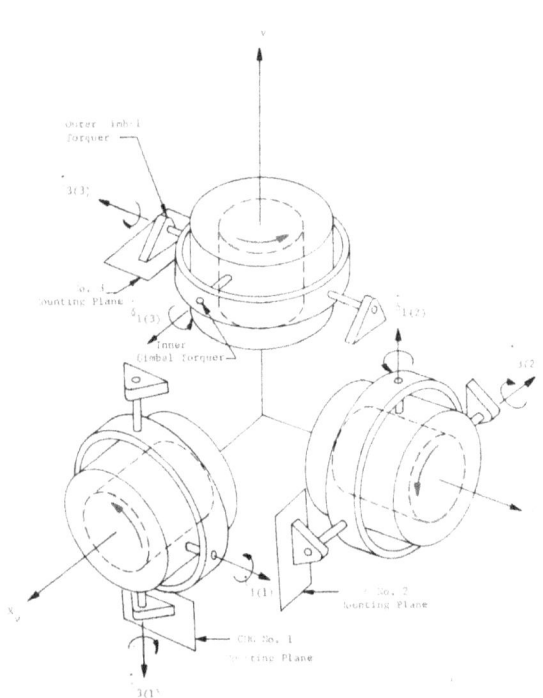

Figure II-65 — CMG Mounting Arrangement

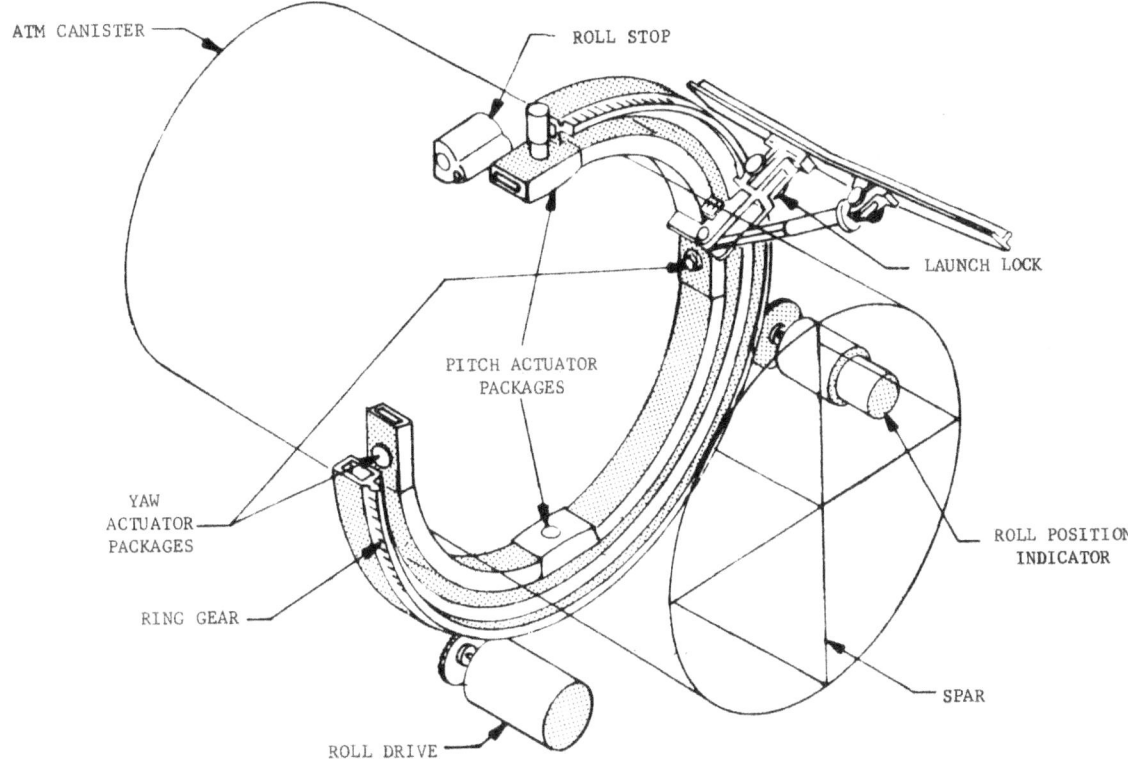

Figure II-66 — Experiment Pointing Control Subsystem

The RPM is commanded by the astronaut via rate switches on the ATM C&D Console. Once the experiment package is positioned, the RPM will hold the location until a repositioning command is received. The astronaut uses the ATM EVA Rotation Control Panel during EVA to reposition the canister for experiment retrieval.

Component Descriptions

The components comprising the APCS, with total units available noted are listed in Table II-10.

Control Moment Gyro (CMG) Assembly — The ATM CMG (Figure II-67) consists of an induction-motor-drive constant-speed wheel, gimbal supported to provide 2° of freedom. Associated with each CMG is an Electronics Assembly (CMGEA) for positioning the gimbals and controlling the gimbal rates, and an Inverter Assembly (CMGIA) for providing power.

The actuator pivot contains a DC torque motor, gear train, output shaft and rate-feedback tachometer. The sensor pivot assembly contains a ball-bearing mounted pivot shaft, a resolver assembly and a flex lead assembly. The resolver assembly provides gimbal position information for gimbal caging, control law computations and momentum management.

Each CMG has an angular momentum storage capability of 3,117 N-m-sec (2,300 foot-pound-seconds). The rotor is driven by two identical double squirrel cage, 3-phase induction motors. A single motor is capable of maintaining rotor operating speed. The CMGs are sized so that vehicle attitude control can be maintained with any two CMGs with minimal use of TACS.

Table II-10 — APCS Components

```
IU/TACS

    IU (ST-124 Stabilized platform, ST-124 Servo Amplifier Box,
    Flight Control Computer, etc.)

    Thruster Attitude Control System (TACS) Thruster Assemblies — 2

    TACS Power and Control Switching Assembly (PCSA) — 2

CMGS/TACS

    Rate Gyro Packages (RGPs) — 9

    Acquisition Sun Sensors (Acq SSs) — 2

    Control Moment Gyro Inverter Assemblies (CMGIAs) — 3

    Control Moment Gyros and Electronic Assemblies
    (CMGs and CMGEAs) — 3/3

    Apollo Telescope Mount Digital Computers/Workshop Computer
    Interface Unit (ATMDCs/WCIU) — 2/1

EPCS

    Rate Gyro Packages (RGPs) — 4

    Fine Sun Sensor (FSS) — 1

    Manual Pointing Controller (MPC) — 2

    Experiment Pointing Electronics Assembly (EPEA) — 1

    Experiment Package Caging and Gimbal Assembly — 1

    Star Tracker (ST) — 1
```

<u>Star Tracker</u> — The ATM Star Tracker function is to provide star position inputs to the ATMDC for calculating the roll reference angle and the orbital plane error. The roll reference angle provides an experiment pointing reference with respect to the solar disk for roll positioning, telemetry and experiment file recording.

The Star Tracker consists of an Optical-Mechanical (OM) Assembly (Figure II-68) and a Star Tracker Electronics (STE) Assembly.

<u>Fine Sun Sensor</u> — The Fine Sun Sensor (FSS) provides the highly accurate attitude information for the X and Y axes of the EPCS. The FSS is comprised of four separate components: an OM Assembly, Preamplifier Electronics Assembly, Control Electronics Assembly and the FSS Signal Conditioner.

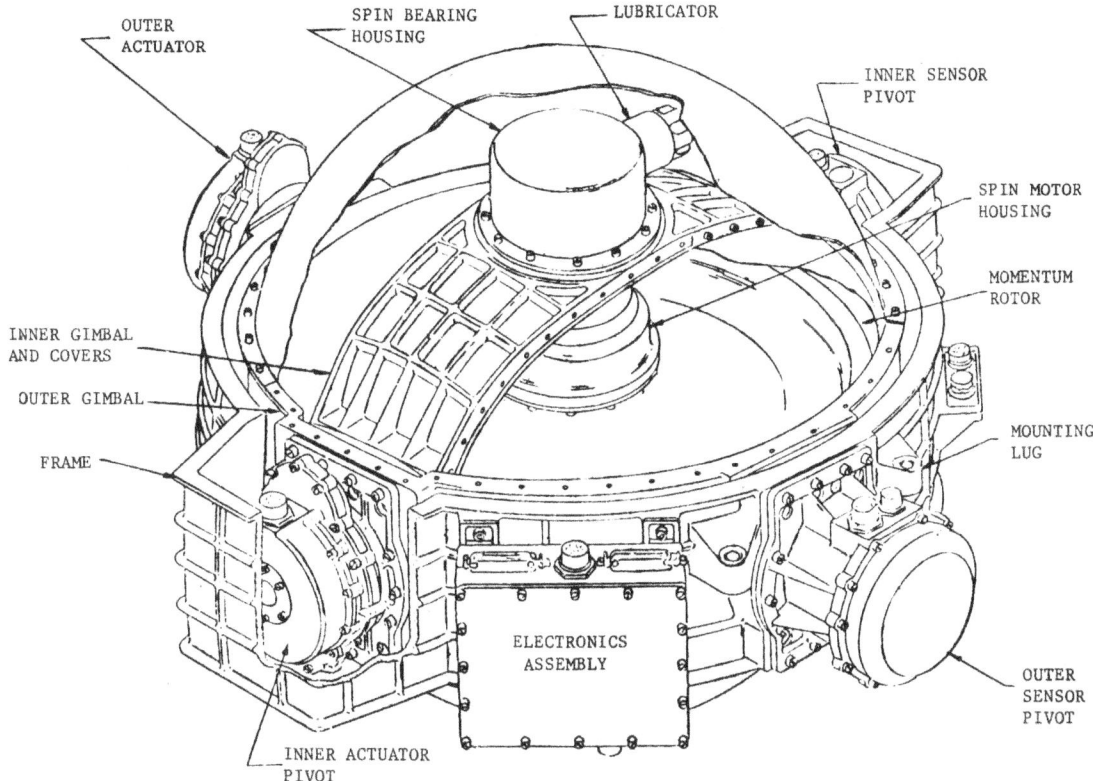

Figure II-67 — Control Moment Gyro (CMG)

The two primary and two redundant Sun sensor channels of the FSS are housed in the OM Assembly. Similarly, redundant electronics are housed in the other three components. If a malfunction occurs in the primary system, the complete redundant system is switched in. An offset pointing capability for each axis is also provided.

Acquisition Sun Sensor — The Acquisition Sun Sensor (consisting of an optical and electronics assembly) provides attitude information for the X and Y control axes of the vehicle. These data are used to update the strapdown computation in the ATM Digital Computer during orbital daytime. Both Acquisition Sun Sensors are active in the vehicle control loop under normal operating conditions but either unit may be selected.

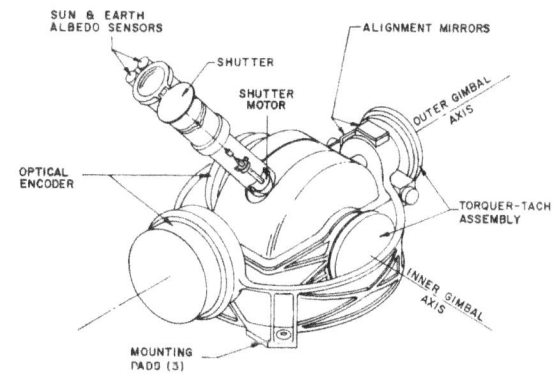

Figure II-68 — Star Tracker

The system provides electrical analog signals proportional to the solar pointing error of the SWS and a Sun presence signal to indicate when the Sun is in the field-of-view of the error sensors.

Rate Gyro Processor — The gyro is a floated rate integrating type gyro. Each component contains a gyroscope and associated electronics. Nine rate gyro processors are used in the CMGS/TACS and two per axis in the EPCS.

Experiment Package Caging and Gimbal Assembly — The ATM experiments are hardmounted and attached to a three degree of freedom caging and gimbal assembly. The EPCS maintains attitude control of the experiment package. The experiment package is provided with a separate control system to isolate it from vehicle disturbances and provide fine pointing.

This electromechanical system, comprised essentially of two large concentric rings, provides the torquing capability for the fine pointing servo system in pitch and yaw. Roll control is provided in an open loop fashion. The astronaut can rotate the experiment package to align optical axes of the experiment as required for data taking or during EVA for film retrieval. A caging system constrains both rings when the experiments are inactive. The launch locks are removed with pyrotechnics once orbit is achieved.

Manual Pointing Controller — The Manual Pointing Controller (MPC) is used by the astronaut to either offset point the ATM instrument package or to manually search for a desired star.

The MPC (Figure II-69) control handle can be tilted from a center null position to a left/right position for yaw, to an up/down position for pitch, or a combination of both positions. The MPC consists of a spring loaded two-axis gimbal control handle and mechanism, two linear transducers and a housing with connectors for the input/output wiring.

Experiment Pointing Electronics Assembly (EPEA) — The EPEA contains the electronic circuits for amplifying, shaping and mixing the outputs of the ATM Experiment package sensors to obtain error signals for driving the Experiment Pointing Control Subsystem (EPCS) actuators (DC torque motors). Interface circuitry between the MPC, the FSS and the Star Tracker is also included in the EPEA.

Caging of the EPCS is done with orbital locks for the pitch and yaw axes and by a brake mechanism for the roll axis.

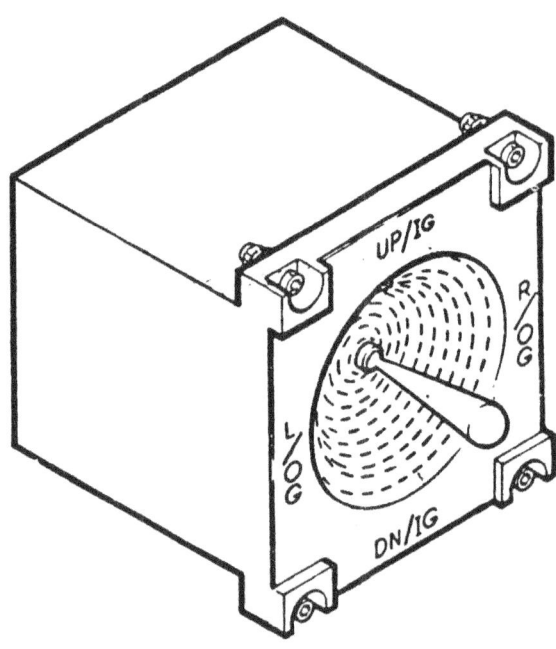

Figure II-69 — Manual Pointing Controller

ATM Digital Computer/Workshop Computer Interface Unit — The ATMDC/WCIU subsystem provides high speed general purpose computing capabilities along with a multipurpose, flexible input/output capability. It accepts from several sources analog and discrete signals which are used to perform calculations under the direction of a stored program. It also provides analog and discrete outputs to several devices. The subsystem consists of two identical ATMDC units and a single WCIU unit.

The computer accepts signals from the vehicle and the ATM experiment package sensors, commands from the ATM C&D Console, including the DAS, and the DCS. The computer sends commands to the system actuators and information to the ATM C&D Console and to the telemetry system.

Memory Load Unit (MLU) — Two ATMDCs are included in the ATM APCS. During the ATM mission, only one ATMDC is powered up and operating at any one time. If the operational ATMDC fails, switchover (automatic or commanded) will occur. The Memory Load Unit makes it possible to load ATMDC flight programs in flight from the onboard tape recorder or from the ground via the ATM RF command uplink.

Thruster Attitude Control Subsystem (TACS) — The OWS mounted TACS consists primarily of:

Cold gas propellant (nitrogen) utilized in a blowdown system (capacity of 61,000 lb-sec minimum).

Twenty-two propellant storage spheres on the thrust structure of the OWS.

Two thruster modules of three thrusters each on the aft skirt of the OWS at position planes I and III.

Quad-redundant valves for each thruster.

Power and Control Switching Assemblies (PCSA).

The nitrogen is stored at 2,130 N/cm² ±69 (3,100 ±100 psia) at 38°C (100°F) in the 0.13 cubic meter (4.5 cubic foot) titanium spheres and plumbed to a common manifold connected to the two thruster modules (Figure I-6 and II-70). Propellant is supplied to each of the six thrusters through quad-redundant (series-parallel) control solenoid valves. The location of the thrusters was chosen to satisfy vehicle control and thermal requirements. A passive thermal control system is used, consisting primarily of insulation over the propellant supply lines.

The Power and Control Switching Assemblies receive command signals from either the Saturn V Instrument Unit (IU) Launch Vehicle Digital Computer (LVDC) via the IU Flight Control Computer (FCC) or the ATM Digital Computer (ATMDC).

II-5

COMMUNICATIONS

The Skylab communications system provides a transfer of voice and instrumentation data between the SWS and the Spaceflight Tracking and Data Network (STDN) during all phases of the mission, in addition to providing an intravehicular communication network between crewmen in the Skylab.

The Skylab communication system is composed of the CSM, AM and ATM communication systems (Table II-11).

The CSM system is basically identical to that of Apollo with slight modifications, primarily in the audio/intercom, antenna and video downlink areas. There will be no high-gain antenna on the CSM, and the omni antennas will all be selectable via ground command. The ground will be able to select video downlinking using the CSM frequency modulation transmitter.

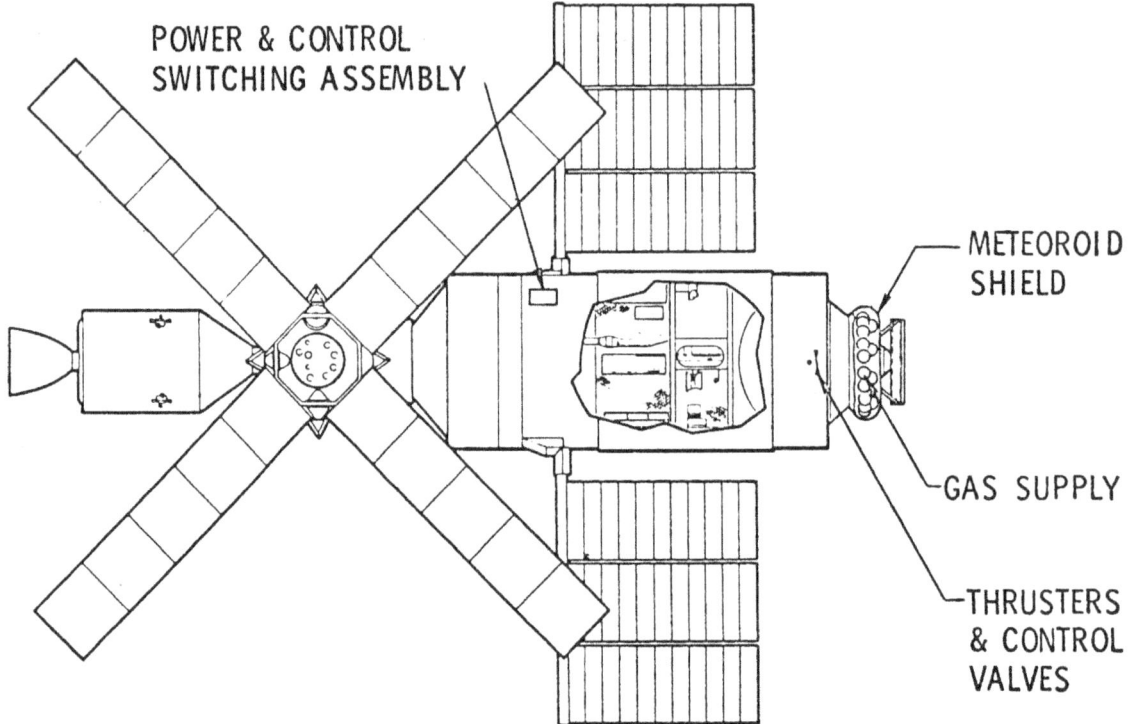

Figure II-70 — TACS Components Locations

The AM communication system consists primarily of modified Gemini equipment. The audio/intercom system will use any three of four VHF transmitters for downlinking real-time data, delayed-time recorded data or recorded voice.

The AM has three tape recorders (replaceable in flight) for recording systems, experiment and voice data. Command capability to the AM will be through redundant UHF receiver decoders.

Modified Saturn hardware is being utilized in the ATM communication system. This data system has two VHF transmitters for downlinking real-time and delayed-time data. It has two non-replaceable recorders and related interface equipment to provide for the recording systems and experiments data.

During the manned phases of the mission, the communication system provides the following:

> Voice communication between crewmen in the Skylab and the spacecraft communicator in the MCC.
> Indications of a caution and warning alert.
> Range information between the CSM and SWS.
> Hard-copy printed messages from MCC to the crew.
> Transmission of television data to STDN.

Operation of the communication system is divided into six subsystems: Audio, television, teleprinting, ranging, caution and warning, and photography.

Audio

The SWS will use the CSM S-band equipment as an integral part of the audio/intercom system. The system has two channels — one to be used for downlinking real-time voice or as an intercom (I/C) channel, the other being used for recording experiment voice data.

Table II-11 — Skylab Communications Links

Data System	Downlink	Uplink	Frequency (MHz)
CSM (USB)	R/T TM R/T VX		2287.5 (All cluster R/T VX via CSM)
	D/T TM D/T VX R/T TV D/T TV		2272.5 (Primary D/T VX via AM)
		CMD VX	2106.4 (All cluster R/T VX via CSM)
CSM (VHF)	Voice	VX	296.8/259.7
AM (UHF)		CMD (Including Teleprinter)	450
AM (VHF)	R/T TM		230.4 and 235.0 and 246.3 (All cluster D/T VX normally via AM)
ATM (UHF)		CMD	450
ATM (VHF)	Real Time R/T TM D/T TM		231.9 and 237.0

Notes: R/T = Real Time   TV = Television
D/T = Delayed Time   VX = Voice
TM = Telemetry   CMD = Command

All real-time and SWS intercom voice will require the use of the CSM S-band system with the CSM VHF equipment serving as a backup. Voice recording will normally be accomplished by using one of the voice channels and the AM recorders.

Voice may also be recorded on the CSM data storage equipment (DSE) and downlinked via the CSM S-band system during later passes. The crewmen, when required, may voice record through any of the 13 speaker intercom assemblies (SIAs) throughout the SWS (Figure II-71).

In addition to the 13 SIAs, the audio subsystem consists of EVA panels, one IVA panel, two load compensators (ALC), three CSM audio centers (A/C) and a CSM speaker box.

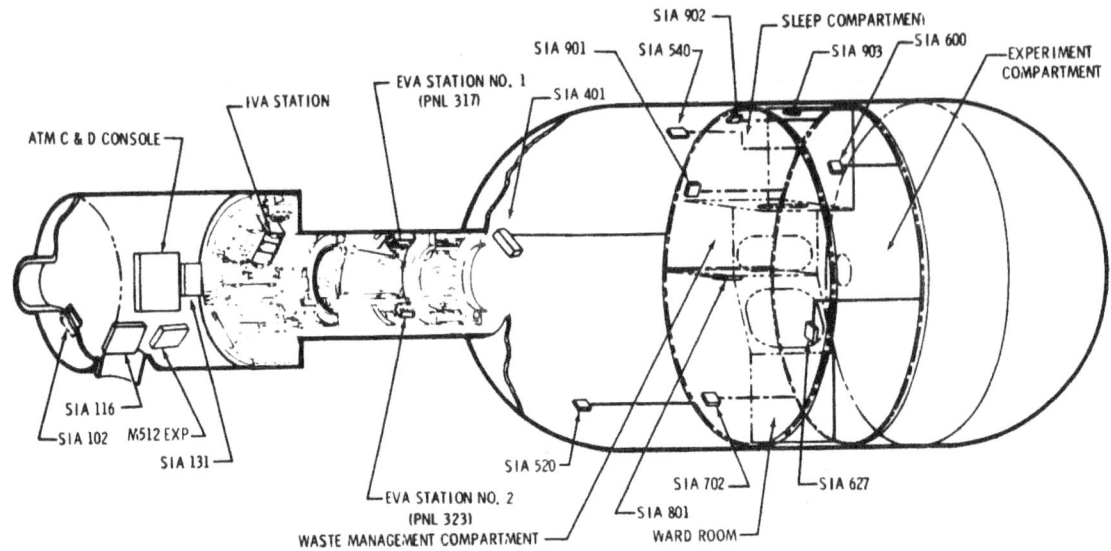

Figure II-71 — Intercom Locations

The SIAs (Figure II-72) provide the voice communication capability for crewmen in an unsuited mode throughout the SWS. This is accomplished via a speaker and microphone or through a headset and Crewman Communications Umbilical (CCU). The intercom system also provides audio tones and visual displays for Caution and Warning.

The two EVA panels and the IVA panel provide the voice communication capability when crewmen are wearing their pressure suits.

The audio load compensators provide automatic compensation for varying audio loads by supplying regulated audio signal levels.

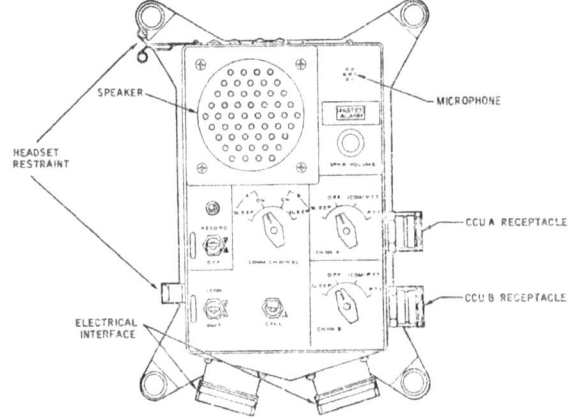

Figure II-72 — Speaker Intercom Assembly

The CSM audio center consists of headsets and communication umbilicals and provides the voice capability when crewmen are unsuited within the CSM. The CSM speaker links a crewman in the CSM with those in the SWS via the internal communication system.

The rescue voice mode does not utilize the CSM equipment but is to be used in the case of a CSM failure. When operating in this mode, audio communication from the vehicle is one-way only via the AM VHF transmitter. STDN must utilize the AM teleprinter for uplink messages. The basic system is designed with redundant components and buses to minimize loss of audio communication.

Television

The Skylab TV Subsystem provides television coverage to Earth of internal and external Skylab scenes during manned and unmanned mission periods. A TV signal from one of five ATM black and white TV cameras or from a portable color TV camera can be selected by the crew for radio transmission to a ground station using the FM transmitter in the docked CSM. When out of contact with STDN, a selected TV signal can be recorded on the Video Tape Recorder (VTR) for subsequent playback during ground station contact. During recorded playback to the ground, real time TV signals can not be transmitted.

The ATM TV cameras are fixed-mounted and are an integral part of the experiment. The cameras provide real time monitoring of the experiments operations. Table II-12 identifies the camera, manufacturer, type, experiment and purpose of each. Sync signals are inserted into the video signal by redundant sync generator and switcher processor elements to provide commercial standard TV signals on two channels, ATM-1 and ATM-2. However, the aspect ratio is 1:1, instead of the standard 4:3.

Table II-12 – ATM TV Cameras

Experiment	Type	Manufacturer	Purpose
H-Alpha One	Vidicon	MSFC	Pointing control of the ATM toward specific points of the Sun.
H-Alpha Two	Vidicon	MSFC	View Sun through interference filter at H-Alpha frequency
NRL XUV	Low Light Level Vidicon	MSFC	View the Sun's image in the 150 to 600 Angstrom spectral region via conversion layer
NRL XUV Slit	Low Light Level Vidicon	MSFC	View the Sun's image directly through 2 x 60 arc-second slit
White Light Coronagraph	Low Light Level Vidicon	MSFC	View the Sun's corona in white light

The portable color TV camera will be used to cover scenes of experiment performance and crew operation and habitation activities, both within and outside of Skylab. This camera, which weighs about 5.4 kg (12 lb) and measures 28 by 15.2 by 10.1 centimeters (11 by 6 by 4 inches), is similar to that used on Apollo missions. The camera controls include focus, iris, zoom and light sensitivity. The camera is used with a small TV monitor to assist the crewman in scene composition and focusing. The camera is connected to the TV subsystem with a 30 ft cable at any one of five Television Input Stations (TVIS). The camera operates on 28 volts power from the TVIS and provides a field sequential color TV signal. This is converted to commercial broadcast format on the ground. The camera may be hand-held by a crewman or fixed-mounted on a universal mount. External viewing is accomplished by connecting the camera at the TVIS in the Airlock for EVA, by mounting on the experiment S191 Viewfinder Tracking System, or by deployment through the Scientific Airlock (SAL) on the T027 Experiment Universal Extension Mount. In this latter mode, the camera focus, iris, zoom and orientation are controlled remotely by crewman from within the OWS.

Three TVISs are in the OWS, one in the AM and one in the MDA (Figure II-73). These stations accept the portable TV camera signal and condition it to modulate the CSM transmitter or be recorded by the VTR. The TVIS gain adjustments are individually set prior to launch.

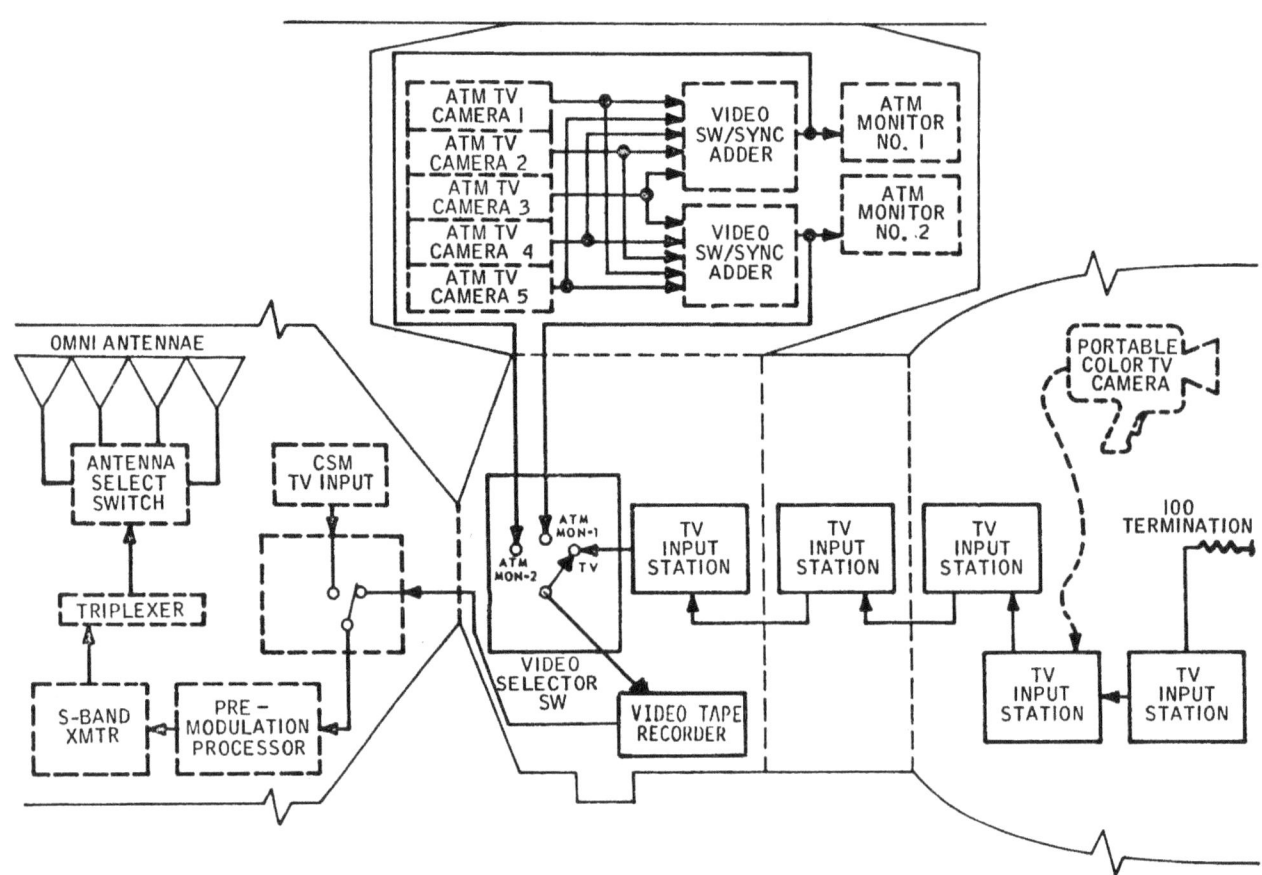

Figure II-73 — TV Subsystem

The Video Selector Switch (VSS) is in the MDA. It permits manual selection of the portable color camera or ATM-1 or ATM-2 TV signals to be transmitted or recorded. The VSS contains electronics to condition ATM-1 and ATM-2 channels to modulate the CSM transmitter.

The VTR in the MDA can be controlled manually or by ground command except for the playback function which can only be initiated from a ground station. The VTR can record up to 30 minutes of TV signals from any source as selected at the VSS. The playback rate is the same as the recording rate. During recording, the VTR can accept voice signals from the Audio Subsystem and multiplex this voice signal with the TV signal being recorded. The voice signal is later recovered after recorder playback by demultiplexing at the ground station. Table II-13 lists major TV subsystem elements and manufacturers.

The docked CSM accepts the conditioned TV signal, either real time or recorded, and routes it to the S-Band FM transmitter. The signal is broadcast while in ground station contact using the one of four CSM omni antennas that provides optimum coverage for that pass.

Table II-13 – TV Subsystem Elements

TV Subsystem Element	Manufacturer
Television Input Station	Martin Marietta Aerospace
Video Selector Switch	Bendix
Portable Color TV Camera	Westinghouse Electric Corporation
Video Tape Recorder	RCA
FM S-Band Transmitter	Motorola

A spare portable color camera, TVIS, VSS and VTR are provided.

Real time coverage of television will be provided by STDN stations at Goldstone, Corpus Christi and Merritt Island. The TV signal will be recorded and routed at the end of the day to MSC over NASCOM. For special events, real-time video will be made available to MCC. Other STDN stations around the Earth will record the signals received routinely and mail the tapes to MSC.

Teleprinting

Teleprinter messages to the Skylab crew will provide updates in detailed flight planning, experiment schedule activity and general mission information. Messages will originate at MCC, and remote sites will pass these messages up during site passes. Daily uplink time will be about 9 minutes; however, when EREP passes are scheduled, this may be lengthened to 12 minutes.

Teleprinter messages are limited to 30 characters per line. Messages with 50 lines or less can be uplinked as one load; those of greater length will require an appropriate number of loads to accommodate the message length (50 lines per load).

The uplinked messages will be received in the teleprinter printout in the SWS. Some 156 rolls of spare paper are aboard the SWS.

Expected messages, their lengths and scheduled transmittal frequency are:
  Summary Flight Plan – 48 lines – Daily.
  Detailed Flight Plan – 105 lines – Daily.
  ATM solar activity PAD – 40 lines – Daily.
  ATM schedule – 36 lines – Daily.
  Next station contact PAD – 48 lines – Daily.
  Block data – 20 lines – Twice daily.
  General messages – 15 lines – Three times daily.
  EREP setup PAD – 40 lines – as required prior to EREP passes.
  Viewfinder tracking system PAD – 25 lines – as required prior to EREP passes.
  EREP operations PAD – 50 lines – as required prior to EREP passes.
  Other experiment data ranges from 5 to 20 lines and is transmitted as required.

Ranging

Tracking adequacy is important to SL-2 launch and rendezvous support. For the SL-2 launch, there are only eight passes over sites which have C-band skin tracking capability (CRO and MIL). Six of these passes have elevation angles less than 10 degrees. Therefore, the CSM command and communications (CCS) USB beacon will be powered to include the entire rendezvous sequence. The SWS ranging subsystem, which utilizes the CSM ranging subsystem with the SWS ranging antenna, VHF transceiver and a range tone transfer assembly, facilitates rendezvous of the CSM with the SWS. The CSM sends a range tone to the SWS. The SWS receives the tone and, through its range tone transfer assembly (RTTA), returns a signal to the CSM. The CSM ranging subsystem uses this to compute and display range and range rate.

The C-band support for SL-3 and SL-4 launch targeting and rendezvous is thought to be adequate since there is a long period of undisturbed coasting flight during which accurate tracking can be established on the SWS.

Caution and Warning

The Caution and Warning (C&W) System (Figure II-74) will monitor the performance of itself (voltage only) and systems and alert the crew to hazards or out-of-limit conditions which constitute or could result in jeopardizing the crew, compromising primary mission objectives, or if not responded to in time could result in loss of a system. Parameters monitored by the C&W System are categorized as either EMERGENCY, WARNING or CAUTION. Criticality and/or crew response will be used to determine the category. The categories are defined as follows:

EMERGENCY − Any condition which can result in crew injury or threat to life and which requires immediate corrective action, including predetermined crew response.

WARNING − Any existing or impending condition or malfunction of a cluster system that would adversely affect crew safety or compromise primary mission objectives. Immediate action by the crew is required.

CAUTION − Any out-of-limit condition or malfunction of a Skylab system that affects primary mission objectives or could result in loss of a system if not responded to in time. Crew action is required although not immediately.

The number of monitored parameters must be consistent with effective monitoring. When any of the monitored parameters reach the predetermined out-of-tolerance level, appropriate visual and acoustical signals will be activated. Parameters and display requirements are enumerated in Table II-14.

OA C&W System − The OA Caution and Warning System consists of C&W Systems installed in both the Saturn Workshop (SWS) and the Command and Service Module (CSM). Each system provides the crew with visual displays and audio tones when selected parameters reach out-of-tolerance conditions. In the docked configuration, the two C&W Systems interface by means of discrete contact closures to provide for cluster-wide monitoring of selected parameters. The C&W System equipment used to monitor these parameters and its location is depicted in Figure II-75.

SWS C&W System − The SWS Caution and Warning System monitors the performance of specified vehicle systems and alerts the crew to hazards or out-of-limit conditions. The SWS C&W System utilizes two independent subsystems, a Caution and Warning Subsystem for monitoring various system parameters and an Emergency Subsystem for detecting fire or rapid loss of pressure.

Fifteen separate panels are provided in the SWS for control, display, operation and testing of the Caution & Warning and Emergency Subsystems. Three of these panels are used for control and display of both subsystems; the remaining twelve are used primarily for control of the fire detection portion of the Emergency Subsystem.

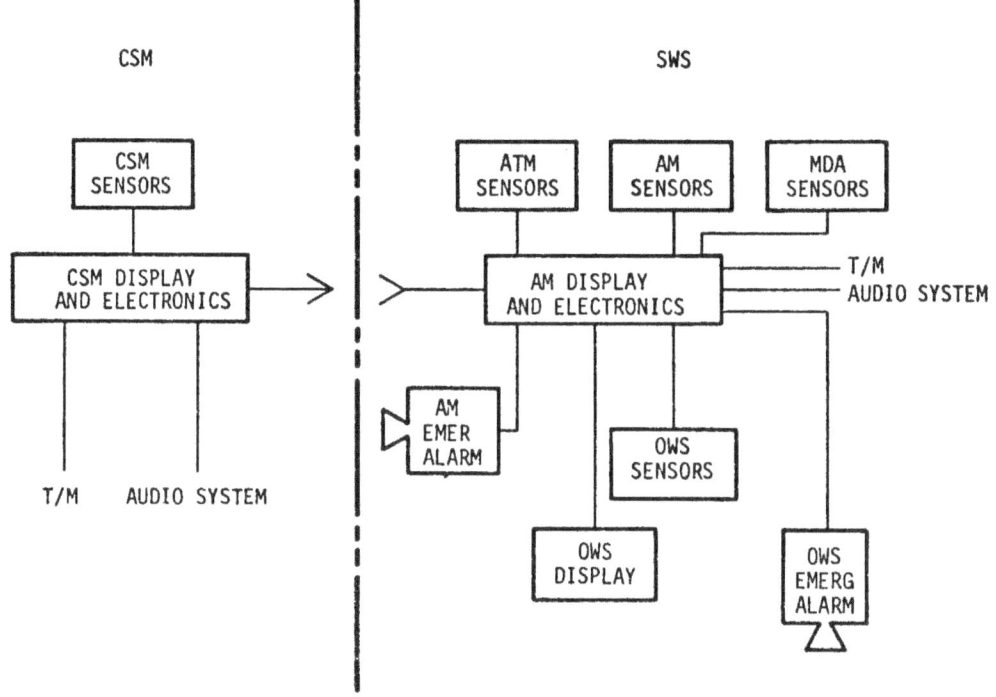

Figure II-74 — Caution and Warning System

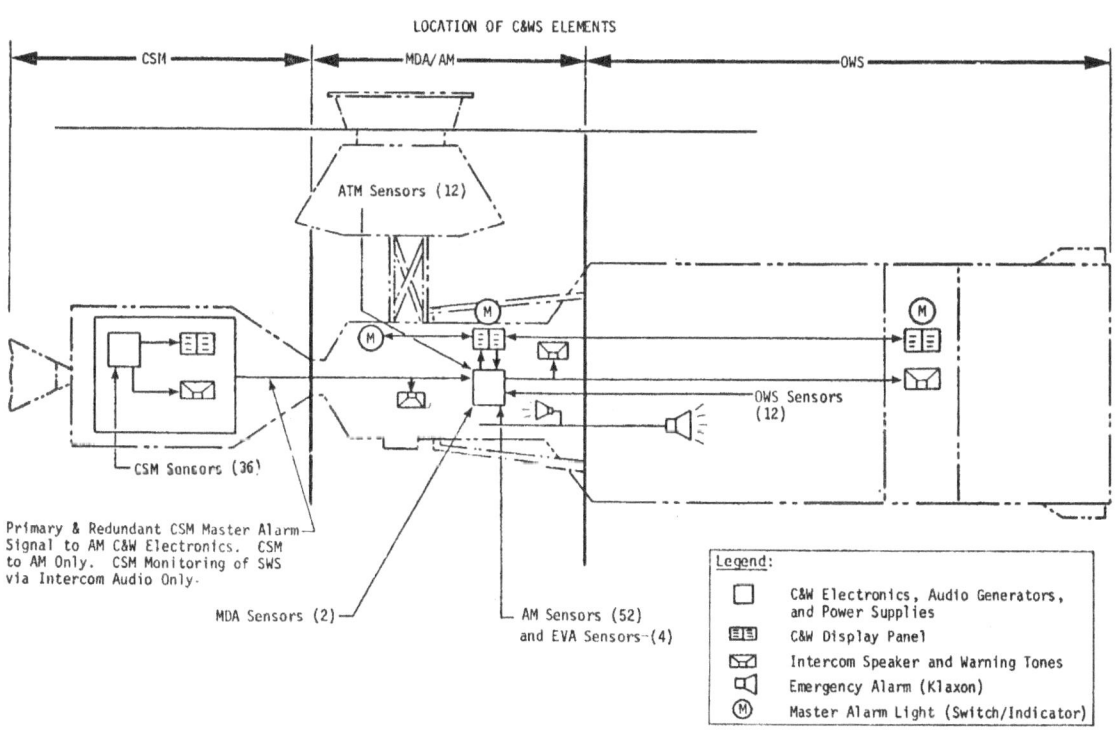

Figure II-75 — C&W System Elements

II-69

Table II-14 — C&W Parameters

Item/Parameter	Monitored Module	Criticality (See Note 2)	No. of C&W System Channels (See Note 3)	Light Labeling
AM–ECS				
1. Sieve A Bed 1/2	AM	C ⎱	1	SIEVE TEMP HIGH
2. Sieve B Bed 1/2	AM	C ⎰		
3. PPO_2 1	AM	W ⎱	1	PPO_2 LOW
4. PPO_2 2	AM	W ⎰		
5. Pri Cool Pump 1	AM	W ⎱		
6. Pri Cool Pump 2	AM	W ⎬	1	PRI COOL FLOW
7. Pri Cool Pump 3	AM	W ⎰		
8. Sec Cool Pump 1	AM	W ⎱		
9. Sec Cool Pump 2	AM	W ⎬	1	SEC COOL FLOW
10. Sec Cool Pump 3	AM	W ⎰		
11. Cluster Pressure	AM	W	1 (R)	CLUSTER PRESS LOW
12. Sieve A PCO_2	AM	C ⎱	1	SIEVE OUT $PPCO_2$ HIGH
13. Sieve B PCO_2	AM	C ⎰		
14. Sieve A Gas Flow	AM	C ⎱	1	SIEVE FLOW
15. Sieve B Gas Flow	AM	C ⎰		
16. Sieve A Timer	AM	C ⎱	1	SIEVE TIMER
17. Sieve B Timer	AM	C ⎰		
18. OWS Gas Flow	AM	C	1	OWS GAS INTER CHG
19. Condensate Tank ΔP	AM	C	1	CNDST TANK ΔP
20. Pri Cool 47° Valve	AM	C	1	PRI COOL TEMP LOW
21. Sec Cool 47° Valve	AM	C	1	SEC COOL TEMP LOW
22. Pri Cool Loop Temp	AM	C	1	PRI COOL TEMP HIGH
23. Sec Cool Loop Temp	AM	C	1	SEC COOL TEMP HIGH
(Subtotal: 23 Inhibit SW's)				
Integrated EPS				
24. Reg Bus 1 Low	AM	W	1	REG BUS 1 LOW
25. Reg Bus 1 High	AM	W	1	REG BUS 1 HIGH
26. Reg Bus 2 Low	AM	W	1	REG BUS 2 LOW
27. Reg Bus 2 High	AM	W	1	REG BUS 2 HIGH
28. ATM Bus 1 Low	ATM	W	1	ATM BUS 1 LOW
29. ATM Bus 2 Low	ATM	W	1	ATM BUS 2 LOW
30. C&W Power Control 1	AM	C ⎫		
31. C&W Power Logic 1	AM	C ⎬ See Note 1		
32. C&W Power Control 2	AM	C ⎬	1	C&W POWER
33. C&W Power Logic 2	AM	C ⎭ See Note 1		
34. C&W Sig Cond Power	AM	C		

Table II-14 (Continued)

Item/Parameter	Monitored Module	Criticality (See Note 2)	No. of C&W System Channels (See Note 3)	Light Labeling
35. Emerg Power Control 1 (See Note 1)	AM	C		
36. Emerg Power Logic 1 (See Note 1)	AM	C	1	EMERG POWER
37. Emerg Power Control 2 (See Note 1)	AM	C		
38. Emerg Power Logic 2 (See Note 1)	AM	C		
39. Emerg Sensor 1	AM	C	1	EMERG SENSOR POWER
40. Emerg Sensor 2	AM	C		
41. OWS Bus 1 Low	OWS	C	1 (R)	OWS BUS 1 LOW
42. OWS Bus 2 Low	OWS	C	1 (R)	OWS BUS 2 LOW
43. Battery 1 70% D.O.D.	AM	C		
44. Battery 2 70% D.O.D.	AM	C		
45. Battery 3 70% D.O.D.	AM	C		
46. Battery 4 70% D.O.D.	AM	C	1	BAT CHARGE LOW
47. Battery 5 70% D.O.D.	AM	C		
48. Battery 6 70% D.O.D.	AM	C		
49. Battery 7 70% D.O.D.	AM	C		
50. Battery 8 70% D.O.D.	AM	C		
(Subtotal: 23 Inhibit Switches) (See Note 1)				
ATM–ACS				
51. ACS–Overate	ATM	W	1	CLUSTER ATT
52. ACS–Thruster Stuck	ATM	W		
53. ACS–CMG Saturate	ATM	C		
54. ACS–Auto TACS Only Option	ATM	C	1	ACS MALF
55. ACS–2nd/3rd Rate Gyro Failure	ATM	C		
56. ACS–Computer Self Test Failure	ATM	C	1	COMPUTER MALF
57. Computer X–Over	ATM	C		
58. ATM Collant Fluid Temp	ATM	C		
59. ATM Coolant Htr Temp	ATM	C	1	ATM CNST THERM
60. ATM Coolant Pump ΔP	ATM	C		
(Subtotal: 10 Inhibit Switches)				

Table II-14 (Concluded)

Item/Parameter	Monitored Module	Criticality (See Note 2)	No. of C&W System Channels (See Note 3)	Light Labeling
Extravehicular Activity				
61. EVA LCG–1 Pump ΔP	AM	W ⎫	1	EVA 1
62. EVA LCG–1 H$_2$O in Temp	AM	W ⎭		
63. EVA LCG–2 Pump ΔP	AM	W ⎫	1	EVA 2
64. EVA LCG–2 H$_2$O in Temp	AM	W ⎭		
(Subtotal: 4 Inhibit Switches)				
Miscellaneous				
65. CSM 1	CSM	W ⎫	1	CSM
66. CSM 2	CSM	W ⎭		
67. Crew Alert 1	AM	W ⎫	1 (R)	CREW ALERT
68. Crew Alert 2	AM	W ⎭		
(Subtotal: 4 Inhibit Switches)				
Emergency				
69. MDA/STS Fire 1	MDA/STS	E ⎫	1(R)	MDA STS FIRE
70. MDA/STS Fire 2	MDA/STS	E ⎭		
71. AM AFT Fire 1	AM	E ⎫	1 (R)	AM AFT FIRE
72. AM AFT Fire 2	AM	E ⎭		
73. OWS Fwd Fire 1	OWS	E ⎫	1 (R)	OWS FWD FIRE
74. OWS Fwd Fire 2	OWS	E ⎭		
75. OWS Exp Fire 1	OWS	E ⎫	1 (R)	OWS EXP FIRE
76. OWS Exp Fire 2	OWS	E ⎭		
77. OWS Crew Qtrs Fire 1	OWS	E ⎫	1 (R)	OWS CREW QTRS FIRE
78. OWS Crew Qtrs Fire 2	OWS	E ⎭		
79. Rapid ΔP 1	AM	E ⎫	1 (R)	RAPID ΔP
80. Rapid ΔP 2	AM	E ⎭		
(Max Subtotal: 12 Inhibit Switches for Parameters)				

Note 1 — 'or' gates utilized to minimize switching complexity.
Note 2 — E = Emergency, W = Warning, C = C = Caution
Note 3 — Brackets denote use of 'or' gates to minimize channel complexity.
 (R) — Denotes items repeated in OWS.

Control and Display Panel 206 — The major power and control switches for the SWS C&W System are on Panel 206 (Figure II-76). The panel is in the STS. The master alarm telelight switch, which is colored aviation red, is illuminated when either a caution, warning or emergency parameter is activated. The memory recall telelight switch has an amber lens and is used to indicate that parameter(s) which activated the C&W Subsystem have been stored in memory. Three power switches are provided for powering the SWS C&W System. One is used to control power to the C&W Subsystem and the other two are used for the Emergency Subsystem. Four test switches are provided for testing the C&W Subsystem electronics, audio tone and visual displays. Three volume controls are also provided for controlling the intensity of the emergency, warning, and caution tones.

Display and Inhibit Switch Panel 207 — The parameter identification lights and inhibit switches are on Panel 207 (Figure II-77) which is in the STS. Forty parameter identification lights are used to help the crew identify which parameter or system has gone out-of-tolerance. Each display has two bulbs for redundancy with each bulb being driven by separate power sources. Each parameter monitored by the C&W System has a corresponding inhibit switch on Panel 207. The inhibit switches are used to disable a malfunctioning circuit or input signal without disabling other active parameter inputs.

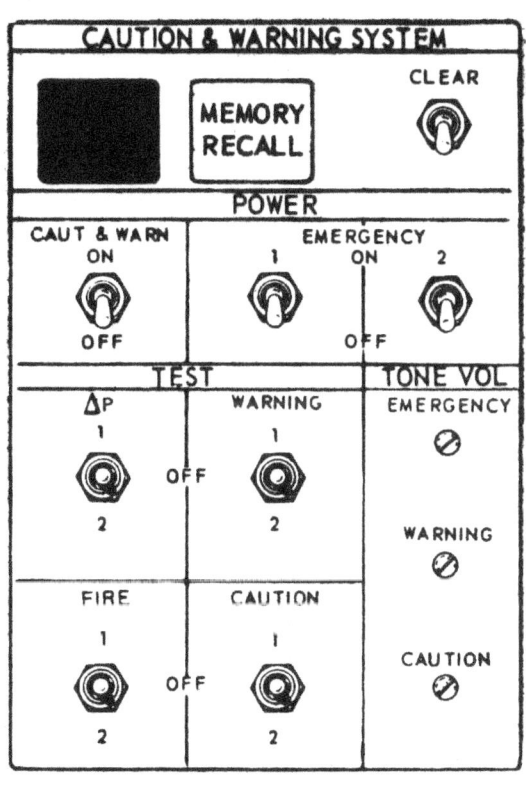

Figure II-76 — C&W Panel 206

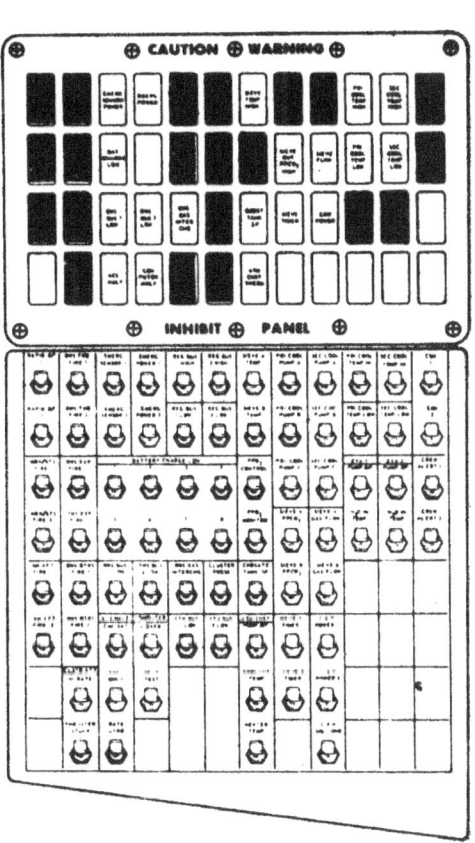

Figure II-77 — C&W Panel 207

OWS Repeater Panel 616 — Panel 616 is in the Experiment Compartment of the OWS. The panel contains one master alarm reset telelight switch (Figure II-78) which contains two bulbs.

Ten parameter identification lights are used in helping the crew identify various parameters or systems that have gone out-of-tolerance. Each display contains two bulbs for redundancy; each drawing current from separate power sources.

Fire Sensor Control Panels — The Fire Sensor Control Panels (Figure II-79) provide the controls for operation and test of the Fire Sensor Assemblies. Each panel has the capability of controlling two sensors. Two power switches are provided, one for each sensor, which allow manual selection of one of two normally energized buses capable of supplying power to the respective sensor. A master alarm reset/test switch is provided for testing the sensor(s) and resetting the SWS C&W System. The bulbs and lenses on the panels and the panels themselves can be replaced in flight. Two panels (complete with lenses and bulbs) and eight lens and bulb assemblies, are stowed in the OWS for inflight replacement.

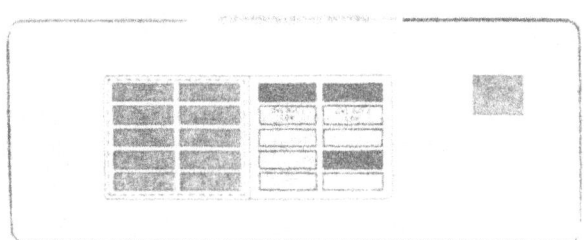

Figure II-78 — C&W Panel 616

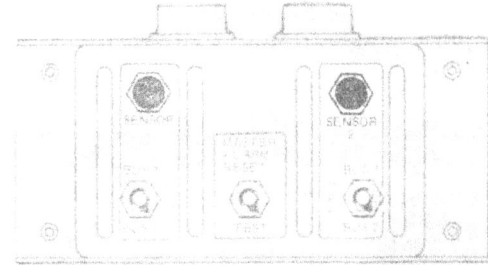

Figure II-79 — Fire Sensor Control Panel

C&W Subsystem Operation — The OA C&W Subsystem (Figure II-80) encompasses the SWS C&W Subsystem and the CSM C&W Subsystem. The CSM C&W Subsystem monitors preselected caution and warning parameters in the CM and SM. An out-of-tolerance condition produces an audio tone and the illumination of visual displays in the CM. The SWS C&W Subsystem monitors preselected parameters in the ATM, AM and OWS and, through the CSM C&W Subsystem, monitors the performance of critical systems in the CSM. An out-of-tolerance condition in the SWS or CSM will produce a tone in the MDA, the AM and the OWS and illumination of display lights on the SWS C&W control and display panels and the SIAs.

Emergency Subsystem Operation — The Emergency Subsystem (Figure II-81) monitors fire and rapid spacecraft pressure loss conditions. An out-of-tolerance condition produces audio tones and visual displays. Also, signals are telemetered to the ground indicating the type of emergency situation. Fire sensors are located throughout the SWS to detect fires. Two rapid change of pressure sensors are in the Structural Transition Section (STS) of the Airlock Module.

Photography

Different cameras are used to record the human, operational, and scientific events occurring during the manned portions of the Skylab missions. This section gives a brief technical description of the cameras and a tabulation (Tables II-15 and II-16) of what cameras and film types are used and Skylab mission numbers.

Pertinent information about cameras used for ATM, Spectrographic and some EREP experiments are in specific experiment sections.

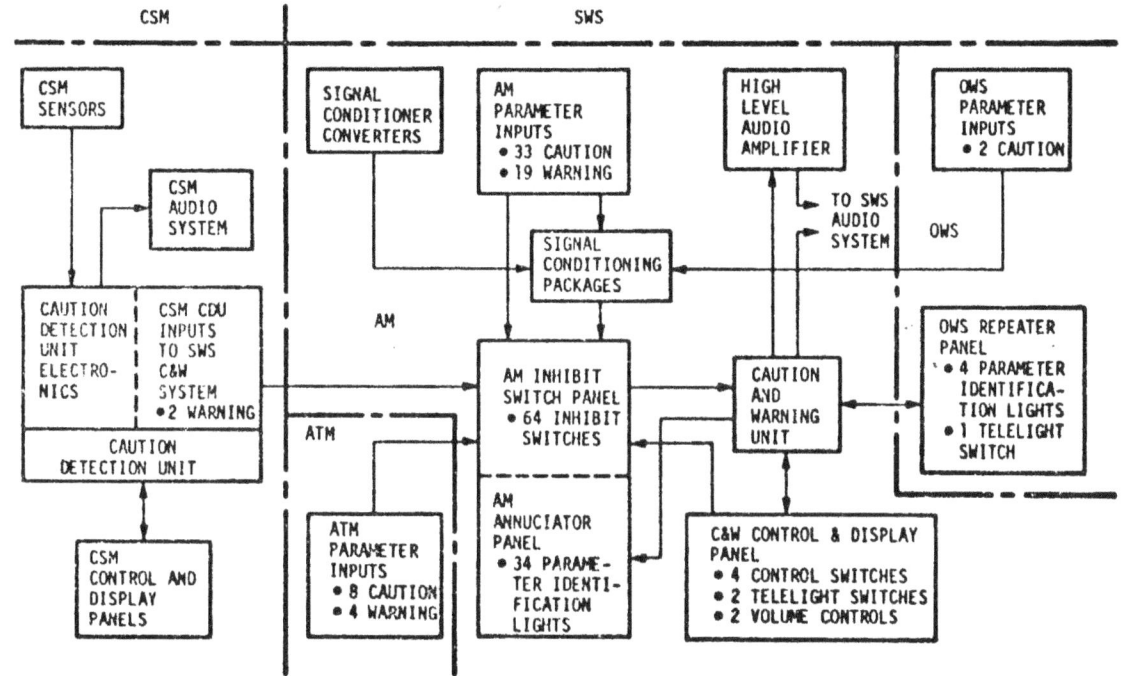

Figure II-80 — C&W Subsystem

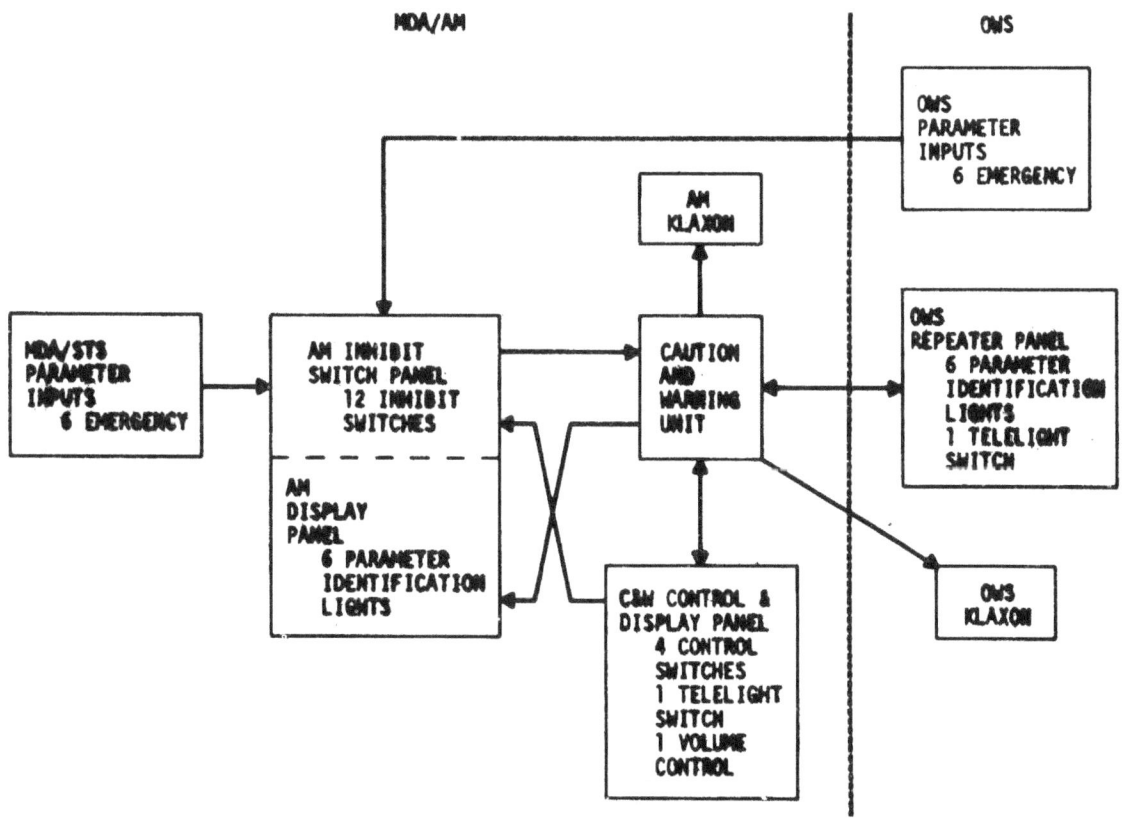

Figure II-81 — Emergency Subsystem

Table II-15 — Cameras and Film

Camera	Experiment	Mission Number	Film Type
16 mm DAC	M131	SL2 and SL3	S0168
	M151	SL2, SL3, and SL4	S0168
	M479	SL3	S0180
	M487	SL2, SL3, and SL4	S0168
	M512	SL2	7242
	Operational	SL2, SL3, and SL4	S0168 S0368
	S073/T027	SL2, SL3, and SL4	2485
	S183	SL3 and SL4	103a-0
	T013	SL3	S0168
	T020	SL3	S0168
	S191	SL2, SL3 and SL4	EK 3401
35 mm Nikon	S063	SL2, SL3, and SL4	2485 S0368
70 mm Hasselblad	Operational	SL2, SL3, and SL4	S0168
	Operational	SL2, SL3, and SL4	S0368
	T020	SL3	S0168
70 mm Itek (6)	S190A	SL2, SL3, and SL4	SO 022 EK 2424 SO 127 SO 356
Earth Terrain Camera, Actron	S190B	SL2, SL3, and SL4	SO 242 EK 3414 EK 3443

Table II-16 — Film Description

S0368	Ektachrome MS; color reversal; ASA 64; daylight application
S0180	Ektachrome Infrared; color reversal; aerial type
S0168	Ektachrome EF; high speed color reversal; ASA 160; daylight, low light, level applications
S0212	Panatomic X; black and white, very high resolution
S0101	Panchromatic with extended red sensitivity; maximum sensitivity at 6560 Angstroms
SC-5	Short Wave Radiation; Kodak-Pathe (French) manufactured
7242	Ektachrome EFB; color reversal; tungsten balanced; ASA 160; low light level applications
EK 3414	Tri-X Aerographic, black and white
SO 022	Panatomic-X; black and white; ASA 40; high resolution terrain photography applications
2485	Panchromatic with extended red sensitivity; very high speed black and white; ASA 6000; low light level applications
2403	Tri-X Aerographic; black and white with extended red sensitivity; low light level applications
026-02	Same as 3400 but with spectral sensitization and enhanced reciprocity response to long exposure times
104-06	Schumann type emulsion for ultraviolet applications in wavelengths shorter than 2200 Angstroms
101-06	Schumann type emulsion for ultraviolet applications in the 50-4000 Angstrom band
103a-0	UV emulsion, 2500 to 5000 Angstrom sensitivity; medium contrast, low resolution
SO 356	Aerial color (High resolution)
EK 2424	Infrared aerographic, sensitized to blue, red and infrared
3401	Black and white, high resolution aerial type, ASA 125
7241	Experimental, in place of SO 168 for Experiment M512
2443	Aerochrome IR, color (4 mil base)
3443	Aerochrome IR, color (2½ mil base)

16 mm Data Acquisition Camera (DAC) — The 16 mm Data Acquisition Camera (DAC) is used to obtain sequential photographic data. Unlike typical movie cameras, this camera has independent shutter speeds and framing rates. The DAC can be hand-held or bracket-mounted, can operate from spacecraft or portable battery power, and can accept various lenses and assorted accessories.

Characteristics:

Manufactured by J. A. Maurer, Inc., Long Island City, New York 11101.

Weight — 0.77 kg (1.7 lbs.)

Size — 15.2 × 9.5 × 6.1 cm (6 × 3.75 × 2.4 in.)

Volume — 885 cm^3 (54 in.^{3})

Power requirements: 28 ±4 VDC at 0.6 amps nominal from spacecraft or DAC Power Pack. DAC incorporates self-resetting overload protection circuit and replaceable power line fuse.

Sequencing frame rate settable to 1 (or 2), 6, 12, or 24 frames per second (fps) and time exposure.

Automatic Modes [1 (or 2), 6, and 12 fps] are initiated by depressing and releasing camera front button and continue uninterrupted even if sequencing rate is changed among automatic modes. Camera operation is stopped by depressing and releasing front button or by switching to the time exposure or 24 fps mode settings. Green operate light will flash at frame rate.

Additional DAC accessories are 140 ft film magazines, six types of lenses (5 mm, 10 mm, 18 mm, 25 mm and 75 mm), right angle mirror, fuse assembly, power cables, film cassette (400 ft) and transport mechanism.

35 mm Nikon Camera — Two motorized 35 mm Nikon cameras, modification of commercial Nikon equipment, are supplied for the S063 and T025 Experiments. The camera body incorporates reflex viewing and through-the-lens coupled light metering along with motorized film advancement. For the experiment operations, a visible lens and an ultraviolet lens are used.

Characteristics:

Manufactured by Nippon Kogaku K.K., Tokyo, Japan, and distributed by Ehrenreich Photo-Optical Industries, Inc., Garden City, N.Y. 11530.

Weight (w/o film or lens) — 16 kg (3.66 lbs)

Size — 16.6 × 15.7 × 7.1 cm (6.55 × 6.16 × 2.8 in)

Volume — 1,850 cm^3 (112.9 in^3)

The included focal plane shutter has the following settings: T (time), 1, 1/2, 1/4, 1/8, 1/15, 1/30, 1/60, 1/125, 1/250, 1/500, and 1/1000 second. A standard connector for X-synchronization with the shutter at 1/60 sec or lower shutter speeds is provided.

Accepts lenses with the commercial Nikon mounting. The 55 mm, f/1.2 visible lens and the 55 mm, f/2 UV lens are intended for use on this camera body.

The visible camera viewfinder shows the full through-the-lens coverage, shutter speed setting, and the light meter needle and matching indicator. The viewfinder can be removed easily to permit waist-level camera operation or replacement.

Nikon camera accessories include two 55 mm lenses (UV and visible), film cassette assemblies and battery assembly.

Hasselblad Data Camera — The Hasselblad Data Camera (HDC) is a rugged version of the commercial electric Hasselblad camera, 500 EL, and is used for medium resolution, photogrammetric photography. This camera incorporates a glass reseau plate positioned immediately in front of the film plane. The reseau plate places a pattern of precision crosses on each photograph to facilitate photogrammetric utilization of the photography.

    Characteristics:

        Manufactured by Victor Hasselblad AB, Goteborg, Sweden, and distributed by Paillard Inc., Linden, New Jersey 07036.
        Weight (with batteries) — 1.4 kg (3.1 lb)
        Size — 14.6 × 9.8 × 12.1 cm (5.75 × 3.86 × 4.77 in.)
        Volume — 1,735 cm^3 (106 in.^{3})

    The HDC has a 4 mm thick glass plate rigidly mounted in the rear opening of the camera. An array of 25 reseau crosses is engraved on the rear surface of the plate to facilitate precision geometrical calibration of the film and of the camera and lens optics. The glass plate also incorporates a fine rim on each vertical edge to provide minimum but positive contact with the film.

The following HDC accessories are stowed aboard the OWS; 70 mm film magazines, 80 mm lens, nickel cadmium cell batteries and a ring sight.

Earth Terrain Camera — The Earth Terrain Camera (ETC) is the major system component and includes an outer lens cone/mount assembly, the lens cone itself comprising the lens and the camera electronics, a control box and the camera body which houses the ETC Magazine.

The ETC is designed to obtain high resolution photographs of the Earth during the performance of Skylab Experiment S190B (EREP). The functional capabilities of the ETC permit automatic operation for overlapping topographic coverage and manual operation for single photographs of selected scenes.

    Characteristics:

        Manufactured by Actron Industries, Inc., Monrovia, California 90106
        Weight (with ETC Magazine, without film — launch configuration) — 35 kg (77 lbs) (with film) — 36.4 kg (82.5 lb)
        Size — 718 × 268 × 342 cm (28.31 × 11.25 × 13.45 in)
        Volume — 70,000 cm^3 (4,280 in.^{3})
        Electrical power requirements — supplied to the ETC through a cable from the spacecraft. AC Power — 115 vac, 400 Hz, 3 phase is for all camera drive motors — shutter, film transport, and forward motion compensation (FMC). DC Power — +28 (+2, -4) is for camera control circuits and relay operation. The camera power ON/OFF switch is located on the camera control box.
        Film format — 114 × 114 mm (4.5 × 4.5 in.) with the use of a 19 × 19 mm (0.75 × 0.75 in.) area at one format corner for data recording.
        Lens type — color corrected, f/4, 457 mm (18 in.) focal length.
        Field-of-view — 14.2° × 14.2°; 20.0° diagonal; for 435 km (235 n.mi.) altitude — 58.7 × 58.7 n.mi.

The ETC camera system also includes film magazines and film canisters.

The Scientific Airlock Assembly provides the interface between the Skylab OWS anti-solar Scientific Airlock (SAL) and the ETC. The assembly incorporates a precision optical window that is matched to the ETC lens optics and maintains the pressure integrity of the SAL.

    Characteristics:

        Manufactured by Actron Industries, Inc., Monrovia, California 91016
        Weight (with covers) — 5.5 kg (12.1 lbs)
        Size 29.6 × 31.3 × 7.6 cm (11.65 × 12.32 × 3 in.)

Volume 7,050 cm³ (430 in.³)

The front (square) side of the Window Assembly installs in the interface flange of the -Z (anti-solar) SAL. The back (round) side of the Window Assembly attaches to the ETC outer lens cone/mount assembly by means of a toggle latch ring (Marman) clamp incorporated on the Window Assembly.

Protective metal covers are attached to the front and the back interface surfaces of the Window Assembly during handling and stowage. The front cover is secured with two self-contained spring latches. The back cover is secured with the Marman clamp of the Window Assembly.

Other crew equipment items include: spotmeter (and batteries), chronograph watches, two speed mechanical timers, ball point pens, mechanical pencils, marker pens, and general and all purpose adhesive tape.

CREW EQUIPMENT

Wearing Apparel

Skylab crews inside the OWS will wear gold colored shirts fabricated from fire resistant PBI (polybenzimidazoic) fabric and trousers and jackets made of fire resistant woven Durette fabric. Enough of these items, including socks and underwear, are provided so that crewmen can change clothing regularly — trousers once a week and socks, shirts and underwear every two days.

The clothing for the three manned flights are in 15 Clothing Modules (Figure II-82), each holding enough clothes to supply a crewman for 28 days. Each crewman on SL-2 will have a 28-day Clothing Module; while crew members on the 56-day missions SL-3 and SL-4 will each have two 28-day modules. Also, two contingency Clothing Modules are stowed in the OWS prior to launch. The contingency modules contain clothing sized for backup crew members.

More than 700 clothing items will be stowed in the OWS prior to launch. Among these items are: 3 jackets, 53 trousers, 58 shirts, 15 boots, 15 pairs of gloves, 4 union suits, 199 T-shirts, 47 half union suits, 34 jockey shorts, 102 boxer shorts, 16 knee shorts, 128 pairs of socks and 7 constant wear garments (Figures II-83 and II-84).

The 28-day Clothing Module is a fabric rucksack made of 14 mil Armalon fabric, sized such that when filled with clothes it will fit in the standard OWS stowage locker (Figure II-85), 41 × 29 × 25 cm (16 × 11.4 × 9.8 in.). It is sectioned into seven main areas for each of identification (through labeling) and access to each specific clothing item.

The Skylab jacket is a contoured, custom fit, waist-length jacket made of fire resistant woven Durette fabric. Three snaps on the waistband attach to the trousers. The jacket has a front, full-length zipper. The American Flag emblem is on the upper left arm, the NASA emblem on the upper right front section and the crew emblem and name tag on the upper left front section.

The trousers are custom fit, waist-to-ankle, and made of fire resistant woven Durette fabric. The legs are removable above the knee by zippers for easy conversion to shorts. The trousers have an adjustable waistband for stability and sizing comfort, and a concealed, zippered front closure.

The Skylab shirt is a short-sleeve pullover of fire resistant knitted PBI fabric. It incorporates a raglan sleeve design and a mock turtleneck collar. The shirttail is a straight cut side-slit design long enough to be kept in the trousers. A utility pocket with zipper is in the upper left front section.

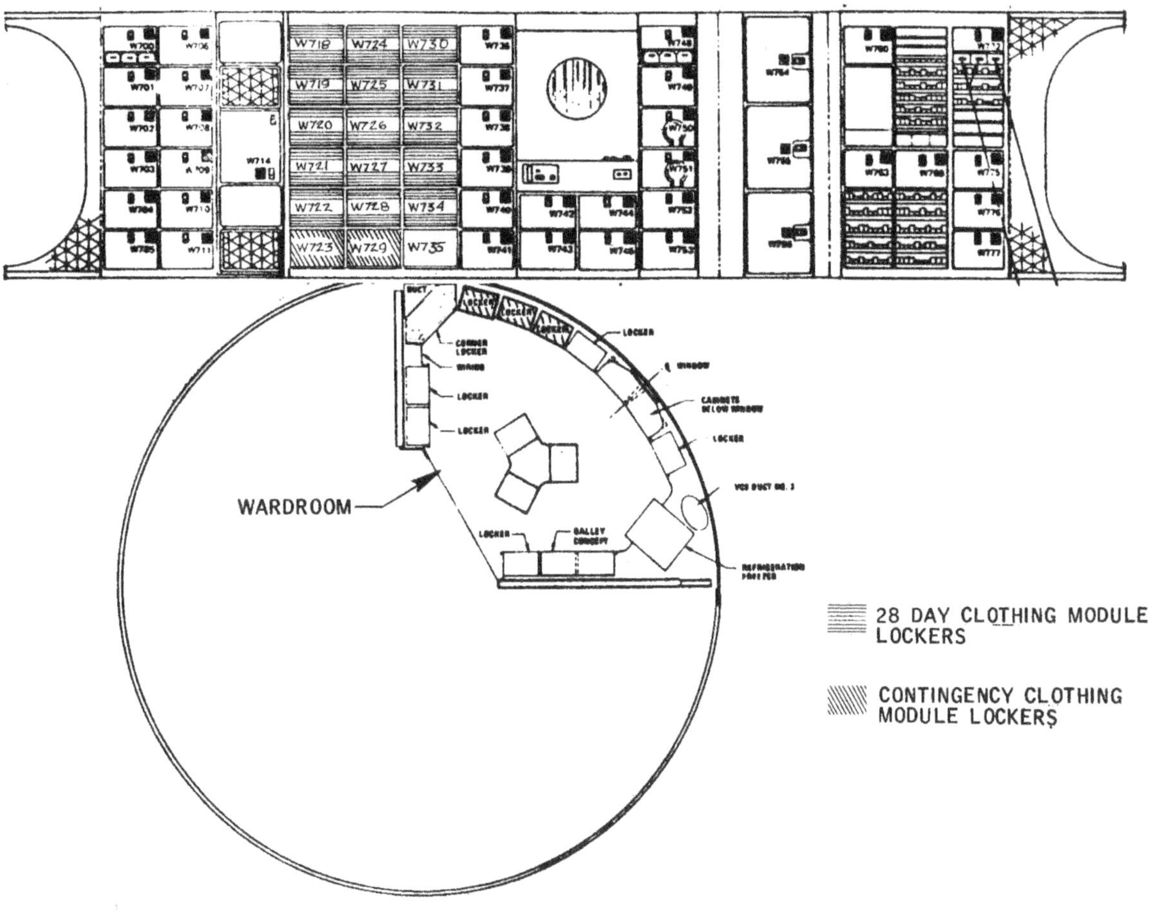

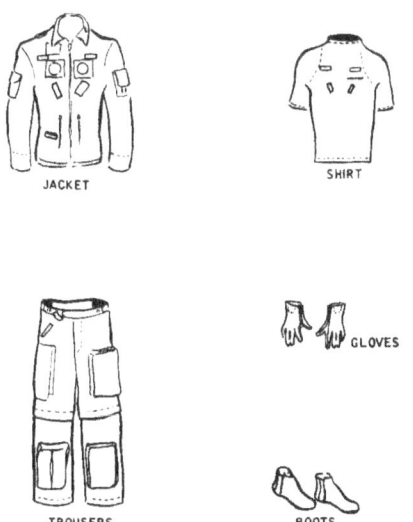

Figure II-82 — Clothing Module Stowage

28 DAY CLOTHING MODULE LOCKERS

CONTINGENCY CLOTHING MODULE LOCKERS

Figure II-83 — Crew Clothing

Shirts are provided in various sizes ranging from extra small to large.

The Skylab boot is ankle high fabricated and comfort-lined with woven fire resistant Durette fabric. The inside sole comfort lining is made from simplex knitted Durette fabric. The boot is lightweight and designed to completely cover the foot and ankle without discomfort from tightness or material bunching and with minimum degradation to mobility.

A pair of wrist-length gloves made of deerskin and fire resistant simplex knitted Durette fabric are provided for each crewman. The glove is five-fingered, unlined and lightweight. The palm and front of the gauntlet, thumb and fingers are seamless deerskin, and the remaining portions are Durette fabric.

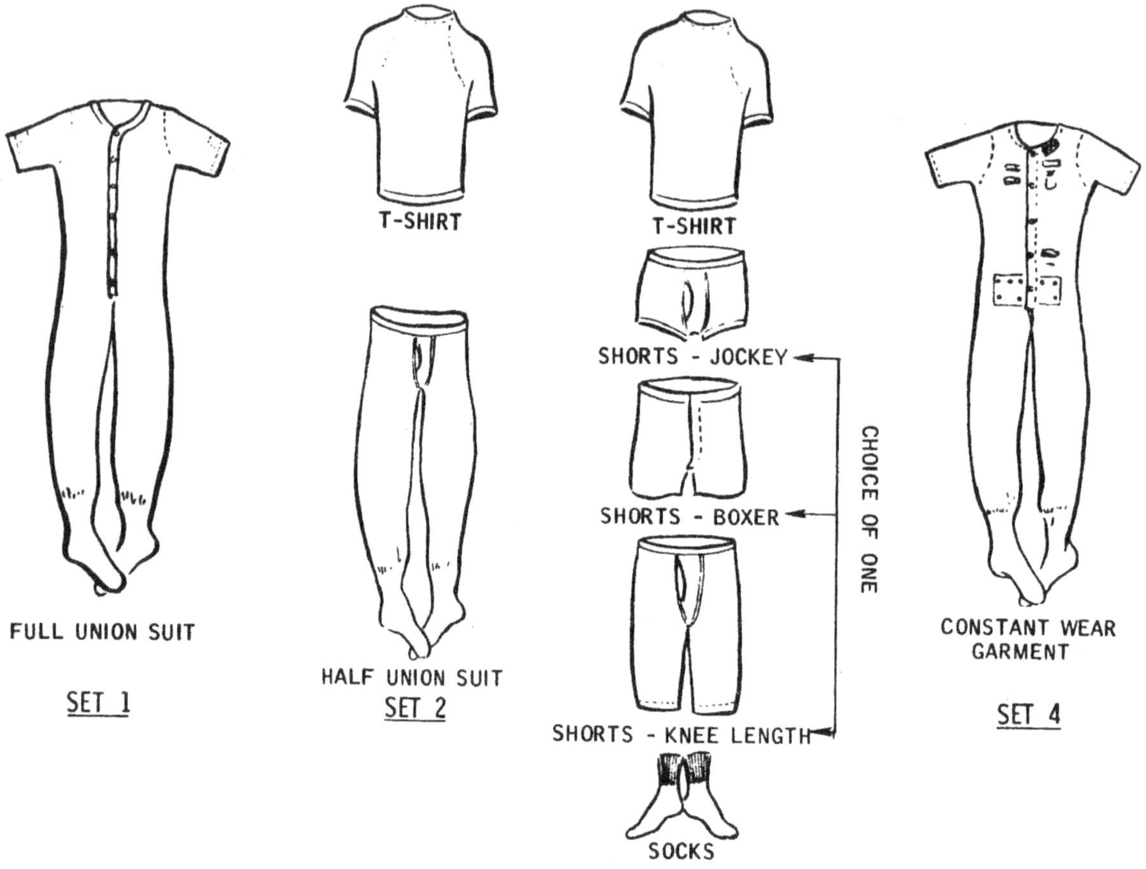

Figure II-84 — Crew Underwear

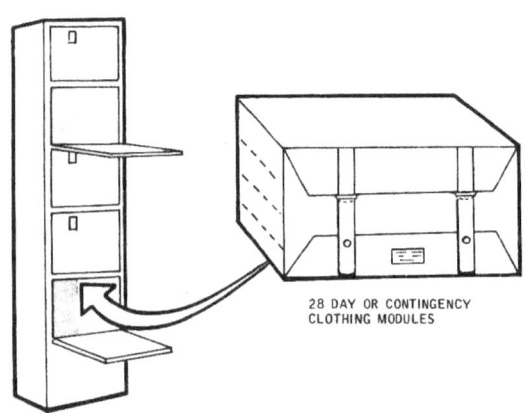

Figure II-85 — Standard Stowage Locker

Four different types of underwear sets are in the 28-day Clothing Modules.

Set 1 is a Full Union Suit with integrated socks and without an opening in the seat. It is a white 100 percent cotton "off-the-shelf" item.

Set 2 is a Half Union Suit ending at the waist. It has integrated socks and no opening in the seat. The other half is a standard pull-over T-shirt. Both are white 100 percent cotton "off-the-shelf" items.

Set 3 is T-shirts, shorts and socks. The short styles include jockey, knee and boxer. The T-shirts and shorts are white cotton. The socks are white cotton with reinforced heels and stretch tops, with one size to fit sizes 10 through 13.

Set 4 is a white cotton Apollo Constant Wear Garment (CWS).

Any combination of clothing articles contained in the 28-day Clothing Module may be packed in the contingency module as long as specified weight and volume requirements are not exceeded. The clothing sizes of the backup crewmen will be compared to the clothing sizes of the prime crewmen. If a backup crewman cannot wear a particular article

II-82

of the prime crewman's clothes, this article, in the backup crewman's size, will be packed into one of the contingency modules.

If the backup crewman can wear all of the prime crewman's clothes, the backup crewman may choose extra clothing subject to the same weight and volume limitations. The two contingency modules will contain all the contingency clothes for the SL-3 - SL-4 backup crew.

Construction of the contingency clothing module is identical to the 28-day rucksack, except that no valet kit is provided and there are only three compartments. It may contain any combination of the following types of items: jackets, shirts, trousers, underwear, boots or gloves.

The 28-day clothing modules are stowed in OWS wardroom lockers W718 through W722, and W730 through W734. Used clothing is placed in the waste tank after use. The two contingency clothing modules are stowed in OWS wardroom lockers W723 and W729.

A7LB Suit

Skylab crew members will use the Extravehicular Mobility Unit (EMU) (Figure II-86) for all operations which require pressure suits. The primary components of the EMU are the Apollo A7LB pressure garment assembly (PGA), the Liquid Cooling Garment (LCG), the pressure helmet, the Skylab Extravehicular Visor Assembly (SEVA), EVA gloves and the Astronaut Life Support Assembly (ALSA) (Figure II-87). Each crewman wears his A7LB suit in the CM at launch.

The PGA provides an oxygen environment for the crewman not only for breathing but also for spacesuit ventilation and pressurization. It is supplied with electrical provisions for bioinstrumentation and communications. After each use, the PGAs are placed in the suit drying station before they are again used.

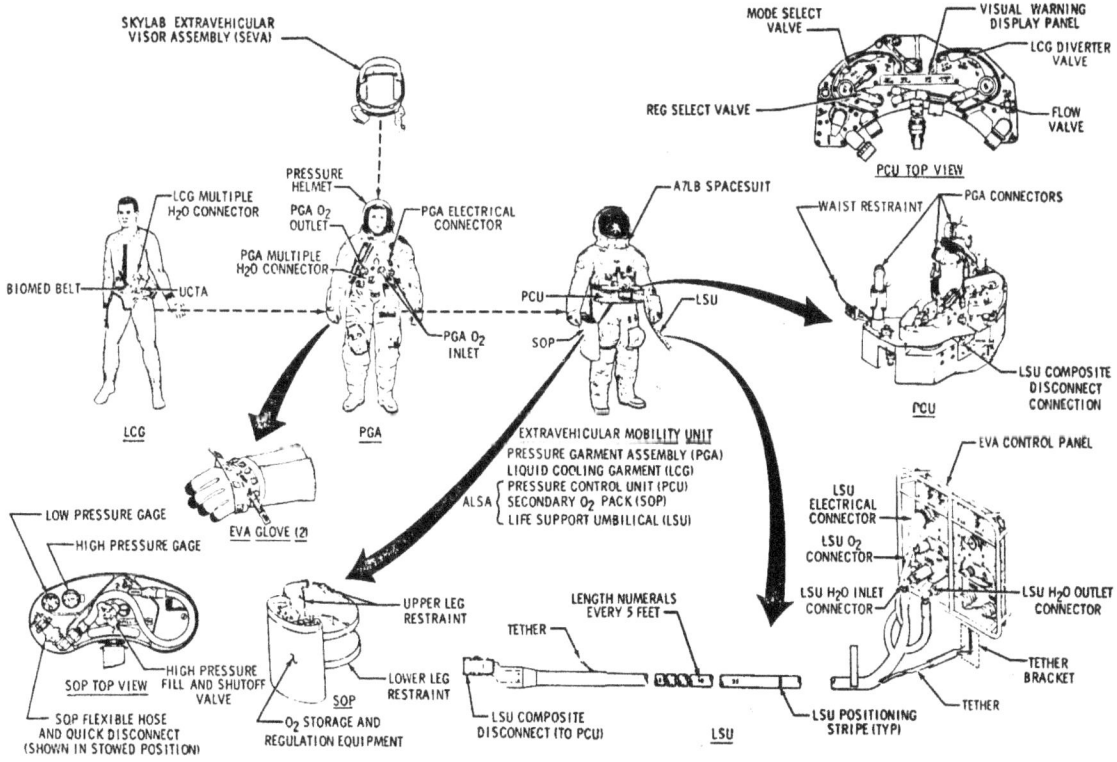

Figure II-86 — Extravehicular Mobility Unit

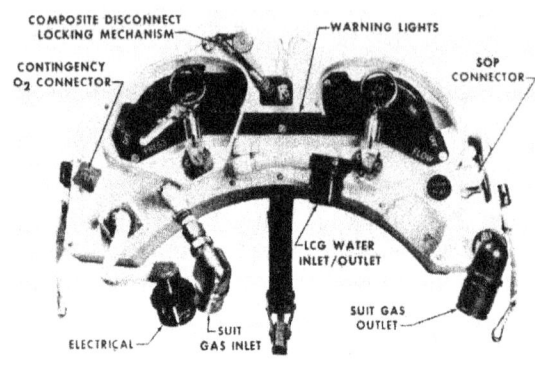

Figure II-87 — Skylab ALSA

The LCG, which is worn under the PGA, is made up of a network of water-carrying tubes that provide body cooling. Cold water is routed through the tubing, circulated about the crewman's body and returned to the SWS for heat rejection.

The helmet is a transparent, detachable enclosure and is supplied oxygen from the PGA to pressurize and ventilate with a provision to diffuse the oxygen about the crewman's head.

The SEVA attached to the helmet provides micrometeoroid and thermal protection while filtering out the harmful light rays of the sun.

EVA gloves are worn during EVA to provide additional thermal and micrometeoroid protection of the hand and forearm area. Three pairs of gloves are launched in each CM and each pair is stowed during flight in the SEVA stowage bag near the suit donning station.

The ALSA receives water, oxygen and electrical provisions through its Life Support Umbilical (LSU) and regulates and/or distributes them to the spacesuit by a Pressure Control Unit (PCU). An additional oxygen supply is furnished through a Secondary Oxygen Pack (SOP).

The LSU is 18.3 meters (60 feet) long and is marked at 1.5-meter (5-foot) intervals to aid in visually tracking the deployed length as the crewman exits the AM. The LSU is connected to a panel which provides the required water, electrical power and oxygen.

The only other activity requiring a pressurized suit besides EVA is a portion of the M509 astronaut maneuvering unit performance. With the exception of launch, undocking and separation the crewmen will be in a "shirt sleeve" mode throughout the mission.

Following rendezvous and docking, the pressure suits will be dried at the drying station in the OWS and then stowed in the CM. The crew will retrieve the suits for EVA. After EVA, the suits will again be dried and returned to stowage in the CM.

Facilities are furnished within the OWS to service the various components of the life support system as well as dry the pressure suits between uses. In addition, stations are designated in the OWS for suit donning and doffing.

The crew members don and doff their suits using the portable foot restraints in the forward compartment near the suit drying station. The drying station suit dryer on the ring section is used to blow air into the PGA to remove the accumulation of water after each use.

The Skylab pressure suit is an Apollo A7LB modified for use in Earth orbit. There are fewer layers of thermal insulation because the Earth orbit environment is not as hazardous as the lunar surface. The special lunar boot is not required, and the boot assembly for Skylab has a rigid sole outfitted with restraint devices to provide the astronaut firmer interface with the various foot restraints on the outside of the vehicle.

Off Duty Activities Equipment

The flight plan provides daily recreation time and regularly scheduled "off duty" days in addition to more than ten hours per day of work activities per crewman. On SL-2 the crew has three days off, the SL-3 crew has seven days and the SL-4 crew is scheduled to have eight days off duty.

The crew has an Off-Duty Activities Equipment module outfitted with exercise gear, reading material, taped music and games. Each crew member had a hand in selecting the type of music and books and in testing the items before flight.

The Off Duty Activities Equipment (ODAE) locker (Figure II-88) is in a corner wall of the Wardroom. The interior of the locker is divided into three stowage areas.

In addition, three tape players are stowed separately, one in each of the sleep compartments in lockers S909, S921 and S931.

The door of the ODAE locker is a rigid hinge-mounted unit to which is affixed the tape player and speakers and shelves holding 60 tapes, four decks of playing cards and 16 ear pieces for use in four stereo headsets. The tape player and speakers are commercial grade and the taped music may be played through the speakers or through the headsets, which may be plugged into special jacks on the locker door.

Power for the player is provided by the OWS power outlet or by dry cell batteries, 112 of which are in the locker door unit.

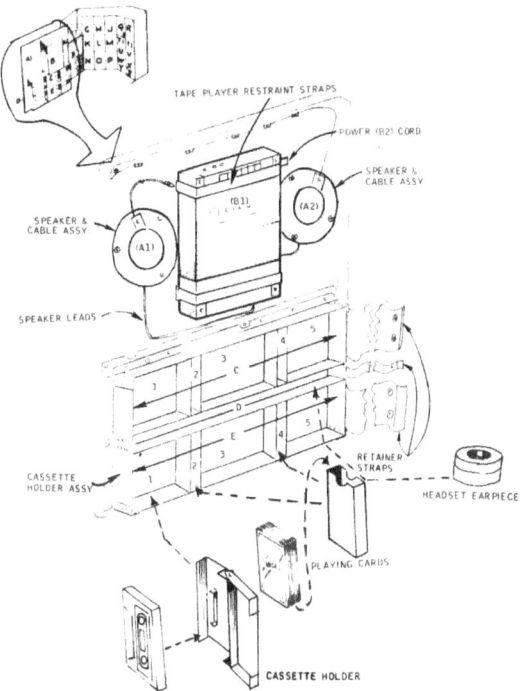

Figure II-88 — ODAE Locker

The four decks of cards are made of fire resistant paper, and the 16 headset ear pieces are foam pyrel. Special flexible magnetized straps and retainers are included to prevent the cards from floating in zero-g.

The recreational unit of the ODAE, an L-shaped unit built into the locker, contains the following: Individually selected off-the-shelf paperback books in eight stowage compartments; three library book covers of fire-resistant Teflon/Beta material; three card deck retainers; three card retainer hand holders; six hand exercisers constructed of Fluorel-coated sponge rubber; and three hand balls of different sizes (one fluorel coated commercial ball, one pyrell Nerf-ball dipped in ammonium-dyhydrogen phosphate and coated with Fluorel, and one commercial toy ball coated with Fluorel).

Three tape player Power Cords for playback of tapes are in the ODAE, as are four stereo headsets for listening to cassette tape playback; twelve darts (six blue, six gold), made of Fluorel with heads covered with Velcro hook and a dart board (Figure II-89) 29 by 38 cm (11½ by 15 inches), the front of which is covered with Velcro pile. The board is mounted in front of the library section.

Other recreational supplies include: three commercial grade Exer-Gyms (Figure II-90); two commercial monaural microphones to be used by the crew for recording their comments on the cassette tapes; and one set of binoculars (10 × 40).

The ODAE is stowed in the Wardroom in corner wall locker W714.

Containers with Velcro pile and hook latches are provided for stowage of equipment. These are kept in place during the mission by spring retainers.

Personal Hygiene Kit

The Personal Hygiene Kits (Figure II-91) provide each crewman with an individual supply of articles for personal hygiene use during the Skylab mission. These items are packed in Personal Hygiene Kits No. 1 and No. 2.

II-85

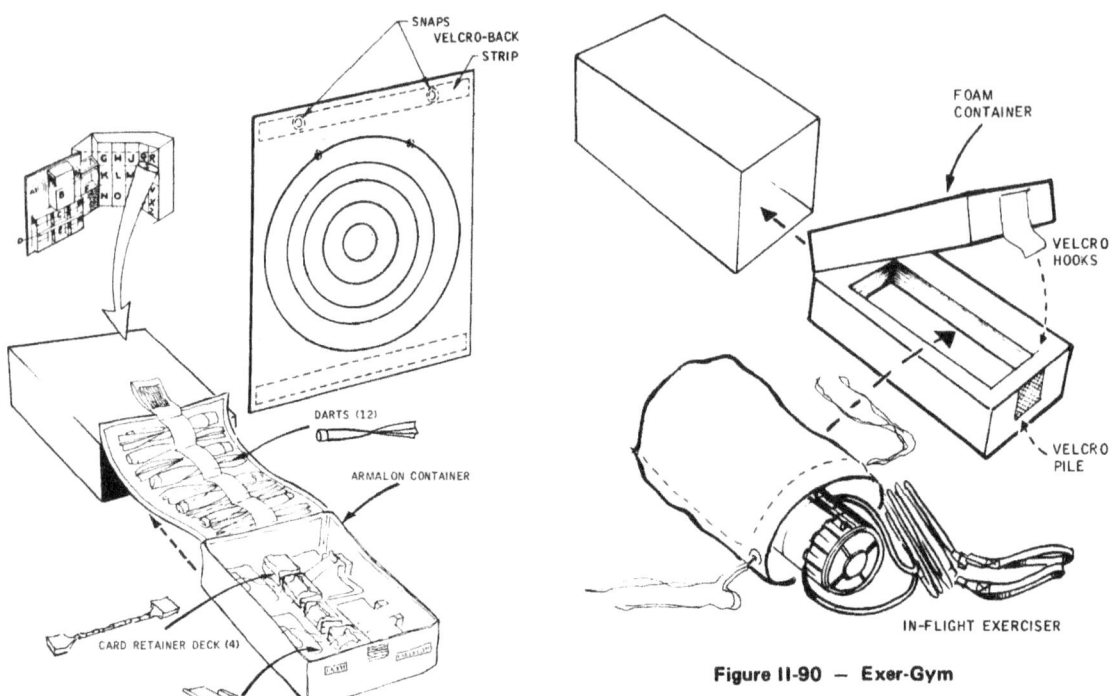

Figure II-89 — Darts and Board

Figure II-90 — Exer-Gym

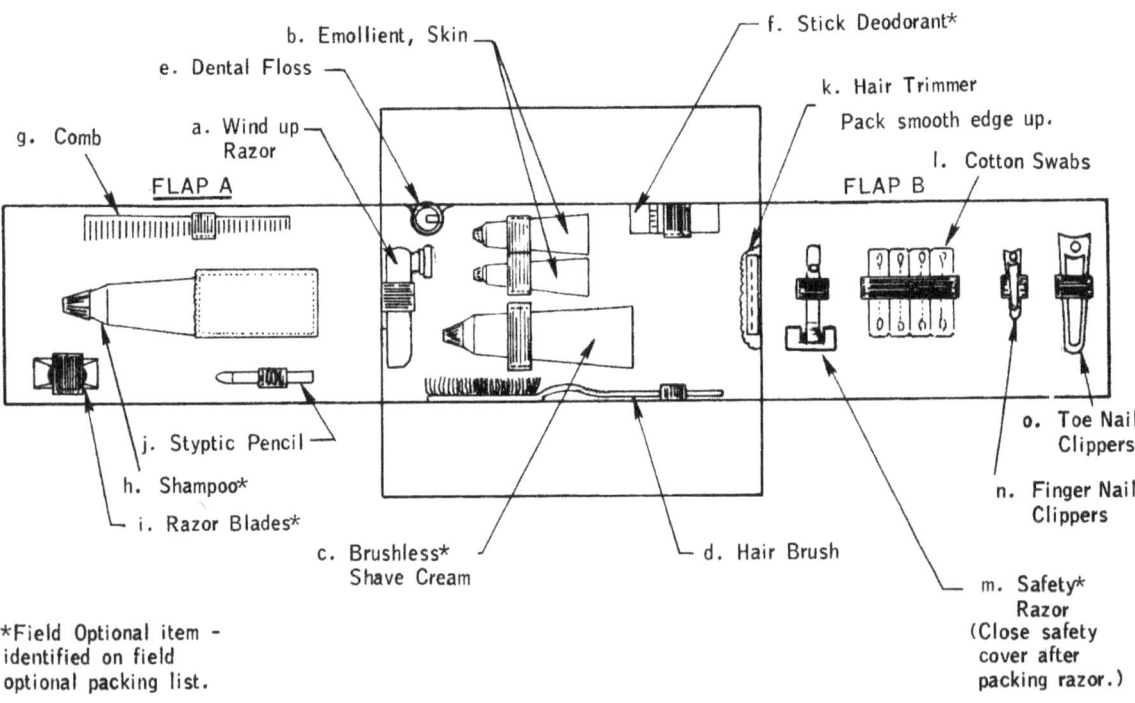

Figure II-91 — Personal Hygiene Kit

One Kit No. 1 is provided for each crewman for the initial 28-day mission. Six No. 2 kits are provided for use by crew members of subsequent missions. One resupply kit (Figure II-92) is furnished for all missions. Each kit bears the name of the crewman to whom it is assigned.

Kit No. 1 is a semi-rigid, rucksack shaped and sized to accommodate the personal hygiene articles for each crewman. The rucksack is five-ply layup construction, consisting of sandwiched layers of Teflon-coated Beta cloth, asbestos aluminum and Armalon. Hinged flaps provide kit closure, and other flaps inside the kit furnish stowage for personal hygiene articles. A pull tab is provided for opening the flaps. A Velcro hook is attached to the base of the kit, when in use, for retention in zero "g." Contents of the kit are held in place by elastic keepers and pockets.

The rucksack for Kit No. 2 is identical in construction to that of Kit No. 1. Kit No. 2 contains the same items as Kit No. 1, except no wind-up razor is provided. The wind-up razor is transferred from Kit No. 1 to Kit No. 2 after the initial 28-day mission. A new head is provided for the razor from the six replacement heads stowed in the Personal Hygiene Resupply Kit.

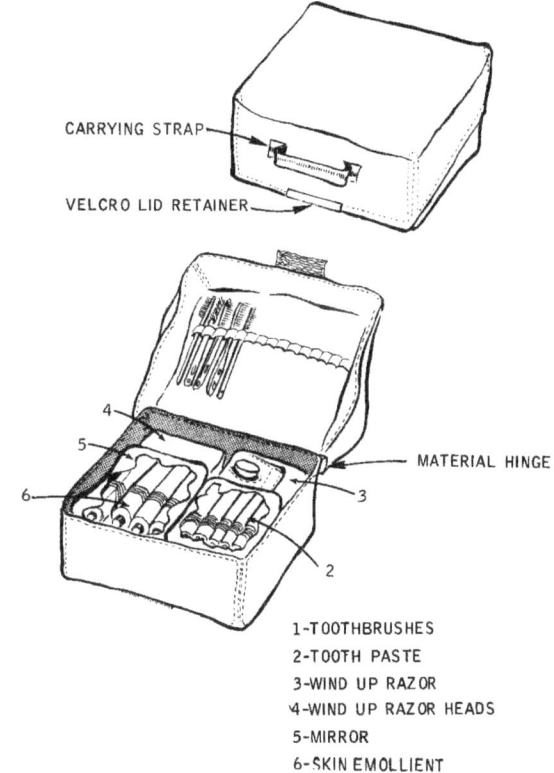

Figure II-92 — Personal Hygiene Resupply Kit

The Personal Hygiene Resupply Kit is a semi-rigid rucksack shaped and sized to accommodate toothbrushes, toothpaste, a resupply of expendable articles and space shaving equipment for Skylab Mission 2, 3 and 4.

Provisions are made for stowing Kit No. 1 in personal lockers assigned to crew members of the initial 28-day mission. One Personal Hygiene Kit No. 1 will be stowed in each of three WMC lockers.

Six Kits No. 2 and one Personal Hygiene Resupply Kit will be stowed in the Sleep Compartment.

Survival Equipment

Survival equipment, intended for use in an emergency after Earth landing, is stowed in two rucksacks (Rucksack 1 has survival equipment and Rucksack 2 contains a life raft) in the right-hand forward equipment bay of the CM. A lifevest for each crew member, which is to be worn during launch procedures, donned prior to reentry, and worn throughout recovery, is also stowed in the CM.

The three-man life raft is composed of a floatation tube with reversible, inflatable baffle and floor, three ballast buckets, and carbon dioxide and oral inflation systems. It is designed to support a load of 340 kg (750 lbs) for an extended period with an initial inflation pressure of 1.4 N/cm² (2 psig).

The raft is initially inflated by pulling a lanyard which simultaneously activates two cylinders, each charged with 375 grams of carbon dioxide. The raft is inflated in about 25 seconds.

The other rucksack contains a beacon transceiver, survival lights, desalter kits, machete, sun-glasses and water cans.

The radio/beacon transceiver is a signal and communications device, (manufactured by Cubic Corp.) about 19.7 cm (7 ¾ inches) high, 10.8 cm (4 ¼ inches) wide and 5.7 cm (2 ¼ inches) thick, which weighs 2.7 kg (6 lbs). The unit operates on a frequency of 243 mHz, and broadcast and reception are line of sight.

The transceiver, when operated continuously for 24 hours with a single Apollo Survival Radio Battery Pack, will give the following performance:

	Altitude of Receiver	Range, km (nm)	Watts Output
BEACON MODE:	3048 meters (10,000 feet)	231 (125)	1.25
	7620 meters (25,000 feet)	370 (200)	
VOICE MODE:			
Transmission	3048 meters (10,000 feet)	185 (100)	0.5
Reception	3048 meters (10,000 feet)	222 (120)	
Transmission	7620 meters (25,000 feet)	277 (150)	
Reception	7620 meters (25,000 feet)	370 (200)	

The combination Survival Light Assembly is a miniature survival kit, the function of which is self-explanatory. Components are: signal mirror, strobe signal light, flashlight, compass, a water-proof receptacle, 14 water purification tablets, a siren whistle, a knife blade, three needles, four fishhooks, a spool, heavy nylon thread, a flint striker and four cotton balls.

The Sunglasses especially designed for the Apollo Program will be used in Skylab. They satisfy the criteria for resistance to breakage, compactness and protection from harmful sunrays and glare.

The sunglasses have a soft fabric frame and are adjustable to head size and face contour. They are held against the face by elastic braids which are fastened together behind the head with hook and pile fastener.

The blade of the machete is high quality stainless steel; the handle is aluminum. The blade is ground and honed to a shaving sharpness. A sawing edge is opposite the cutting edge. It is sheathed.

Three Water Containers with 2.3 kg (5 lbs) capacity are in Rucksack No. 1. Two are similarly configured; the third is shaped for neat interface with the spacecraft stowage compartment. Each has an aluminum body and an aluminum cap with a neoprene gasket.

The Sun screen is a protective cream that may be rubbed on the exposed portions of the body to prevent sunburn.

Two survival knives are provided. They are standard three-bladed, all metal knives developed by the U.S. Navy.

The Utility Netting is 1.5 by 1.3 meters (60 by 52 inches) and is made of a standard fine mesh nylon material. The netting is intended primarily for protection from insects, but can be used as a fishing net if required.

Three pieces of Nylon-Mylar material 1.5 by 1.07 meters (60 by 42 inches) are provided. The material can be used for thermal protection and for signaling. The color of the inner surfaces is International Orange. The outer surface is aluminized.

PROVISIONS AND STOWAGE

Outfitting and provisioning for 420 man days of living and working in the Skylab has resulted in the biggest packing job NASA has undertaken. More than 20,000 items have been stowed throughout Skylab (Figure II-93).

Management of the Skylab "stowage list," has produced a Skylab Workshop Stowage numbering system with assigned items in each of the vehicles (Figure II-94). Each of the stowage locations in the AM, MDA and OWS is again subdivided into series. Command Module Stowage is shown in Figure II-95.

Each stowage location has a label that contains the assigned stowage number, the items contained and their quantities. A label kit and marking pen permit crewmen to write numbers on the items to maintain control.

The items stowed aboard SL-1 range from 1,200 aspirin and 952.6 kg (2,100 lbs) of frozen and unrefrigerated foods to more than 150 rolls of teleprinter paper and 27 kg (60 lbs) of maintenance tools.

MDA Stowage

In the MDA (Figure II-96) are four stowage compartments and four film vaults. The items stowed in the MDA include the film for the ATM cameras, lithium hydroxide replacement cartridges for the CSM, mission and data files, crewman communications equipment, mol sieve solids traps and CSM and SWS power and signal transfer cables.

The CSM LiOH cartridges, communication equipment, SWS activation equipment, flight data files and mol sieve trap spare parts are stowed in MDA lockers M125, M126, M157 and M168. All but M168 are permanently located on structures in the MDA. Stowage compartment M168 is launched in the AM and later transferred to the MDA.

The MDA film vaults (M124, M141, M143 and M152) provide stowage for the ATM film magazines and cameras. Each vault features aluminum shielding to supply radiation protection for the stored film. There are more than 175 film magazines stowed in the SWS.

AM Stowage

Sixteen stowage compartments are in the AM (Figures II-97 and II-98), five in the STS, four in the forward compartment, two in the lock compartment and five in the aft compartment. Stowage in the STS is assigned (M)200 series numbers, while the AM forward, lock and aft compartments stowage is assigned (M)300 series numbers.

Supplies for the teleprinter, life support umbilicals, portable timers and spare parts are in compartments M201, M202, M301, M305, M310 and M311. All of the compartments are removable through use of Calfax fasteners. A compartment labeled M208 is reserved for inflight stowage of flight data material.

The M168 compartment is transferred to the MDA following crew activation.

The life support umbilicals (LSU) are mounted on the exterior of the lock compartment. One 18-meter (60-foot) LSU is in each spherical enclosure.

There are also two spare tape recorders, a speaker intercom assembly and a spare condensate tank module mounted on the walls of the AM.

OWS Stowage

Stowage in the Orbital Workshop (Figures II-99 through II-104) is provided on the Forward Compartment ring and in the Forward Compartment and Crew Quarters in the form of various sized compartments, dispensers, refrigerated and ambient temperature food stowage facilities, refrigerated urine stowage and a film vault. The OWS stowage and series

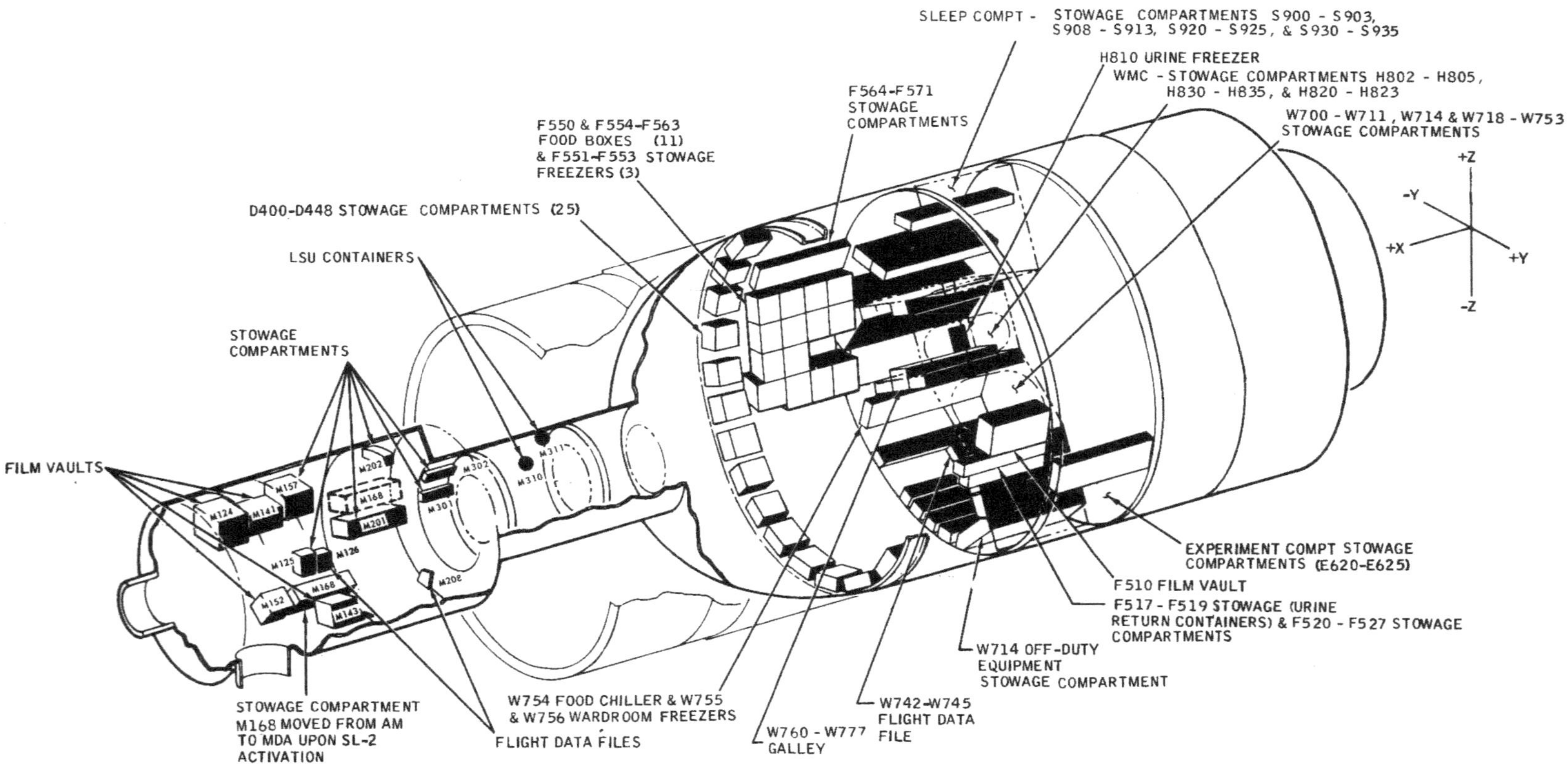

Figure II-93 — Skylab Stowage

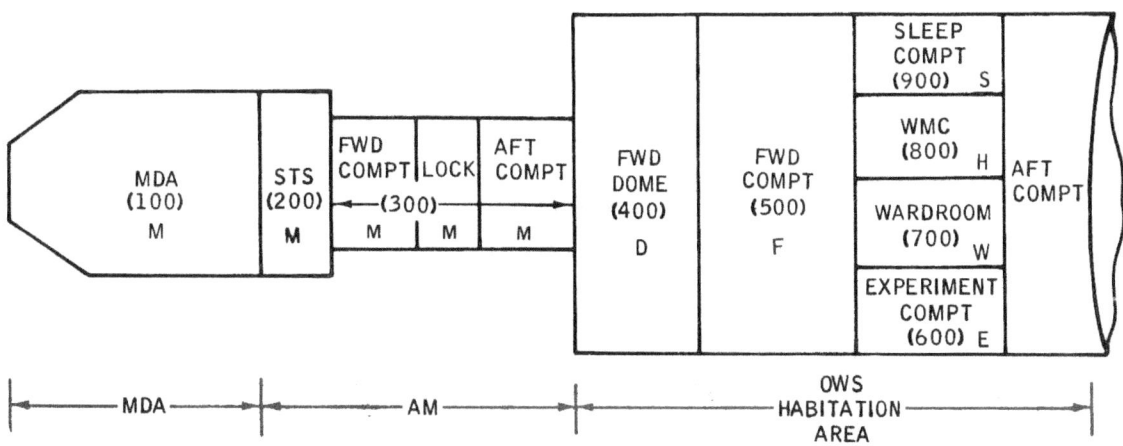

Figure II-94 — Stowage Numbering

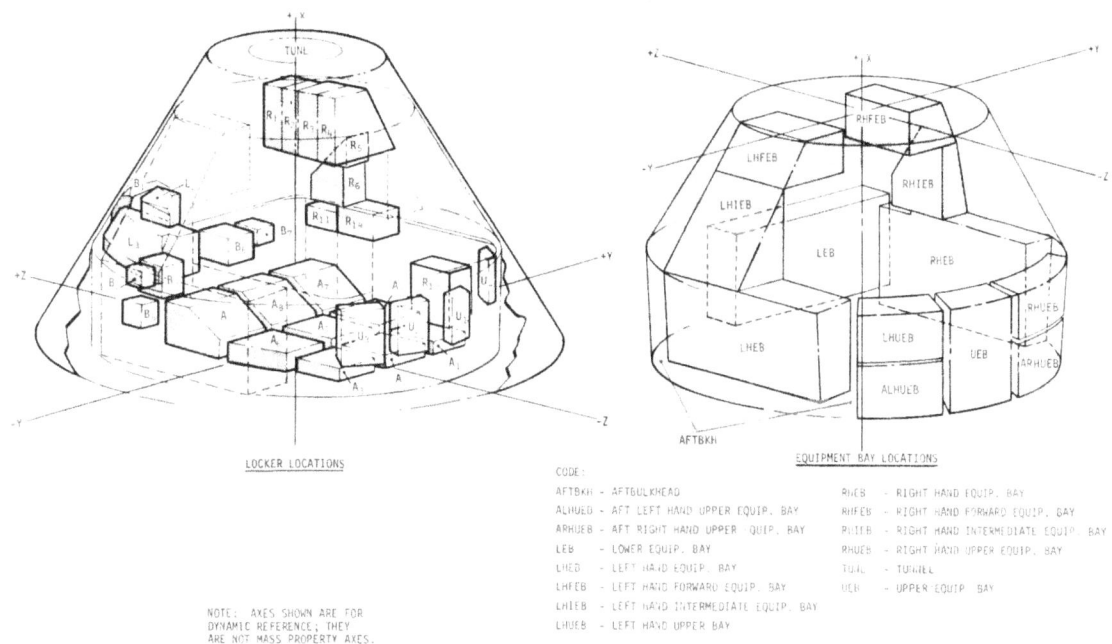

Figure II-95 — Command Module Stowage

numbers are: Forward Compartment F500, experiment E600, Wardroom W700, Waste Management Compartment H800, Sleep Compartment S900, and forward dome stowage (25 containers) D400 series. Summary of the OWS stowage items are: dispensers, food boxes, urine freezer, film vault, wearing apparel, EVA support and workstation provisions, off-duty equipment, trash and waste management, personal hygiene, maintenance equipment and spare parts, fire suppression equipment and Inflight Medical Support System.

Dispensers are stowed for waste management collection, bags, towels and trash containers. The waste management bag dispensers are stowed in H833 and are readily available to the crew in the WMC.

II-91

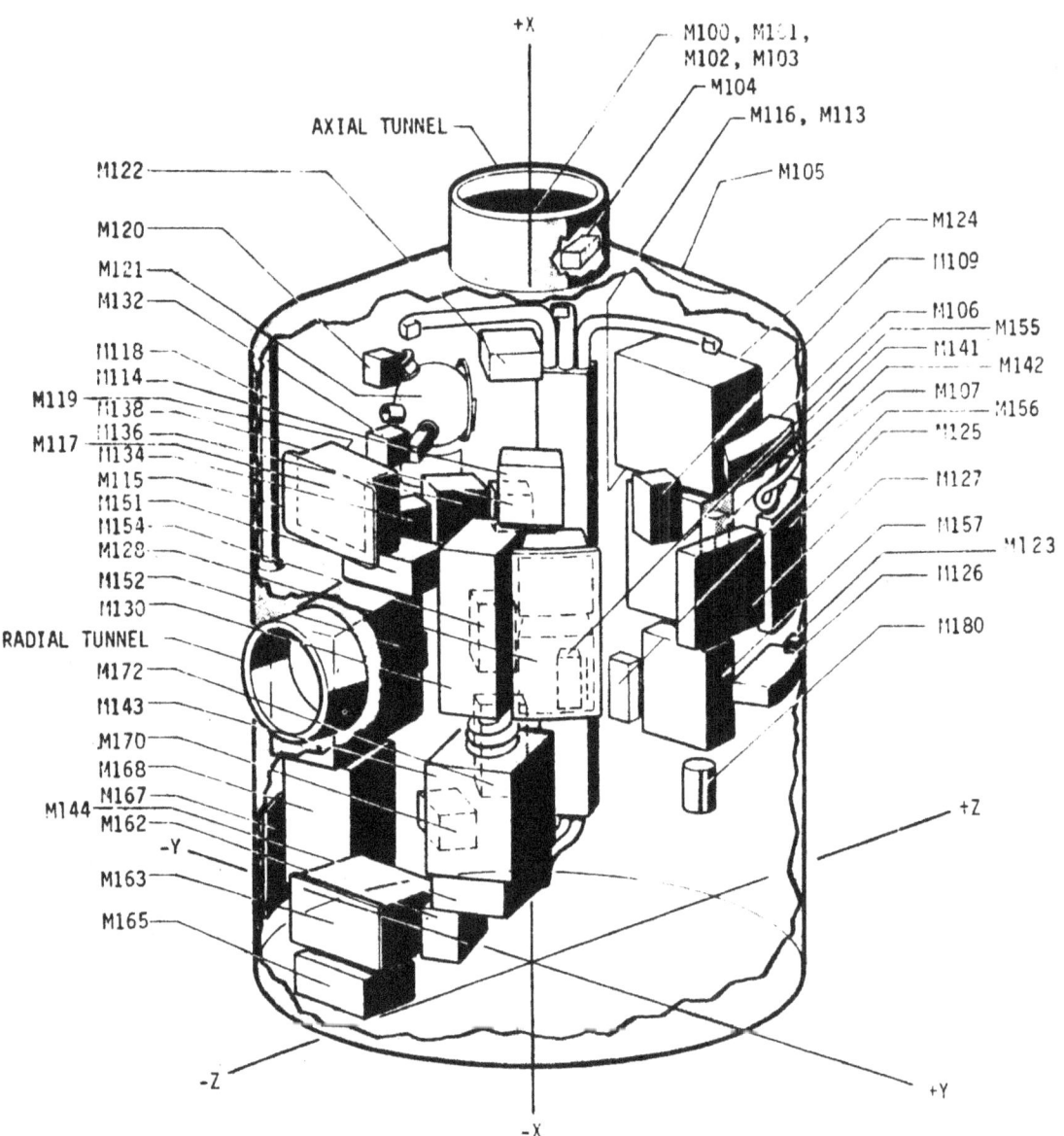

Figure II-96 — MDA Stowage

Towels are stowed in five compartments in the WMC. Each towel dispenser contains 18 towels.

Trash containers are in certain stowage compartments of the OWS. Trash is inserted directly into the bag through the opening in the trash container door.

Eleven food boxes provide launch and on-orbit stowage of ambient temperature foods for the SL-2, SL-3 and SL-4 missions. Five of the 11 boxes are permanently mounted in the OWS Forward Compartment on the grid above the Wardroom. The remaining six food boxes are dispersed on the forward compartment floor to ensure structural integrity during launch. Upon SL-2 activation, the six dispersed foodboxes are unbolted from their launch positions and placed in permanent locations.

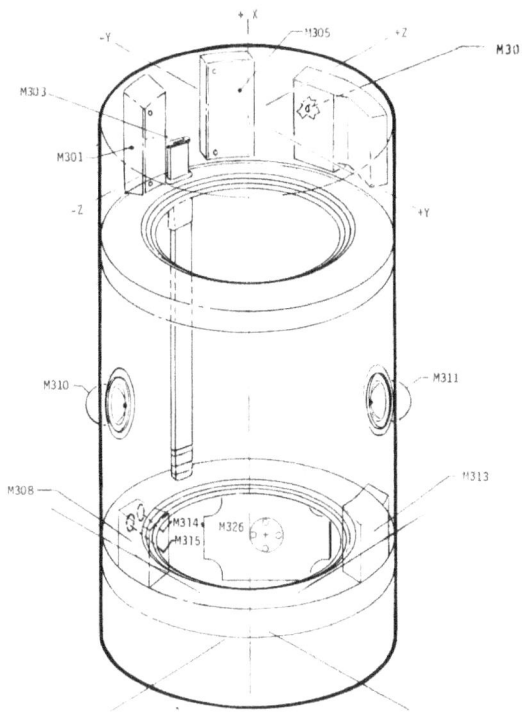

Figure II-97 — AM Tunnel Stowage

The entire mission supply of refrigerated foods is stowed in a three-chambered refrigerated unit in the OWS Forward Compartment and in one in the Wardroom. The stowage freezer is divided into three chambers with each chamber containing a 28-day supply of frozen foods.

Food Stowage

Food is stored in cans and beverage packs which are grouped in menu form in food overcans. The overcans are stowed in bundles in food boxes and the freezers. The food trays in the Wardroom food table permit the temporary stowage of food when preparing meals and managing leftovers.

The food in, in two forms — ambient and frozen. The ambient consists of dehydrated food and beverages, thermostabilized foods, dry bites and puddings. The frozen foods consist of thermostabilized food, some of which must be heated prior to consumption.

Ambient temperature foods (excluding beverages) and frozen foods are vacuum packed in single meal portions in food cans which come in types — large, small and pudding. The food is prepared in the can, all of which have pull-top lids, and the can is used as the dish.

The food overcans contain 12 food cans or beverage packs which are packed and identified according to the menu for the particular crewman. The overcans, containing a mixture of large and small overcans, are stored in bundle form, two deep.

The overcans are stored in the 11 food lockers. Every seven days, a week's supply of food overcans is transferred from a food overcan bundle in one of the food boxes to the galley stowage trays.

The galley in the Wardroom accommodates 22 galley trays, a week's supply of food readily available for meal preparation. The trays consist of five galley trays per crewman for their individual menu, one galley tray for the weekly pudding supply, one galley tray per crewman for snacks (dry bites) and one galley tray per crewman for beverages.

Each galley tray slides out on a track and may be removed from the galley. Each tray holds 20 items: large and small food cans, pudding cans or beverage packs in partitioned segments.

Urine Freezer

A urine freezer in the WMC stores and preserves up to a 56-day accumulation of urine samples from the three crewmembers. Urine samples are in portable urine trays. The freezer can stow up to four urine trays simultaneously.

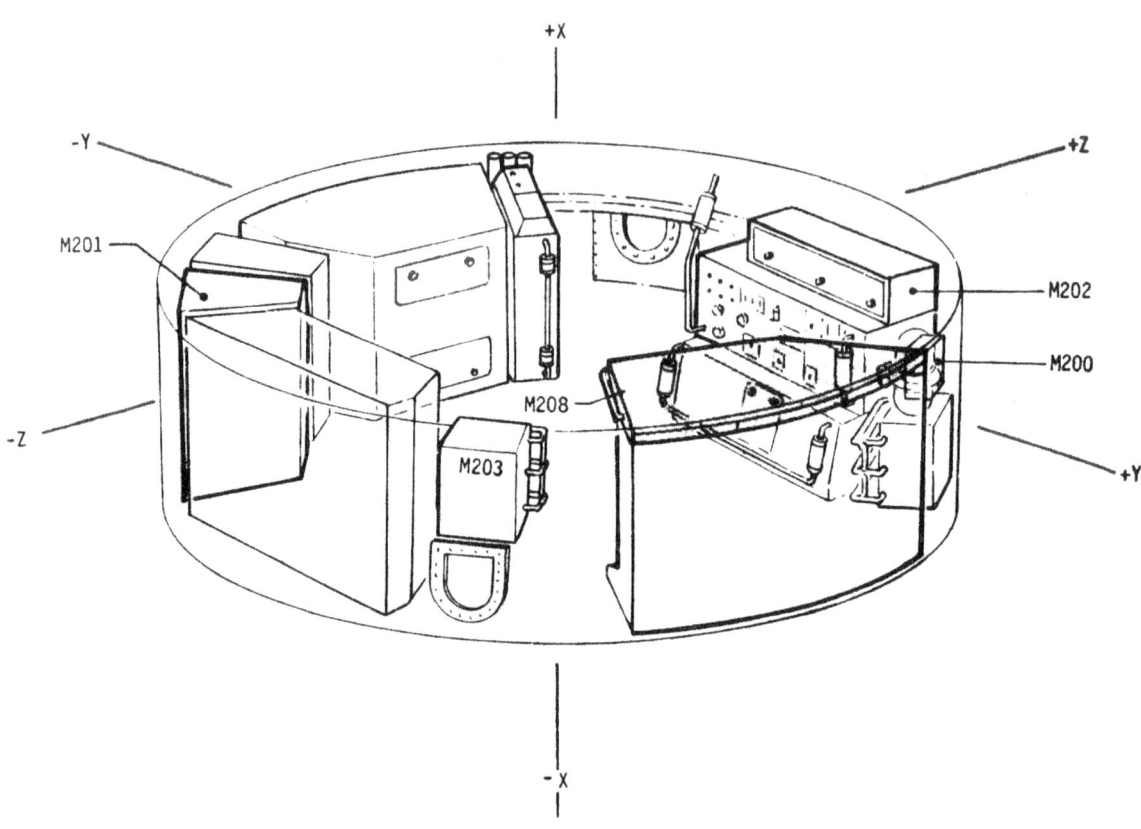

Figure II-98 — AM STS Stowage

Film Vault

The OWS film vault, F510, is a shielded, drawered vault in the Forward Compartment which stores hand-held camera film cassettes used for experimentation. On-orbit the doors are secured closed with two dial latches. The doors and vault walls contain aluminum shielding to protect the unexposed film from radiation.

Skylab Flight Data File

Skylab requires a comprehensive assembly of documents, charts and data for crew use in operating the spacecraft systems and experiments. The data collectively is known as the Flight Data File and consists of: flight plans (crew timelines), experiment and system checklists (condensed procedures), systems data, logs, maps and charts. Using this group of books the crewman knows not only what to do and when to do it, but also is provided supplemental reference data with provisions for real-time updating.

The Flight Data File for Skylab is formidable when compared to the FDF of Apollo. The Apollo FDF weighed 15.8 kg (35 lbs). The Skylab FDF consists of 54 kg (120 lbs) launched in the workshop and 18 kg (40 lbs) launched in each CM. Basically, the workshop FDF consists of books pertinent to all three missions whereas the CM books are peculiar to the particular mission in question. Data file items that can be used again for subsequent missions are left stored in the SWS during the unmanned mission phase. Other books are either discarded in the SWS or returned in the CM for postflight use.

Flight Data Files are stored in four basic areas: the CM (R1, R2 and R3), the MDA (M126), the AM (M208) and the OWS workroom (W742, W743, W744, and W745). Special beta bag provisions are provided for large bulky maps and charts and for inflight stowage of books adjacent to experiment hardware.

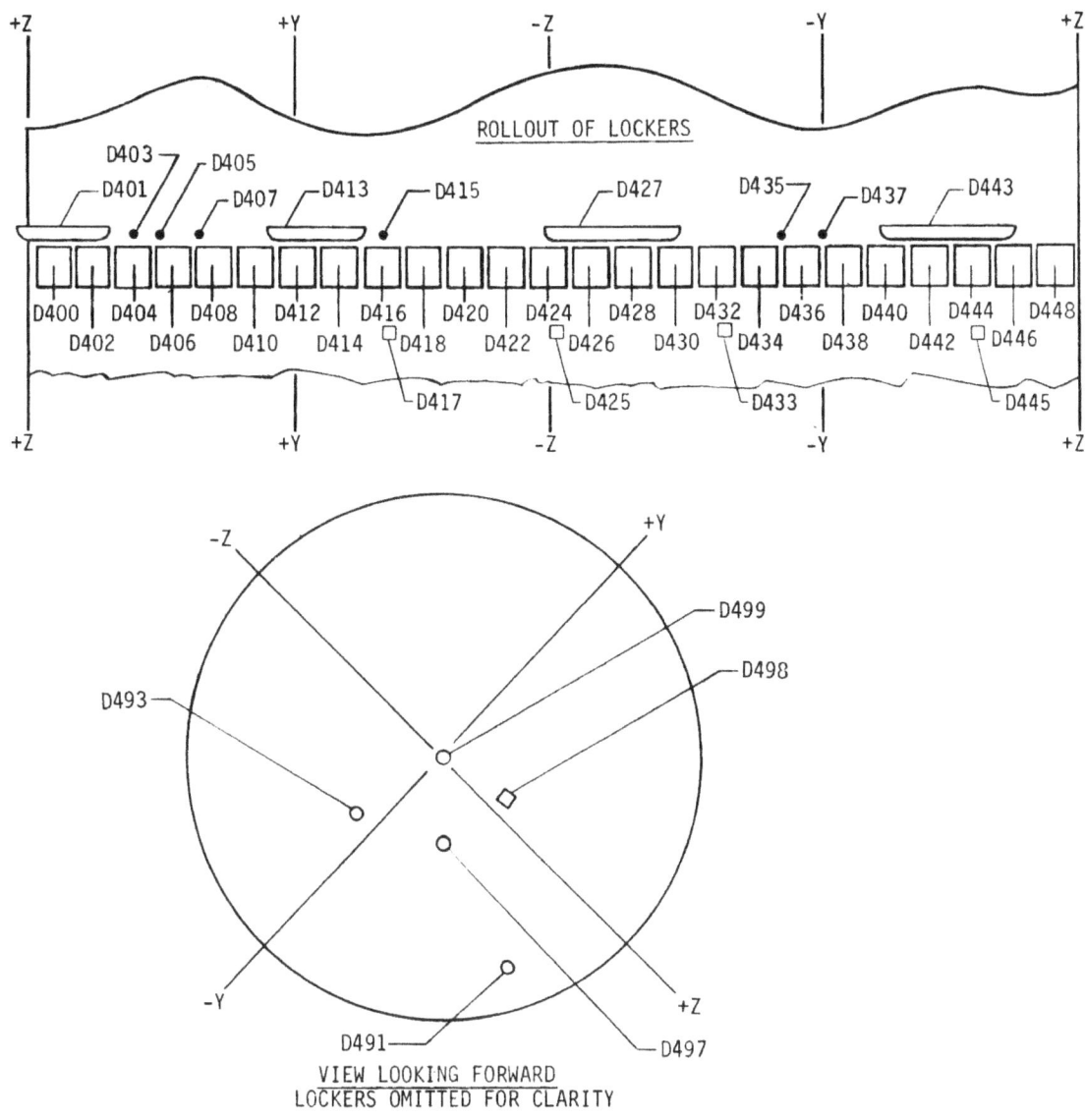

Figure II-99 — OWS Stowage Ring

Tools

Tool box number one has five drawers containing common usage tools, i.e., hammer, clamps, pinch bar, vise, standard slot screw drivers, phillips screw drivers, allen wrenches, 3/8 drive socket wrenches (deep well and standard), ratchet handles, speeder handles, torque handles (5-150 inch pounds), universal joint, ratchet extensions, open end/box wrenches, crow-foot wrenches, retrieval tools and mechanical fingers (Figure I-16).

Tool box number two has five drawers and a storage bin containing common usage tools and special repair items, i.e., channel pliers, needle-nose pliers, slip joint pliers, pocket knife, large crowfoot wrenches, tweezers, O-ring extractors, torque wrench (0-600 inch pounds) scissors and tape.

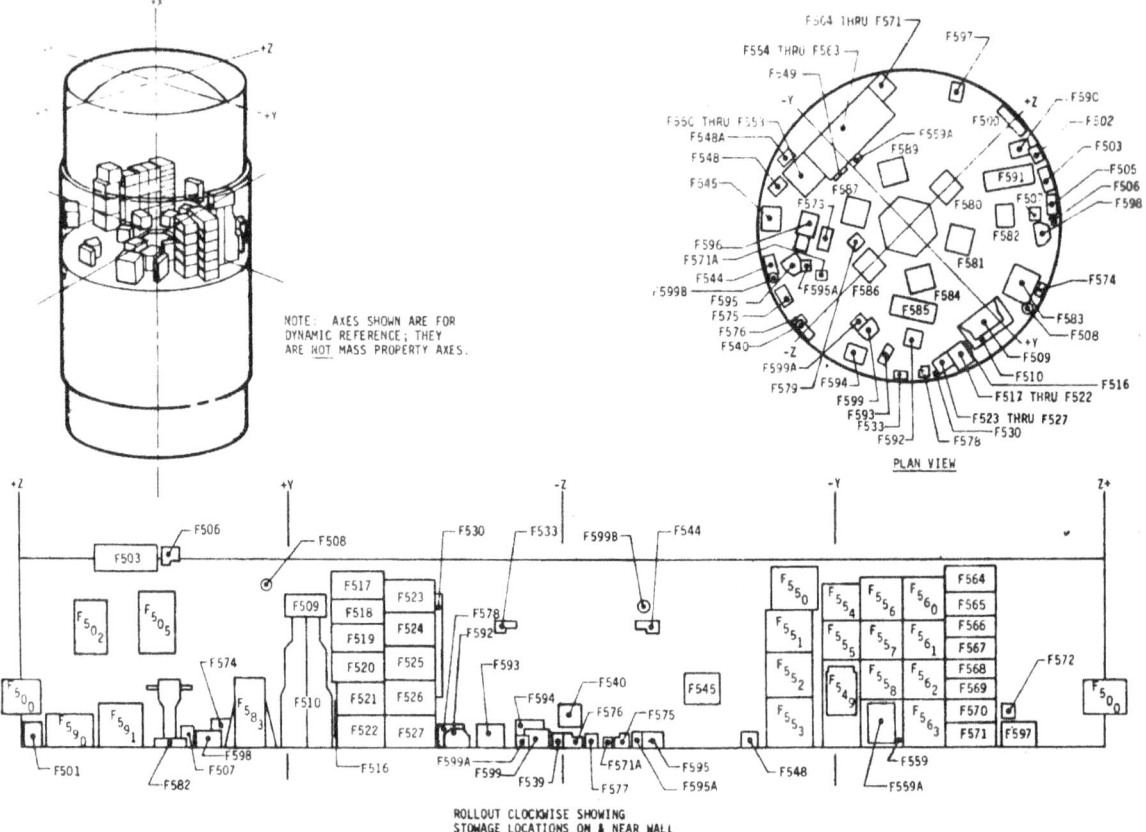

Figure II-100 — OWS Forward Compartment Stowage

The repair kit contains five drawers of special patching materials, i.e., flat and blister patches, teflon tape, sealant putty, Velcro strips, snip scissors and a leak detector. Adhesive tape, scissors and tweezers are used in repairing fabric material damage. Punctures in the pressurized structure caused by inadvertent damage are repaired using pliable, pressure sensitive blister and flat patches which are sized for specific leaks. These patches are made of aluminum sheet backed with a layer of polymeric compound to flow into the puncture. The aluminum provides a rigid surface upon which cabin pressure acts in applying the force necessary to maintain a pressure-tight seal.

Tool box drawers are designed to be removed by the astronauts and transferred to the maintenance area and attached to the structure using a universal mounting technique.

The tool caddy is attachable to a crewman's utility belt by four snaps. The caddy has elastic bands and pocket restraints for holding tools. It also has two boxes with clear windows on the front and flexible sides with slits. Small articles, such as nuts and washers, can be inserted through the slot and kept in view until needed.

Spare parts carried in the experiment compartment include: Power module collections, experiment M093 VCG cable, electrode kit, subject interface box and experiment M171 absorbtion cartridge.

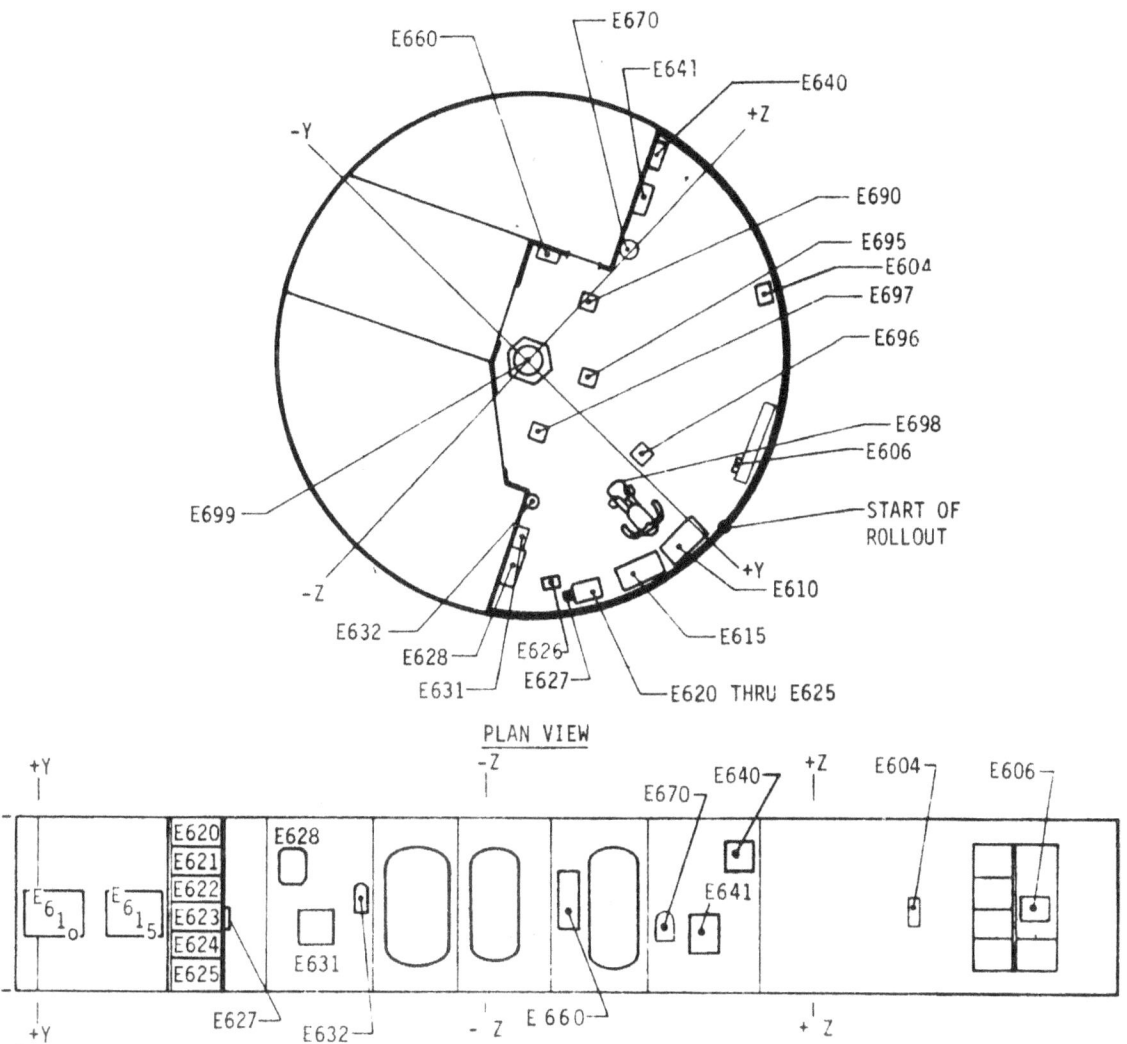

Figure II-101 — OWS Experiment Compartment Stowage

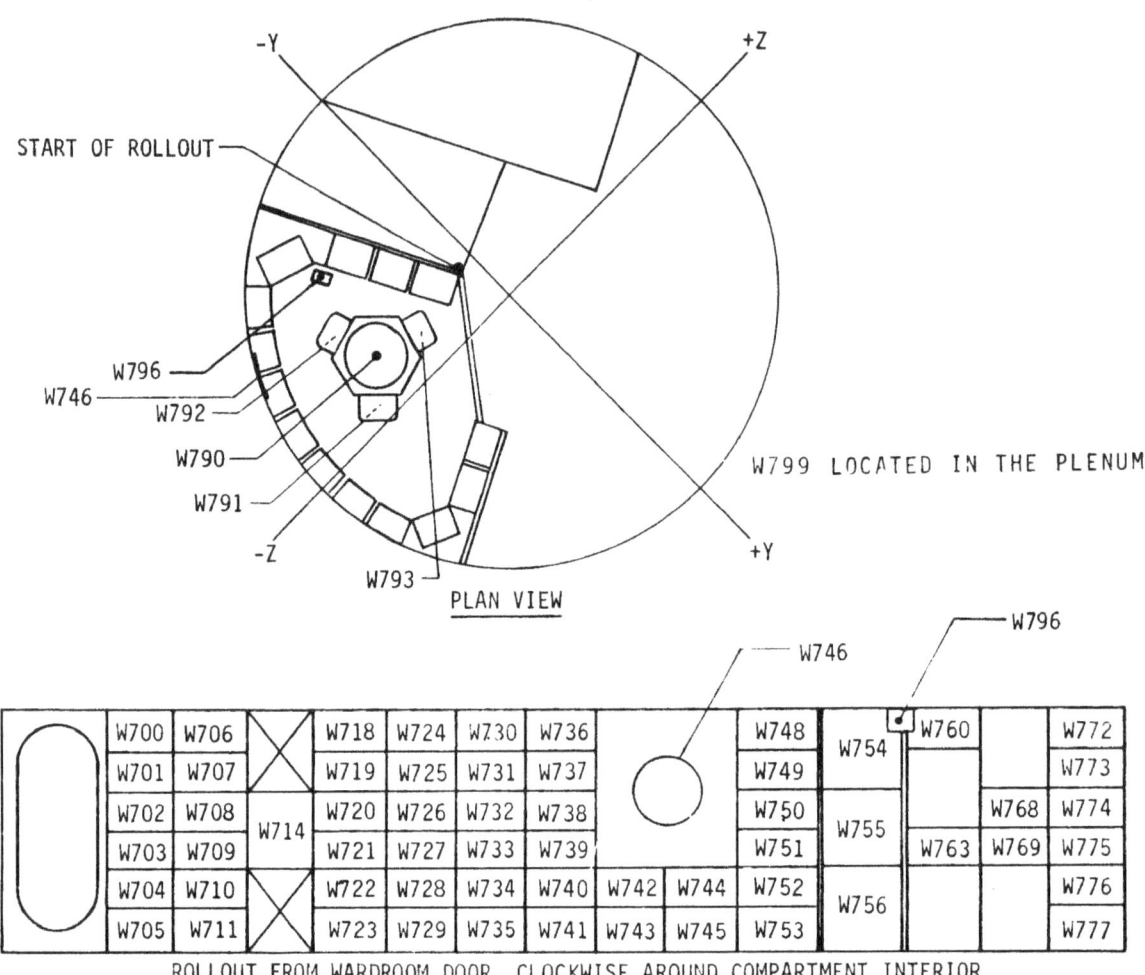

Figure II-102 — OWS Wardroom Stowage

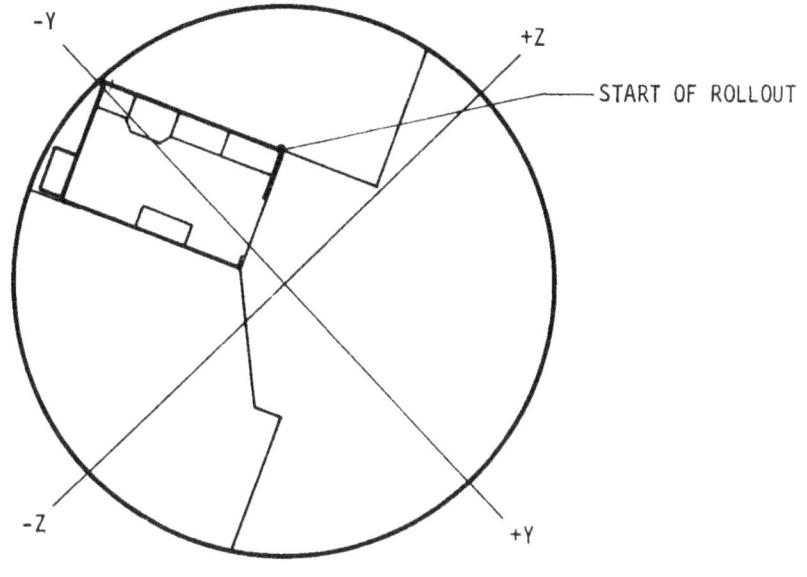

PLAN VIEW

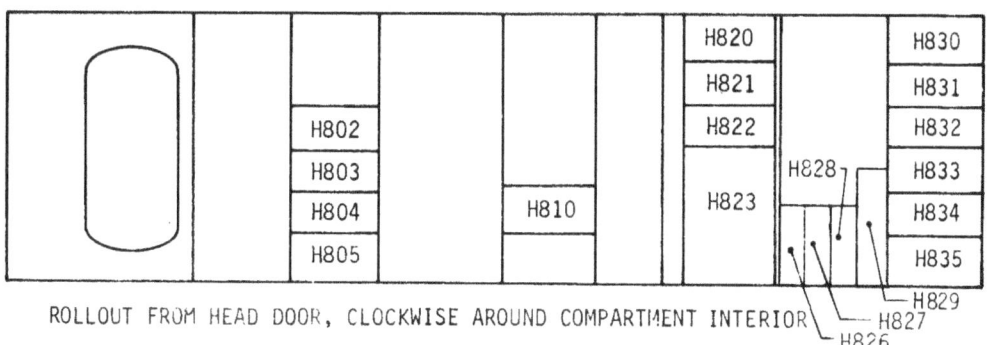

ROLLOUT FROM HEAD DOOR, CLOCKWISE AROUND COMPARTMENT INTERIOR

Figure II-103 — OWS WMC Stowage

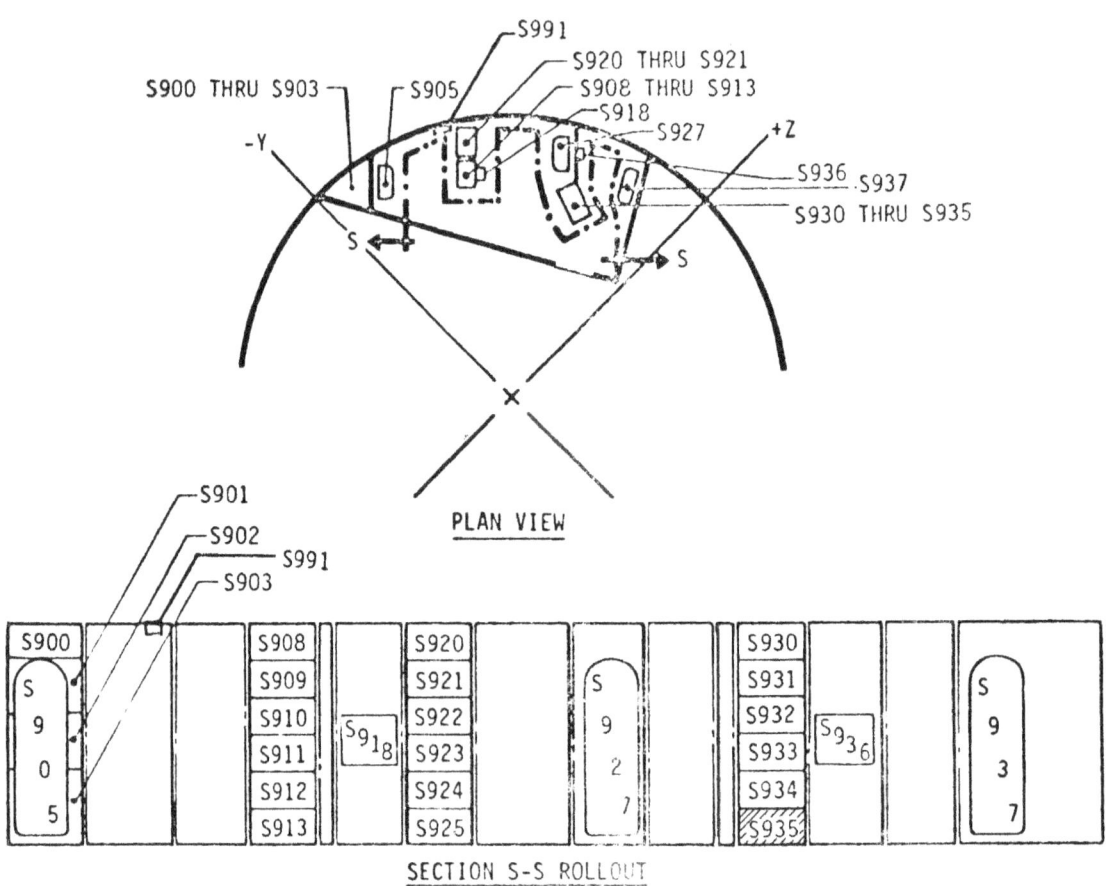

Figure II-104 — OWS Sleep Compartment Stowage

SECTION III

EXPERIMENTS

The Skylab experiment program consists of more than 270 scientific and technical investigations representing virtually every field that has been recognized as being able to benefit from operations in near-Earth orbit. The instruments, sensors and other equipment for conducting these experiments and investigations are located in various parts of Skylab, some inside and some outside.

The experiment payload has been developed through the joint cooperation of engineers and scientists from both foreign and domestic educational institutions, private industry and government agencies. Included are 19 experiments proposed by high school students. The 54 items of experiment hardware include available off-the-shelf equipment whenever practicable to reduce costs. On the other hand, sophisticated hardware not used before on manned spacecraft was developed when required to meet certain experiment objectives.

The major experiment areas are:

Life Sciences – 19 experiments will support some 28 investigations dealing with the effects on men and animals of long duration in the space environment.

Solar Physics – Nine solar instruments will provide unprecedented observations for 45 investigations of solar phenomena.

Earth Observations – Six remote sensing instruments will support more than 140 individual investigations for the study of the Earth from space.

Astrophysics – 14 instruments will make observations to support 24 studies of the solar system and beyond.

Material Science – The properties of orbital weightlessness will be exploited to investigate the advantages of materials processing in space – 18 experiments.

Engineering and Technology – To advance the knowledge for design and operation of future space systems – 13 experiments.

Student Experiments – 19 investigations.

Refer to "Skylab Experiments" published in August 1972 by the National Aeronautics and Space Administration for further details. See Table III-1 for the list of experiments.

LIFE SCIENCES

From the beginning of manned space flight, there has been concern about the ability of man to survive and perform satisfactorily in the space flight environment. Concern has centered around the exposure of the human body to the launch accelerations, adaptation to weightlessness, the ability to withstand reentry loads and the readaptation to full gravity following the return to Earth, the latter two emerging as real concerns.

The medical thrust of the first decade of manned space flight has been directed primarily to qualifying man for Apollo lunar landing missions. In the Gemini VII mission, limited inflight medical experiments were conducted to study man's physiological reactions to a two-week mission. The balance of the Gemini and Apollo flights have been used to study the physiological effects on man primarily through pre- and postflight medical investigations.

Table III-1 – Skylab Experiments

LIFE SCIENCES

M071	Mineral Balance
M073	Bio-Assay of Body Fluids
M074	Specimen Mass Measurement
M078	Bone Mineral Measurement
M092	Lower Body Negative Pressure
M093	Vectorcardiogram
M111	Cytogenic Studies of the Blood
M112	Man's Immunity, in Vitro Aspects
M113	Blood Volume and Red Cell Life Span
M114	Red Blood Cell Metabolism
M115	Special Hematologic Effect
M131	Human Vestibular Function
M133	Sleep Monitoring
M151	Time and Motion Study
M171	Metabolic Activity
M172	Body Mass Measurement
S015	Effects of Zero G on Single Human Cells
S071	Circadian Rhythm – Pocket Mice
S072	Circadian Rhythm – Vinegar Gnat

SOLAR PHYSICS

S020	X-Ray/UV Solar Photography
S052	White Light Coronagraph
S054	X-Ray Spectrographic Telescope
S055	UV Scanning Polychromator Spectroheliometer
S056	Dual X-Ray Telescope
S082	XUV Spectrograph/Spectroheliograph
	XUV Coronal Spectroheliograph (S082A)
	Chromospheric XUV Spectrograph (S082B)
Hα-1,-2	Hydrogen-Alpha Telescopes

EARTH OBSERVATIONS

S190	Multispectral Photographic Facility
	Multispectral Photographic Cameras (S190A)
	Earth Terrain Camera (S190B)
S191	Infrared Spectrometer
S192	Multispectral Scanner
S193	Microwave Radiometer/Scatterometer/Altimeter
S194	L-Band Microwave Radiometer

ASTROPHYSICS

S009	Nuclear Emulsion
S019	UV Stellar Astronomy
S063	UV Airglow Horizon Photography
S073	Gegenschein/Zodiacal Light
S149	Particle Collection
S150	Galactic X-Ray Mapping
S183	Ultraviolet Panorama
S228	Trans-Uranic Cosmic Rays
S230	Magnetospheric Particle Composition

MATERIALS SCIENCE AND MANUFACTURING IN SPACE

M512	Materials Processing in Space
M551	Metals Melting
M552	Exothermic Brazing
M553	Sphere Forming
M555	Single Crystal Growth
M479	Zero Gravity Flammability
M518	Multipurpose Electric Furnace
M556	Vapor Growth of II-VI Compounds
M557	Immiscible Alloy Compositions

Table III-1 – (Concluded)

M558	Radioactive Tracer Diffusion		**STUDENT EXPERIMENTS**
M559	Microsegregation in Germanium		
M560	Growth of Spherical Crystals	ED11	Atmospheric Heat Absorption*
M561	Whisker-Reinforced Composites	ED12	Volcanic Study*
M562	Indium Antimonide Crystals	ED21	Libration Clouds*
M563	Mixed III-V Crystal Growth	ED22	Objects Within Mercury's Orbit*
M564	Metal and Halide Eutectics	ED23	UV From Quasars*
M565	Silver Grids Melted in Space	ED24	X-Ray Stellar Classes*
M566	Copper-Aluminum Eutectic	ED25	X-Rays from Jupiter*
		ED26	UV From Pulsars*
	ENGINEERING AND TECHNOLOGY	ED31	Bacteria and Spores**
		ED32	In Vitro Immunology**
D008	Radiation in Spacecraft	ED41	Motor Sensory Performance**
D024	Thermal Control Coatings	ED52	Web Formation**
M415	Thermal Control Coatings	ED61	Plant Growth**
M487	Habitability/Crew Quarters	ED62	Plant Phototropism**
M509	Astronaut Maneuvering Equipment	ED63	Cytoplasmic Streaming**
M516	Crew Activities/Maintenance	ED72	Capillary Study**
T002	Manual Navigation Sightings	ED74	Mass Measurement**
T003	In-Flight Experiment Aerosol Analysis	ED76	Neutron Analysis**
T013	Crew/Vehicle Disturbances	ED78	Liquid Motion in Zero "G"**
T020	Foot-Controlled Maneuvering Unit		
T025	Coronagraph Contamination Measurements	*	Data Only
T027	Contamination Measurements	**	Hdwe. Fab. Required

Before NASA can embark on future major manned programs of exploration and orbital operations, man's ability and usefulness in space over long periods must be assured. This can be done ultimately only through a careful quantitative study of man's physiological, psychological and social adjustments as they occur in flight. Overall limitations of the crew at a given time during flight must be established, and an accurate time profile of the adaptation of men to space conditions must be developed. It must be determined whether the long-term adjustments a man makes in space eventually lead to a new stable level or continual adjustments cause him to eventually exceed his reserve capacity for meeting stress. Even if man does successfully adapt to space conditions, the return to Earth involves an additional adaptive change about which more must be learned.

The Skylab Program offers the first opportunity to study these questions in depth. The 28- and 56-day missions are long enough to study acute effects which could threaten man's safety as well as to observe slower biological processes. The biomedical experiments for the Skylab Program have been designed to study the suspected changes and to investigate the basic mechanisms involved in these changes. The experimental investigations are much more comprehensive than previous medical safety monitoring techniques. The latter function will be performed operationally by known and fully tried bioinstrumentation and medical techniques and procedures.

The Skylab medical program is an intensive study of normal, healthy men and their reactions to the stresses of space flight. Seldom has such a comprehensive examination been performed in ground-based studies, and never under the unusual stresses of prolonged space flight. In addition, by preparing for and conducting these multi-man extended missions, advances in Earth-based medical applications in such areas are non-invasive bio-sensors, continuous long term monitoring of physiological processes and biotelemetry will have a significant impact on medical diagnosis and treatment.

A basic set of biomedical data has been collected as a safety monitoring procedure on all the manned flights of the Mercury, Gemini and Apollo programs. The parameters recorded have been heart and respiration rates, body temperature and blood pressure. These were supplemented by a variety of pre- and postflight measurements of such factors as exercise capability, cardiovascular stress response, hematological-biochemical changes, immunology studies and microbiological evaluations. In the Gemini program, medical experiments of limited scope were conducted in flight to investigate the time course of the changes which had been noticed on previous missions.

The following physiological effects of space flight on man have been observed: Loss of body weight; a small and inconsistent loss in bone calcium and muscle mass; and generally a reduction in orthostatic tolerance upon return to Earth.

These effects completely reversed themselves within a few days after return to Earth and so far have shown no consistent relation to flight duration (up to 14 days). However, some concern remains that continued effects in extended mission could significantly reduce man's effectiveness in space and increase the danger of re-adapting to the gravity conditions on Earth.

Each manned mission in the United States space program was built upon the cumulative experience of preceding flights. Skylab will fly a total of nine men, three at a time, in a larger spacecraft with more varied activities and for longer times than any previous flight. It will provide the test contitions under which the biomedical effects observed to date can be studied more extensively than has previously been possible.

The Skylab biomedical program consists of four parts:
> The actual stay of nine men in space, with the associated operational medical monitoring and the observations of crew performance in a wide variety of scientific and operational tasks.

The medical experiments designed to investigate in depth the physiological effects and their time courses which were revealed in previous flights.

The biology experiments designed to study fundamental biological processes affected by the space environment.

The biotechnology experiments directed toward advancing the effectiveness of man-machine systems in space operations and improving the technology of space-borne bio-instrumentation.

The knowledge and experience gained from all four parts of the program will be used to establish criteria for incremental increases in the duration of manned missions after the 28- and 56-day Skylab flights.

Mineral Balance (M071)

The purpose of this experiment is to collect data for a predictive understanding of the effects of space flight on the muscle and skeletal systems by measuring the day-to-day gains or losses of pertinent biochemical constituents.

The data to be collected in support of M071 are: Daily body weight, accurate food intake (quantity and composition), accurate fluid intake, volume of 24-hour urine output, samples of pooled 24-hour urine output, determine the mass, process and store all feces and vomitus, if any, (all collected and processed inflight for return and postflight analysis), and preflight, inflight and postflight blood samples taken for analysis.

Urine will be analyzed for calcium, phosphorus, magnesium, sodium, potassium, chlorine, nitrogen, urea, hydroxyproline and creatinine. Feces will be analyzed for calcium, sodium, phosphorus, magnesium, potassium and nitrogen. Blood will be analyzed for calcium, phosphorus, magnesium, alkaline phosphotase, sodium, potassium, total protein, glucose, hydroxproline, creatinine, chloride and electrophoretic pattern.

All hardware used in M071 is a part of other systems and will be described in the appropriate sections. Hardware used in this experiment and the systems of which they are a part includes:

Urine Measurement and Collection System (a part of the Habitability Support System).
Fecal Collection System (a part of the Habitability Support System).
Specimen Mass Measurement Device (a part of M074).
Body Mass Measurement Device (a part of M172).
Food system.
Inflight blood collection equipment.

The Principal Investigator is Dr. G. D. Whedon, National Institute of Health, Co-investigator is Dr. Leo Lutwak, Cornell University, and principal coordinating scientist is Dr. Paul C. Rambaut, Manned Spacecraft Center. (Development center MSC, integration center MSFC.)

Bio-Assay of Body Fluids (M073)

The purpose of this experiment is to assess the effect of space flight on endocrine-metabolic functions including fluid and electrolyte control mechanisms (Figure III-1).

The data to be collected in support of M073 are: Daily body weight, accurate food intake (quantity and composition), accurate fluid intake, volume of a 24-hour urine output, samples of pooled 24-hour urine output (collected and processed inflight for return and postflight analysis), and preflight, inflight and postflight blood samples taken for analysis.

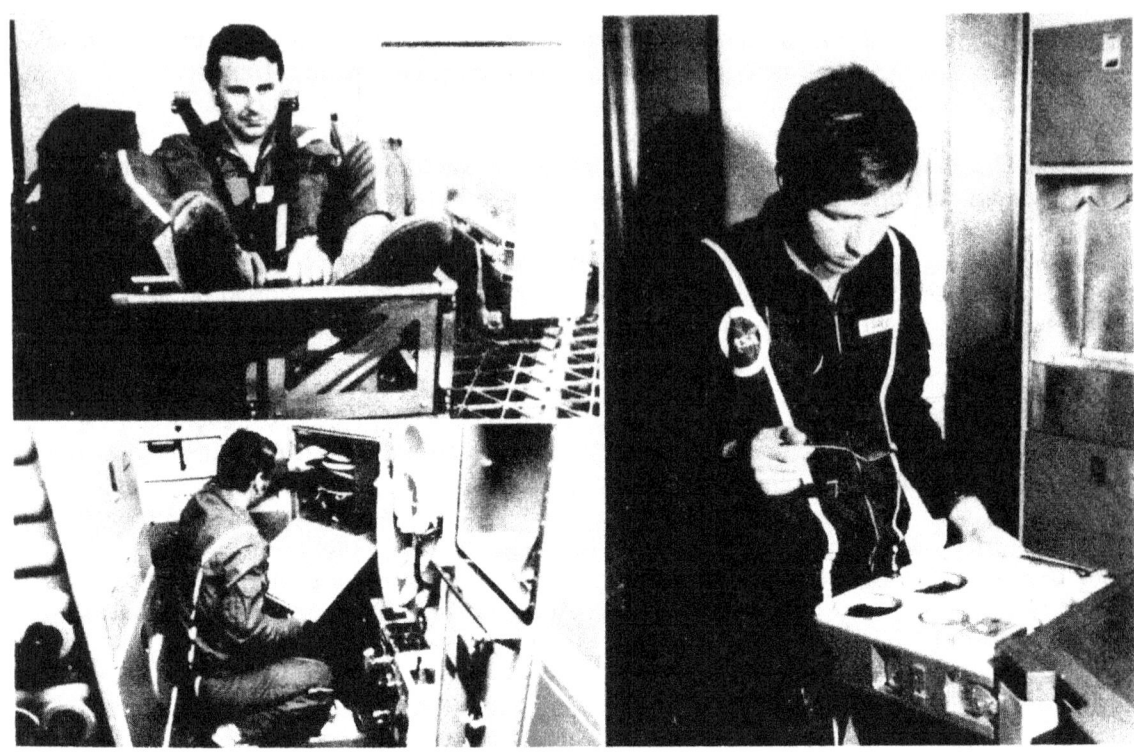

Figure III-1 — M071-M073 Experiments

Urine will be analyzed for sodium, potassium, aldosterone, epinephrine, norepinephrine, antidiurectic hormones (ADH), urine osmolality, hydrocortisone, total body water and total and fractional Ketosteroids. Blood will be analyzed for renin, sodium, potassium, chloride, plasma osmolality, extracellular fluid volume (ECF), parathyroid hormone, thyrocalcitonin, thyroxine, adrenocorticotropic hormone (ACTH), hydrocortisone and total body water.

All hardware used in M073 is a part of other systems. Hardware used in this experiment, along with the systems of which they are a part, includes:

Urine Measurement and Collection System (a part of the Habitability Support System).

Specimen Mass Measurement (a part of M074).

Body Mass Measurement (a part of M172).

Principal investigator is Carolyn S. Leach, PhD, Manned Spacecraft Center. (Development center MSC, integration center MSFC.)

Specimen Mass Measurement (M074)

The primary purpose of this experiment is to demonstrate the capability of accurately weighing 50 to 1,000 gm masses in a null gravity environment. The secondary purpose is to provide a means of accurately determining the mass of feces, vomitus and food residue generated in flight. These mass measurements provide data for M071 and M073 experiments.

The data to be collected in support of M074 are: Preflight calibration of the Specimen Mass Measurement Device (SMMD), measurement of known masses three times during each Skylab mission, and Skylab environmental temperature. The SMMD will also be used to determine the mass of feces, vomitus and food residue.

The SMMD is a device that utilizes the inertial property of mass in lieu of a gravity field to determine mass (Figure III-2). Basically, the SMMD consists of a spring mounted tray. The oscillatory period of the spring is a function of the amount of mass on the tray. The spring's period is measured electro-optically, and this measurement is electronically converted to a direct mass read out on board.

Principal investigator is Dr. William E. Thornton, astronaut at the Manned Spacecraft Center, and co-investigator is Col. John W. Ord, Medical Corps, USAF Hospital, Clark AF Base, Philippine Islands. (Development center MSC, integration center MSFC, contractor Southwest Research Institute.)

Bone Mineral Measurement (M078)

The purpose of this experiment is to assess the effects of the spaceflight environment on the occurrence and degree of bone mineral changes.

The data to be collected are preflight and postflight Gamma-Ray scans of the heel bone and right radius of the forearm. Comparison of these Gamma-Ray scans will give a comparison of bone density (a measure of bone calcification) before and after flight (Figure III-3).

No inflight equipment is required because the experiment is to be conducted preflight and postflight only.

Principal investigator is Dr. J. M. Vogel, U.S. Public Health Service, San Francisco, Calif., and co-investigator is John R. Cameron, PhD, of the University of Wisconsin Medical Center. (Development center MSC.)

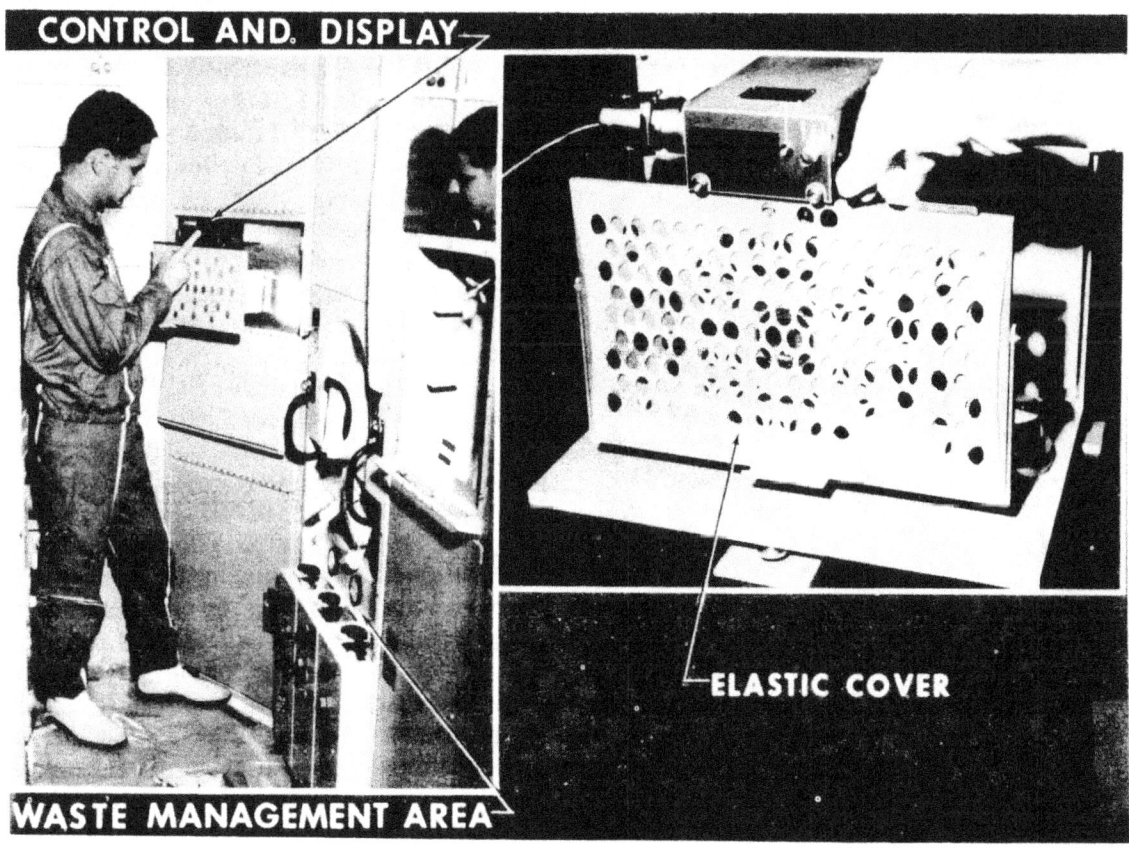

Figure III-2 — M074 Experiment

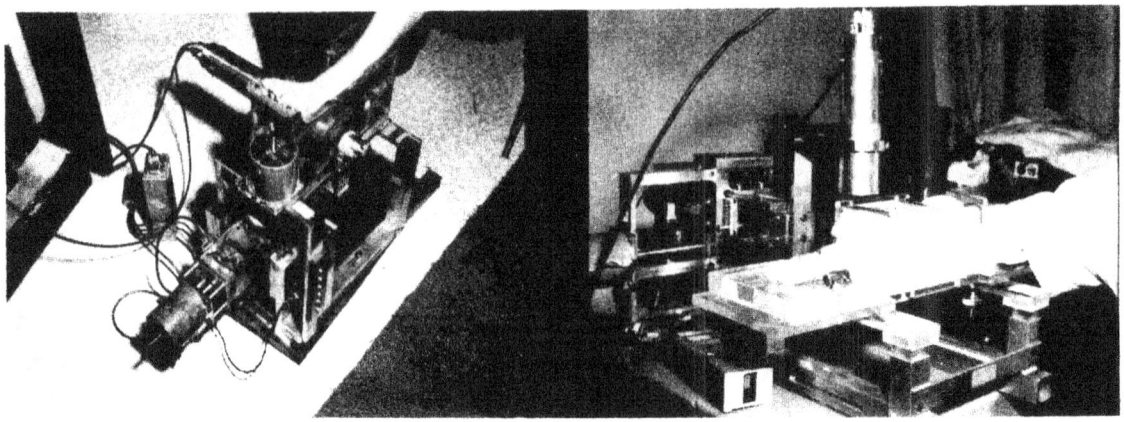

Figure III-3 — M078 Experiment

Lower Body Negative Pressure (M092)

The purpose of this experiment is to provide information concerning the time course of cardiovascular adaptation during flight, and to provide inflight data for predicting the degree of orthostatic intolerance and impairment of physical capacity to be expected upon return to Earth environment.

The data to be collected in support of M092 are blood pressure, heart rate, body temperature, vectorcardiogram, LBNPD pressure, leg volume changes, heart rate and body weight.

The Lower Body Negative Pressure Device (LBNPD) (Figure III-4) consists of a cylinder which encloses the lower half of the subject. A diaphragm forms an air seal around the subject's waist. Provisions are made to lower the pressure in the cylinder thus expose the lower body to a series of negative pressures. This negative pressure simulates the effects of the normal hydrostatic pressure of the blood in the cardiovascular tree, or a person standing erect on a one-g field.

Two Leg Volume Plethysmographs (one for each leg) are required. These devices are capacitance gauges for measuring the expansion of the legs on exposure to the LBNPD vacuum. The amount of expansion is a measure of the amount of blood pooling in the legs.

The Blood Pressure Assembly consists of a pressure cuff affixed around the upper arm, a microphone to pick up the Korotkoff's sounds, and signal conditioners. These interface with a programming unit to cycle the pressure cuff automatically, the electronic circuitry for blood pressure decisions and displays, and calibration circuitry in the Experiment Support System (ESS).

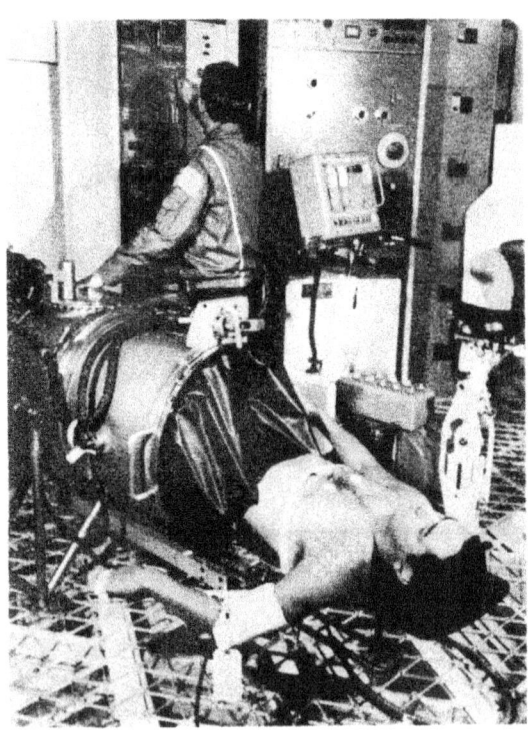

Figure III-4 — M092 Experiment

Principal investigator is Dr. Robert L. Johnson, Manned Spacecraft Center, and co-investigator is Col. John W. Ord, Medical Corps, Clark Air Force Base. (Development center MSC, integration center MSFC, contractors MSFC and Martin-Marietta Aerospace.)

Vectorcardiogram (M093)

The purpose of this experiment is to measure the vector cardiographic potentials of each astronaut during the preflight, inflight and postflight periods so that flight-induced changes in heart function can be detected.

The VCG will also be used to determine crewmen heart rates during the M092 and M171 experiments.

The data to be collected in support of M093 is a readable VCG taken at regular intervals throughout the mission while the crewmen are at rest and before, during and after specific exercise periods.

The VCG system consists of an eight-electrode input harness, a Frank Lead Network, calibration and timing circuits, and three ECG signal conditioner channels. The VCG system (Figure III-5) presents three normalized, amplified ECG signals and an analog heart rate signal to the spacecraft telemetry system. A bicycle ergometer (part of the M171 system) will be used to provide a specific exercise profile with which to compare the VCG data.

Principal investigator is Dr. Newton W. Allebach, USN Aerospace Medical Institute, Pensacola, Fla., and co-investigator is Dr. Raphael F. Smith, School of Medicine, Vanderbilt University. (Development center MSC, integration center MSFC, contractor Martin-Marietta Aerospace.)

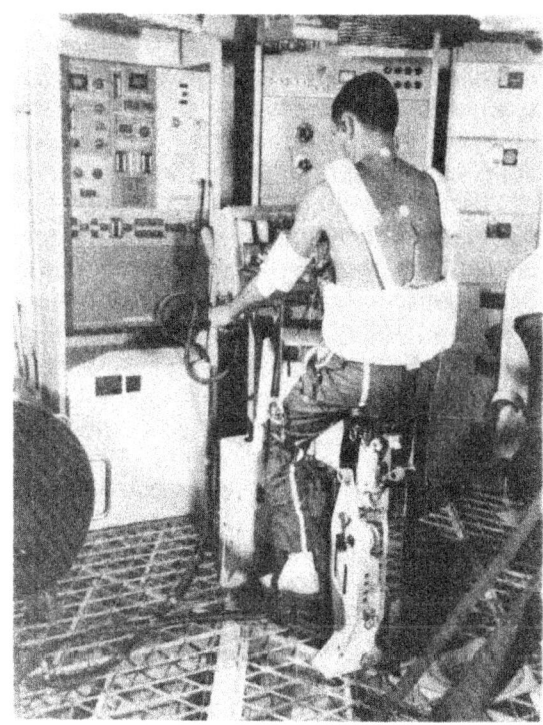

Figure III-5 — M093 Experiment

Cytogenetic Studies of the Blood (M111)

The objectives of this experiment are to make preflight and postflight determinations of chromosome aberration frequencies in the peripheral blood leukocytes of the Skylab flight crewmen and to provide "in vivo" radiation dosimetry. Another objective is to acquire data that will add to the findings of other Skylab cytologic and metabolic experiments to determine the genetic consequences of long duration space travel on man.

Periodic blood samples will be taken before and after the flight, beginning one month before and ending three weeks after recovery. The leukocytes will be placed in a short terminal culture. During the first cycle of mitotic activity in the "in vitro" cultures, standard chromosome preparations of the leukocytes will be prepared.

The leukocytes from the cell culture will be removed during metaphase and "fixed." A visual analysis will be performed which involves counting the chromosomes, the number of breaks, the types where possible, and then making a comparison between the identifiable chromosome forms with groups of chromosomes comprising the normal human complement.

The data from this experiment will consist of the chromosome aberration frequencies which appear postflight for nine men. An estimate of the radiation dose experienced by each man will be made based on the number of chromosome breaks.

Principal investigator is Dr. Lillian H. Lockhart of the University of Texas Medical Branch. Co-investigator is P. Carolyn Gooch of Brown & Root-Northrop. (Development center MSC.)

Man's Immunity – in Vitro Aspects (M112)

The objective of this experiment is to assay changes in humoral and cellular immunity as reflected by the concentrations of plasma and blood cell proteins, blastoid transformations and synthesis of ribonucleic (RNA) and desoxyribonucleic (DNA) acids by the lymphocytes.

The experiment will obtain preflight baselines (21, 7 and 1 days before launch), which will be indications of normal metabolism, from the crewmen and a ground control group composed of three men physically similar to the crewmen. The group will serve as ground controls during the flight. Inflight blood samples will be taken four times from each crewman during the SL-2 mission and eight times from each crewman during the SL-3 and SL-4 missions. Seven days and 21 days after recovery, samples will be taken from each crewman. Information will be compared with prelifght baselines, inflight profiles and with data from the control group to detect any significant deviations.

Blood will be analyzed for kinetics of lymphocyte RNA and DNA, RNA and DNA distribution in lymphocytes, observation of blastoid formation, lymphocyte morphology and antigen response, lymphocyte functional response to antigen, quantitation of plasma constituents, presence of immunoglobulins, albumin and globulin concentration and total plasma protein. The Inflight Blood Collection System will provide the capability to draw venous blood and centifuge the samples for preservation. The blood samples are frozen and returned to Earth for postflight analysis.

Principal investigator is Dr. Stephen E. Ritzmann and co-investigator is Dr. William C. Levin, both of University of Texas Medical Branch. (Development center MSC, integration center MSFC, contractor MDAC-ED.)

Blood Volume and Red Cell Life Span (M113)

The objective of this experiment is to determine the effect of Earth orbital missions on the plasma volume and the red blood cell populations with particular attention paid to the changes in red cell mass, red cell destruction rate, red cell life span and red cell production rate.

This experiment has four parts; in each, a different radioisotope tracer will be injected into crewmen's veins and into a control group with similar physical characteristics on the ground.

The site of red blood cell (RBC) production in the mature adult is the marrow of membranous bones (e.g. sternum and vertebrae) with the rate of production dependent on metabolic demands and the current red cell population. The rate of RBC production will be measured quantitatively by injection of a known quantity of a radioactive iron tracer into crew members.

Since the rate of RBC production acts with RBC loss to increase or decrease the total RBC mass present at a given time, any changes in the rates of RBC production and destruction will be necessarily reflected in the red cell mass. Such changes in red cell mass will be measured and analyzed in the flight crew members by injection of radioactive chromium (in the form of sodium chromate) tagged red cells.

To determine the selective age dependent erythrocyte destruction and mean red cell life span, carbon fourteen labelled glycine will be injected into a superficial arm vein of each crew member and control subject.

Finally, plasma volume changes will be measured by adding a known amount of radioiodinated human serum albumin to each crew member's blood.

Blood samples are frozen and returned to Earth for postflight analysis.

Blood samples of each crewman will be taken preflight (21, 20, 14, seven and one days before launch), inflight (four times for SL-2, eight times for SL-3) and post-flight (recovery day, one, three, seven, 14 and 21 days after recovery).

Principal investigator is Dr. Phillip C. Johnson, Jr., Baylor University College of Medicine. (Development center MSC, integration center MSFC, contractor MDAC-ED.)

Red Blood Cell Metabolism (M114)

The objective of this experiment is to determine if any metabolic and/or membrane changes occur in the human red blood cell as a result of exposure to the space flight environment.

This experiment will assess the influence of the space flight environment on the metabolic processes which support crewmen's erythrocytes. The experiment is designed to complement Experiment M113.

Blood samples of each crewman will be taken preflight (21, seven and one days before launch), inflight (four times during SL-2, eight times during SL-3), and postflight (recovery day, one and 14 days after recovery).

Blood will be analyzed for methemoglobin, glyceraldehyde-6-phosphate dehydrogenase, phophglyceric acid kinase, reduced gluthathione, adenosine triphosphate, glutathione reductase, lipid peroxide levels, acetylcholinestecase, phosphofructokinase, 2, 3-diphosphoglycerate, and hexokinase.

Blood samples are frozen and returned to Earth for postflight analysis.

Principal investigator is Dr. Charles E. Mengel, University of Missouri School of Medicine. (Development center MSC, integration center MSFC, contractor MDAC-ED.)

Special Hematologic Effect (M115)

The primary objective of this experiment is to examine critical physiochemical blood parameters relative to the maintenance of a stable state of equilibrium between certain blood elements and to evaluate the effects of space flight on these parameters. A secondary objective is to provide essential data on blood which will assist in interpreting other hematology/immunology experiments.

Blood studies made on Gemini and Apollo astronauts have shown that changes in red cell mass, blood constituents and in the fluid and electrolyte balance can be expected as a result of the space environment.

Blood samples of each crewman will be taken preflight (21, 14, seven and one day before launch), inflight (four times on SL-2, eight times on SL-3), and postflight (recovery day, one, three, seven, 14 and 21 days after recovery).

Blood will be analyzed for sodium, potassium, single cell hemoglobin, red blood cell hemoglobin, RNA, protein distribution, hemoglobin characterization, electrophoretic mobility, red blood cell age profile, red blood cell electrolyte distribution, membrane and cellular ultrastructure, acid and osmotic fragility, critical volume, volume distribution, red blood cell count, white blood cell count, differential white cell count, micro hematocrit, platelet count, hemoglobin, and reticulocyte count.

Blood samples are frozen and returned to Earth for postflight analysis.

Principal investigator is Stephen L. Kimsey, PhD, MSC, co-investigator is Dr. Craig L. Fischer, Eisenhower Memorial Hospital. (Development center MSC, integration center MSFC, contractor MDAC-ED.)

Human Vestibular Function (M131)

The purpose of this experiment is threefold: To test the crew's susceptibility to motion sickness in the Skylab environment; to acquire data fundamental to an understanding of the functions of human gravity receptors in the prolonged absence of gravity; and to test for changes to the sensitivity of the semicircular canals.

The data to be collected in support of M131 are: Threshold perception of rotation; motion sickness symptoms caused by out-of-plane head motions while being rotated; and ability of crewman to determine his orientation with respect to spacecraft reference points with visual cues. Data will be collected before, during and after flight.

The inflight equipment used for M131 includes (Figure III-6):

Rotating Litter Chair (RLC) — This chair is a framed seating device which is convertible for operation in either a rotating or tilt litter mode.

Drive Motor for Chair Rotation — This motor has the capability of rotating the seated subject within the limits of 1 to 30 rpm and with an accuracy of ±1%.

Control Console — The console contains mode selector, speed selector, tachometer, indicators, timers, and other deivces for control and a response matrix for coding a subject's response to the rotational tests.

Otolith Test Goggle — This device is used to measure the visual space orientation in two dimensions. It provides the visual target for the oculogyrol illusion test.

Custom Bite Boards — The bite boards are used to hold the otolith test goggle precisely and confortably in position over the observer's eyes.

Reference Sphere and Magnetic Pointer/Readout Device — These

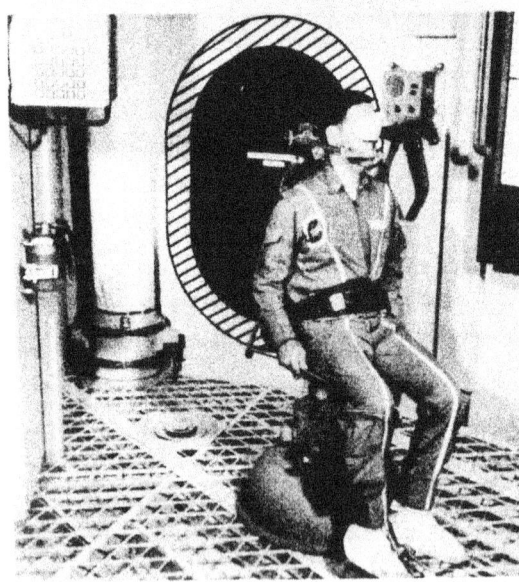

Figure III-6 — M131 Experiment

devices are used for measiring spatial localization using nonvisual clues. A magnetic pointer is held against the sphere and moved by the subject to determine the subject's judgments of his orientation. The position is measured by the three-dimensional readout device.

Principal investigator is Dr. Ashton Graybiel and co-investigator is Dr. Earl Miller, both of Naval Aerospace Medical Research Laboratories, Pensacola, Fla. (Development center MSC, integration center MSFC, contractor Applied Physics Laboratory.)

Sleep Monitoring (M133)

The purpose of this experiment is to evaluate objectively the quantity and quality of inflight sleep through an analysis of electroencephalographic (EEG) and electro-oculographic (EOG) activity.

It has been demonstrated that disrupted patterns of sleep are associated with modifications in performance capabilities. Therefore, objective investigative data regarding sleep in the space environment is of practical significance in terms of learning more of man's capabilities and limitations in the performance of space missions.

The data to be collected in support of M133 are: Preflight baseline EEG and EOG data on a crewman for three consecutive nights of sleep; periodical inflight EEG and EOG data throughout a crewman's sleep period; and postflight sleep EEGs and EOGs on approximately one, three and five days after recovery.

The M133 equipment (Figure III-7) consists of a wholly self-contained device which will record the EEG and EOG on magnetic tape and also provide TM data in near real time. Electrodes are incorporated into a fitted cap that crewmen will wear during sleep periods to detect the EEG and EOG signals. The cap is also fitted with accelerometers to detect head movement to keep this movement from influencing the sleep-stage determination systems.

The EEG and EOG tapes are returned at the conclusion of the mission for postflight analysis.

Principal investigator is Dr. J. D. Frost, Jr., Baylor School of Medicine, Houston, Tex. (Development center MSC, integration center MSFC, contractor Martin-Marietta Aerospace.)

Time and Motion Study (M151)

The purpose of M151 is to study the adaptation of man in prolonged periods of zero gravity by comparing, through use of time and motion determination, identical activities performed by astronauts during ground-based training and inflight (Figure III-8).

The only inflight data required for this experiment is to have the crew photograph themselves as they perform flight-planned activities required of them during execution of the Skylab missions. Films made by the crewmen will be returned and analyzed postflight. These data are compared to baseline data taken during crew training.

Flight-planned tasks selected for analysis will satisfy the following functional objectives:

> To study the locomotion of crewmen as they translate in the zero gravity environment with and without "loads."
> To study the fine and gross motor activities of crewmen in performing operations with and without the use of restraints.
> To study crewmen performing tasks which require visual, tactile, auditory feedback or combinations thereof.
> To study IVA and EVA activities.
> To study repeated activities performed early, middle, and late in the missions which will show adaptation to the zero gravity environment.

Principal investigator is Joseph F. Kubis, PhD, of Fordham University, and co-investigator is Edward J. McLaughlin, PhD, of NASA-OMSF. (Development center MSC, integration center MSFC.)

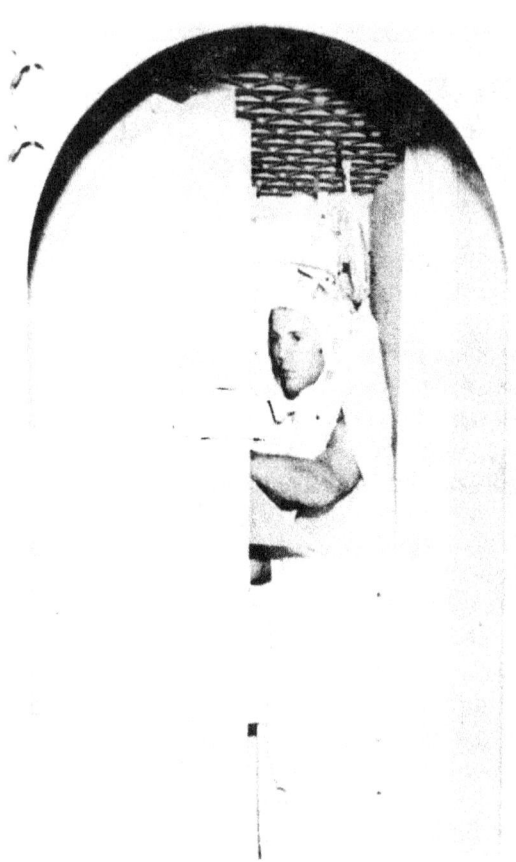

Figure III-7 — M133 Experiment

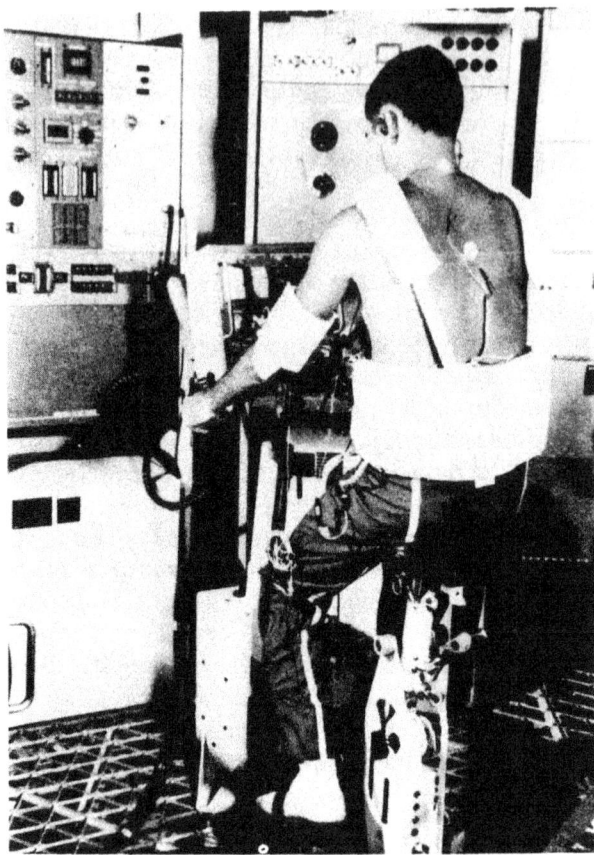

Figure III-8 — M151 Experiment

Metabolic Activity (M171)

The primary purposes of this experiment are to determine if man's metabolic effectiveness in doing mechanical work is progressively altered by exposure to the space environment. The secondary purpose is to evaluate the bicycle ergometer as an exerciser for long duration missions.

The data to be collected in support of M171 are: Ergometer workrate, Ergometer RPM, oxygen uptake, carbon dioxide output, minute volume, vital capacity, respiratory quotient, heart rate, blood pressure, vectorcardiogram, body weight, body temperature and Skylab environmental parameters.

A metabolic analyzer is used to measure oxygen uptake, carbon dioxide output and minute volume. These determinations are made through the use of a spirometer and mass spectrometer.

The bicycle ergometer is used to provide a calibrated work load. It will have an adjustable load selector and tachometer (Figure III-9).

Principal investigator is E. L. Michel and co-investigator is Dr. J. A. Rummel, both of Manned Spacecraft Center. (Development center MSC, integration center MSFC, contractors MSFC, Martin Marietta Aerospace and Perkin Elmer.)

Body Mass Measurement (M172)

The Body Mass Measurement experiment will demonstrate body mass measurement in a null gravity environment, validate theoretical behavior of this method, and support those medical experiments for which body mass measurements are a requirement.

The data to be collected in support of M172 are: Preflight calibration of BMMD and measurement of known masses up to 100 kilograms (220 pounds) three times during each Skylab mission. The BMMD will also be used for daily determination of the crewmen's weight, which will be manually logged and voice recorded for subsequent telemetered transmission.

The self-contained instrument (Figure III-10) consists of a spring/flexure pivot mounted chair, the method of operation being similar to that for the Small Mass Measurement (M074). A photo optical pickup is mounted beneath the chair to determine zero crossing, and its output is timed and converted to a direct mass readout.

Principal investigator is Dr. W. E. Thornton of Manned Spacecraft Center. (Development center MSC, integration center MSFC, contractor Southwest Research Institute.)

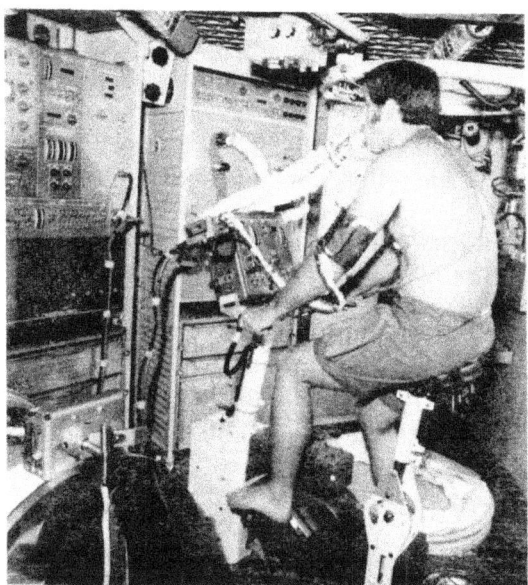

Figure III-9 — M171 Experiment   Figure III-10 — M172 Experiment

Effect of Zero G on Single Human Cells (S015)

The purpose of this experiment is to determine the effect of zero gravity on living human cells in a tissue culture.

The data to be collected in support of S015 are: Time lapse cinematographic films of inflight tissue culture growth and crew logs. Also, the tissue cultures will be returned for postflight analysis.

The flight hardware for Experiment S015 consists of a microscope-camera assembly and a growth curve module subsystem, both enclosed in a single hermetically sealed package.

Microscope-Camera Assembly — Two phase-contrast microscopes, one 20 power and one 40 power, each focused on its own specimen chamber, supply a magnified image to the two 16 mm time-lapse cameras. The cameras are automatically cycled by a built-in timing mechanism; each runs at 5 frames per minute for 40 minutes twice per day for the entire 28 day mission of SL-2.

A separate specimen chamber for each microscope provides temperature controlled environments in which to grow the cell cultures. Each chamber has its own independent media exchange assembly to provide fresh nutrients to the cultures twice each day.

Growth Curve Module — This system consists of two operationally independent assemblies, each capable of maintaining living cells in nine chambers. At preprogrammed times a fixative will be injected into eight chambers, one at a time. The cells will be returned for post-flight analysis.

Principal investigator is Dr. P. O. Montgomery and co-investigator Dr. J. Paul, both Dallas (Tex.) County Hospital, Dr. P. Kruse, Jr., Noble Foundation, and Dr. L. Hayflick, Stanford University. (Development and integration center MSC, contractor Dallas County Hospital.)

Circadian Rhythm — Pocket Mice (S071)

The purpose of the experiment is to determine if the daily physiological rhythms of mammals are altered in space flight. If the rhythms are changed or affected, the indication is that physiological rhythms are timed by some factor not found in space. Changed or affected rhythms alter the basic control of metabolism. It is important that normal biological rhythms in man be maintained for his well being and effectiveness during space flight. If normal physiological rhythms are maintained, the conclusion can be made that space flight does not impose bio-rhythm restriction and that man can work in space without degrading his performance due to bio-rhythm disturbance.

The experiment consists of six pocket mice placed in a completely dark cage having 15°C (60°F) temperature, relative humidity of 60% and an atmospheric pressure equivalent to sea level.

Three weeks before the mission, the mice are placed in the cage. Body temperature and activity level are automatically monitored to establish the natural period, phase and stability of the rhythms. The cage is installed in the Service Module prior to the launch. The same measurements are made during the flight. Data is automatically recorded and telemetered to Earth for interpretaiton.

Principal investigator is Robert G. Lindberg, PhD, of Northrop Corporate Laboratories. (Development center is Ames Research Center, integration center MSC, contractor is Northrop Electronics.)

Circadian Rhythm — Vinegar Gnat (S072)

The purpose of the experiment is to determine if the daily emerging cycle of the vinegar gnat (drosophila) pupae is altered during space flight.

Extensive experiments have shown that even though gnats in the pupal stage develop at different rates depending on temperature, they will not emerge from the pupae as adult gnats until some kind of internal signal is given off. This triggering signal is somehow timed to occur at an exactly fixed time delay after a flash of light, and it occurs at the same daily time interval thereafter, regardless of the temperature.

The experiment will measure the emergence times of four groups at 20°C (68°F) to find out whether space flight conditions change the mechanism which keeps the rhythm constant despite changes in temperatures.

Each group is divided by one-half. A synchronizing flash of light is used to initiate the pupae at two different times.

If the delayed group shows the same rhythms of emergence response as the earlier group, probably no external factor contributes to the rhythm and the rhythms are internally synchronized.

This experiment is conducted in conjunction with the pocket mice experiment (S071). If rhythms of both experiments are disrupted or altered during space flight, it can be assumed that space flight disrupts or alters the common basic rhythm mechanisms, and man's biological rhythm mechanism is probably affected.

Experiment S072 is an automatic operation. After the initial white light flash is used for synchronizing the pupae, a dim red light is turned on every 10 minutes and 180 pupae photocells are scanned electronically. If a gnat has emerged the pupa is transparent and its photocell activated.

Emergence data is stored and later telemetered to Earth.

Principal investigator is Colin S. Pittendrigh of Stanford University. (Development center is Ames Research Center, integration center MSC, contractor Northrop Electronics.)

Experiment Support System

The Experiment Support System (ESS) (Figure III-11) provides the maximum practical support of common and special requirements of Medical Experiments M092, M093, M131 and M171. The system concept is one of providing a central unit from which to provide efficient support to those experiments. The ESS provides regulated power, spacecraft power, controls and displays, data management, event time, pressurized gas sources, and calibration commands. The necessary support hardware is integrated into a frame assembly suitable for mounting in the OWS near the experiments. Other hardware within the ESS frame assembly are the M092 blood pressure measurement system module and the plethysmograph electronics, and the M093 vectorcardiogram module.

The major subassemblies are the ESS Frame Assembly, Distributor Assembly, two Power Supply Assemblies, Power Panel Assembly, Meter Panel Assembly, Experiment Control Panel Assembly, Pressure Transducer Assembly, Nitrogen plumbing, M092 Blood Pressure Assembly, M093 Blood Pressure Assembly, M093 VCG Electronics Module, and M092 LVMS Electronics.

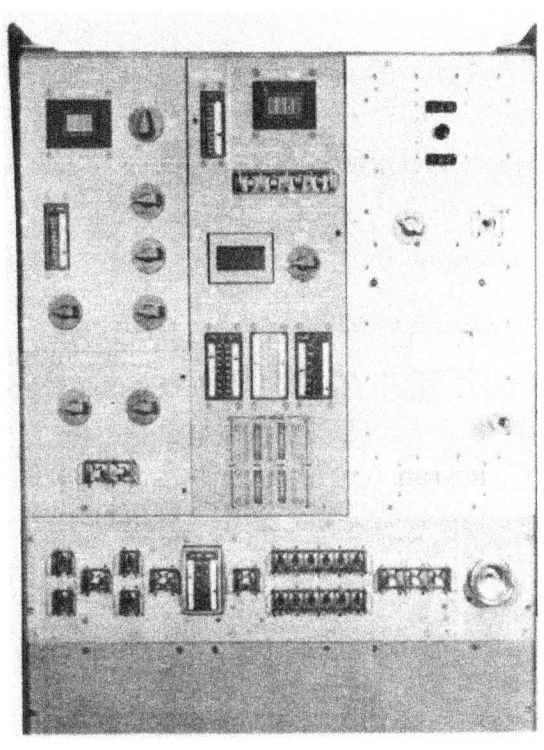

Figure III-11 — Experiment Support System

(Integration center and contractor MSFC, development center MSC.)

Medical Definitions

Aldosterone — The principal electrolyte-regulation steroid secreted by the adrenal complex (steroid is a group name for compounds that resemble cholesteral chemically, and some of the substances in this group include sex hormones and bile acids; check below for definition of electrolyte).

Angiotensin — A vessel-constricting substance present in the blood, and formed by the action of renin (enzyme involved in changing proteins in other products) on a globulin (a class of proteins characterized by being insoluble in water, but soluable in saline solutions.)

Antidiuretic hormone (vasopressin)	A hormone that suppresses the secretion of urine. Vasopressin also stimulates the contraction of the muscular tissue of the capillaries and arterioles.
Antigen	A substance, as a toxin, enzyme, or any of certain constituents of blood corpuscles or of other cells, which when introduced into the body stimulates the production of an antibody; with possible exceptions, antigens are proteins.
Bioassay	Determination of the active power of a sample of a drug by noting its effect on animals, as compared with the effect of a standard preparation.
Cardiotachometer	An instrument for counting the total number of heart beats over long periods of time.
Cardiovascular	Heart and vessel system.
Circadian rhythm	Pertaining to a period of about 24 hours. Applied especially to the rhythmic repetition of certain phenomena in living organisms at about the same time each day.
Cytogenetics	The branch of genetics devoted to the study of the cellular constituents which are concerned with heredity, that is, the chromosomes and genes. Also, clinically, it is the scientific study of the relationship between chromosomal aberrations (abnormal) and pathological conditions.
Dehydrogenase	An enzyme which mobilizes the hydrogen of a substrate (the base on which an organism lives) so that it can pass to a hydrogen acceptor.
Electrocardiogram	A photographic record of the heart's action, made by electrocardiograph, an instrument for recording the changes of electrical potential occurring during the heartbeat.
Electrolyte	An electric conductor in which passage of current is accompanied by liberation of matter at the electrodes; also, a substance, as an acid base, or salt, that becomes such a conductor when dissolved in a suitable solvent, or fused. The current is carried by charged particles, ions.
Enzyme	An organic compound, frequently a protein, capable of accelerating or producing by catalytic action some change in a substrate for which it is often specific. An activating enzyme activates a given amino acid by attaching it to the corresponding transfer of ribonucleic acid.
Epinephrine concentrations	Concentrations of adrenaline (compound occurring naturally as the adrenal hormone. In certain concentrations the compound causes an increase in blood pressure and in the sugar content of the blood.)

Ergometer	A device for measuring energy expended or work done.
Fibrin	A whitish, insoluble protein which forms the essential portion of the blood clot.
Fibrinolysis	The splitting up of fibrin by enzyme action.
Fibrinolytic	Pertaining to, characterized by, or causing fibrinolysis.
Glutathione	Co-enzyme of glyoxalase and acts as a respiratory carrier of oxygen.
Glyceraldehyde	A compound formed by the oxidation of glycerol, a mixture of glycerin and acetanilid powder.
Hematocrit	An instrument for determining the relative amounts of plasma and corpuscles in blood, generally some form of centrifugal apparatus.
Hemolysis	The dissolution of red blood corpuscles with liberation of their hemoglobin, an iron-containing protein respiratory pigment occurring in the red blood cells.
Hemostasis	Stagnation of blood, arrest of a hemorrhage, as by the tieing or binding of arteries.
Hexokinase	An enzyme that catalyzes the transfer of a high-energy phosphate group of a donor to D-glucose, producing D-glucose-6-phosphate.
Humoral	Pertaining to fluid or semifluid substances in the body.
Immunology	The medical, bacteriological, and chemical science treating of the phenomena and causes of immunity.
Kinase	An enzyme that catalyzes the transfer of a high-energy group of a donor to some acceptor in respect to organic compounds in the body.
Leukocyte chromosome	A leukocyte is any colorless amoeboid cell mass, applied especially to one of the formed elements of the blood, while chromosomes are rod-shaped bodies which appear in the nucleus of a cell at the time of cell division and contain the genes, or heridatary factors.
Metabolic changes	Changes in the processes concerned in the building up of protoplasm and its destruction incidental to life; or chemical changes in living cells, by which the energy is provided for the vital processes and activities, and new material is assimilated to repair the waste.

Methemoglobin	A soluble brownish-red, crystalline compound from which the oxygen cannot be removed in a vacuum. It is formed by the spontaneous decomposition of blood and also by the action on blood of various oxidizing reagents, as oxone, etc.
Morphology	Branch of biology dealing with the form and structure of animals and plants; the science of structural organic types; broadly, it includes anatomy, histology and organography, and also the nonphysiological aspects of cytology and embryology.
Norepinephrine	A hormone secreted by the adrenal medulla in response to stimulation in the viscera, and stored in granules that stain strongly with chromium salts. They are released predominantly in response to hypotension (diminished tension or lower blood pressure.)
Oculogyral illusion	An illusion developed by the movement of the eye about the anteroposterior axis.
Plasma osmolality	Osmolality is defined as a property of a solution, in this case plasma. Osmolality depends on the concentration of the solute (a dissolved substance) per unit of solvent (a substance capable of dissolving another substance).
Plasma renin	Plasma is the fluid portion of the blood in which the corpuscles are suspended. Renin is an enzyme involved in changing proteins into other products.
Plethysmograph	An instrument for determining and registering variations in the size of an organ part or limb and in the amount of blood present or passing through it for recording variations in the size of parts and in the blood supply.
Pyruvate kinase	Pyruvate is a salt or pyruvic acid, while kinase is an enzyme that catalyzes the transfer of a high-energy group of donor to some acceptor.
Reductase	An enzyme that has a reducing action on chemical compounds.
Roentgenologic visualization techniques	Using X-rays for medical diagnosis.
Thorax	That part of the body between the neck and the abdomen.
Thymidine	A compound isolated from the liver.
Thyroxin	The white, crystalline, active principle of the thyroid gland. It has been synthesized, and is used for goiter, cretinism, obesity, etc.

Urine osmolality	Osmolality is defined as a property of a solution, in this case urine, Osmolality depends on the concentration of the dissolved substance per unit of solvent (a substance capable of dissolving another substance).
Vectorcardiograph	An instrument for taking a graphic record of the magnitude and direction of the electrical forces of the heart.
Vestibule	A space or cavity at the entrance to a canal; for instance, a small space within the left ventrical at the root of the aorta.

SOLAR PHYSICS

The Sun is one of the most widely studied objects in the sky but much about it remains a mystery. Such questions as the origin of solar flares, the development and decay of active regions and the temperature characteristics of the corona remain. Until recently it was possible to observe solar emissions only at wavelengths which could penetrate the Earth's atmosphere — visible and radio emission. Vital ultraviolet and X-ray regions of the solar spectrum are cut off from Earth-bound viewing. Also, the daytime atmospheric scattering of visible light causes the sky to be much brighter than the solar corona. Thus rare solar eclipses are the only opportunities to view the extended solar corona.

The Sun is the ultimate source of all energy on the Earth, and all terrestrial life depends on it. The Sun is also the nearest star. An understanding of the stars depends on an understanding of the Sun. The Sun is an astrophysical laboratory close at hand. By using the Sun, studies can be made of atomic, nuclear and plasma physics, aerodynamics, hydrodynamics and magneto-hydrodynamics.

A better understanding of the following solar physics problems is awaiting use of the instruments in space to study the Sun:

How is the corona heated?
What is the nature of the atmospheric structural detail?
What do coronal streamers look like in space?
What is the relationship between these streamers and surface features?
What are solar flares?
How do active regions evolve?

Several basic features of the Sun's emission are of interest to solar physicists. The first is the hydrogen alpha emission, a red light emitted by the hydrogen gas present near the Sun's surface. The total visible emission of the Sun changes very slightly during solar flares; however, the hydrogen alpha emission is enhanced greatly and has been a fundamental feature of flare and active region studies for some time. The hydrogen alpha images will be available to the astronauts on Skylab to assist them in finding interesting portions of the solar disk to study with Skylab instruments. Two H-alpha telescopes will be on the ATM.

Several instruments will study a region of the solar atmosphere (chromosphere). This region is from about 800 to 9,650 kilometers (500 to 6,000 miles) about the Sun's surface. Temperature in this region increases from about 5,000 to 1,000,000°C. The much higher temperatures of the chromosphere give rise to the ultraviolet radiation. UV radiation will be studied in terms of particular emission features at specified wavelengths. These studies will reveal the types of atoms present in this region under various phases of solar activity and possibly shed some light on the mechanism which supplies the heat to this region. Three different UV detector experiments are on the ATM and an additional UV detector is aboard for operation through an airlock.

The solar corona begins about 9,650 kilometers (6,000 miles) above the surface and continues far into space. The density of matter in the corona is quite low, but temperatures vary from almost a million degrees in quiet regions to tens of millions of degrees in certain regions during solar flares. These high temperatures cause the ions and electrons in the corona to radiate X-rays. X-ray telescopes on Skylab are equipped with cameras to photograph the X-ray corona of the Sun.

Because the corona is so large and hot, many free electrons are available to scatter the white visible light radiated from the surface of the Sun. This scattered light is very much weaker than the radiation from the solar surface. A coronagraph is an instrument capable of studying the faint corona without viewing the bright solar surface. Since the intensity of scattered light is a measure of the electron density in the corona, photographs taken out to 4.8 million kilometers (3,000,000 miles) from the surface will provide the first look at the corona over an extended period.

Experiment S020, X-Ray/UV Solar Photography, is the one experiment to be conducted using an airlock in the sunlit side of the OWS. All other solar physics experiments are mounted in the ATM.

X-Ray/Ultraviolet Solar Photography (S020)

The objective of S020 is to record on photographc film the detailed energy spectrum of X-ray and ultraviolet radiation from normal and explosive areas in the solar atmosphere.

The 10 to 200 Angstrom region of the solar spectrum is rich in emission lines of highly ionized atoms. Many of these emission lines are weak and require instruments of high sensitivity for their observation. S020 will take observations by taking advantage of the long exposure times and film return capability of the Skylab spacecraft.

The S020 spectrograph will be mounted in the solar SAL (Figure III-12) and will be operated at the same time the ATM experiments are in operation.

It is planned to obtain 10 preplanned exposures of the "quiet" Sun ranging in exposure from five to 60 minutes. Photography of the active Sun is to be scheduled when solar flare activity has been predicted. During this period, 10 exposures will be allocated varying in period from one to several minutes.

The spectral data associated with solar flares is of interest in solar flare prediction and the effects of solar flares on the quality of radio communications. The ability to accurately predict solar flares will be useful in planning manned space flights. The data from this experiment will also be useful to persons with interests in the fields of astrophysics and plasma physics.

Principal Investigator is Dr. Richard Toussey, U.S. Naval Research Laboratory, Washington, D.C. (Development center MSC, integration center MSFC, contractor Naval Research Laboratory.)

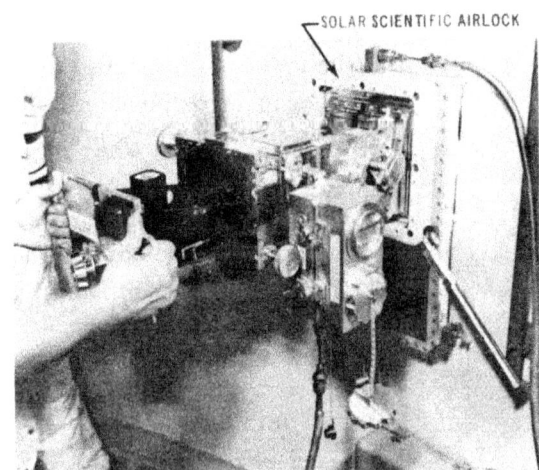

Figure III-12 – S020 Experiment

White Light Coronagraph (S052)

The White Light Coronagraph is designed to obtain photographs of the solar corona out to 3 million miles (six solar radii) in visible light. These photographs, taken twice daily or at rates up to one every 13 seconds during periods of high limb activity, will provide new information related to the rapidly moving matter, sometimes traveling at relativistic speeds, which is transported outward from the Sun due to solar events occurring at the solar limb. Systematic changes, to the extent that the corona will be apparent over several solar rotations, will allow correlations to be made with surface features as they move into proper position on the limb of the Sun. Existing Earth based coronagraphs are hampered because Earth sky brightness is much greater than the corona and prevents acquisition of detailed corona information.

The experiment is approximately 3 meters (10 feet) long, 0.46 meters (1.5 feet) in diameter, and weighs 142 kilograms (314 pounds). It includes an externally mounted disk system to occult (block) the brilliant solar surface to allow viewing of the faint corona radiation. The experiment data are presented on the TV system, which is displayed on the C&D video monitors or transmitted by telemetry to ground, or recorded on film (Figure III-13).

One loaded film camera, which contains 35mm film for approximately 8000 data frames, is placed in the experiment prior to launch, and three replacement film cameras are stored in the MDA. Film retrieval and replacement is accomplished by astronaut EVA from the center work station with exposed film being returned to Earth by CM for evaluaiton.

Dr. Robert MacQueen of the High Altitude Observatory, Boulder, Colo., is principal investigator. (Development and integration center MSFC, contractor HAO with subcontractor Ball Brothers Research Corporation.)

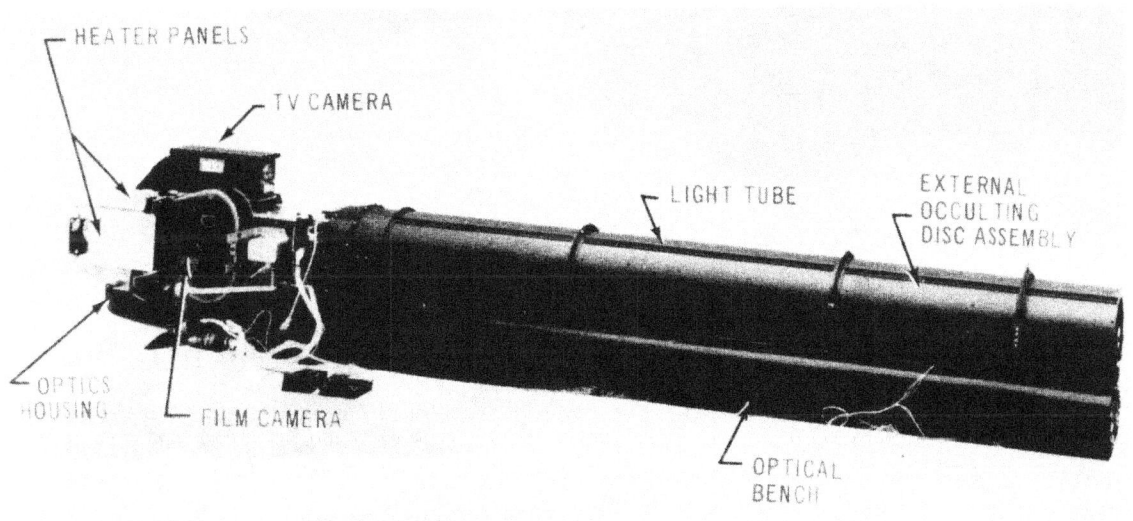

Figure III-13 — S052 Experiment

X-Ray Spectrographic Telescope (S054)

The X-Ray Spectrographic Telescope will take sequential photographs of X-ray producing events (flares and active regions) for determining corona temperatures and energetic particle densities. This information will be correlated with similar data from other experiments.

The X-ray telescope consists of two concentric mirrors of highly polished metal alloy to intercept the X-radiation and focus it at grazing incidence. Filters of beryllium, aluminized mylar and other materials with varying thickness will select the X-ray wavelength band to be photographed. A transmission grating will also be used in conjunction with the filters to obtain information on the spectral features of the X-ray emission.

The telescope weighs more than 136 kilograms (300 pounds) and is three meters (10 feet) long and almost half a meter (1.5 feet) in diameter (Figure III-14).

A small 7.6 centimeter (3 inch) grazing incidence instrument placed in the unused central portion of the larger telescope is used to provide a "live" picture of the Sun in X-rays for the astronaut to view. This aid will complement the H-Alpha images in TV and will assist the astronaut in getting the best possible data from the ATM.

Dr. Riccardo Giacconi of American Science and Engineering Inc., is principal investigator. (Development and integration center MSFC, contractor American Science and Engineering.)

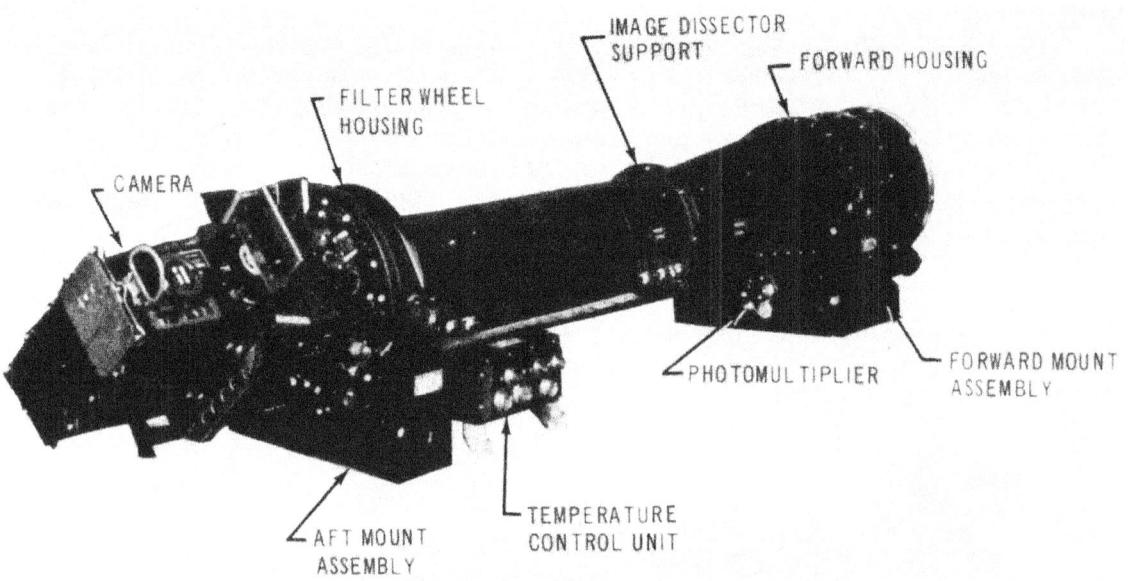

Figure III-14 — S054 Experiment

UV Scanning Polychromator Spectroheliometer (S055)

The purpose of this experiment is to observe temporal changes in the ultraviolet (EUV) radiation emitted by several types of solar regions. The EUV region of the solar spectrum is generated in the chromosphere and lower corona. The instrument operates photoelectrically and requires no film. All data are recorded electronically. The instrument is capable of accurately measuring the strength of certain emission features of elements with high time resolution in various stages of ionization. It observes seven emission lines in the wavelength region from 300 to 1,350 angstroms. Simultaneous observations of chromospheric and coronal layers of flares will be obtained, and the energy radiated in selected emission lines in the EUV region will be measured.

Radiation from the Sun enters the instrument and is reflected by a mirror which is movable along both axes. The mirror is adjusted to place the desired square segment of the solar surface on the spectrometer grating. The rest of the light is reflected back and out of the instrument. The radiation is broken up into its spectral components and the EUV portion is received by seven detectors. The eighth detector is the zero order position and, hence, sees light at all wavelengths. The zero order detector indicates whether or not the solar disk is being viewed.

The instrument is 3 meters (10 feet) long and 0.6 by 0.6 meter (2 by 2 feet) in cross section. It weighs 156.5 kilograms (345 pounds) (Figure III-15).

Dr. E. M. Reeves of Harvard College Observatory is principal investigator. (Development and integration center MSFC, contractor Harvard College Observatory with subcontractor Ball Brothers Research Corporation.)

Figure III-15 — S055 Experiment

X-Ray Event Analyzer/X-Ray Telescope (S056)

The objective of the S056 experiment (Figure III-16) is to gather solar radiation data in the X-ray region of the solar spectrum which will provide information regarding physical processes occurring within the solar atmosphere, with special emphasis on obtaining data of active solar phenomena such as solar flares. The evaluation of this information will lead to an increased knowledge of the influence of the Sun's magnetic field on flare development and a more definite understanding of the relationships between Sun spots and solar flare formation. The experiment consists of two separate and independently operated instruments, the X-Ray Event Analyzer and the X-Ray Telescope, which obtain complementary X-ray data.

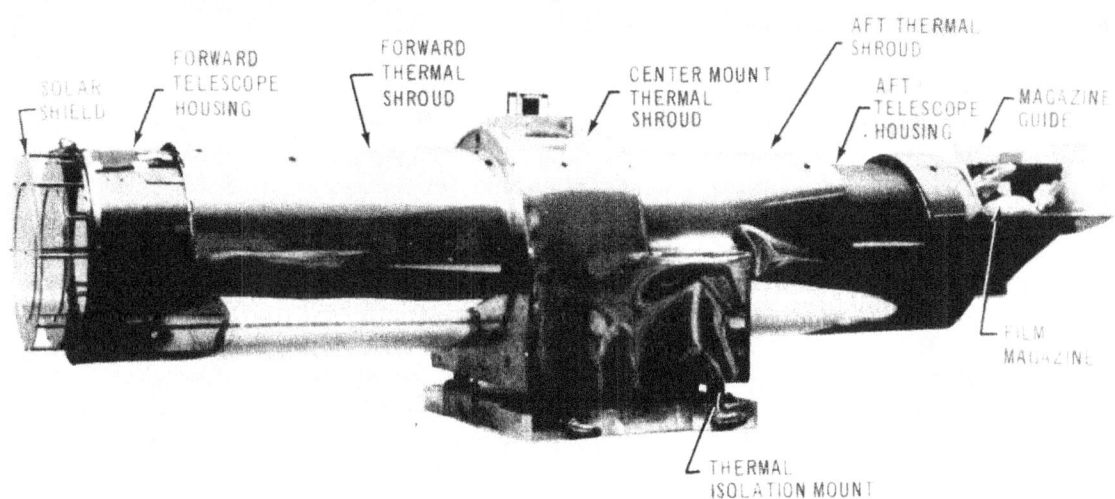

Figure III-16 — S056 Experiment

The X-Ray Event Analyzer (X-REA) provides spectral data (photon intensity as a function of wavelength) in 10 bands from 2 to 20 angstroms. These spectral data serve two purposes: 1) their real time in-orbit display on the C&D Console gives the astronaut solar X-ray and microwave flux information necessary for pointing and operating the X-Ray Telescope, and 2) their analysis will provide detailed flare temperature, density and chemical abundance information.

The X-Ray Telescope (X-RT) records solar images in the form of X-Ray filtergrams (images viewed in narrow wavelength intervals) in five band widths from 5 to 33 angstroms and one in visible light. The filtergrams will provide temporal (time oriented) and spectral (position oriented) variations of the spectral data in flare regions.

The X-REA consists of two gas filled proportional counters, pulse height analyzers and associated electronics components. The proportional counters produce linear outputs proportional to the solar energy levels detected. The pulse height analyzer sorts the proportional counter outputs, with respect to amplitude, into six energy levels in one output channel and four energy levels in the other. These data are then transmitted via telemetry on STDN to the Earth and/or displayed on the counter or activity history plotter of the ATM Control and Display Console for real time in-orbit astronaut information display.

The X-RT consists of a telescope using glancing incident optics to obtain solar images on photographic film. Film retrieval and replacement is accomplished by astronaut EVA with three replacement film canisters, each containing film for about 7,200 data frames, stored in the MDA. Exposed film canisters are returned to Earth in the CM for data evaluation.

Operation of the X-RT camera and the X-REA is by the astronaut at the ATM Control and Display Console. The level of X-RT operation will depend upon solar activity with the astronaut selecting patrol mode with a quiet Sun or active modes with an active Sun which require more photographs per unit of time. The astronaut will also be capable of acquiring targets to photograph using experiment pointing control of the ATM canister which is provided from the C&D Console.

James E. Milligan of the Marshall Space Flight Center is principal investigator. (Development and integration center MSFC.)

<u>XUV Spectrograph/Spectroheliograph (S082)</u>

These experiments will sequentially photograph the Sun over long periods required for proper studies in selected ultraviolet wavelengths. Resulting pictures (spectroheliograms) will show specific emission features greatly enhanced over photographs of the solar disk in white light. Therefore, the Sun will appear quite "blotchy" with much of the emission confined to active regions. The spectroheliograph ("A" instrument) will cover the wavelength region from 150 to 650 angstroms (EUV regions). The spectrograph ("B" instrument) will take data highly resolved into wavelengths in the middle and near ultraviolet region. This instrument can be pointed anywhere on the solar disk to obtain detailed emission characteristics of a region only 1,600 kilometers (1,000 miles) wide.

The astronaut will take photographs of the Sun with the spectroheliograph. He will select the wavelength range to be studied and the exposure time. The "B" instrument will be used to take spectra at various portions of the limb or solar disk. The astronauts will select the mode of operation and the wavelength region to be covered. The XUV monitor provides a display of the 150-650 angstrom activity and gives an indication of XUV images being taken photographically by the "A" instrument.

The "A" instrument (Figure III-17) consists of a concave grating which separates the UV light into its various wavelength components from 150 to 650 angstroms and forms images of the solar disk on film at specific positions corresponding to wavelength. The "A" instrument weights 114.3 kolograms (252 pounds) and is about 3 meters (10 feet) long, one meter (three feet) wide and half a meter (1.5 feet) thick.

The "B" instrument (Figure III-18) consists of a mirror and entrance slit which will select portions of the solar disk or limb to be viewed. A set of two gratings will spread the UV region from 970 to 3,940 angstroms onto photographic film. An XUV monitor allows the Sun to be viewed by the astronauts on TV in the XUV regions. The instrument weighs 169 kilograms (373 pounds) and is essentially the same size as the "A" instrument.

Dr. Richard Tousey of the Naval Research Laboratory is principal investigator. (Development and integration center MSFC, contractor NRL with subcontractor Ball Brothers Research Corporation).

H-Alpha Telescopes

The H-Alpha Telescopes provide the primary means for the boresight pointing of the ATM experiment package. There are two telescopes sensitive to the red hydrogen alpha light of the Sun in the H-Alpha package. One will be equipped with a beam splitter for simultaneous photographic and television pictures. The other telescope will be operated in the TV mode only. Both telescopes will be equipped with a Fabry-Perot filter to make precise observations at the desired wavelength. A zoom capability will allow specific portions of the solar disk to be viewed in detail. These telescopes will take TV and photographic pictures of the solar disk showing flare activities as tremendously enlarged H-Alpha emission sources, which are the primary mode of classifying the size of a flare region.

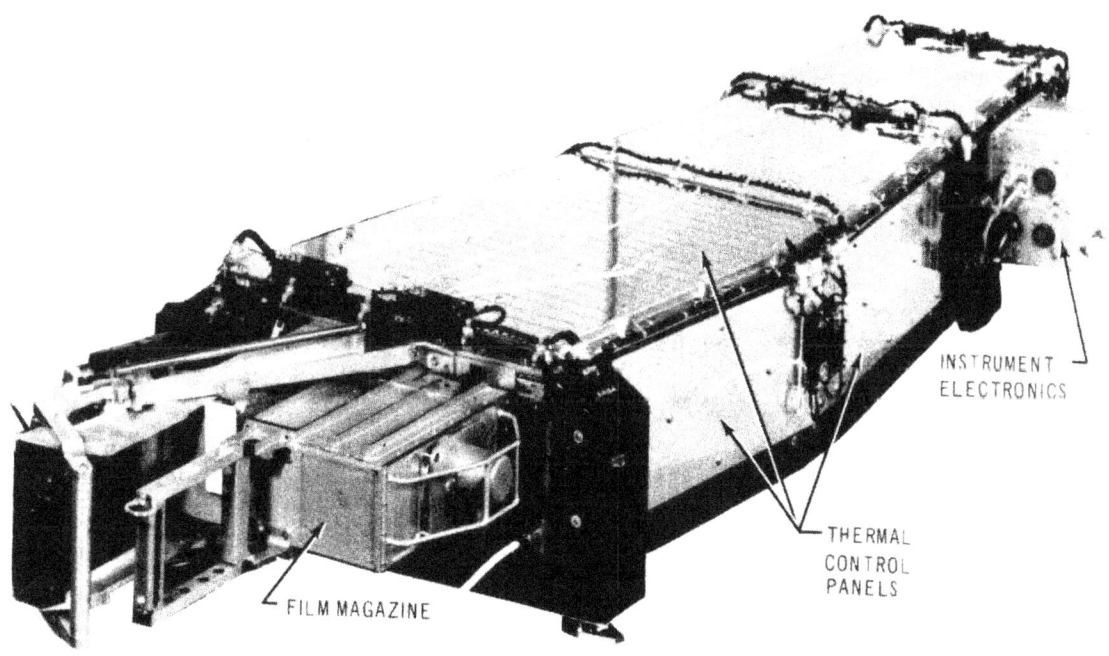

Figure III-17 – S082 Experiment Instrument "A"

Figure III-18 — S082 Experiment Instrument "B"

The H-Alpha telescope will be one of the "eyes" of the astronaut. Active regions will be followed as they traverse the solar disk. When flares are observed, the amount of H-alpha emission will be correlated with emission intensities in other energy regions.

One telescope weighs 86 kilograms (190 pounds) and is 2.7 meters (9 feet) long and approximately 0.3 meter (1 foot) in diamaeter. The second telescope which is not used for photographic purposes is 1.5 meters (5 feet) long, 0.3 meter (1 foot) in diameter and weighs 50 kilograms (110 pounds) (Figure III-19).

Dr. E. M. Reeves of Harvard College Observatory is the principal investigator. (Development and integration center MSFC, contractor Harvard College Observatory with subcontractor Perkin-Elmer.)

Figure III-19 — H-Alpha Telescope

EARTH OBSERVATIONS

Remote sensing of the Earth from orbital altitudes has the potential of yielding information which is of fundamental importance for effective use and conservation of natural resources in both underdeveloped and technologically advanced nations.

Photography from orbital altitudes in the visible and near-infrared spectral regions has already proven to be invaluable for standard synoptic mapping of geographic features over large areas. Systematic use of multispectral remote sensing techniques over an extensive wavelength region has the potential of greatly extending the scope of this capability to include mapping of terrestrial resources and land uses on a global scale. Resources amenable to study are: crop and forestry cover; health state of vegetation; types of soil; distribution of snow pack; surface or near-surface mineral deposits; sea-surface temperature; and the location of likely feeding areas for fish. Comprehensive surveys of such resources will help cope with developing world-wide problems of such accelerating urgency as food supplies, mineral shortages, energy needs, environment pollution and expanding patterns of human settlements.

Many of the environmental features requiring study are in remote regions of the Earth and are highly variable in time. Space Systems can offer the following distinct advantages over conventional aircraft: a broad field of view afforded by the increased altitude; periodic coverage of the same area; and coverage of areas otherwise not easily accessible (Figure III-20).

The Earth resources program objectives are to gether natural and cultural resource data and to minitor environmental and ecological relationships. Major areas of application are agriculture, forestry, hydrology, geology, oceanography, geography, meteorology and ecology. Some of the specific applications in these areas are:

Agriculture/forestry: Improve planning and marketing with current crop census and yield estimates; increase yield by determining soil characteristics and optimizing water management; and reduce losses by early identification of disease, infestation, etc.

Oceanography: Improve fishing productivity by locating cold water upwellings, biologically rich areas, optimum thermal conditions; improve ship routing by measurement of sea state, detection of navigation hazards, and monitoring of sea ice; and improve development of continental shelves by mapping submarine topography and locating oil seeps.

Hydrology: Inventory of water sources for optimum water management; identify new sources of fresh water; monitor health and other characteristics of lakes; identify, monitor and evaluate pollution; and predict and assess flood damage.

Geology: Identify geologic features related to mineral resources such as faults, folds and lateral changes in rock beds; and monitor dynamic features such as volcanic eruptions, landslides and coastal and river sedimentation changes.

Geography: Inventory and classify man's activities through production of thematic maps; and understand physical geography to improve rural and urban development.

Currently, Earth resources data are being collected by ocean ships and buoys, sounding rockets, aircraft flights and spacecraft. Each method has advantages and disadvantages. In general, direct, or ground sensing, methods provide greater accuracy than remote sensing methods. Sensor resolution may be the limiting factor in remote sensing. However, much of the desired Earth resources data can be provided satisfactorily with present sensors and those currently being developed.

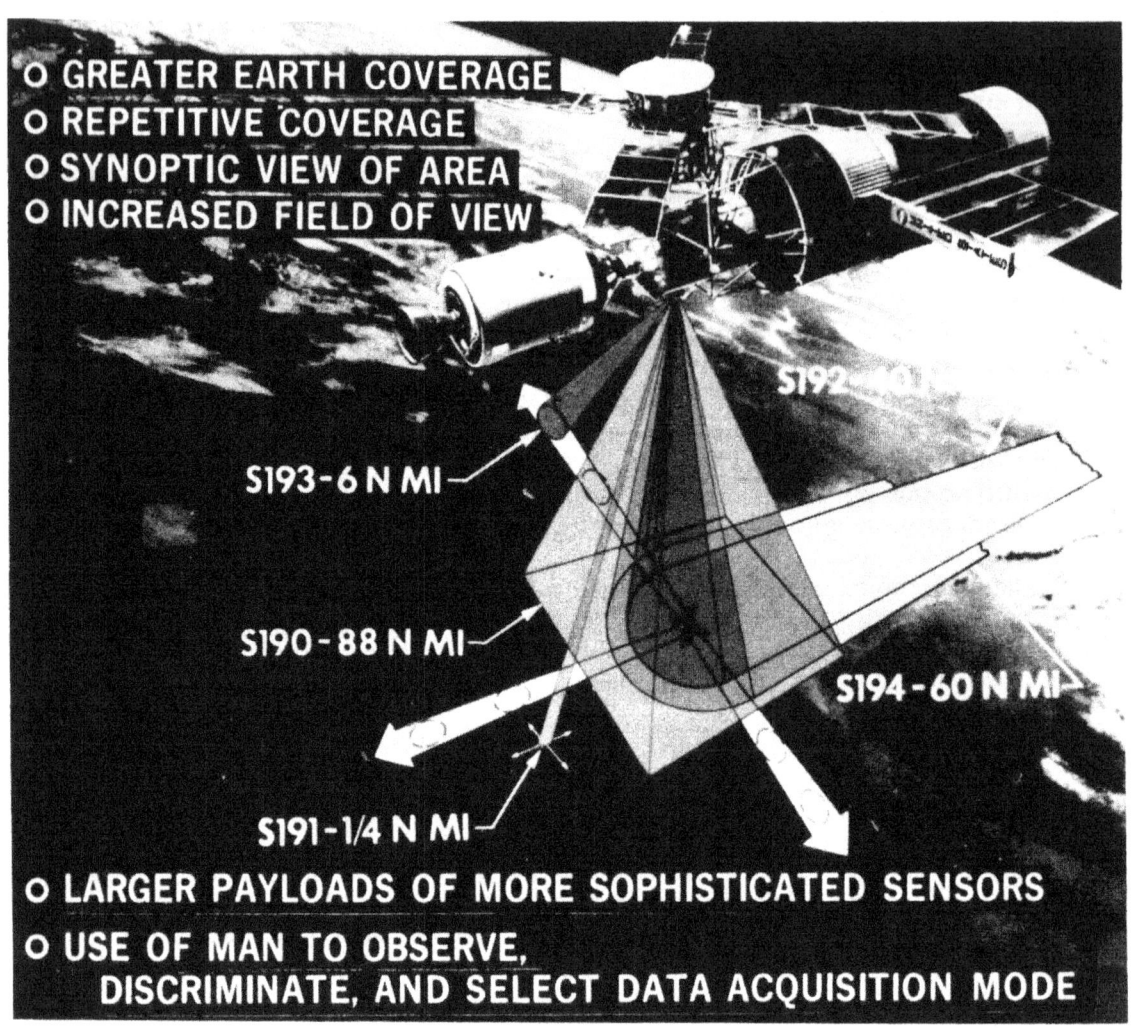

Figure III-20 — Advantages of Skylab EREP

The Skylab EREP, composed of six remote sensing systems, is designed as a spaceborne facility for use as a part of, and in support of, the already existing broad-base international studies on the techniques and application of Earth remote sensing. These studies encompass multispectral sensing (Figure III-21) at ground level, by aircraft and by unmanned spacecraft in addition to Skylab.

The Skylab EREP will provide additional and more precise data on spacecraft sensing capabilities, allowing a more thorough evaluation of sensor techniques and returned-data correlation and application. Also, the manned Earth resources satellite offers unique features not presently possible with automated systems. These are the ability to evaluate test site conditions, to acquire and track uniform, small test sites off the ground track, and to vary the data-acquisition activities as a system conditions warrant.

The EREP sensors can be operated individually or as a group, depending on the scientific requirements and other factors such as weather and vehicle capability. Data will be recorded on tape and film and returned to the NASA Manned Spacecraft Center for

processing. Total tape and film utilizations for the three missions will be as follows:
- 16 rolls of 5-inch film (450 frames per roll)
- 78 rolls of 70-mm film (400 frames per roll)
- 9 rolls of 16-mm film for V/TS (5,600 frames per roll)
- 25 reels of 28-track magnetic tape (7,200 ft. per reel)

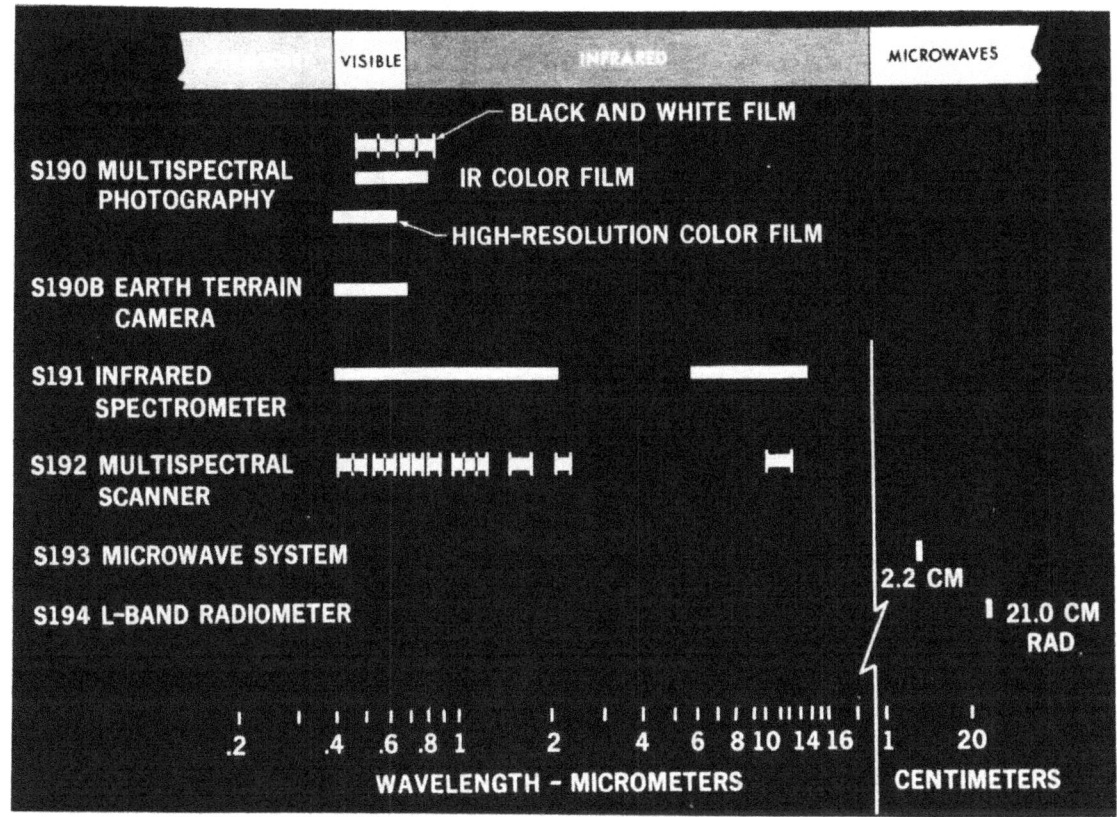

Figure III-21 — EREP Spectral Coverage

Each EREP experiment consists of the data collecting instrument and the necessary support hardware for equipment mounting and stowage. The experiments are centrally controlled from the EREP control and display panel. EREP data are recorded on magnetic tape by the tape recorder. These data are supplemented by voice annotation and MDA housekeeping data.

The EREP hardware located on the exterior of the MDS is shown in Figure I-20. External doors are opened on the S190, S191 and S192 experiments to permit optical viewing of the Earth's surface. The internal arrangement of EREP in the MDA is shown in Figure III-22.

The EREP experiments can be operated individually or as a group, depending on scientific requirements and other factors, such as weather and vehicle capability. The data recorded on tape, S190 film and S191 16mm DAC films are returned to Earth for processing after each mission (Table III-2).

Principal investigators numbering more than 100 from the United States and 40 from other nations have been selected for Earth observation experiments to be performed on Skylab next year.

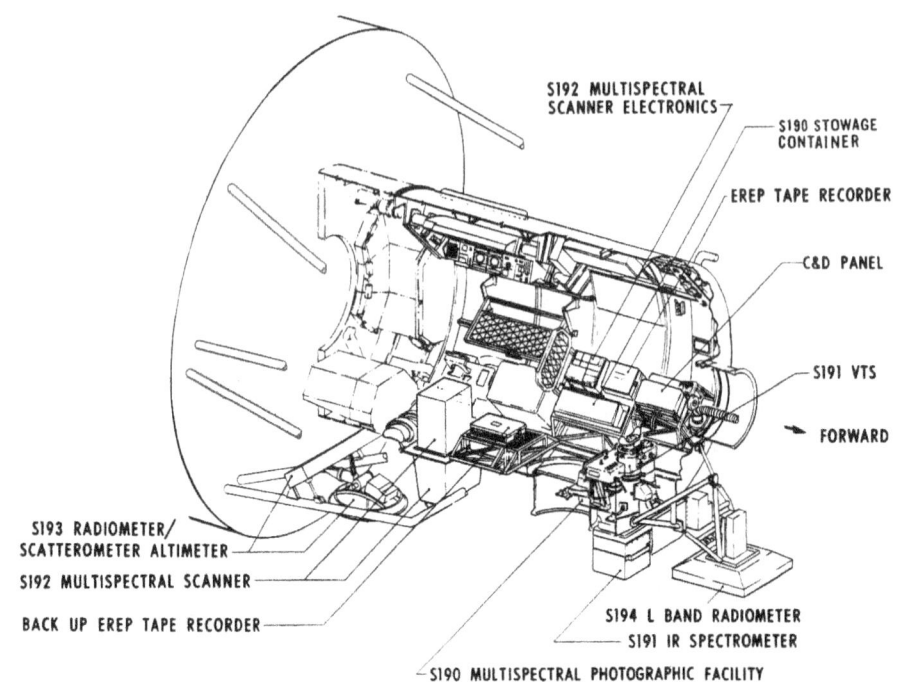

Figure III-22 – EREP Arrangement In MDA

Table III-2 – EREP Data

Associated Experiment	Data Media	Quantity	Total/Mission		
			SL-2	SL-3	SL-4
S190A	IR color film type S0127 Color film, type EK S0-356 IR Aerographic B&W film, type EK 2424 Panotomic-X aerial B&W film type S0-022	13 cassettes 13 cassettes 26 cassettes 26 cassettes	3 3 6 6	5 5 10 10	5 5 10 10
S190B	IR color film, type EK 3443 Color film, type EK S0-242 or High-Resolution B&W film type EK 3414	16 cassettes total	4	6	6
S191	140 feet of 16-millimeter B&W film, type EK 3401	9 reels total	2	4	3
S190A S191 S192 S193 S194	Magnetic tapes 1 in. × 1 mil × 7200 ft.	25 reels	8	11	6

The investigators will use data acquired from the Earth Resources Experiment Package (EREP) which consists of six sensors especially developed for Earth observations.

Data from EREP will be correlated with information acquired from aircraft and ground observations and sensors on the NASA Earth Resources Technology Satellite (ERTS). More than 850 scientists submitted proposals in April-June 1971 for investigations using ERTS and EREP data.

The EREP data-use investigations were selected jointly by NASA's Office of Applications and Office of Manned Space Flight, and NASA negotiated the scope and funding of the EREP experiments selected from U.S. proposers. Those proposed by foreign scientists will be funded by the scientists' own country.

Multispectral Photography Facility (S190)

S190 has been designed to photograph regions of the Earth's surface, including oceans, in a range of wavelengths from near infrared through the visible. The facility is in two parts: the Multispectral Photographic Camera (S190A), six-channel 70 mm cameras (Figure III-23) that simultaneously photograph the same area, each viewing a different wavelength; and the Earth Terrain Camera (S190B), a single lens camera (Figure III-24).

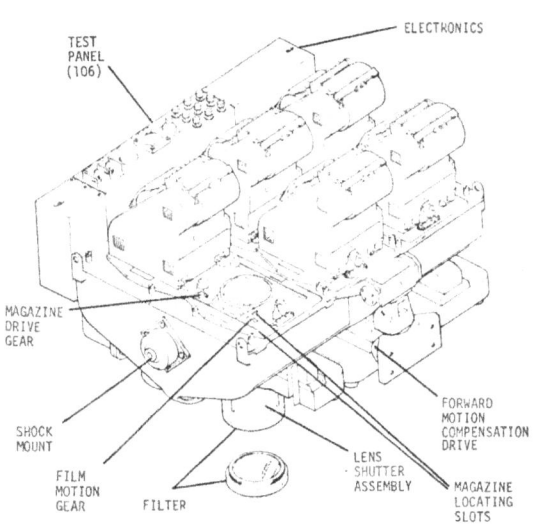

Figure III-23 – S190 A Cameras

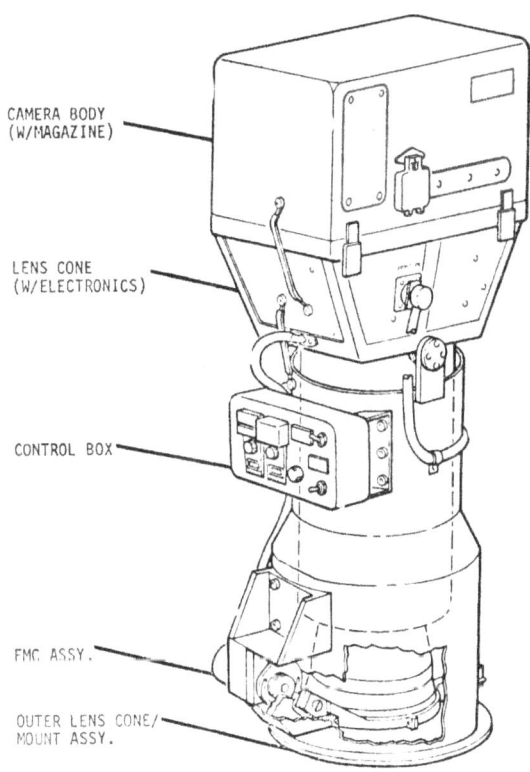

Figure III-24 – S190 B Camera

The S190A experiment uses a six channel, high precision 70 mm camera system. The matched distortion and focal length camera array contains forward motion compensation to correct for spacecraft motion. The six-inch focal length lenses have a field of view of 21.1 degrees across flats providing a square surface coverage of about 169 kilograms (88 nautical miles) from the expected 435 kilogram (235 nautical mile) altitude. The system is designed for the following wavelength/film combinations:

III-33

.5 - .6	Micrometers	Pan-X B & W
.6 - .7	Micrometers	Pan-X B & W
.7 - .8	Micrometers	IR B & W
.8 - .9	Micrometers	IR B & W
.5 - .88	Micrometers	IR Color
.4 - .7	Micrometers	HI RES Color

The spectral regions designated were selected to separate the visible and photographic infrared spectrum into the bands that are expected to be most useful for multispectral analysis of Earth surface features. Further spectral refinements are made by using different filter combinations.

The S190B portion utilizes a single 18-inch focal length lens with 5-inch film. Its field of view of 14.2 degrees across flats provides square surface coverage of about 112 kilograms (59 nautical miles) from orbit. This camera is designed to use high resolution color film and will be operated from the OWS SAL window.

The camera compensates for spacecraft forward motion through programmed camera rotation. Shuttle speeds are selectable at 5, 7, and 10 milliseconds with a curtain velocity of 110 inches per second.

Areas of the Earth surface shall be photographed that are of particular interest to EREP investigators. Before each pass the crew will receive ground update for settings, number of exposures, etc.

Project scientist is K. Demel of the Manned Spacecraft Center. (Development center MSC, integration center MSFC, contractors Itek Corporation for Multispectral Photographic cameras and Actron Industries for Earth Terrain camera.)

Infrared Spectrometer (S191)

The primary goal of Experiment S191 is to make an evaluation of the applicability and usefulness of sensing Earth resources from orbital altitudes in the visible through near-infrared and in the far infrared spectral regions. Another specific goal is to assess the value of real time identification of ground sites by an astronaut.

The S191 is a dual spectral band system, the short wavelength band, 0.4 to 2.4 micrometers, and the long wavelength spectral band, 6.2 to 15.5 micrometers.

The field of view of the system is one milliradian with a spectral resolution of 1 to 5 percent. The system includes a Viewfinder Tracking System (Figure III-25) which a crewman uses in acquiring and tracking desired sites for S191 use, providing the ability to look at relatively small ground targets about 0.44 kilogram (¼ nautical mile) in size. A 16 mm camera is used to photograph these sites.

The primary data will be recorded on magnetic tape along with data from other sensors in the EREP. The magnetic tape and the film from the viewfinder camera will be returned with each crew rotation.

External Components are shown in Figure III-26.

Project Scientist is Dr. Thomas Barnett, NASA Manned Spacecraft Center. (Development center MSC, integration center MSFC, contractors Block Engineering Company for IR Sensor, Martin Marietta Aerospace for Viewfinder/Tracking System.)

Multispectral Scanner (S192)

The primary goal of Experiment S192 is to assess the feasibility of multispectral techniques, developed in the aircraft program, for remote sensing of Earth resources from space. Specifically, attempts will be made at spectral signature identification and mapping of ground truth targets in agriculture, forestry, geology, hydrology and oceanography.

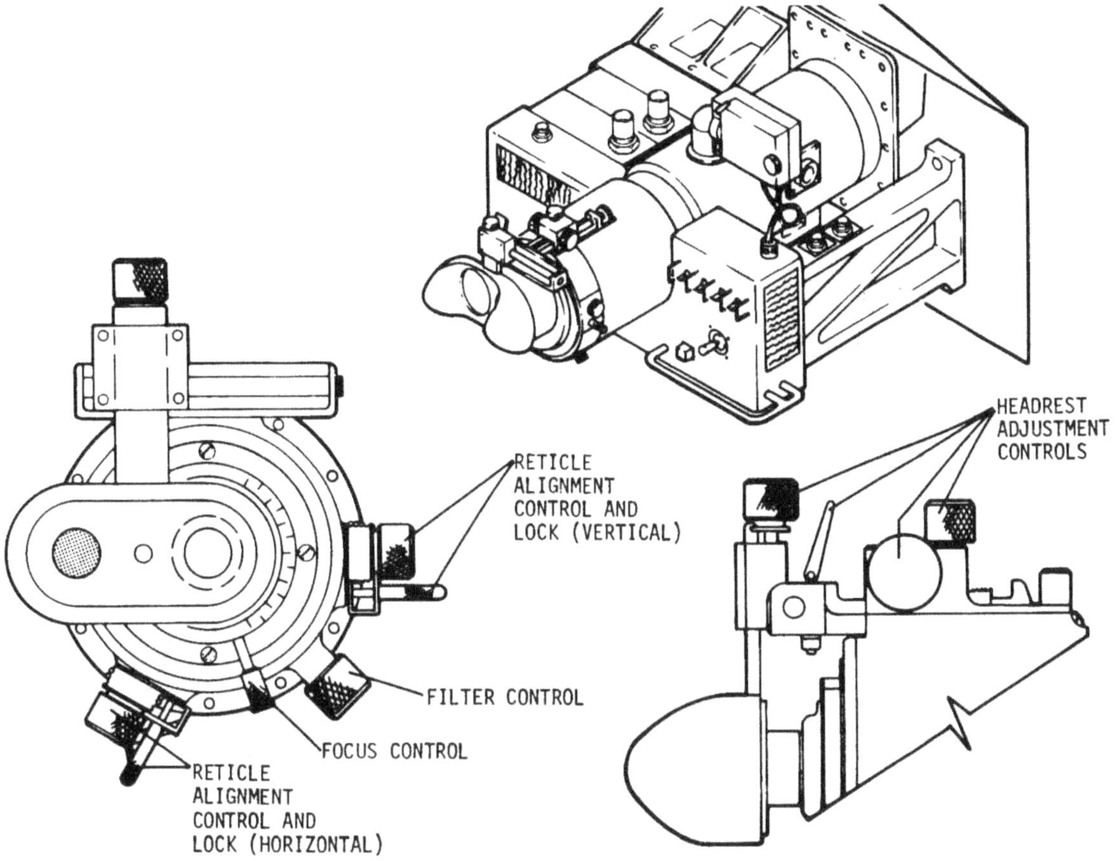

Figure III-25 — Viewfinder Tracking System (S191)

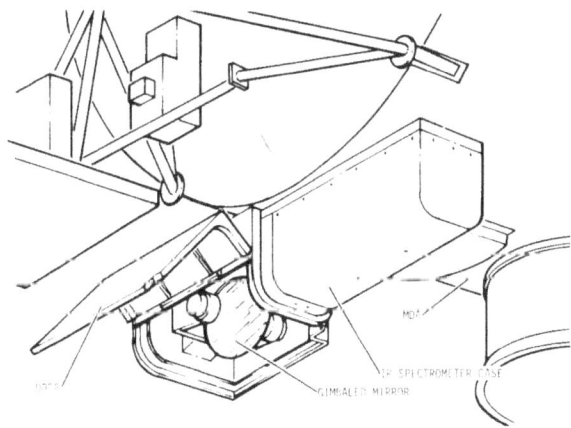

Figure III-26 — S191 External Components

The S192 instrument has 13 spectral bands from 0.4 to 12.5 micrometers in the visible, near infrared and thermal infrared regions. The system gathers quantitative high-spatial-resolution line-scan imagery data on radiation reflected and emitted by selected ground sites in the U.S. and other parts of the world.

The motion of the sensor is a circular scan (Figure III-27) with a radius of 41.8 kilometers (22.6 nautical miles). Data of ground scenes are recorded as the scan sweeps a track 75 kilometers (37 nautical miles) wide in front of the spacecraft, yielding a 260-foot resolution at an altitude of 435 kilometers (235 nautical miles). The S192 optical mechanical scanner utilizes a 30 cm (12 inch) reflecting telescope with a rotating mirror. The telescope and mirror are mounted outside the MDA.

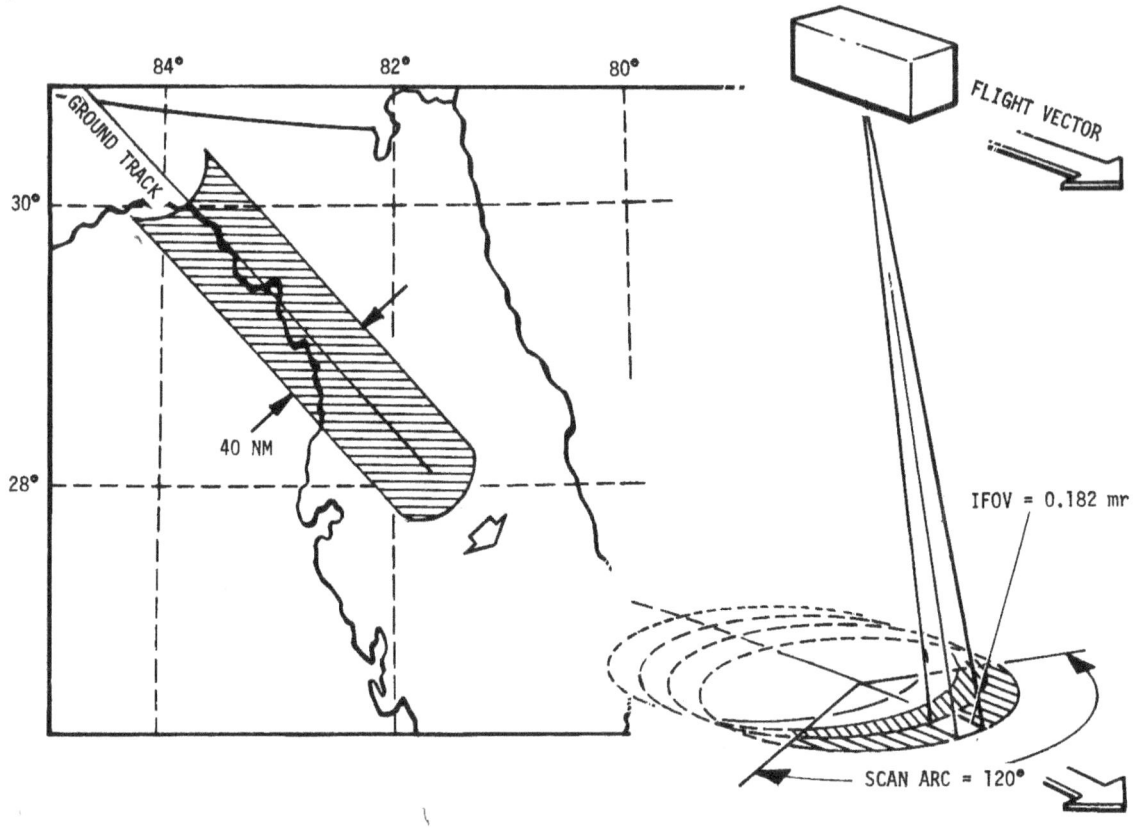

Figure III-27 — S192 Footprint

the S192 and the data are recorded on the EREP tape recorder. Since the high data rate of S192 is not compatible with the standard recording speed of the EREP recorder, the recorder speed is increased 60 inches per second during operation of the scanner.

Project Scientist is Dr. Charles Korb, Manned Spacecraft Center. (Development center MSC, integration center MSFC, contractor Honeywell Radiation Center.)

Microwave Radiometer/Scatterometer/Altimeter (S193)

The objectives of the experiment are the near simultaneous measurement of the radar differential back-scattering cross section and the passive microwave thermal emission of the land and ocean on a global scale and to provide engineering data for use in designing space radar altimeters.

The S193 will be useful in studying varying ocean surfaces, wear conditions, sea and lake ice, snow cover, seasonal vegetational changes, flooding, rainfall and soil types. The sensor generally will operate over ocean and ground areas where ground truth data are available; however, additional targets of opportunity, such as hurricanes and storms, will be viewed if the opportunity arises.

S193 incorporates a radiometer, scatterometer and radar altimeter, all operating at the same frequency of 13.9 gigahertz (GHz). The equipment shares a common gimballed antenna mounted on the outside of the MDA (Figure III-28). The scatterometer measures the back-scattering coefficient of ocean and terrain as a function of incidence angle ranging from 0 to 52 degrees. The radiometer is a passive sensor which measures the brightness temperature, from a cell on the Earth's surface, as a function of incidence angle from the surface.

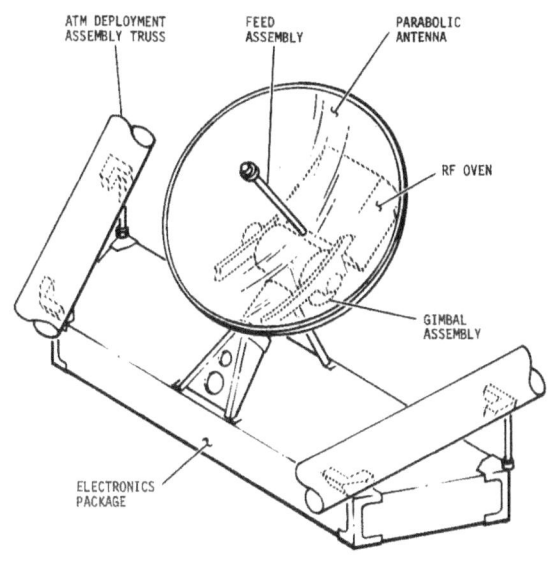

Figure III-28 — Experiment S193

The S193 ground coverage is 48 degrees forward and 48 degrees to either side of the spacecraft ground track.

All data will be recorded on magnetic tape on one digitized channel. The radiometer/scatterometer data are recorded at 5.33 kilobits per second; the altimeter data at 10 kilobits per second.

Project Scientist is Dallas Evans, Manned Spacecraft Center. (Development center MSC, integration center MSFC, contractor General Electric.)

L-Band Microwave Radiometer (S194) – The experiment objective is to supplement experiment S193 in measuring brightness temperature of the Earth's surface along the spacecraft track, which would provide ocean surface features, meteorology winds and Earth surface features.

The S194 experiment is a passive microwave sensor utilizing a fixed planar array antenna (Figure III-29). S194 records the brightness temperature of the Earth in the L-band range with a digital output giving an absolute antenna temperature to an accuracy of one degree Kelvin. The system utilizes a built-in calibration scheme that samples known sources. The spatial characteristics are a half power beam width of 15 degrees, first null beam width of 37 degrees (97 percent of power) and having a circular footprint of approximately 111 kilometers (60 nautical miles) diameter (half power) at the expected 435 kilometer (235 nautical mile) altitude.

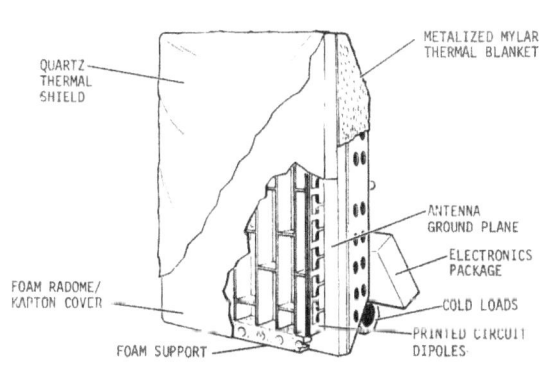

Figure III-29 — Experiment S194

All data will be recorded on magnetic tapes. The data output is at 200 bits per second.

Principal scientist is Dallas Evans, Manned Spacecraft Center. (Development center MSC, integration center MSFC, contractor Airborne Instrument Laboratory.)

EREP Aircraft Support for Skylab

Underflights of remote sensor-equipped aircraft will support the EREP experiments by acquiring data to correlate with the spacecraft data. Each Earth resources aircraft underflight of Skylab will be coordinated with the actual Skylab flight plan. In many cases simultaneous coverage will be provided with spacecraft coverage of a given area.

Over 200 underflights will be flown by NASA and University of Michigan aircraft. The flights will be about equally high-altitude and medium-altitude in support of the EREP experiments. No foreign sites will be covered by the NASA aircraft.

The high-altitude flights will be conducted by two WB57F aircraft staging initially out of Ellington Air Force Base near MSC and other locations throughout the United States, and by two aircraft staging out of Moffett Field near Ames Research Center and Wallops Island, Va.

The low-altitude flights will be conducted by C-130B, P3A, OV1-C and C-47 aircraft and an H47-G helicopter. The C-47, a University of Michigan aircraft, will stage out of various locations in the United States. The C-130B and P3A will stage out of EAFB and other locations throughout the United States. The OV1-C will stage out of EAFB. The H-47-G helicopter will stage out of EAFB and White Sands Missile Range, New Mexico.

During Skylab some 87,000 data miles are scheduled to be flown by the nine aircraft in support of the EREP experiments. MSC and ARC will be responsible for processing and disseminating data acquired by their respective aircraft.

EREP Data Handling

The EREP data that are returned to Earth in the CMs will consist of 78 cassettes of 70 millimeter color and B&W film from S190A, 16 cassettes of five-inch color and B&W film from S190B, and nine reels of 16 millimeter film from the viewfinder of S191. Also, twenty-five 7,200-foot reels of one-inch wide magnetic tape, each with 28 tracks of recorded data, will be returned.

Upon arrival at MSC, the film will be processed through the NASA bonded storage facility for inventory and inspection. Then, development and duplication of the film will be performed in the Photographic Technology Laboratory. The original (flight) films will be placed into archival storage. Quick-look screening and science analysis will be performed with duplicate film.

The tape reels will be processed through the NASA bonded storage facility for inventory and inspection. The reels then will be forwarded to the Earth resources production processing facility for duplication and reformatting and the original tapes will be sent to archival storage. The duplicates will be used for further processing.

No EREP sensor data will be telemetered to the Earth in real time or during the active mission. All data will be recorded on the on-board EREP recorder or on film and returned to MSC after each mission.

Earth Resources Processing Facility

Processing of EREP electronic sensor data as necessary to provide data products for investigator use will be performed in the MSC Earth Resources Production Processing Facility.

The EREP tape reproducer/reformatter will be required to convert the original tapes to compatible formats for data processing because the original flight tapes are incompatible with existing ground-data processing facilities. After original flight tapes are reformatted and reproduced on 14-track analog master tapes, the tape masters containing data from experiments S190A (supporting data), S191, S192, S193 and S194 will be decommutated and processed on telemetry processing stations and computers.

All EREP sensor data will be cataloged and these catalogs will be issued formally to federal outlets of the Department of Interior's EROS Data Center at Sioux Falls, S.D., and the Department of Commerce's NOAA Environmental Data Service.

Photographic Technology Laboratory – The S190 and S191 film will be processed in the MSC Photographic Technology Laboratory (PTL). From the three Skylab missions, at least 10,400 frames of color and 20,800 frames of B&W 70 millimeter film will be returned to Earth from S190A; 7,200 frames of color and B&W film will be returned from S190B; and 50,400 frames of 16 millimeter B&W film will be returned from S191.

ASTROPHYSICS

Until recently, progress in astronomy has been made by theoretical and observational advances using instruments associated with ground based optical telescopes. Skylab experiments will identify advantages and disadvantages of man-attended space observations and obtain information needed for planning future, more advanced instruments. Most of the observations performed will be of a type not possible from the surface of Earth because of the absorbing and light scattering effects of the atmosphere surrounding Earth.

Nine astrophysical experiments are included on Skylab. Three experiments (S063, S073 and S149) will investigate the outer atmosphere of the Earth and the interplanetary medium. The other six experiments (S009, S019, S150, S228, S230 and S183) will study objects external to our solar system.

Nuclear Emulsion (S009)

The objective of this experiment is to record the cosmic ray flux incidence outside the Earth's atmosphere, more specifically, the relative abundance of high energy primary heavy nuclei.

Theories of nucleogenesis predict the relative abundance of nuclei that would be produced in the thermonuclear reactions occurring in possible sources such as neutron stars. It is therefore of great interest to study the relative abundance of the nuclei reaching the earth. To obtain as accurate a measure as possible of the relative abundance of various nuclei in the primary cosmic ray flux, nuclear emulsions must be carried into space.

The Skylab nuclear emulsion experiment will provide a long exposure to determine the abundance of the rarer heavy nuclei.

The instrument (Figure III-30) consists of two adjacent stacks of nuclear emulsion strips. This emulsion differs from regular photographic emulsions being considerably thicker and containing a much higher density of grain material to improve the detection of tracks left by charged particles. The stacks are hinged together like the two sides of an open book and contain several layers of different emulsion types.

The emulsion stacks are mounted inside the Skylab Multiple Docking Adapter, separated from space by a thin section of the spacecraft wall. During exposure the "book" is open, allowing high energy particles which have passed through the wall to enter the front surface of both emulsion stacks.

The exposed emulsion will be returned to the earth and peeled apart in thin strips which are numbered, developed and scanned for tracks. By measuring the variations in thickness and direction of the tracks and tracing their entire path through the strips, the energy and charge of the cosmic rays can be determined.

The relative abundances of various nuclei observed from the primary cosmic radiation provide crucial information from which one can learn something about the physical conditions where the nuclei were formed, the time that has elapsed since they were formed, and the nature of their interactions with interstellar material in transit.

Maurice M. Shapiro, PhD, Naval Research Laboratory, is the principal investigator. (Development and integration center MSFC, Contractor Naval Research Laboratory.)

Ultraviolet Stellar Astronomy (S019)

Experiment S019 is designed to take UV photographs of large areas of the Milky Way in which young, hot stars are abundant. About 50 star fields, each 4 by 5 degrees in area, will be photographed during each Skylab mission, two to three exposures on each field. Exposure times are 30, 90 and 270 seconds, depending on the brightness of the stars being photographed.

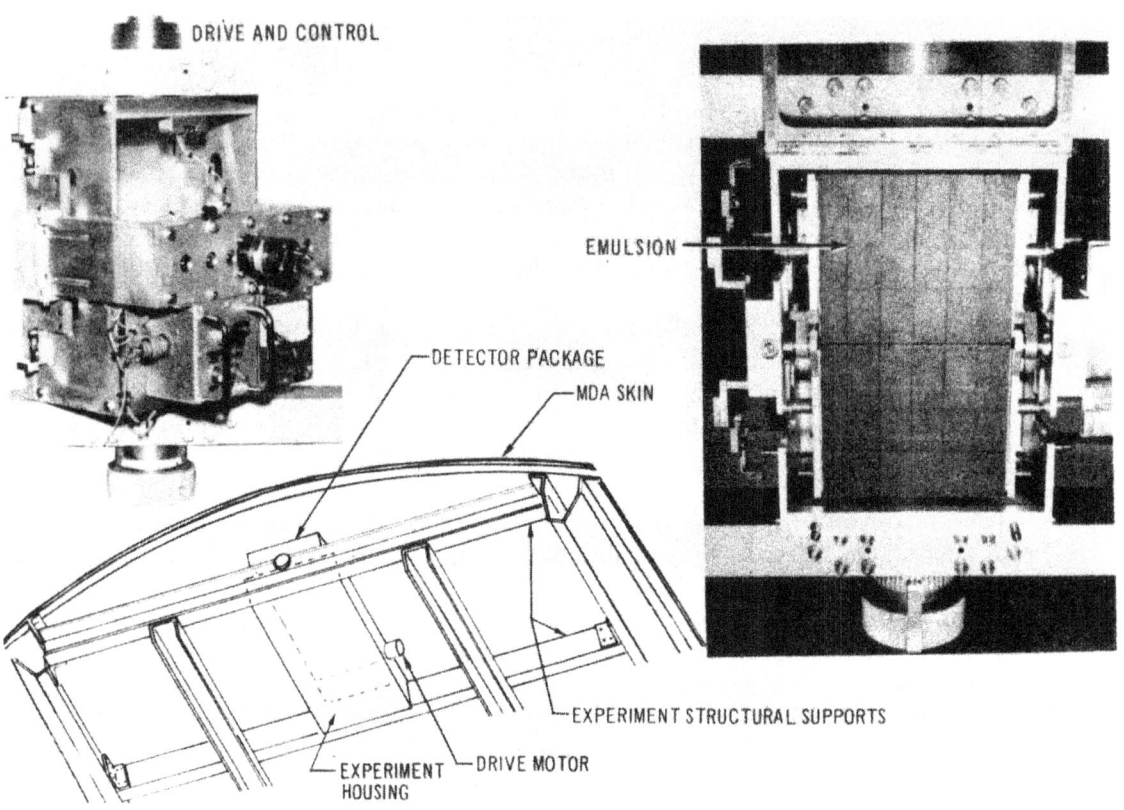

Figure III-30 — Experiment S009

A magazine holding 164 frames of special UV sensitive film will be carried on each mission. When OWS stability permits, a fourth exposure without using a spectral widening mechanism may be made. The astronaut, observing spacecraft drift by viewing star field motion through a small seven-power telescope, will decide when such as exposure can be made.

On 270-second exposures, UV spectra will be obtained on early type stars as faint as the sixth magnitude, the faintest visible to the unaided human eye. On "widened" special exposures, stars 10 times fainter should be reached.

The S019 experiment consists of a 6-inch reflecting telescope and a moveable mirror. The telescope will be operated through one of the scientific airlocks in the OWS (Figure III-31). The rotating mirror is first extended through the SAL by the extension mechanism (Figure III-32) to allow the telescope line of sight to be pointed over a large area of the sky. The telescope is operated manually by the astronaut.

The telescope is capable of reording ultraviolet radiation down to a wavelength of 1350 angstrons. Its main objective will be to study the differences from star to star in the several strong spectrum lines known to exist in the 1350 and 2000 anstrom region of the spectrum.

Some of these lines are formed by the atmosphere of the star while other lines are formed by the interstellar gas between Earth and the stars. One special advantage of the telescope is its ability to photograph a 4 by 5 degree region of the sky and to record the spectra of all the stars in that region on one photograph. It is expected that spectra of more than a thousand stars will be obtained by this technique.

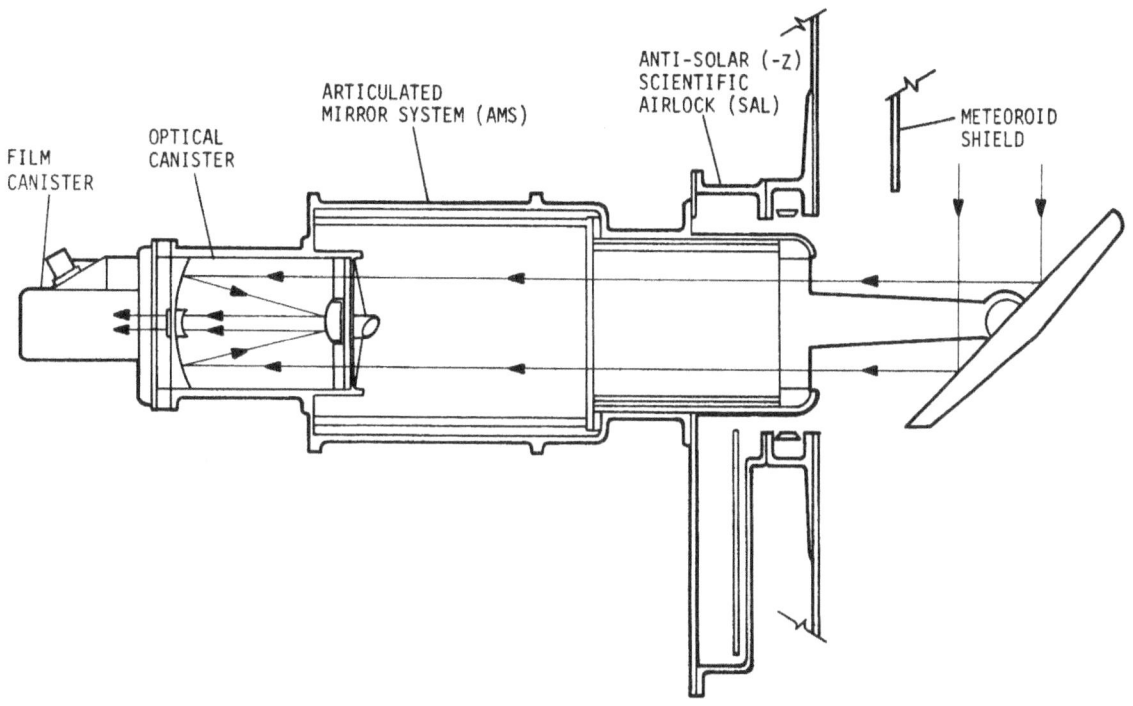

Figure III-31 — Experiment S019 In SAL

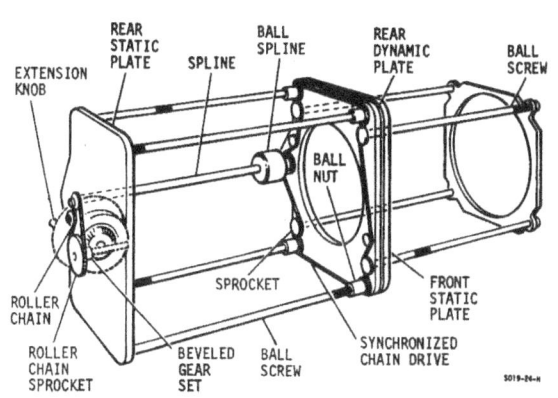

Figure III-32 — AMS Extension Mechanism

The quality of these spectra (two angstrom resolution) is between those of the first and third Orbiting Astronomical Observatories.

The principal investigator is Dr. Karl G. Henize, a scientist-astronaut at the Manned Spacecraft Center. A team of astronomers headed by Dr. James D. Wray at the University of Texas will measure the spectra and participate in their analysis. This team was formerly based at Northwestern University in Evanston, Ill., where the program was conceived and equipment was designed. Equipment construction was by Cooke Electric Company and by Boller and Chivens Division of Perkin Elmer Corporation. (Development center MSC, integration center MSFC.)

UV Airglow Horizon Photography (S063)

The Ultraviolet Airglow Horizon Photography — Experiment S063 — is a photographic experiment with two separate experiment assemblies. One experiment assembly performs ozone photography, the other assembly performs twilight airglow photography.

The behavior of ozone is an important factor in the thermal balance of the atmosphere. The amount of absorption can be determined by taking two series of simultaneous photographs. One series, using a 35mm camera with ultraviolet filters, will record the varying amounts of absorption of ultraviolet illumination, indicating varying densities of ozone. The other series, using a second 35mm camera without UV filters, will be aimed at the same target but will obtain color photographs of atmospheric and ground features such as water, mountains and clouds (Figure III-33).

The UV camera (Figure III-34) will be mounted at the antisolar SAL and the color camera in the Wardroom window. Three-hundred exposures (150 UV and 150 color) are planned.

The twilight airglow experiment will photograph the glow occurring in the upper atmosphere caused by chemical reactions in the ozone, oxygen and other gases when they are simulated by the sun's radiation. The upper atmosphere will be photographed at twilight against the dark sky of space (Figure III-35).

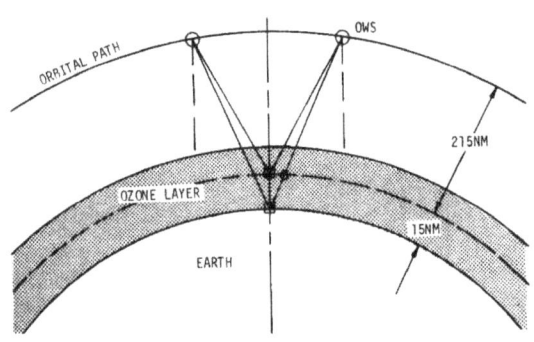

Figure III-33 – S063 Ozone Photography

The twilight airglow camera (Figure III-36) will be attached to the solar SAL and will photograph the airglow, together with the Earth limb and background starfields. Three-hundred exposures are required.

Principal Investigator is Dr. Donald M. Parker, Naval Research Laboratory, Washington, D.C. (Development center MSC, integration center MSFC, contractors NRL and Martin Marietta Aerospace.)

Gegenschein/Zodiacal Light (S073)

The purpose of this experiment is to measure the brightness and polarization of the visible background of the sky as seen from the Skylab above the earth's atmosphere.

Photographs and light level readings taken from the ground, rockets and satellites have not yet been sufficiently accurate to distinguish between models of the sources contributing to the faint visible background light of the sky outside the atmosphere. Hand-held photographs, taken from Gemini spacecraft, of the Gegenschein, or anti-solar enhancement, have established that it is extraterrestrial in origin rather than a phenomenon occurring in the Earth's atmosphere. An Apollo lunar orbit experiment attempted to photograph the Gegenschein from the dark side of the lunar orbit to determine by triangulation whether or not it is due to a "zero phase" enhancement of sunlight scattered off the interplanetary medium or a cloud of dust maintained in the gravitational null point 1,609,000 kilometers (one million miles) from the Earth, opposite the Sun. The Skylab study will be a complete survey of the sky from above the variable interference of visible airglow emission and atmospheric extinction.

There is no dedicated hardware for this experiment. The T027 photometer and 16mm operational data acquisition camera are used.

The content of the magnetic tape is telemetered to the ground. The film is returned to the ground in the CM. After subtracting the signal caused by stars in the telescope field-of-view, the photographs and light level readings are combined to make a map of the background brightness of the sky.

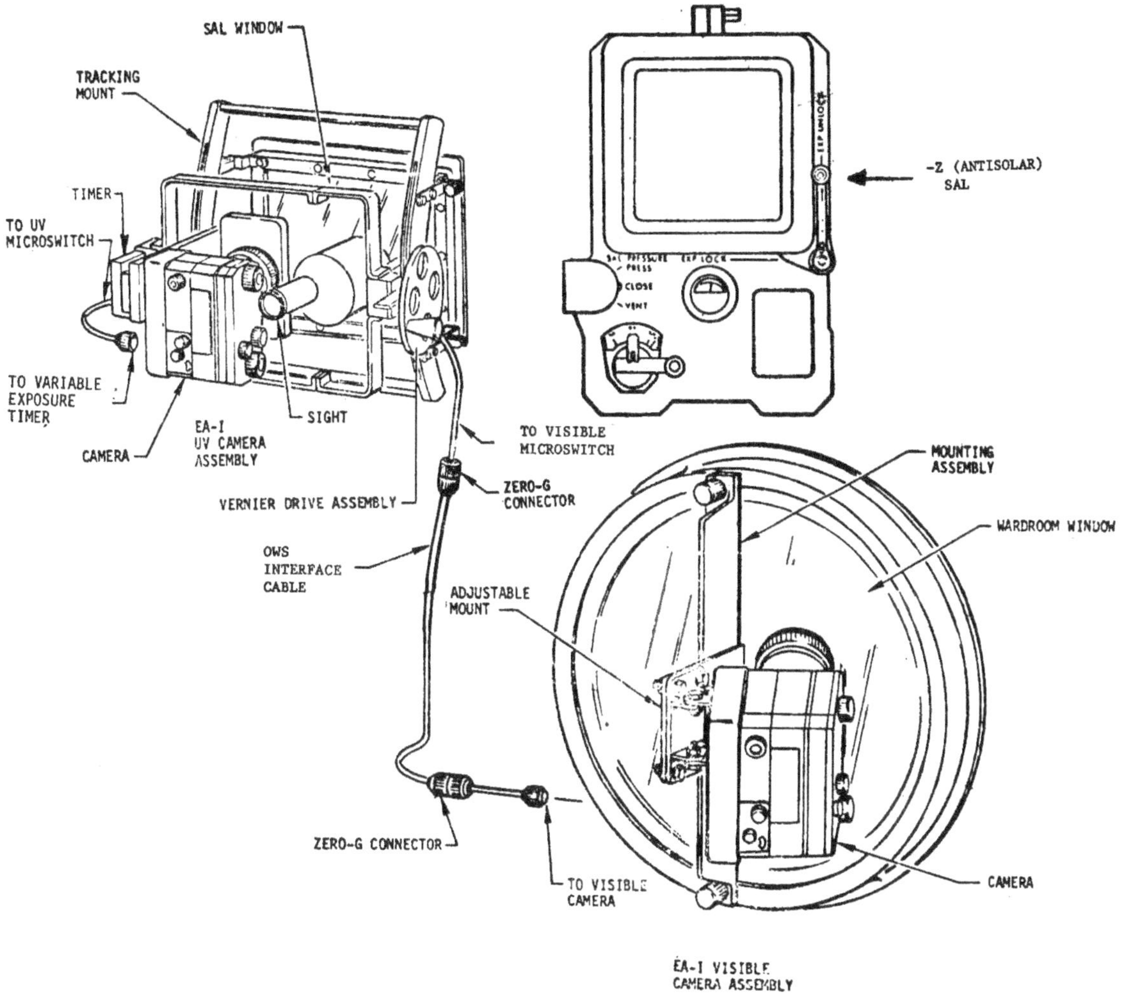

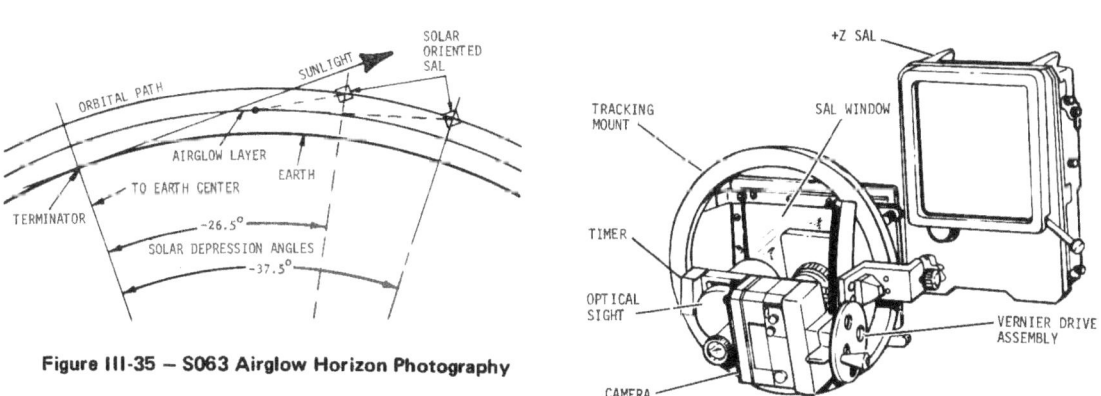

Figure III-34 – Ozone Photography Equipment

Figure III-35 – S063 Airglow Horizon Photography

Figure III-36 – Airglow Photography Equipment

III-43

By comparing the amount of light at different colors and polarizations with the spectrum of the Sun one can indirectly obtain information about the sizes, shapes, composition and numbers of the dust particles traveling in interplanetary space which reflect the sunlight and produce the zodiacal light. In addition, the variation of the brightness relative to the direction of the Sun and the ecliptic plane will enable the interplanetary contribution to background light level to be distinguished from the interstellar background and any contribution from a hypothetical dust cloud associated with the Earth and Moon.

J.L. Weinberg, PhD, Dudley Observatory, Albany, N.Y., is the principal investigator. (Development and integration center MSFC, contractors Dudley Observatory and Martin-Marietta.)

Particle Collection (S149)

The purpose of the S149 Particle Collection experiment is to study the nature and distribution of interplanetary dust by exposing specially prepared surfaces to space. Impacts of cosmic dust at high speeds produce impact craters and in some cases a portion of the profile remains at the impact site. The controlled, exposed surface may also be examined to determine the rate of surface contamination of the Skylab environment. The S149 instrument consists of gold covered smooth plates 15 centimeters (six inches) square and layers of film mounted inside resealable cassettes.

Sets of four plates will be exposed for a 72-hour period during the manned missions and also during the two unmanned storage periods. After being exposed to the space environment, the first set of cassettes (four) will be retrieved and returned to Earth for analysis.

This experiment will also help to better define the hazard to space travel posed by dust particle erosion of critical optical surfaces and windows.

The experiment hardware will be positioned by deployment of the T027 extension device through the Scientific Airlocks (Figure III-37).

Principal investigator for S149 is Dr. Curtis L. Hemenway of Dudley Observatory. (Development center MSC, integration center MSFC, contractor Dudley Observatory.)

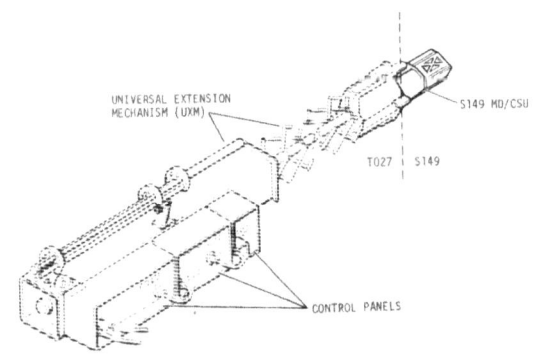

Figure III-37 – S149/T027 Interface

Galactic X-Ray Mapping (S150)

The objective of this experiment is to conduct a survey of the sky for faint X-ray sources.

X-radiation has been observed from more than 40 stellar sources over the past 10 years. Most of these studies have been conducted in the energy region from 1-10 KeV using rockets with a viewing time of only 3 minutes. Satellites such as SAS-A, launched December 1970, have completed the survey of stellar sources in the 1-10 KeV region. The Skylab experiment provides a sky survey in the 0.2-12 KeV energy range.

The instrument consists of a set of proportional counters which will cover the spectral region from .2 to 12 KeV. It is physically mounted on the launch vehicle (Figure III-38). Accordingly, the life time of the experiment is limited to 4-5 hours. During this time detectors with a 20° field-of-view will determine the location of X-ray sources to within 20 arc minutes. Only about 1/2 of the sky will be viewed, however. Due to daylight X-ray fluorescence, no data below .7 KeV will be available during the daytime half of the orbit.

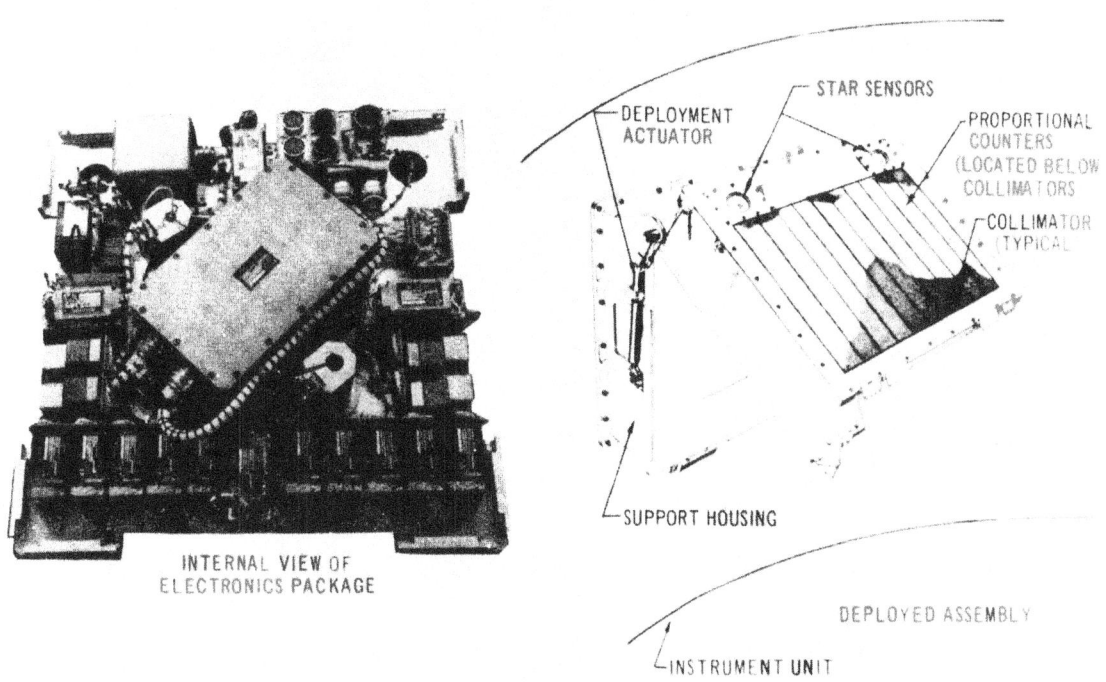

Figure III-38 — Experiment S150

Data will be recorded and transmitted back by telemetry. The spectral data will have an energy resolution of 50 percent at 1/2 KeV and 25 percent at 10 KeV.

Because most X-ray sources are not recognizable from the ground in visible light, it is necessary to develop a catalog listing the precise positions of as many sources as possible, to guide later, more detailed studies. The results of the Skylab survey will provide a catalog of faint X-ray (.2-12 KeV) sources including their strength and spectral characteristics.

William Kraushaar, PhD, University of Wisconsin, is the principal investigator. (Development and integration center MSFC, Contractor University of Wisconsin.)

Ultraviolet Panorama (S183)

The objective of this experiment is to measure ultraviolet brightness of a large number of stars.

Rocket experiments have obtained high resolution spectra of individual bright stars and the OAO-A2 telescope has obtained images of many star fields in four spectral bands.

The Skylab experiments will provide a photographic survey in three bands with fine spatial and photometric resolution of a number of star fields previously unavailable.

The instrument, a spectrographic assembly (Figure III-39) uses the same movable mirror as S019. Total weight is about 175 pounds. The spectrographic assembly is mounted in the anti-solar airlock in the wall of the Skylab with the moveable mirror extended outward to permit viewing in different directions.

The assembly includes a grating spectrograph which collects the ultraviolet light from the spectrum of the stars in the field-of-view into 600 Å wide bands centered at 1800 Å and 3100 Å. Stars as faint as 7th magnitude can be recorded with 7 arc-minutes of angular resolution over a 7 by 9 degree field-of-view. It also includes a small Schmidt camera which records the field-of-view in a 600 Å wide band of UV radiation centered at 2500 Å. angular resolution over a 7 by 9 degree field-of-view.

III-45

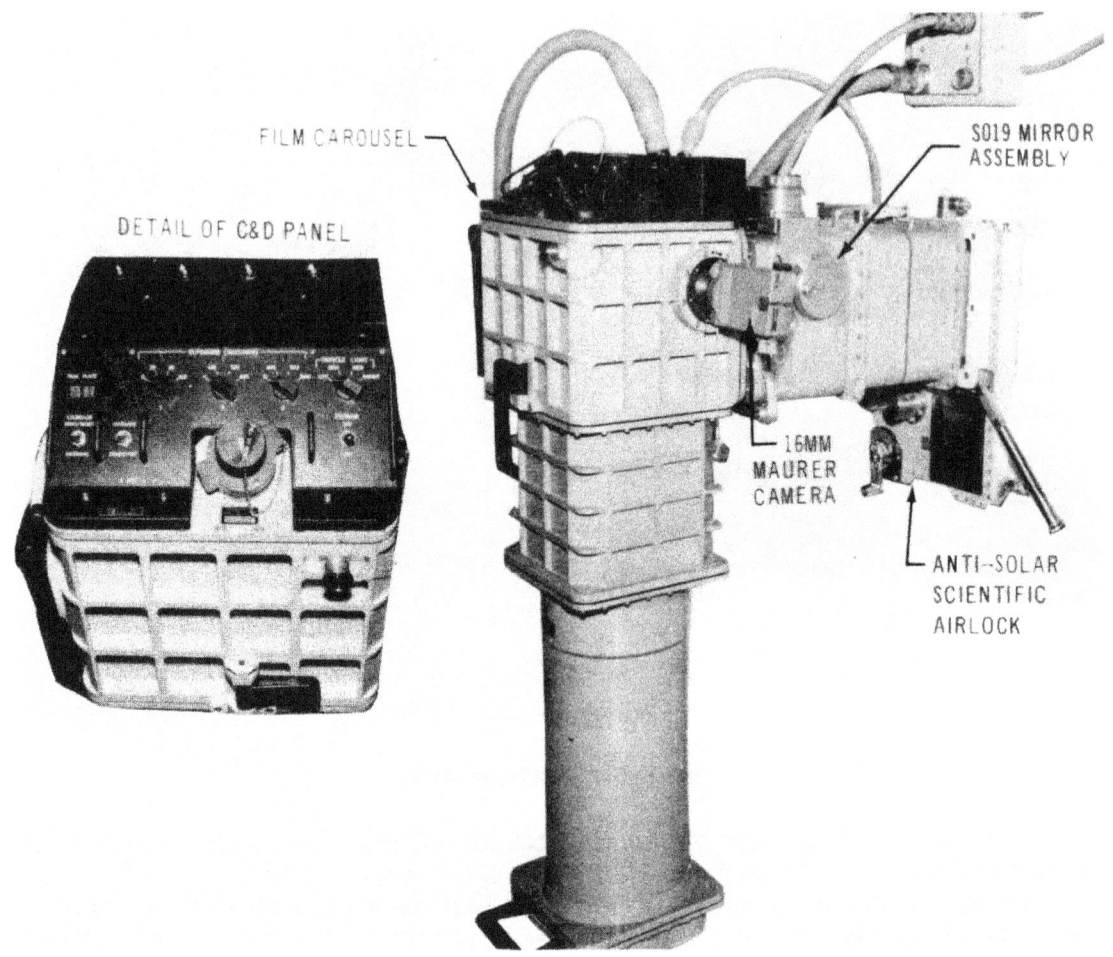

Figure III-39 — Experiment S183

The photographs are returned to the Earth for processing. From the film images, the amount of light emitted by stars in the three ultraviolet bands can be determined. These values can be compared with theoretical ultraviolet spectra and the spectra of brighter stars obtained by OAO and the S019 experiment to determine average ultraviolet colors or differences between stars of the same type. In addition, the variation with wavelength of observation of distant stars due to interstellar dust can be used to study the distribution and composition of this dust. The overall average ultraviolet color of a group of stars such as a cluster or a galaxy will be compared with the visible color of unexpected discrepancies.

Dr. Georges Courtes, Director of the Laboratoire d'Astronomie Spatiale du CRNS, Marseille, France, is the principal investigator. (Development and integration center MSFC, Contractor CRNS, France.)

Trans-Uranic Cosmic Rays (S228)

The objective of this experiment is to provide a detailed knowledge of the relative abundances of nuclei with Z (atomic map number) greater than 26 in the cosmic radiation, specifically to observe and identify as many trans-uranic nuclei as possible. This experiment will also observe and help determine upper limits on the flux or super heavy cosmic rays with Z greater than 110, and determine simultaneously the energy spectrum of cosmic rays with $Z = 26$, $Z > 60$, $Z > 85$, from about 1500 to 1500 MeV/nucleon.

The experiment will utilize plastic (Lexan) detectors mounted in the OWS and exposed to cosmic radiation (see Figure III-40). These cosmic rays will penetrate the detector packages streaking the plastic sheets within. Subsequent to return to Earth these plastic sheets will be chemically etched and the cosmic ray tracks will be measured with an optical microscope. With this technique, both the atomic number and energy of each particle can be determined.

Dr. P. Buford Price of the University of California, Berkley Campus, is the principal investigator. (Development and integration center MSFC.)

Magnetospheric Particle Composition (S230)

The objective of S230 is to measure the abundances of heavy, rare ions in the Earth's magnetosphere, principally the isotopes of the noble gases, and to compare these to the abundance ratios previously measured in the solar wind composition experiment used in the Apollo program. Analysis of experiment results will help to determine the origin of magnetospheric particles through careful study of the isotopes.

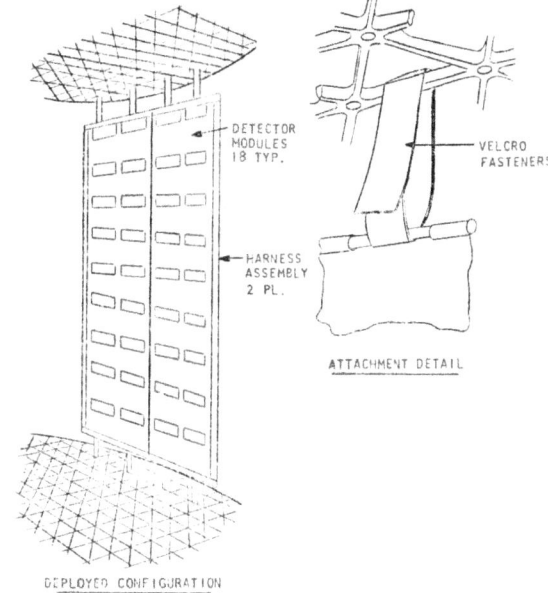

Figure III-40 — Experiment S228

The experiment consists of sheets of collecting foil of aluminum, aluminum oxide, and platinum mounted on the ATM Deployment Assembly truss. Each piece of foil, a rectangle of about 35 by 48 centimeters (14 by 19 inches) is mounted on a flexible backing material and formed into a cuff that is wrapped around one of two spools on the DA truss. Two cuffs are used per spool with one cuff being covered by the other.

Two cuffs from one spool are retrieved during the first and third EVA of SL-3 and two cuffs from the other spool are retrieved during the first and second EVA of SL-4. The cuffs will be stored in an envelope in the OWS food freezer until return to Earth for analysis.

Co-investigators are Dr. Don Lind, astronaut at the Manned Spacecraft Center, and Dr. Johannes Geiss, University of Bern, Switzerland. Development and integration center is MSFC.

MATERIALS SCIENCE AND MANUFACTURING IN SPACE

The condition of weightlessness in orbital flight makes it possible to conduct operations in materials processing that could not be done easily on Earth, if at all. Melting and mixing without the contaminating effects of containers, suppression of convection and buoyance in liquids and molten material, control of voids, and the ability to use electrostatic and magnetic forces otherwise masked by gravitation open the way to new knowledge of material properties and processes and ultimately to valuable new products for use on Earth. These potential products range from composite structural materials with highly specialized physical properties to large highly perfect crystals with valuable electrical and optical properties to new vaccines that could not be produced by conventional means on Earth.

NASA has been interested for some years in the properties of welding in zero gravity, information necessary for assembly in space. Several studies indicated that electron beam welding would be the best method of joining operations. The Marshall Space Flight Center began development work in 1963 on compact electron beam welders. Considerable interst was developing at that time in manufacturing products in space for use on Earth. The investigations described in this section will probe the feasibility of several specific processes and will help select the most promising processes and products for use in providing sound design criteria for future large scale facilities.

The progress of Space Shuttle planning has raised the prospect that vehicle capabilities sufficient to support large-scale experiment programs and limited commercial manufacturing operations will be made available. Practical experience developing the materials for Skylab has already proved of great value in concept planning of an improved and enlarged facility for the Space Shuttle program. Evaluation of Skylab results will help engineers finalize the design of Shuttle equipment.

Materials Processing Facility (M512)

The objectives of the experiments to be performed in the M512 facility are to demonstrate and evaluate the merits of molten metal phenomena for manufacturing in a space environment, to gain experience in molten metal characteristics in space for future application of constructure, assembly, and maintenance outside the Earth environment, and for the manufacture and retrieval of valuable products for use on Earth.

The facility will provide a basic apparatus and a common spacecraft interface for a group of metallic and non-metallic materials experiments and test and demonstrate a system approximating the "facility approach" projected for future space experimentation, where common hardware will be used to perform multiple experiments.

The M512 facility will demonstrate in space the feasibility of joining metallic materials by applying heat through an electron beam and an exothermic source, respectively. The experimental hardware will also be utilized as a common facility for materials processing experiments.

The M512 facility (Figure I-19), hard mounted in the MDA, consists of a vacuum work chamber with associated mechanical and electrical controls, an electron beam subsystem and a control and display panel. The vacuum chamber is a 40 centimeter (16.25 inch) sphere with a hinged hatch for access. It is connected to the space environment by a 10 centimeter (4 inch) diameter line containing two butterfly poppet valves. The electron beam subsystem is mounted to the chamber so that the beam traverses the sphere along a diameter parallel to the plane of the hatch closure. The chamber wall contains a cylindrical well accommodating the small electric furnaces used for the M518 and M555 experiments. Auxiliary provisions include ports for a floodlight and the 16mm data acquisition camera, a bleed line, a repressurization line and a port for a vacuum cleaner. A subsystem is also provided for spraying water into the chamber during some runs of the M479 experiment for quenching.

The electron beam operates nominally at 20 kV and 80 mA, and is provided with focusing and deflection coils that can be operated from the control panel.

The control panel has controls and displays for all of the experiments to be performed in the facility, including a gauge for the vacuum chamber, voltage and current meters for the electron beam and a thermocouple temperature indicator.

Data from the experiments will be comprised of the samples, those parts of the apparatus that are returned, motion picture records of the two electron beam experiments and M479, plus comments by the operating crewmen. The returned samples will be studied in comparison with control samples produced on Earth.

The experiments to be conducted using the M512 facility, the objectives of each and the principal investigators are:

Metals Melting (M551)
The objectives of Experiment M551 are to:
>Study the behavior of molten metals in micro-gravity.
>Characterize the structures formed in metals melted and rapidly solidified in zero gravity.
>Test means of joining and cutting metals by electron beam welding in zero gravity.

Richard M. Poorman of the Marshall Space Flight Center is principal investigator.

Exothermic Brazing (M552)
The objectives of Experiment M552 are to:
>Test and demonstrate a method of brazing components in space repair and maintenance operations.
>Study surface wetting and capillary flow effects in weightless molten metals.

J.R. Williams of the Marshall Space Flight Center is principal investigator.

Sphere Forming (M553)
The objective of the experiment is to demonstrate the effects of zero gravity on fundamental solidification phenomena. In particular, high purity nickel, a Ni-12% Sn alloy, Ni-1% Ag alloy, and Ni-30% Cu alloy, will be melted on stings and resolidified in both the free flowing and captive conditions.

These materials will permit study of the effects of low-gravity solidification phenomena which could apply equally to the majority of the more complex alloys and solidification procedures which are of commercial importance. The four materials selected allow investigation of solidifications which:
>Include metals which have invariant, narrow and wide melting ranges, yet all melt between 1225°C and 1455°C, simplifying the experimental procedure.
>Include a pure metal and alloys which have small density differences between the constituents.
>Will allow investigation of the effects of low-gravity on magnetic properties.

E.A. Hasemeyer of the Marshall Space Flight Center is principal investigator.

Single Crystal Growth (M555)
The objective of this experiment is to grow single crystals of gallium arsenide from solution in order to:
>Produce material of exceptionally high chemical and crystalline perfection.
>Have better doping homogeneity.
>Have more homogeneous starting melts.
>Have uniform growth.

Dr. M. Rubinstein of the Westinghouse Company and M. Davidson at Marshall Space Flight Center are the principal investigators.

Experiment M479 – Zero Gravity Flammability
The objective of Experiment M479 is to ignite various materials in a five psia oxygen/nitrogen mixture to determine:
>Extent of surface flame propagation and flashover to adjacent materials.
>Rates of surface and bulk flame propagation under zero convection.
>Self-extinguishment.
>Extinguishment by vacuum and water spray.

The combustion chamber and controls for this experiment are provided by Experiment M512. Individual flammability tests last from a minimum of 10 seconds to a maximum of four minutes. Each test will be photographed in its entirety so that combustion rates can be determined by post-flight analysis.

Additional data will be available from the voice comments of the astronaut performing the experiment. The astronaut will be invaluable in observing several aspects of the tests that may be missed by photography. He should be better able to note drift rates of detached fuel specimens, and sublimation products and overall energy profiles as well as trouble-shoot malfunctions of ignitors, cameras, etc., and report environmental changes exterior to the combustion chamber. Water spray patterns can also be reported.

Principal investigator is J.H. Kimzey of the Manned Spacecraft Center.

Multipurpose Electric Furnace (M518)

The Multipurpose Electric Furnace system provides a means to perform experiments on solidification, crystal growth and other processes involving phase changes in materials. The furnace system will be used to perform experiments involving phase changes at elevated temperatures in systems comprising selected combinations of solid, liquid and vapor phases. Because of the near zero gravity aboard Skylab, the liquid and vapor phases will be essentially quiescent and phases of different density will have little or no tendency to separate.

The system will consist of three main parts: the furnace, designed to interface with the M512 Materials Processing Facility; a programmable electronic temperature controller which will control the temperature levels in the furnace; and experiment cartridges which will contain the sample materials. The furnace will have three specimen cavities so that three material samples can be processed in a single run. The furnace is constructed to provide three different temperature zones along the length of each sample cavity, as follows:

> A constant temperature hot zone at the end of the sample cavity where temperature up to 1,000 degrees C (1,832 degrees F) can be reached.
>
> A gradient zone next to the hot zone where temperature gradients ranging from 20 degrees C (63 degrees F) to 200 degrees C (392 degrees F) per centimeter can be established in the samples.
>
> A cool zone in which heat conducted along the samples will be rejected by radiation to a conducting path that carries the heat out of the system.

Each sample of material will be enclosed in a cartridge in which the actual temperature distribution applied to the sample will be controlled by the thermal design of the cartridge.

The control package will provide active control of the furnace temperature. It can be set to any specified temperature within the furnace's capability by the astronaut operating the system. Two timing circuits in the controller will enable the astronaut to program the soak time spent at the set temperature and the cooling rate of the furnace at the end of the soak period. Active temperature control will continue during programmed cooling.

Once the specimens are installed in the furnace and the system activated by the astronaut, the system will operate automatically except for complete system shutdown. The material cartridges will be returned to Earth for examination.

The experiment equipment (Figure III-41) is comprised of the furnace, control package and 33 cartridges (11 experiment sets), cables to inter-connect the system and to connect the system to power and data outlets, two cartridge containers, a tube of thermal grease and a cartridge extraction tool.

The 11 experiments which will use the Multipurpose Electric Furnace, the objectives and principal investigators are:

Vapor Growth of II-VI Compounds (M556) — To determine the degree of improvement that can be obtained in the perfection and chemical homogeneity of crystals grown by chemical vapor transport under weightless conditions in space. Dr. Harry Wiedemeier, Rensselaer Polytechnic Institute, Troy, New York.

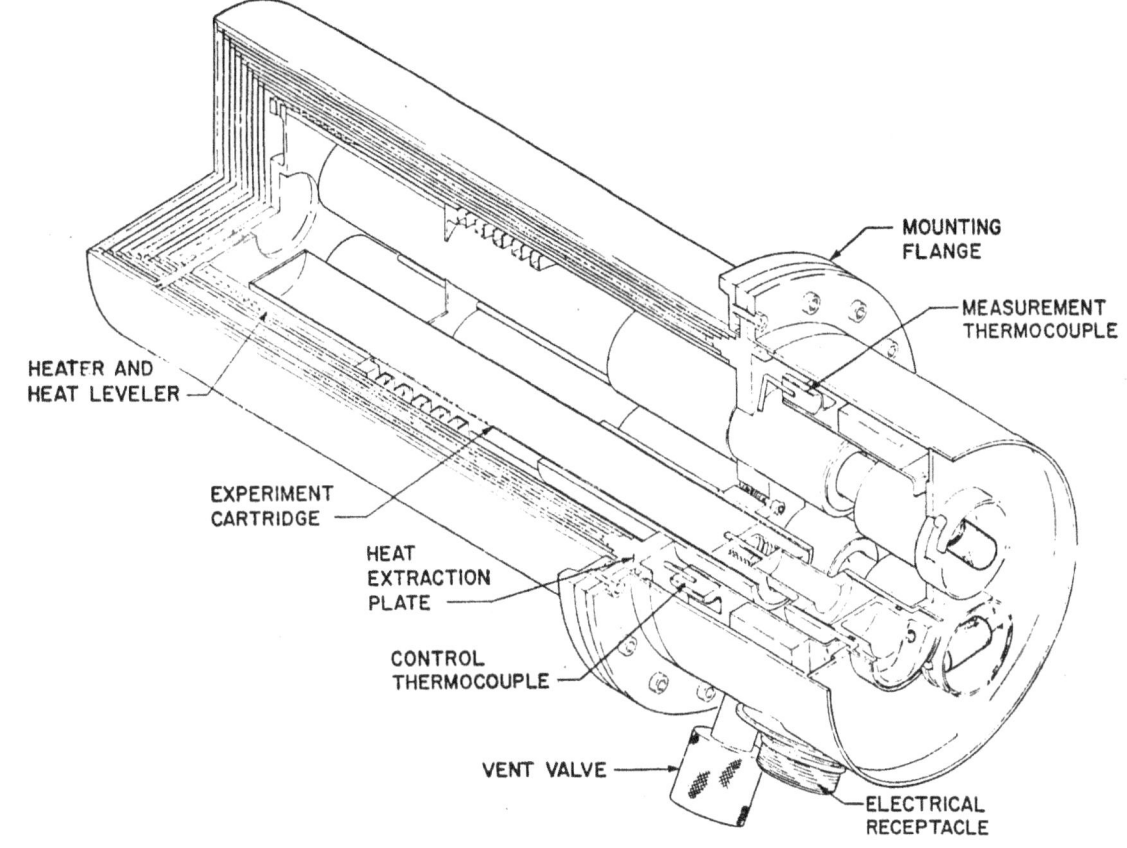

Figure III-41 — Experiment M518

<u>Immiscible Alloy Compositions (M557)</u> — To determine the effects of near-zero gravity on the processing of materials compositions which normally segregate on Earth. J.O. Reger, TRW Systems Group, Redondo Beach, California.

<u>Radioactive Tracer Diffusion (M558)</u> — To measure self-diffusion and impurity diffusion effects in liquid metals in space flight and characterize the disturbing effects, if any, due to spacecraft acceleration. Dr. Anthony O. Ukanwa, Marshall Space Flight Center.

<u>Microsegregation in Germanium (M559)</u> — To determine the degrees of microsegregation of doping impurities in germanium caused by convectionless directional solification under conditions of weightlessness. Dr. Francois A. Padavani, Texas Instruments Corporation, Dallas.

<u>Growth of Spherical Crystals (M560)</u> — To grow doped germanium crystals of high chemical homogeneity and structural perfection and study their resulting physical properties in comparison with theoretical values for ideal crystals. Dr. Hans Walter, University of Alabama at Huntsville.

<u>Whisker-Reinforced Composites (M561)</u> — To produce void-free samples of silver reinforced with oriented silicon carbide whiskers. Dr. Tomoyoski Kawada, National Research Institute for Metals, Tokyo.

Indium Antimonide Crystals (M562) — To produce doped semiconductor crystals of high chemical homogeneity and structural perfection and to evaluate the influence of weightlessness in attaining these properties. Dr. Harry C. Gatos and Dr. August F. Witt, Massachusetts Institute of Technology, Cambridge, Mass.

Mixed III-V Crystal Growth (M563) — To determine how weightlessness affects directional solidification of binary semiconductor alloys and, if single crystals are obtained, to determine how their semi-conducting properties depend on alloy composition. Dr. William R. Wilcox, University of Southern California, Los Angeles.

Metal and Halide Eutectics (M564) — To produce highly continuous controlled structures in samples of the fiber-like sodium fluoride-sodium chloride and plate-like bismuth-cadmium and lead-tin eutectics, and to measure their physical properties. Dr. Alfred S. Yue, University of California, Los Angeles.

Silver Grids Melted in Space (M565) — To determine how pore sizes and pore shapes change in grids of fine silver wires when they are melted and resolidified in space. Dr. A. Deruytherre, Catholic University of Belgium, Reverlee, Belg.

Copper-Aluminum Eutectic (M566) — To determine the effects of weightlessness on the formation of lamellar structure in eutectic alloys when directionally solidified. E.A. Hasemeyer, Marshall Space Flight Center. (This experiment replaces M554 — Composite Casting.)

III-6

ENGINEERING AND TECHNOLOGY EXPERIMENTS

The Engineering and Technology experiments will provide data which is important in the development of future space systems for exploration and the conduct of scientific experimentation. The result will be a better understanding of how man performs in space, what tools he needs to accomplish his tasks, and what his influence is on the space environment.

A number of the experiments are particularly oriented toward the interaction of man with his new zero-gravity environment. In this category are:
        Habitability and Crew Quarters, M487
        Astronaut Maneuvering Equipment, M509
        Crew Activities/Maintenance, M516
        Manual Navigation Sightings, T002
        Crew/Vehicle Disturbance, T013
        Foot-Controlled Maneuvering Unit, T020

The Astronaut Maneuvering experiments (M509 and T020) are closely allied. They investigate several different techniques for future use by man for extravehicular activity (EVA). In the Skylab program, these maneuvering units will be operated inside the SWS working volume.

Several experiments designed to study the spacecraft environment, both natural and induced are:
        Radiation in Spacecraft, D008
        Thermal Control Coatings, D024
        Thermal Control Coatings, M415
        Inflight Aerosol Analysis, T003
        Coronagraph Contamination Measurement, T025
        ATM Contamination Measurements, T027

The two thermal control coating experiments are complementary. The M415 experiment investigates the effect of the launch environment — Earth's atmosphere, retrorockets, etc. — on spacecraft surfaces. The D024 experiment investigates the long-term effects of the space environment, particularly sunlight, on spacecraft surfaces.

Radiation in Spacecraft (D008)

The purpose of Experiment D008 is to test advanced radiation instruments and techniques for determining the radiation effects on man and to provide correlative data for radiation hazard prediction methods for long duration manned spaceflight mission planning.

The D008 experiment (Figure III-42) is composed of a moveable tissue-equivalent dosimeter, a linear energy transfer spectrometer, and five passive dosimeters. The passive dosimeters integrate the dose received during the entire mission.

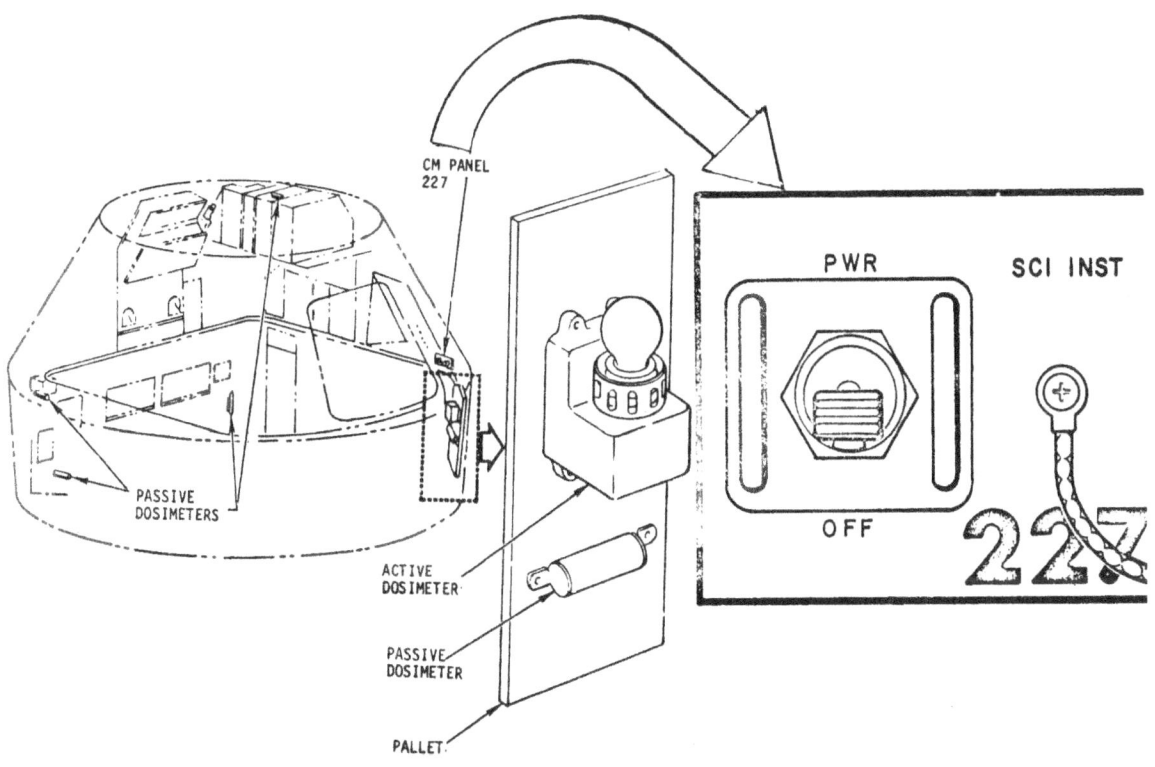

Figure III-42 — Experiment D008

The total weight of the D008 experiment hardware is 3.6 kilograms (8 pounds) and it occupies less than 0.03 cubic meter (one cubic foot).

D008 is scheduled for flight aboard SL-2. The experiment hardware remains in the CM for the course of the mission. The following schedule has been established for D008: Perform two active dosimeter surveys with crew participation in the South Atlantic Anomaly and two while centered around the most northern magnetic latitudes of the orbit. Outputs from the active dosimeters will be recorded on the Data Storage Equipment (DSE) for six consecutive passes per day for 14 days. This includes five passes per day through the SAA and one pass per day at the most northerly latitudes. No crew participation is required.

If a solar flare radiation event occurs, active dosimeter surveys with the crew participation will be conducted.

Principal Investigator for the D008 experiment is Captain Andrew D. Grimm, and Co-investigator is Joseph F. Janni, both of Kirtland Air Force Base, New Mexico. (Development and integration center MSC, contractor AVCO Corp.)

Thermal Control Coatings (D024)

This experiment, consisting of exposing material samples to the space environment has the following objectives:

> Determine the effects of near-Earth space environments on selected experimental thermal control coatings which have been extensively investigated in the laboratory.
>
> Correlate the effects of the space environment on these coatings with measured effects of ground-based simulated space environments.
>
> Gain new understanding of the mechanisms of degradation of thermal control coatings caused by actual space radiation.

This experiment is a companion to Experiment M415, which will determine the effects of the launch environments on thermal control coatings.

Experiment D024 will provide the first opportunity to examine in detail coating samples that have been chemically or physically unaltered since retrieval from space. Test results will enable a better prediction of how currently available coatings degrade in space.

The experiments package (Figure III-43) consists of four panels, two of which contain 36 thermal control coating samples. The samples are 2.54 cm (1 inch) diameter discs coated with various selected thermal control coatings. The other pair of panels contains strips of polymeric plastic five mils thick. The panels are square plates, about 17 centimeters (6½ inches) on a side and 0.6 centimeter (1/4 inch) thick. Each has a flexible handle to prevent contamination of the samples while handling.

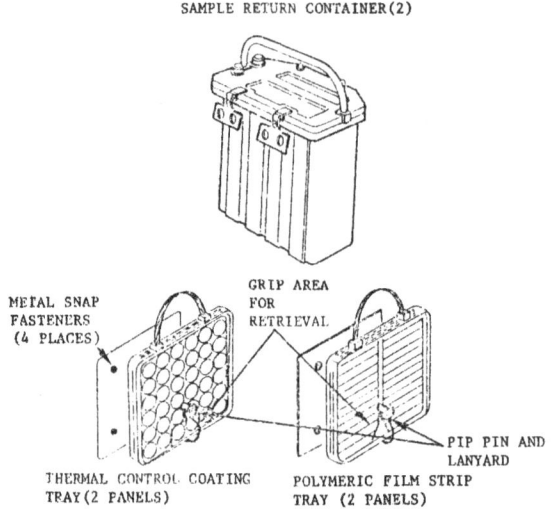

Figure III-43 – Experiment D024

Panels will be attached with snap fasteners to the AM truss assembly. Here they will receive no cluster shadowing in the solar inertial attitude held during most of the mission. Protective covers will be removed from the panels no later than 24 hours before launch. Protected by the payload shroud during launch, the samples will not be affected by the launch environment.

One of each type of the thermal control sample panels will be retrieved and returned to Earth on Skylab flight SL-2 and the other set of panels will be retrieved and returned on flight SL-3. During an EVA an astronaut will retrieve the panel and place it in a return sample container. This hermetically sealed container will maintain a vacuum for the samples until they reach a ground-based laboratory where they will be placed in a vacuum chamber and a vacuum established before removal. Spectral reflection and all other measurements will be made in the chamber.

Dr. William L. Lehn of the Air Force Materials Laboratory, Wright-Patterson Air Force Base, is Principal Investigator. (Development center Wright-Patterson AFB; integration center MSFC.)

Thermal Control Coatings (M415)

The objective of Experiment M415 is to determine the degradation effects of pre-launch, launch and space environments on the thermal absorption and emission characteristics of various coatings commonly used for passive thermal control.

The principal elements of this experiment (Figure III-44) consists of two panels, each containing 12 thermal sensors arranged in four rows of three. Three different thermal control coating samples are mounted on the sensors in each row. All of the coating samples are thermally isolated from surrounding structures. Unlike Experiment D024, detailed spectral reflection measurements cannot be made since the coatings will not be retrieved. Thermal properties will be measured by temperature sensors and the data telemetered to the ground.

Eugene C. McKannan, Marshall Space Flight Center, is Principal Investigator. (Development and integration center MSFC.)

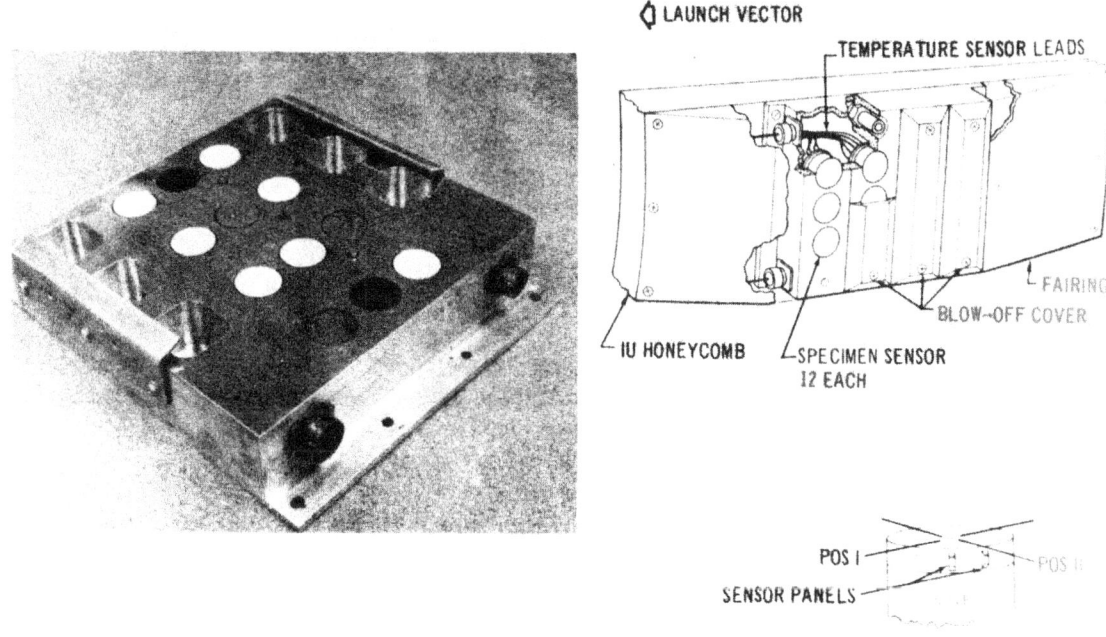

Figure III-44 — Experiment M415

Habitability/Crew Quarters (M487)

Throughout the three manned visits to the Skylab flight crews will be asked to evaluate everyday spacecraft type activities such as sleeping, eating and the ease or difficulty of getting around. The crew will document with film and tape-recorded comments the habitability features of the workshop as part of Experiment M487 — Habitability/Crew Quarters, the objectives of which are to provide data useful in the design of future manned spacecraft.

Habitability features such as architecture, environmental elements and communications techniques will affect every day spacecraft activities and crew performance. Objective and subjective data will be obtained on OWS environment, internal architecture, adequacy of mobility aids and restraints, food and water, garments and personal accouterments, personal hygiene, housekeeping, internal communications, and subjective data on the adequacy of the off-duty activity provisions.

Instruments including a measuring tape, portable thermometers, surface temperature digital thermometer, sound meter and frequency analyzer, velometer and force gauge are provided for the experiment.

Principal Investigator for this experiment is Caldwell C. Johnson, Jr., of the Manned Spacecraft Center. (Development and integration center MSFC, contractor McDonnell Douglas Astronautics Co., Western Division.)

Astronaut Maneuvering Equipment M509

The objectives of Experiment M509 are to: demonstrate Astronaut Maneuvering Unit flying qualities and piloting capability; test and evaluate system response; and relate the data and experience gained to ground based analysis, future AMU design requirements and projected EVA capabilities.

The astronaut maneuvering equipment (Figure III-45) consists of two jetpowered AMUs, a back-mounted hand-controlled unit called the automatically stabilized maneuvering unit (ASMU or backpack) and a hand-held maneuvering unit (HHMU).

The ASMU has a rechargeable/replaceable high pressure nitrogen propellant tank and battery. Control moment gyro and reaction jet stabilization modes are provided. A third mode allows the pilot to fire the reaction jets directly through the hand controllers. The ASMU provides propellant and instrumentation for evaluation of the HHMU mode.

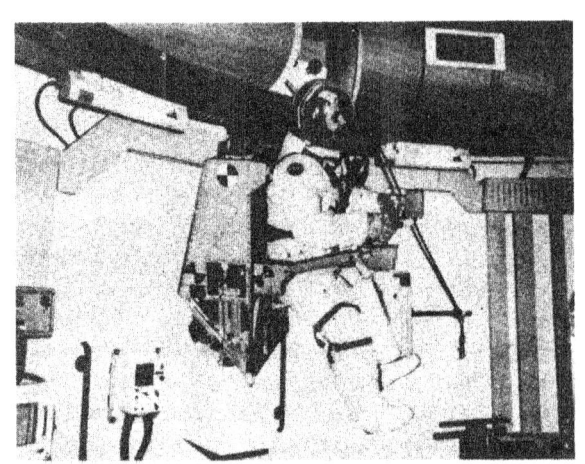

Figure III-45 -- Experiment M509

The ASMU is maneuvered by 14 fired thrusters located in various positions on the backpack. The thrusters are controlled by two hand-controllers mounted on arms extending from the backpack. The left hand controls translation forward, backward, up, down, and sideways; and the right hand, using an aircraft-type handgrip, controls rotation in any direction.

The HHMU is a handgrip unit with a pair of thrusters that pull the astronaut forward, a single thruster that pushes him backwards, and thruster controls.

Skylab crewmembers will fly the units in a shirt sleeve and pressure suit mode inside the forward dome of the workshop. Four experiment runs are planned with two crewmen performing the operation, one serving as pilot and the other as observer.

The observer will operate the cameras, cue the pilot on test operations procedures, and analyze and describe the test progress over the voice communication system. Comments made by the pilot and observer during and immediately after each experiment run (varying from 50 to 80 minutes duration) will be taped and dumped to tracking stations at the scheduled intervals. A total of three crewmen will fly the units, one pilot on each Skylab mission.

Principal Investigator for M509 is Major C.E. Whitsett Jr., assigned to the Air Force Space Transportation System office at the Manned Spacecraft Center. (Development center MSC, integration center MSFC, contractor Martin-Marietta Aerospace.)

Crew Activities/Maintenance (M516)

This experiment investigates crew performance in zero gravity, long duration missions, primarily through observations of normal Skylab tasks. It is related closely to M151 and M487. The experiment calls for:

Systematic documentation of man's performance during prolonged weightless space flight.

Acquisition and evaluation of inflight maintenance data.

Evaluation of data relative to design criteria for Skylab and future missions.

Evaluation and report of findings in terms useful to future manned mission planners.

Performance data will be gathered in the areas of manual dexterity, locomotion, mass handling and transfer, and maintenance.

M516 will be handled in the following phases:

Preflight — Crew performance data acquired during preflight simulations and training sessions will be used to establish baseline data for comparison with data acquired inflight.

Inflight — M516 will use film coverage provided by various other experiments and operational activities which relate to crew performance activities.

Postflight — The crewmen will provide subjective and technical comments during debriefing regarding crew performance activities.

Principal Investigator is R.L. Bond of the Manned Spacecraft Center. (Development and integration center MSC, contractor Martin-Marietta Aerospace.)

Manual Navigation Sightings (T002)

The objective of Experiment T002 is to investigate the effects of the space flight environment (including long mission time) on the navigator's ability to take space navigation measurements through a spacecraft window using hand-held instruments.

Previous data obtained with the use of simulators, aircraft, and the Gemini spacecraft has already demonstrated that man, in a space environment, can make accurate navigation measurements using simple hand-held instruments. This, together with already developed techniques for reducing the data to a position determination, means that a technique is available for man to navigate in space using simple instruments and without a computer. The intent of this experiment is to determine whether long mission duration appreciably affects the capability of man to obtain accurate measurements.

The equipment for this experiment (Figures III-46 and III-47) consists of two hand-held instruments, a sextant and a stadimeter. The sextant, similar to a navigator's sextant, will measure the angles between two stars, and between single stars and the edge of the Moon. The stadimeter, also an optical device, determines spacecraft altitude by measuring the apparent curvature of the horizon.

Data returned will be logbook entries of the sextant and stadimeter readings, supplemented as required by crew comments on the voice tape recorder.

Robert J. Randle, ARC, and Maj. Stanley Powers, USAF, are co-investigators. (Development center ARC; integration center MSFC; Contractor Kollsman Instrument Co.)

In-Flight Experiment Aerosol Analysis (T003)

The objective of Experiment T003 is to measure the sizes, concentration and composition of particles in the atmosphere inside the Skylab as a function of time and location.

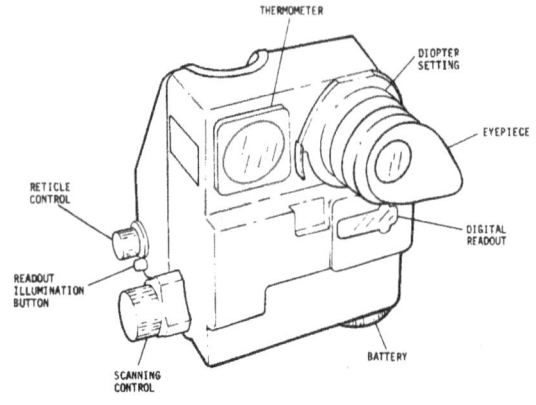

Figure III-46 — Experiment T002 Sextant

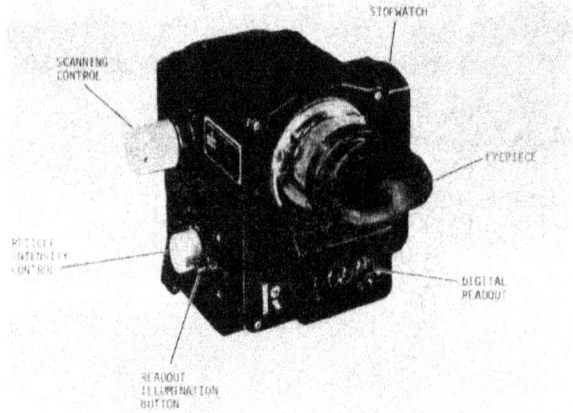

Figure III-47 — Experiment T002 Stadimeter

An aerosol analyzer has been designed for Experiment T003 which is capable of separating particulates into three size ranges, accumulating the total particles in the three size intervals and displaying the results immediately to the astronaut. The instrument also contains a particulate collection system so that post-flight analysis can be used to ascertain the shape and composition of the individual particles.

The instrument for Experiment T003 is a multi-channel, battery operated particle counter capable of sorting aerosol particles into three size groups: 1 to 3 microns, 3 to 9 microns, and 9 to 100 microns.

The experiment equipment (Figure III-48) consists of a Stowage Container, Aerosol Analyzer, Protective Container Assemblies, Filter Inserts and Log Cards and Clips.

The instrument is hand-held by an astronaut at the desired point of measurement. Representative locations throughout the spacecraft will be tested.

Data will be a written log and returned filter.

William Z. Leavitt, Ph.D., Department of Transportation, is Principal Investigator. (Development and integration center MSFC.)

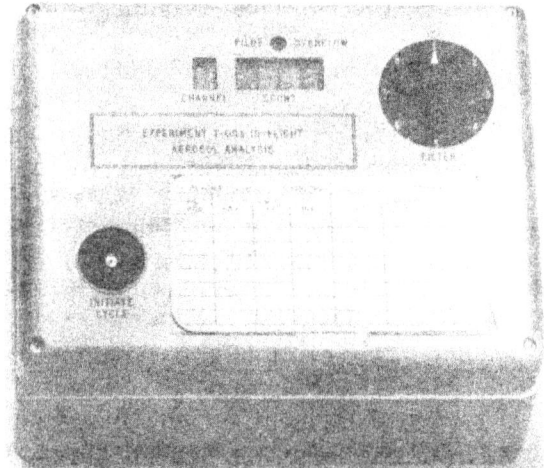

Figure III-48 — Experiment T003

Crew/Vehicle Disturbances (T013)

The objectives of this experiment are to: measure the effects of various crew motions on the dynamics of manned spacecraft, specifically the torques, forces, and vehicle motions produced by the astronaut's body motions; to verify information obtained from ground simulation programs; and to determine the effects of astronaut motion on the attitude and control of the vehicle.

Experiment T013 is designed to resolve uncertainties about crew motion effects on spacecraft and provide system designers with accurate models of crew motion distrubances.

Body motion of the astronaut will be measured by the Limb Motion Sensing System (LIMS) (Figure III-49) which is a skeletal structure incorporated into a suit, with pivots at the major body joints. Each pivot is monitored by a linear potentiometer which provides a continuous measurement of body limb position as the subject astronaut performs the assigned task. On-board motion picture photography, using two 16mm Data Acquisition Cameras, will be used concurrently with the LIMS.

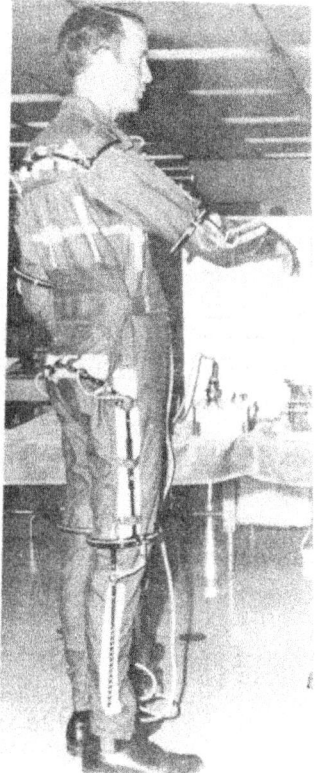
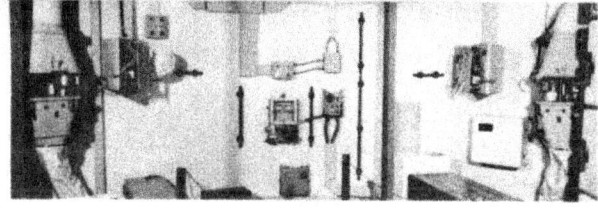

Figure III-49 — Experiment T013

A Force Measuring System (FMS), consisting of two Force Measuring Units (FMUs), will be used to measure the forces and moments applied to the OWS structure during the assigned task, which includes soaring between the FMUs.

The measurement data of the LIMs and FMS are processed and telemetered to the ground along with real time transmission of the applicable ATM Pointing Control System data. The 16mm film will be returned in the CM.

Bruce A. Conway, Langley Research Center, is Principal Investigator. (Development center LaRC; integration center MSFC; Contractor Martin-Marietta.)

Foot-Controlled Maneuvering Unit (T020)

This experiment will evaluate an astronaut maneuvering device that does not require use of the astronaut's hands. Both the ASMU (Automatically Stabilized Maneuvering Unit) and the HHMU (Hand-Held Maneuvering Unit) used in Experiment M509 required the astronaut to use his hands to control the unit. The foot-controlled maneuvering unit (FCMU) is a foot-operated propulsion device that is straddled by the operator as if riding a bicycle. The unit is propelled by high pressure nitrogen contained in the detachable propellant tank used in the M509 experiment (Figure III-50).

Donald E. Hewes, Langley Research Center, is the Principal Investigator. (Development center LaRC, integration center MSFC, Contractor Martin-Marietta.)

Coronograph Contamination Measurements (T025)

The primary objective of Experiment T025 is to visually and photographically observe and record the amount of light scattered by particles from thruster firings and waste dumps. One purpose of T025 is to determine the extent and nature of the induced contaminant and to assess its effect on other optical experiments on the spacecraft. Another objective is to look through the Earth's upper atmosphere to determine the type and amount of particulate matter.

The T025 experiment hardware consists of a modified 35-mm Nikon camera attached to a coronagraph which is fitted into the solar Scientific Airlock.

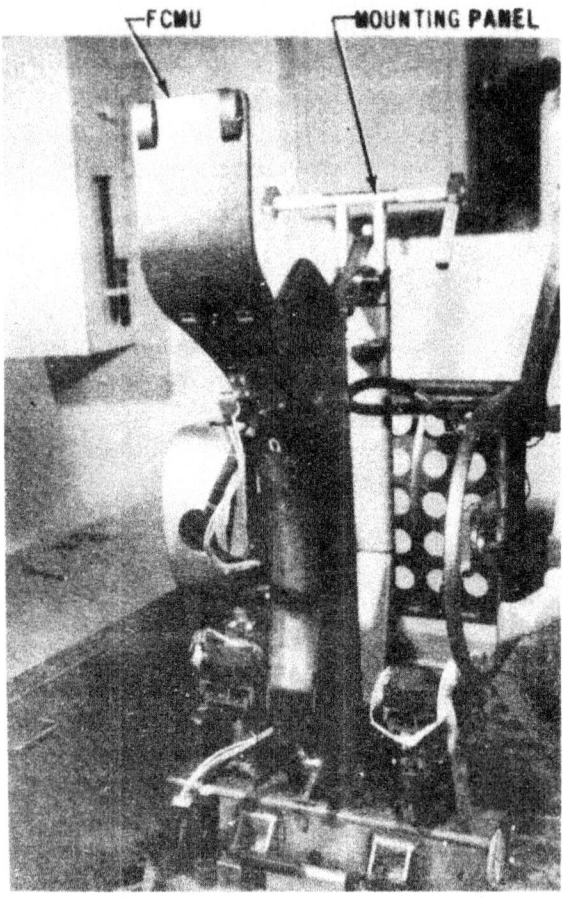

Figure III-50 -- Experiment T020

The coronagraph will be placed in the SAL, the occulting discs will be extended to occult the solar disk and the SWS will be oriented so that only scattered light from particulate matter is recorded on the photographic film. Two photographic sequences (48 exposures per sequence) will be conducted by the crew when there is a minimum of contaminates being vented to the atmosphere surrounding the SWS. One sequence will be conducted as early and the other as late in the mission as possible. Five additional sequences will be performed on SL-2 during periods of high contamination. A total of approximately 210 exposures are required to satisfy the experiment objectives.

Included in the 210 exposures will be sequences made while the M512 materials processing experiment is venting contaminates into space and others while trash is being dumped into the waste tank.

Principal Investigator for this experiment is Dr. Mayo Greenberg of Dudley Observatory. Co-investigator is George Bonner of the Manned Spacecraft Center. (Development Center MSC, integration center MSFC, contractor Martin Marietta.)

Contamination Measurements (T027)

This experiment consists of two separate pieces of hardware, T027 Sample Array (Figure III-51) and the T027/S073 Photometer System (Figures III-52 and III-53), each with its own objectives. The Universal Extension Mechanism, part of the Photometer System, is also used to deploy the portable TV and Experiment S149 through the SAL. Photometer system hardware is also used to meet the objectives of Experiment S073.

Figure III-51 — T027 Sample Array

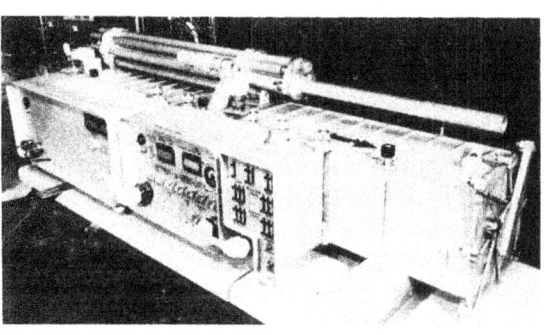

Figure III-52 — T027 Photometer System

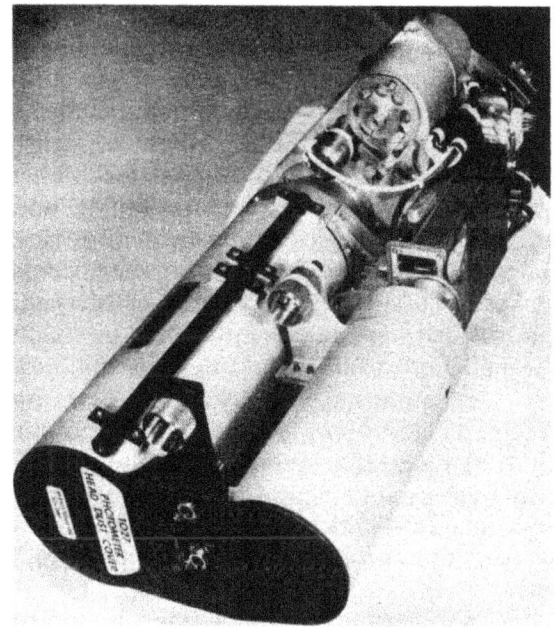

Figure III-53 — T027 Photometer Head

<u>T027 Sample Array</u> — The basic objective of the T027 Sample Array experiment is to obtain controlled data on the degradation effects of contaminants associated with Skylab on the optical properties of various windows, mirrors and diffraction gratings. A secondary objective is to obtain near-real-time data on contamination rates during sample exposure by use of Quartz Crystal Microbalances.

Window contamination on Gemini and Apollo flights have interfered with star sightings and lunar surface photography experiments. Sources of contaminant depositions have been found to be thruster firings and molecular evaporations from various materials on the spacecraft.

III-61

The Sample Array System will expose optical samples to the space environment for controlled periods. The Sample Array will be deployed outside the spacecraft on a boom. A total of 248 samples of different types will be exposed. The samples consist of window materials, mirrors, gratings and other optical surfaces suitable for various wavelength regions.

When the experiment is completed, the Sample Array will be sealed in space and returned to Earth for analysis and study.

T027 Photometer System — The objective of the T027 Photometer System is to measure the brightness and polarization of the scattered sunlight from the solar illumination of the particulate contaminant cloud surrounding Skylab. Variations of the contaminant cloud with respect to time and location will also be measured.

The Photometer System will measure three parameters which fully characterize the radiation from the skyglow and from the Skylab corona; i.e., brightness of the total and of the polarized components, and orientation of the plane of polarization. Measurements pertaining to the skyglow (zodiacal light, Gegenschein, starlight, F-region airglow) are best performed on the dark side. Measurements on the sunlit side and at the terminator will be used to characterize the contaminant cloud and to provide information on the skyglow.

Of particular importance is the measurement of the amount of polarization, which increases in proportion to the amount of light scattered off particulate matter surrounding Skylab. The optimum time for discrimination of the Skylab corona with respect to the nightglow (primarily zodiacal light) occurs just before the Skylab leaves, or enters, the Earth's shadow.

The principal method of collecting photometric data which will satisfy experiment objectives is to scan the areas under study with a photoelectric polarimeter. The data obtained from this system, which includes a rotating polarizer and a DC detection system with photomultiplier tube, will subsequently be used in a computer program to provide synchronous detection by digital techniques. Auxiliary data includes photographs of the observed areas, plus adequate timing and crew comments to correlate the recorded photometric data with the position information determined from the photographs.

The photoelectric photometer is deployed through the Scientific Airlock (SAL) by means of the Universal Extension Mechanism. The Photometer Head is mounted on a gimballed system at the end of the mechanism to permit scanning in elevation and azimuth through limits of 0 to 112.5 degrees and 0 to 354 degrees, respectively.

The photometer head contains an optical train with a polarizing disk, ten selectable filters, a field-of-view system and a photomultiplier tube to sense and analyze the integrated light entering the system. A radioactive calibration source is also provided to allow automatic system calibration. A data acquisition 16 mm camera is included to provide film data for post-flight analysis. Sun shields are provided for both the photometer and camera systems to limit extraneous light and allow data acquisition to within 16 degrees of the Sun line.

An automatic programmer, in addition to the manned control assembly, is included to permit pre-programmed automatic measurement sequences with the photometer system.

During stowage, the extension mechanism and photometer head are retracted into the experiment canister. A flight stowage container is supplied to house and protect the photometer unit when it is not installed in the SAL.

Data is returned by telemetry during operation of the experiment and CM return of film and Data Log Books.

Principal Investigator is Joseph A. Muscari, Ph.D., of Martin-Marietta Aerospace. (Development and integration center MSFC, contractor Martin-Marietta Aerospace.)

STUDENT EXPERIMENTS

NASA and the National Science Teachers Association in October 1971 announced a national competition for students in grades 9 through 12 to propose an experiment for Skylab. NSTA administered the competition in cooperation with the NASA Educational Program Division and Skylab Program Office. More than 3,400 proposals were received, evaluated and judged by the NSTA. Of these 3,400 proposals, 301 were selected as regional winners and 25 national winners were announced in March 1972. Of these 25 national winners, 11 were selected for development of hardware for Skylab, eight others will use existing hardware and six could not be accommodated as proposed because of Skylab performance requirements or schedule constraints. Four of the 6 that could not be accommodated will be provided data from existing Skylab experiments which either are similar to the students' proposal or satisfy an alternate interest of the student. The remaining two students will be associated with NASA researchers in areas closely related to their proposed subject of interest.

All student experiments are being handled in a manner very similar to the mainline Skylab experiments with the Student Investigator assuming the role of Principal Investigator under the guidance and with the assistance of a Science Adviser from the Marshall Space Flight Center. Consulting Science Advisers have also been assigned from the Manned Space Flight Center in Houston.

The 25 students and experiments selected for participation in the Skylab program are:

JOE B. ZMOLEK, 1914 Hazel Street, Oshkosh, Wisconsin, 54901.
   "Atmospheric Heat Absorption" –ED-11
   Lourdes High School, William L. Behring, Teacher/Sponsor.

Joe's objective is to determine the attenuation, due to the Earth's atmosphere, of radiant energy in the visible and near IR regions over both densely populated and sparesely populated sections of the Earth. High resolution photographs of the actual target areas will enable identification of the portions of Earth surveyed by two, on-board, spectrometer systems which measure the energy radiated from the Earth.

EREP (Earth Resources Experiment Program) and ground truth data over the target sites will be utilized to determine the incident and reflected energy at the Earth's surface.

TROY A. CRITES, 736 Wynwood Drive, Kent, Washington, 98031.
   "Volcanic Study" – ED-12
   Kent Junior High School, Richard C. Putnam, Teacher/Sponsor.

The aim of this experiment is to determine the feasibility of predicting volcanic activity based on remotely sensed thermal infrared surveys. These surveys will be carried out using on-board Earth Resources Experiment sensors. The Skylab data will be correlated with ground truth data from instrumented volcanic sites to establish a relationship between volcanic activity and thermal contours and patterns.

ALISON HOPFIELD, 183 Hartley Avenue, Princeton, New Jersey, 08540.
"Libration Clouds" – ED-21
Princeton Day School, Normal Sperling, Teacher/Sponsor.

Alison's experiment will make use of one of the ATM experiments telescopes to observe the two zero-force regions in the Earth-Moon System where it is expected that small space particles will accumulate. Photographs of these two regions, one ahead of the Moon and the other following it in its path, should exhibit greater brightness than surrounding regions.

DANIEL C. BOCHSLER, Route 2, Box 75, Silverton, Oregon, 97381.
"Objects Within Mercury's Orbit" – ED-22
Silverton Union High School, John P. Daily, Teacher/Sponsor.

This experiment will attempt to identify a planetary body (or any other identifiable object) which may orbit the sun at a radius substantially less than that of Mercury's orbit. This identification will be attempted by a detailed analysis of some 30,000 photographs obtained with a Skylab solar telescope system.

JOHN C. HAMILTON, 12 Honu Street, Aiea, Hawaii, 96701.
"UV From Quasars" – ED-23
Aiea High School, James A. Fuchigami, Teacher/Sponsor.

In this experiment, selected photographs obtained by the ultraviolet stellar astronomy equipment will be analyzed. Photographs of target areas in which quasars have been identified will be studied to obtain spectral data in the ultraviolet region to augment existing data in the radio and visible ranges.

JOE W. REIHS, 12824 Wallis Street, Baton Rouge, Louisiana, 70815.
"X-Ray Stellar Classes" – ED-24
Tara High School, Helen W. Boyd, Teacher/Sponsor.

The primary aim of this experiment is to make observations of celestial regions in X-ray wave lengths in an attempt to relate X-ray emissions to other spectral characteristics of observed stars. Due to the fact that the Skylab X-ray telescopes were designed primarily for solar observation, it is predicted that the Stellar X-ray data will be marginal. If stellar X-ray data is inadequate, Joe will be supplied solar X-ray data for analysis.

JEANNE L. LEVENTHAL, 1511 Arch Street, Berkeley, California, 94708.
"X-Rays From Jupiter" – ED-25
Berkeley High School, Harry E. Choulett, Teacher/Sponsor.

Jeanne's experiment will attempt to detect X-rays from the planet Jupiter and establish the correlation of X-ray emission with solar activity and decametric radio emission. Since the Skylab X-ray telescopes were designed for solar observation, their threshold sensitivity is deemed inadequate to detect normal Jovian X-ray emission. However, in the event of a solar flare, the X-ray emission from Jupiter is expected to increase enough that detection will be feasible. If no solar flares occur during the Skylab mission, Jeanne will share in analysis of the SCO X-1 X-ray data under the planned stellar observation program.

NEAL W. SHANNON, 2849 Foster Ridge Road, Atlanta, Georgia, 30345.
"UV From Pulsars" — ED-26
Fernbank Science Center, Dr. Paul H. Knappenberger, Teacher/Sponsor.

Neal's experiment is to attempt to measure the radiation from known Pulsars in the UV spectral region to determine whether or not the UV data correlates with known existing Pulsar Spectral data. Photographs of selected celestial regions, including the Crab nebula, will be made using the Skylab Ultraviolet Stellar Astronomy equipment.

The preceding eight experiments make use of existing Skylab hardware. Each of the above eight students will work closely with Skylab Principal Investigators or Project Scientists in analyzing and interpreting their data.

The following 11 student experiments involve the development of unique hardware capable of operating in the Skylab environment.

ROBERT L. STAEHLE (Figure III-54), Huntington-Hills, North, Rochester, New York, 14622.
"Bacteria and Spores" — ED-31
Harley School, Alan H. Soanes, Teacher/Sponsor.

Colonies of selected, non-pathogenic bacteria will be incubated at a specified temperature in the Skylab weightless environment. The resulting colonies will be photographed periodically during their development cycle to assess the differences in survival, growth and mutations when compared with a similar group of earth environment spores. Both the photographs and the bacterial colonies (inhibited from further growth) will be returned to Earth for further study.

TODD A. MEISTER (Figure III-55), 33-04 93rd Street, Jackson Heights, New York, 11372.
"In Vitro Immunology" — ED-32
Bronx High School of Science, Vincent G. Galasso, Teacher/Sponsor.

This experiment attempts to determine the extent to which the absence of gravity affects the in vitro demonstration of the immune-response mechanism. The technique employed is a precipitin reaction (radial immune diffusion) in vitro, using various concentrations of human antigen to be reacted against an agar suspension of specific antibody under spacecraft conditions.

The reaction between antigen and antibody forms a precipitin ring, the diameter of which is a function of the reaction between antigen and antibody.

The results are observed and photographed for subsequent comparison to control samples run on Earth.

KATHY L. JACKSON (Figure III-56), 18618 Capetown Drive, Houston, Texas, 77058.
"Motor Sensory Performance" — ED-41
Clear Creek High School, Mary K. Kimzey, Teacher/Sponsor.

This experiment utilizes a standardized eye-hand coordination test apparatus to measure the changes in fine, manipulative capabilities of a crew member due to extended exposure to the Skylab environment.

Figure III-54 — MSFC Advisor Steven Hall (left) and Robert L. Staehle

Figure III-55 — MSFC Advisor Dr. Robert Allen (left) and Todd Meister

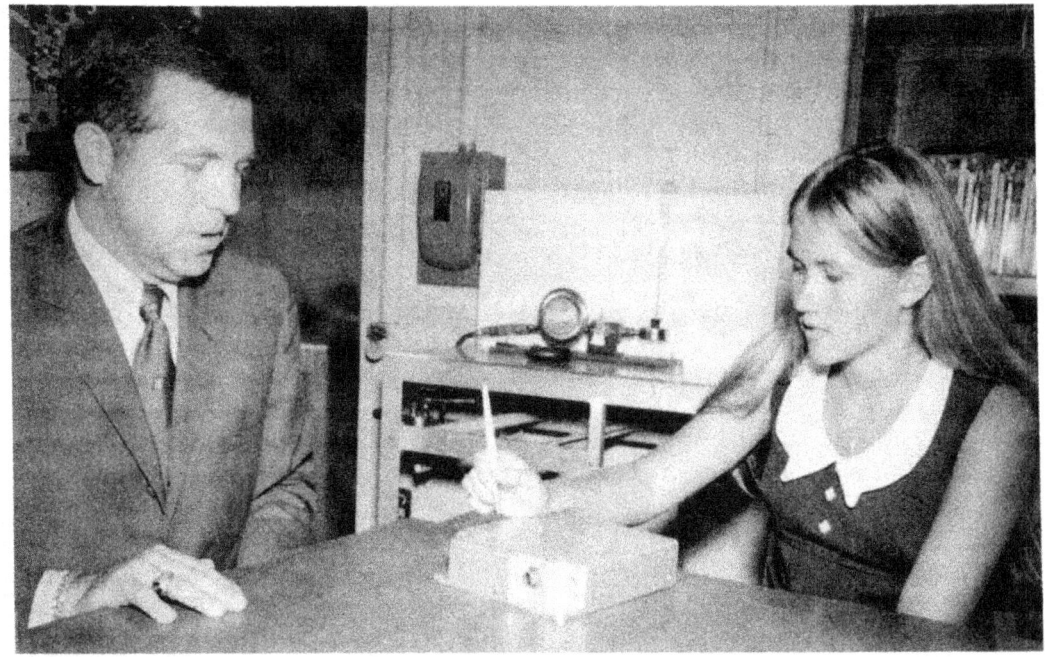

Figure III-56 — MSFC Advisor Dr. Robert Allen and Kathy Jackson

JUDITH S. MILES (Figure III-57), 3 Dewey Road, Lexington, Massachusetts, 02173.
"Web Formation" — ED-52
Lexington High School, J. Michael Conley, Teacher/Sponsor.

This experiment will observe the web-building process and the detailed structure of the web of the common cross spider (Araneus diadematus) in a normal environment and in a Skylab environment. Analysis of experiment results will be similar to analysis of web structure experiments performed by the Research Division of the North Carolina Department of Mental Health, Raleigh, N.C.

JOEL G. WORDEKEMPER (Figure III-58), 810 East Sherman Street, West Point Nebraska, 68788.
"Plant Growth" — ED-61
Central Catholic High School, Lois M. Schaaf, Teacher/Sponsor.

DONALD W. SCHLACK (Figure III-58), 9217 Appleby Street, Downey, California, 90240.
"Plant Phototropism" — ED-62
Downey High School, Jean C. Beaton, Teacher/Sponsor.

These two experiments have been combined into a single joint experiment whose objectives are:

To determine the differences in root and stem growth and orientation of rice seeds in specimens grown in zero gravity and on Earth under similar environmental conditions.

To determine whether light can be used as a substitute for gravity in causing the roots and stems of rice seeds to grow in the appropriate direction in zero gravity, and to determine the minimum light level required.

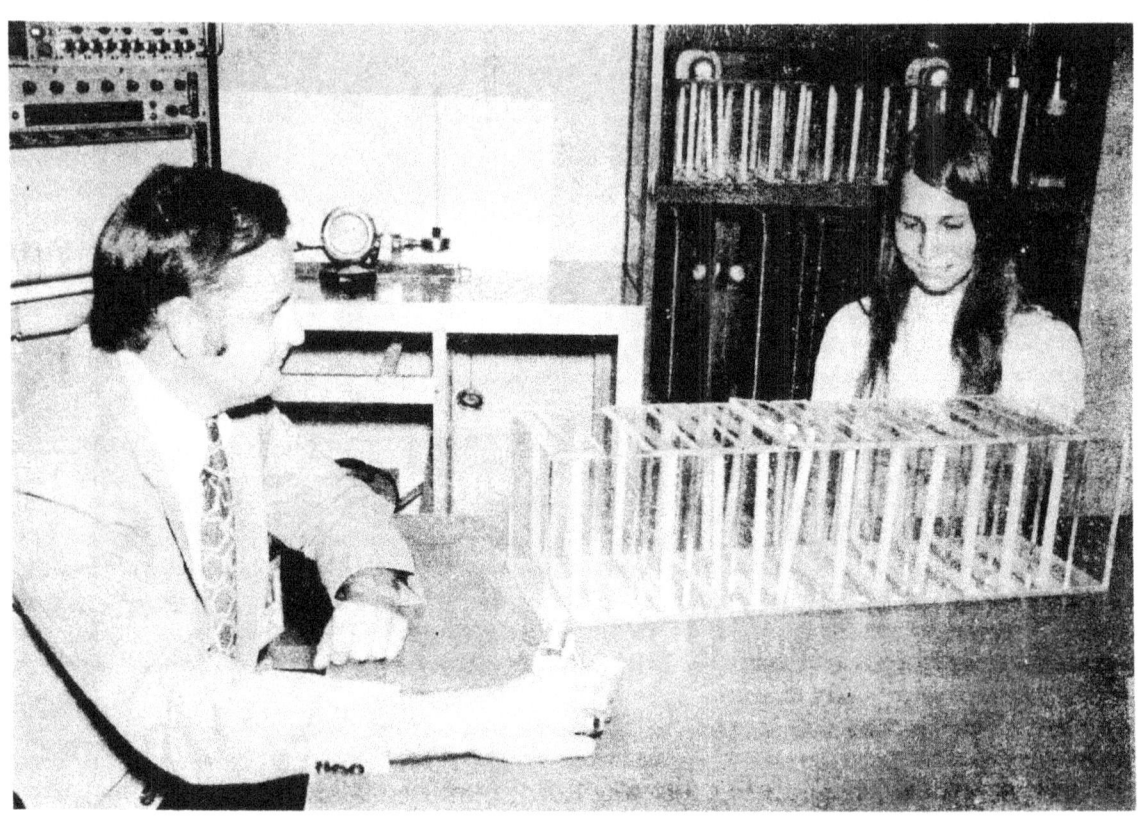

Figure III-57 — MSFC Advisor Dr. Raymond Gause and Judith Miles

Figure III-58 — MSFC Advisor Loren Gross (left), Joel Wordekemper (center) and Donald Schlack

CHERYL A. PELTZ (Figure III-59), 7117 S. Windermere, Littleton, Colorado, 80120.
"Cytoplasmic Streaming" – ED-63
Arapahoe High School, Gordon B. Scheele, Teacher/Sponsor.

The aim of this experiment is to perform microscopic observation of leaf cells of elodea plants in zero gravity to determine if there is any difference between the intracellular cytoplasm motion compared with cytoplasmic motion of similar leaf cells on Earth.

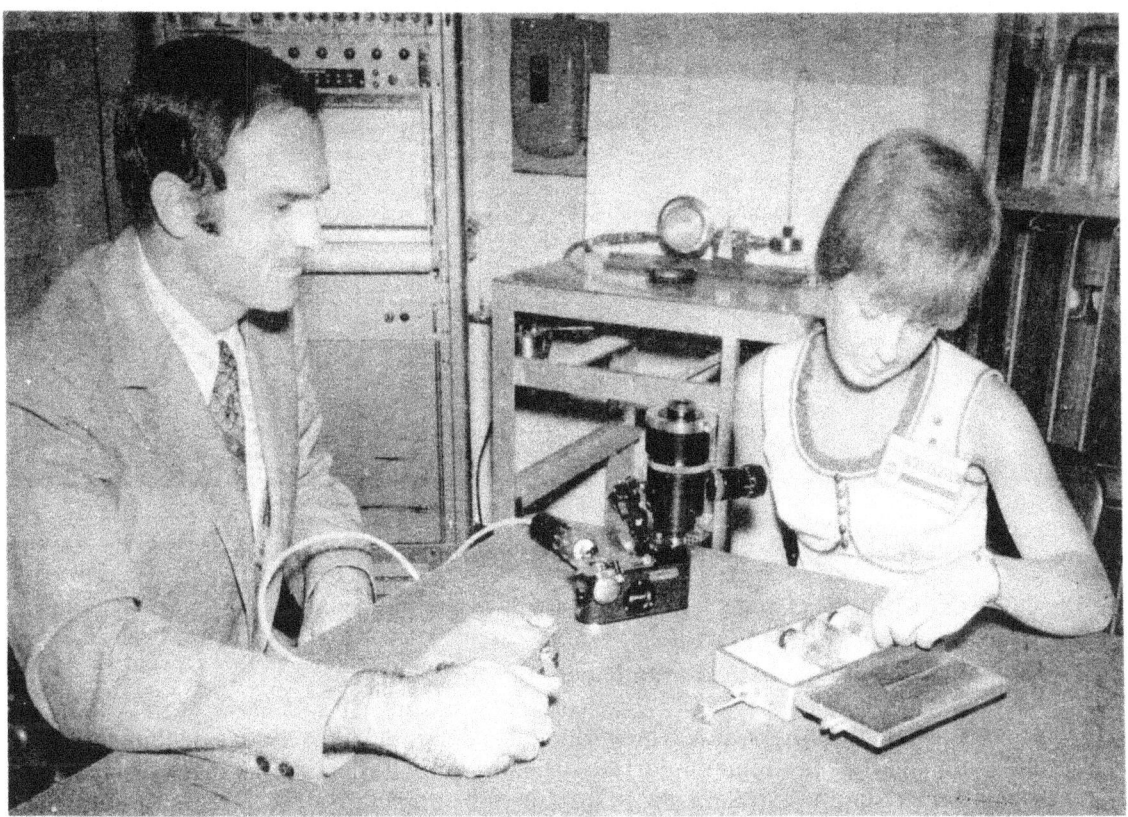

Figure III-59 – MSFC Advisor Charles Cothran and Cheryl Peltz

ROGER G. JOHNSTON (Figure III-60), 1833 Draper Drive, St. Paul, Minnesota, 55113.
"Capillary Study" – ED-72
Ramsey High School, Theodore E. Molitor, Teacher/Sponsor.

The aim of this experiment is to determine if the zero gravity environment induces changes in the characteristics of capillary and wicking action from the familiar Earth-Gravity characteristics.

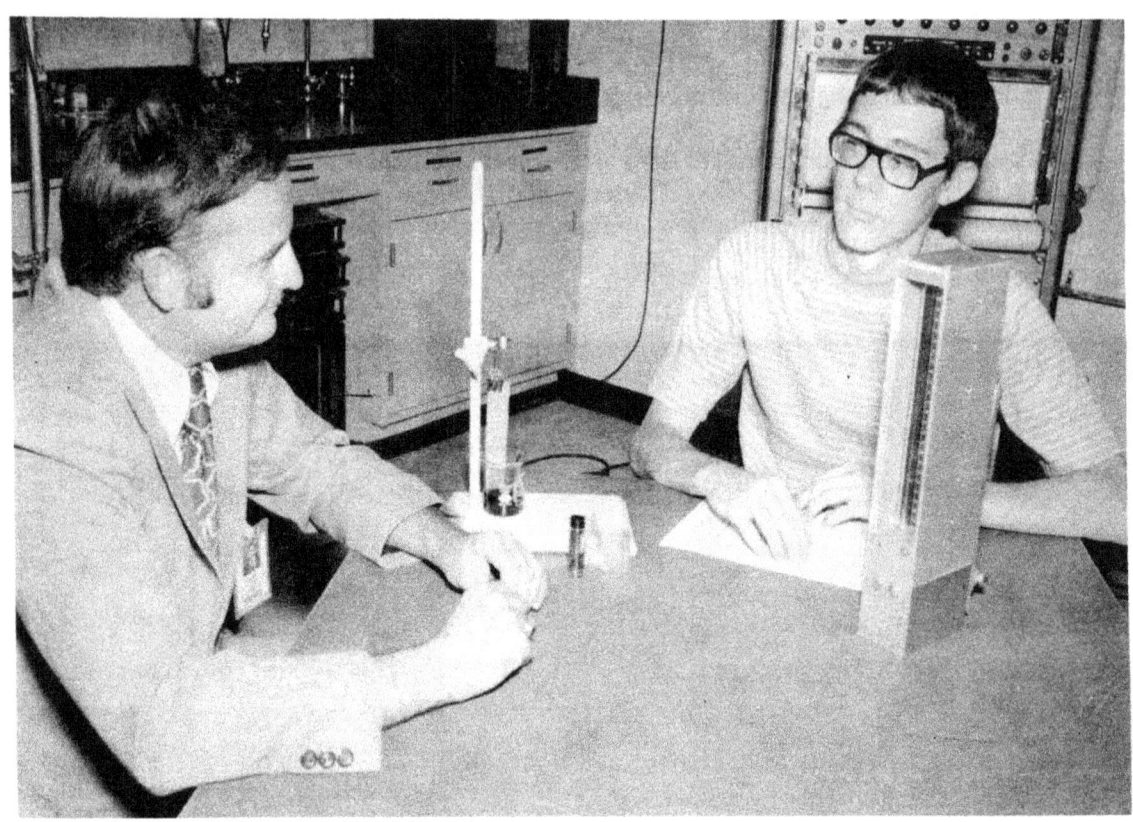

Figure III-60 — MSFC Advisor Dr. Raymond Gause (left) and Roger Johnston

VINCENT W. CONVERSE (Figure III-61), 1704 Roosevelt Road, Rockford, Illinois, 61111.
"Mass Measurement" — ED-74
Harlem High School, Mary J. Trumbauer, Teacher/Sponsor.

This experiment demonstrates the principle of the existing Skylab specimen mass and body mass measurement devices, utilizing the classical spring-mass mechanical oscillator. A simple cantilevered beam (spring) is loaded at the free end with the mass to be measured. The frequency of oscillation, or equivalently the period of the oscillation is a direct indication, by means of an empirical calibration curve, of the mass loading the beam.

TERRY C. QUIST (Figure III-62), 3818 Longridge Drive, San Antonio, Texas, 78228.
"Neutron Analysis" — ED-76
Thomas Jefferson High School, Michael Stewart, Teacher/Sponsor.

In this experiment, detectors inside Skylab record impacts of low energy neutrons. The detectors mounted on the inboard faces of water tanks will be able to discriminate between neutrons in four energy spectra. The neutrons, which have been moderated by their passage through the water in the tanks, impact the detectors and produce fission particles which in turn disrupt the polymer chains in the solid dielectric recording medium. Chemical treatment, on the ground, etches out the track (disrupted regions) and permits microscopic identification of the tracks.

Figure III-61 — Vincent Converse (left) and MSFC Advisor Dr. Robert Head

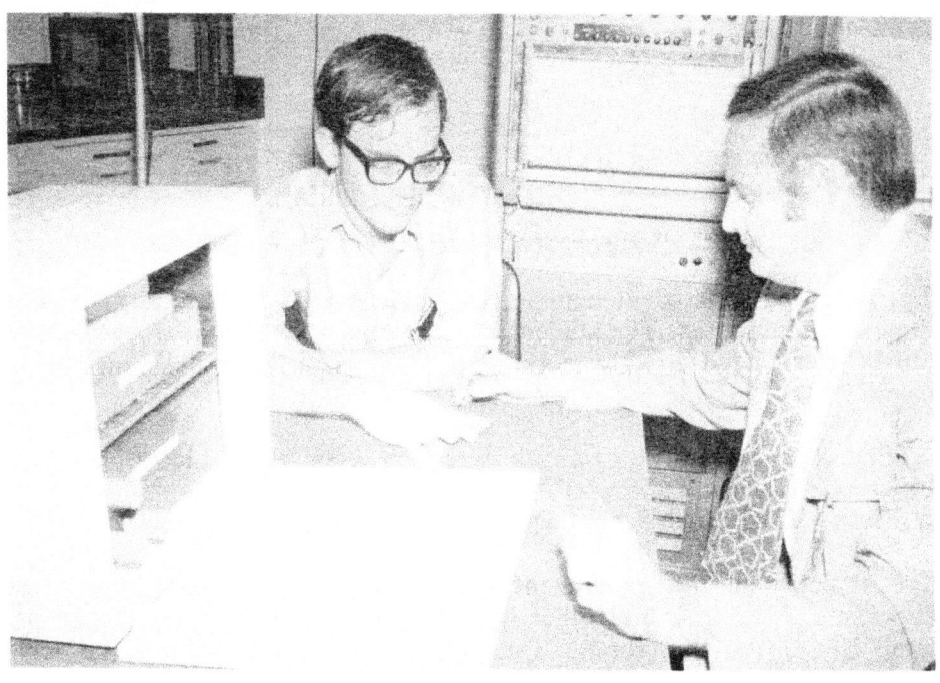

Figure III-62 — Terry C. Quist (left) and MSFC Advisor Dr. Raymond Gause

W. BRIAN DUNLAP (Figure III-63), 6695 Abbot Avenue, Youngstown, Ohio, 44515.
"Liquid Motion in Zero 'G' " — ED-78
Austintown Fitch High School, Paul J. Pallante, Teacher/Sponsor.

The aim of this experiment is to observe and photograph the motion of a liquid-gas interface (gas bubble surrounded by a liquid) subjected to an impulsive force.

Figure III-63 — W. Brian Dunlap (left) and MSFC Advisor Dr. Robert Head

The following six student experiments could not be accommodated in the Skylab Program as proposed. The student, his experiment as proposed, the reasons for its not being included directly in the Skylab Program and the alternatives are listed below:

KEITH STEIN, 2167 Regent Court South, Westbury, New York, 11590.
"Microorganisms in Varying 'G'."
W. Tresper Clarke High School, Dennis Unger, Teacher/Sponsor.

The objective of this experiment was to subject numerous different species of bacteria, ciliated cells and other microorganisms to a complex regime of varying levels of gravitational forces. The varying "G" levels were to be achieved by mounting the specimens at different radii on a centrifuge.

The development of a centrifuge qualified for manned space flight could not be accomplished in the limited time available. Keith will be provided data on microorganisms in zero "G" from the Skylab Microbiology Detailed Test Objective.

KENT M. BRANDT, 11380 Grand Oak Drive, Grand Blanc, Michigan, 48439.
"Chick Embryology."
Grand Blanc Senior High School, Charles E. Martel, III, Teacher/Sponsor.

Kent's experiment involved the launching of a number of fertile chicken eggs, incubating the eggs in orbit. At stated intervals in the incubation cycle, the development of one or more eggs was to be terminated and the egg preserved. At least one egg would have been carried to full term and, hopefully, hatched. The embryonic eggs and the live chick were to be returned for analysis and observation of psycho-motor behavior.

After considerable effort, it was concluded that the hardware involved placed too great a demand in terms of weight, volume and crew time to enable launching in the Command Module. A further problem arose in developing and building suitable, flight-qualified hardware to meet the stringent temperature and humidity requirements to achieve a sufficiently high probability of success in the allowable time. Keith has been associated with Dr. John Lindberg, the Principal Investigator for S071 – Circardian Rhythm-Pocket Mice.

KEITH MCGEE, 122 Sunflower Street, Garland, Texas, 75041.
"Colloidal State."
South Garland High School, Ann Patterson, Teacher/Sponsor.

The stated objective of this experiment was to determine the effect of a zero-g environment on a series of colloidal suspensions, solutions and gels, as well as electrophoretic processes.

The successful performance of this experiment requires a highly stable platform. The normal vibration levels in Skylab, together with the required attitude changes and maneuvers, preclude achieving the required stability.

Keith will be associated with the NASA researchers who were involved in the Apollo 14 and 16 electrophoresis demonstrations.

KIRK M. SHERHART, 2144 Earlmont Road, Berkley, Michigan, 48072.
"Powder Flow."
Berkley High School, Helen Politzer, Teacher/Sponsor.

The objective of this experiment was to study the parameters involved in achieving the flow of powdered or granulated materials as opposed to liquids.

Detailed studies revealed that significant development problems existed that precluded production of hardware within the allowable time.

Kirk will be affiliated with the NASA researchers on material flow in a zero gravity environment.

GREGORY A. MERKEL, 153 Ashland Avenue, Springfield, Massachusetts, 01119.
"Brownian Motion."
Wilbraham and Monson Academy, Solon Economou, Teacher/Sponsor.

The objective of this experiment was to investigate the effect of zero gravity on the Brownian progression of a solute through its solvent. This experiment required a highly stable platform for time periods of up to a month. The Skylab is not capable of providing the required degree of stability.

Gregory has indicated a strong interest in the field of astronomy. Thus he will be affiliated with Dr. Karl Henize, the PI for S019-UV Stellar Astronomy.

JAMES E. HEALY, 84 South Gillette Avenue, Bayport, New York, 11705.
"Universal Gravity."
St. Anthony's High School, Dr. Paul E. Mottl, Teacher/Sponsor.

This experiment proposed developing a space qualified Cavendish balance to measure the mass attraction force (universal gravity).

The forces being measured were found to be at least three orders of magnitude less than the forces induced by the Skylab motions and thus were incapable of measurement.

James exhibited an interest in the effect of crew motion on the attitude stability of the Skylab. Thus he will be affiliated with Bruce Conway, the PI on T013.

Figure III-64 — Skylab Student Experimenters: Left to right, rear row — Robert L. Staehle, Keith L. Stein, Joel G. Wordekemper and Joe B. Zmolek; third row — Judith S. Miles, Cheryl A. Peltz, Terry C. Quist, Joe W. Reihs, Donald W. Schlack, Neal W. Shannon and Kirk M. Sherhart; second row — Alison Hopfield, Kathy L. Jackson, Roger C. Johnston, Jeanne L. Leventhal, Keith D. McGee, Todd A. Meister and Gregory A. Merkel; first row — Daniel C. Bochsler, Kent M. Brandt, Vincent W. Converse, Troy A. Crites, W. Brian Dunlap, John C. Hamilton and James E. Healy

SL-1 AM/MDA at KSC

SECTION IV

SKYLAB MISSIONS

The Skylab mission is a long duration flight of approximately eight months with three separate visits by three-man crews totaling 140 days. The first manned visit to the Skylab space station will be 28 days, which will then be followed by two separate stays of 56 days each. All will be launched from the NASA-Kennedy Space Center.

The first Skylab mission, scheduled for the first half of 1973, will consists of two launches one day apart. The first launch (SL-1) will be that of the unmanned Skylab spacecraft atop a two-stage Saturn V launch vehicle. The configuration will include the Payload Shroud, Apollo Telescope Mount, Multiple Docking Adapter, Airlock Module, Instrument Unit and Orbital Workshop. The Payload Shroud is automatically jettisoned in flight and the Apollo Telescope Mount is deployed prior to arrival of the first crew.

One day later a Command and Service Module with its crew of three astronauts is launched by a two-stage Saturn IB (SL-2). The CSM will rendezvous and dock with the Multiple Docking Adapter of the Skylab. The crew will activate the laboratory and perform experiments during the mission that will last up to 28 days. Then they will undock from the Skylab, deorbit and land in the Pacific Ocean west of San Diego, Calif.

The second manned Skylab mission (SL-3) will be launched about 80 days after the SL-2 launch. Its crew of three astronauts will dock with the Skylab cluster which has been circling 435 kilometers (235 nautical miles) above the Earth unoccupied since the departure of the first crew. The crew will reactivate Skylab, perform experiments and, after a scheduled 56 days, deactivate the cluster, deorbit and land in the Pacific Ocean west of California.

The third manned Skylab mission (SL-4) will follow the same general mission plan as the previous flight. The experiment activities will be continued, but the crew will land in the western Pacific near Hawaii.

In each of the three visits, the CSM is powered down to a quiescent condition after docking and Skylab activation. The Skylab cluster provides power, life support and attitude control for the orbital stay times.

During unmanned periods, the Skylab systems, although powered down, will be monitored at the Mission Control Center, Houston, on a daily basis, and limited operation of certain experiments will be performed. The unmanned periods average about 50 days between undocking of the first crew (SL-2) and the visit by the second crew (SL-3) and 40 days between the departure of the second crew (SL-3) and the final visit of the third crew (SL-4).

Flight Crews

Flight crews for Skylab consist of a commander, a science pilot, and a pilot. Prime crews, listed in that order, are:

First mission: Charles Conrad, Jr., Dr. Joseph P. Kerwin and Paul J. Weitz.
Second mission: Alan L. Bean, Dr. Owen K. Garriott and Jack R. Lousma.
Third mission: Gerald P. Carr, Dr. Edward G. Gibson and William R. Pogue.

Backup crewmen for the first mission are Russell L. Schweickart, Dr. Story Musgrave and Bruce McCandless II. Serving as backup crew for both the second and third missions are Vance D. Brand, Dr. William E. Lenoir and Dr. Don L. Lind.

Astronauts Kerwin, Garriott, Gibson, Musgrave and Lenoir are scientist astronauts; the remaining Skylab crew members are pilot astronauts.

Due to specialized operational and scientific abilities required for Skylab, crew members are not totally cross trained in all areas, which has been the case in Apollo. Major crew responsibilities are divided as follows:

Commander - overall mission, command module and service module systems and flight planning.

Science pilot - Apollo Telescope Mount and medical experiment systems.

Pilot - earth resources experiment systems and cluster systems.

First Skylab Mission SL-1/SL-2

The SL-1/SL-2 mission is directed primarily toward accomplishing a series of group related experiments which include medical, solar astronomy, Earth resources and technological experiments. During the 28 days the first crew will be on board, biomedical investigations will be emphasized, and the habitability of Skylab will be evaluated and experiments associated with solar astronomy, Earth resources and scientific and technical research will be performed.

Mission Objectives

The objectives for the SL-1/SL-2 mission, assigned by the Office of Manned Space Flight, NASA Headquarters, are:

Establish the SL-1 in Earth Orbit — Operate Skylab as a habitable space structure for up to 28 days after the SL-2 launch; obtain data for evaluating the performance of the Orbital Assembly; and obtain data for evaluating crew mobility and work capability in both intravehicular and extravehicular activity.

Obtain Medical Data on the Crew for Use in Extending the Duration of Manned Space Flights — Obtain medical data for determining the effects on the crew which result from a space flight of up to 28 days duration; and obtain medical data for determining if a subsequent Skylab mission of up to 56 days is feasible and advisable.

Perform In-flight Experiments — Obtain solar astronomy data for continuing and extending solar studies beyond the limits of Earth-based observations; obtain Earth resources data for continuing and extending multisensor observation of the Earth from low Earth orbit; and perform the assigned scientific, engineering, technological and Department of Defense (DOD) experiments.

The interval between SL-1 and SL-2 will be about 24 hours with an "early rendezvous" to implement required medical experiments. Another constraint in launch sequence is based on a maximum use of 317.5 kilograms (700 pounds) of SL-2 launch vehicle propellant for yaw steering.

The SL-1 circular orbit of approximately 435 kilometers (235 nautical miles) was chosen to meet requirements of the Apollo Telescope Mount and to provide acceptable subsequent launch opportunities and enhance rendezvous capabilities. The 50-degree orbital inclination was chosen to enhance EREP ground coverage.

SL-1 LAUNCH

The launch of the Skylab (SL-1) is planned for 1:30 p.m. E.D.T. from Pad 39A, Kennedy Space Center. The Saturn V will be launched on a flight azimuth of 40.88 degrees east of north; and, after approximately 9 minutes and 48 seconds into flight, the Skylab will be inserted into a 435 kilometer (235 nautical mile) orbit, the plane of which is inclined 50 degrees to the equator.

After insertion, the S-II stage is separated by retrorockets, and the Skylab is pitched down by the Thruster Attitude Control Subsystem (TACS). As Skylab passes through the nose-down attitude, the Payload Shroud is jettisoned. The TACS orients the spacecraft to a solar inertial attitude while the ATM and its solar array are deployed. The Workshop solar array is deployed and activation of the Control Moment Gyro Subsystem (CMGS) is begun. The solar inertial attitude is one that keeps the ATM's solar arrays pointed toward the Sun.

Launch Windows

The time of day of the Workshop launch establishes the framework for the launch sequence on all subsequent missions. The Skylab payload can be launched between 6:15 a.m. and 3 p.m. E.D.T. and still provide SL-2 the required abort recovery lighting conditions the following day.

SL-2 LAUNCH

Prior to the manned launch, the unmanned Skylab will be checked out by telemetry to assure that all launch commit criteria are validated for the first manned visit.

The launch of a Saturn IB — carrying the first crew — is scheduled for the following day at 1 p.m. E.D.T. The exact launch time was selected to provide a favorable five revolution rendezvous with the Workshop.

The SL-2 launch timing provides a maximum slip of only 10 minutes for rendezvous. Anything beyond a 10 minute slip requires a 24-hour delay. A launch time delayed by one day provides only a 7 minute window with further delays causing a recycle to day 6 or 7 for the manned launch vehicle.

Rendezvous

The CSM and second stage of the Saturn IB are inserted into a 150 by 222 kilometer (81 by 120 nautical mile) orbit inclined 50 degrees to the equator. Orbital insertion occurs shortly before 10 minutes into the flight.

Six minutes after insertion, the CSM will separate from the S-IVB with a posigrade maneuver that increases spacecraft velocity by three feet per second using the Service Module reaction control system thrusters. This maneuver at 16 minutes GET, results in a 150 by 223 kilometer (81 by 121 nautical mile) orbit.

The first phasing maneuver will be performed as the CSM nears its second apogee. This maneuver will be performed by firing the Service Propulsion System (SPS) engine. One and one-half revolutions after the first phasing maneuver, the second phasing maneuver is performed, again by the SPS engine.

A corrective combination maneuver will be performed when the CSM and Workshop are 363 kilometers (196 nautical miles) apart. A coelliptic maneuver will be performed after the corrective combination maneuver to place the CSM 18 kilometers (10 nautical miles) below Skylab.

Terminal Phase

The terminal phase initiation maneuver will be performed at 6 hours, 47 minutes, 44 seconds GET by the SPS when the two vehicles are about 41 kilometers (22 nautical miles) apart. The braking approach will begin when the CSM is about 1.8 kilometers (1 nautical mile) from Skylab. Stationkeeping and docking will follow with the CSM scheduled to dock with the forward (axial) port of the Multiple Docking Adapter (MDA).

Docking

Docking is achieved by maneuvering the CSM close enough to the MDA so that the extended CSM probe engages the MDA drogue. When the probe is engaged through the capture latches, the probe retract system is activated to pull the MDA and CSM together.

Upon retraction, the MDA tunnel ring activates the 12 automatic latches and effects a pressure seal between the modules through the two seals in the CM docking ring face. After docking, the pressure in the tunnel is equalized from the CM through an equalization valve and the CM hatch is removed, and the crew verifies that all of the 12 latches are locked in place. Latches not automatically actuated can be manually actuated by the crew.

Ingress and Activation

The SL-2 astronauts begin the second day in space inside the CM. After breakfast in the CM, the crew begins preparations to enter the Skylab through the MDA tunnel.

When proper pressure checks of the vehicles are confirmed, the pilot (PLT) begins preparations to enter the MDA. The crew removes the CM probe and drogue assembly, stows it in the CM, opens the MDA hatch and crawls into the MDA through the tunnel.

An inspection of the MDA is made by the PLT. He is then followed into the MDA by the scientist pilot (SPT). The two then begin a series of systems checks and housekeeping chores in the MDA and AM.

One power and two control umbilicals are connected to the respective connectors in the CM docking ring and MDA tunnel. The next step is for the crew to install a transfer duct which is used to transfer atmosphere from the MDA into the CM.

Meanwhile, the commander (CDR) is configuring the CM to a powered down mode. The Caution and Warning system, an elaborate network system of some 90 sensors which monitor critical parameters of the vehicles, is activated and tested. (See Section II-5.)

Groundtrack

The groundtrack repeats on a 71-revolution (118-hour) or approximately a 5-day cycle. For example, revolution 77 retraces the path of revolution 6.

The Skylab Space Station passes through the South Atlantic Anomaly, a region of trapped radiation on the Earth's upper atmosphere, approximately nine times per day in which the duration of transit is 10 minutes. Two passes per day are extensive; i.e., revolutions 18 and 19 encounter the northern trapped electron belt containing both trapped electrons and protons.

Revolutions 20 through 23 are moderately long passes, and 24 through 26 much shorter. The time of passage through these areas of trapped radiation are of concern when scheduling radiation experiments such as D008, EVA activities and film handling.

Controlled-Orbit Maneuvers

The Skylab orbit of about 435 kilometers (235 nautical miles) and 93-minute period is designed to provide a repeating groundtrack, which is an advantage in scheduling many ground-related activities such as launch window planning, flight control PAD message preparation, and such EREP preparations as aircraft support. Once a repeating pattern is established, the effects of atmospheric drag can be negated to within 3.7 kilometers (2 nautical miles) cross range error by adding trimburns every 20 days.

If the Skylab insertion is nominal, Trim 1 is not performed. Trim 2, scheduled at about 8 days CSM GET, is an RCS posigrade burn of 6.3 seconds designed to counteract drag by increasing velocity by 0.12 meter (0.5 foot) per second. (A posigrade burn propels the spacecraft forward and increases its velocity.)

Trim 3, scheduled at about 24 days CSM GET, is about a 40-second posigrade burn designed to increase velocity by about 0.8 meter (2.72 feet) per second to compensate for drag and control the location of the groundtrack for the second mission EREP passes. Over this period of time, the groundtrack moves to the west of the reference pattern by about 37 kilometers (20 nautical miles).

Deactivation

The SL-2 crew will begin deactivation of the Skylab about three days before the end of the scheduled manned mission. This includes transfer to and stowage of samples, EREP data and film in the CM. The crew will perform such tasks as securing food and water management areas; turning off crew quarters heaters and circulating fans; configuring the power distribution and control panel; and deactivating the environmental control system in the OWS and the aft end of the AM.

Deactivation of the MDA/AM will include deactivation of the environmental system in the AM Structural Transition Section (STS); configuring the STS power control panel and ATM control and display console; and preparing for closeout of the MDA.

Undocking and Separation

The entire sequence of docking events is reversed in preparing for undocking and deorbit. The umbilicals and interchange ducts are removed, the MDA hatch is closed, the drogue and probe are installed, the 12 docking latches are released, the CM tunnel hatch is installed to seal the CM cabin, the tunnel pressure is vented, the probe extension latch and capture latches are electrically released, and the probe spring extends the probe separating the CSM from the MDA.

Deorbit

The deorbit sequence is initiated by Inertial Measuring Unit (IMU) alignments, undocking, fly-around inspection and then a separation maneuver. The IMU alignments occur prior to undocking to verify spacecraft attitude and velocity and to provide a permanent attitude reference point.

The SL-2 CSM will undock and the undocking impulse will initiate a Skylab circumnavigation sequence which will allow the crew to photograph and inspect the cluster.

A 1.5-meter (5-foot) per-second retrograde firing by the SM-RCS will be done to evade the Skylab and reduce CSM velocity.

Two deorbit systems are available in the CSM. The primary one is the Service Propulsion System (SPS), and the backup system is the SM's Reaction Control System (RCS).

IV 3

SL-3 LAUNCH

Prior to launch of the second crew (SL-3), the SWS will be checked by telemetry to be sure it is ready for reuse.

The second mission is directed primarily toward accomplishing the onboard experiments, medical and technical and others associated with solar astronomy and Earth resources. It is planned for a duration of up to 56 days from the day of launch.

Launch

The launch of SL-3 (Saturn IB) is scheduled to occur in the afternoon E.S.T. from Pad 39B at Kennedy Space Center. The exact launch time will be based on the most favorable conditions for rendezvous.

It is mandatory that rendezvous and docking occur early enough in this mission to allow first-day medical samples to be placed in the Workshop freezer within 24 hours of collection.

Rendezvous Phase
Rendezvous and docking maneuvers will be performed much like the SL-2 mission.

Skylab Reactivation
Reactivation of Skylab will immediately follow docking.

The crew will enter the SWS in shirtsleeves and perform the following major activation tasks:

Open the MDA hatch and store it; turn on the MDA and AM interior lights and make a visual inspection; connect the CSM/MDA power and communications umbilical; activate the molecular sieves; activate the Star Tracker; activate the ATM controls and displays; activate the Caution and Warning system; process and stow M071 and M073 urine samples in the freezer; and activate food and management systems.

Experiments Phase
From habitation of Skylab to preparation for unmanned orbital storage, mission activities other than normal crew functions and systems operations will consist primarily of experiments work.

There will be 28 EREP passes on the second mission (compared with 15 on SL-2) over sites in the U.S. and surrounding coastal waters, plus extensive coverage of foreign sites. The 28 passes include the United States and foreign geographical areas (Central and South America, Europe-Middle East, S. America-Africa-Europe, S.E. Asia-Japan, India-S.E. Asia-Australia).

Deorbit
The deorbit sequence is similar to the first mission — IMU alignments, docking, fly-around inspection, and then a separation maneuver. The IMU alignments occur prior to undocking and the remaining operations begin at daybreak and occupy one daylight pass.

SL-4 LAUNCH
The launch of the third crew aboard a Saturn IB is scheduled to occur in late 1973.

The third manned mission, also planned for 56 days, is directed toward continuing the medical, scientific, technical and student experiments. It will include 22 EREP passes.

SKYLAB RESCUE MISSION AND RESCUE KIT (SL-R)
If a malfunction prevents a Command Service Module docked to the Skylab cluster from being used to return the astronauts, a rescue can be made by launching another CSM. Malfunctions elsewhere in the Skylab cluster should not endanger the safe return of astronauts. A rescue mission will be launched only in case there is complete loss of CSM capability for return to Earth. Malfunctions endangering the safe return of the astronauts are: loss of entry or Earth landing capability; loss of all propulsion; loss of all electrical power; inability to undock from the MDA; loss of environmental control; and loss of precise attitude stabilization.

If it is necessary to launch a rescue mission, the succeeding Skylab CSM on the launch pad will be modified by installation of portions of a rescue kit to receive extra passengers (Figure IV-1). The orbiting crew will continue to perform Skylab mission activities until they are rescued. The life support capability of the Workshop has an overall total mission duration of about 150 days.

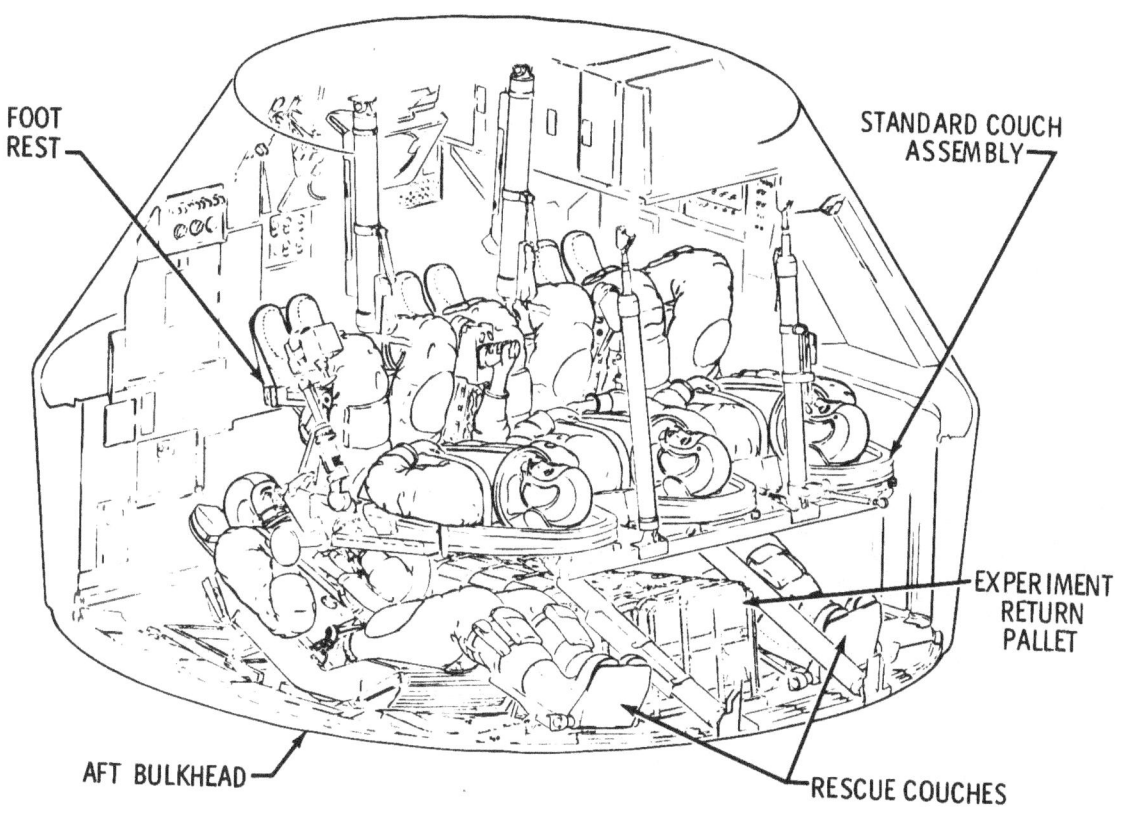

Figure IV-1 — Rescue CM

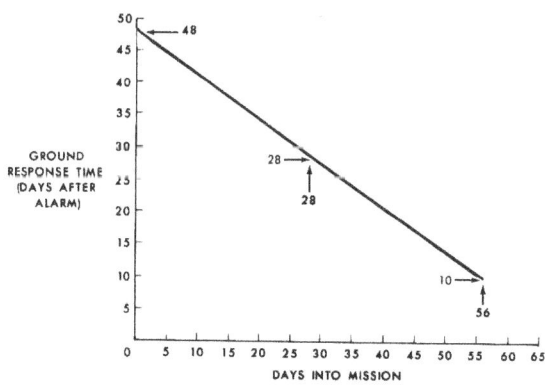

Figure IV-2 — Rescue Response Time

Response time for a rescue mission depends mainly on the time required to prepare the launch and rescue vehicles. Launch window opportunities could delay the launch no more than five days, at most. Rendezvous and crew transfer times are of secondary importance (Figure IV-2).

For an early rescue mission, such as in response to a malfunction just after docking, 48 days of accelerated preparation will be required with 22 days needed to refurbish the launch umbilical tower. The rescue kit can be installed in eight hours. As the mission progresses, time to prepare the rescue launch decreases. The succeeding launch vehicle and tower, being prepared for a normal launch, require less time to get ready for a rescue mission. For a late emergency, time to prepare a rescue launch can be shortened to 10 days.

Backup hardware will be used if a rescue of the third crew is necessary. Items designated for this eventuality are Saturn IB vehicle AS-209 and command module 119.

The rescue kit modifies a normal three-man CSM to carry five men. Also, provision is made for a limited experiment return capability. The modifications include: Removal of storage lockers on the aft bulkhead and installation of two couches in that space; life support and communications umbilicals modified to accommodate two extra crewmen.

For a rescue mission, two crewmen are launched with the center couch empty. The returning CM will then carry five men.

A modified drogue and a mechanical triggering device for the mission CSM docking probe are included in the Skylab portion of the rescue kit. The crewmen to be rescued can install the triggering device on the mission probe, and then use the probe and modified drogue to eject the disabled CSM. This will allow the rescue CSM to dock at the forward (axial) port, while keeping the possibility open for future revisits. If, for some reason, it is not feasible to eject the disabled CSM, the rescue CSM can dock at the side port of the MDA. However, a future revisit mission would not be possible due to the limited stay time allowable for side docking.

IV-6

CREW BIOGRAPHIES

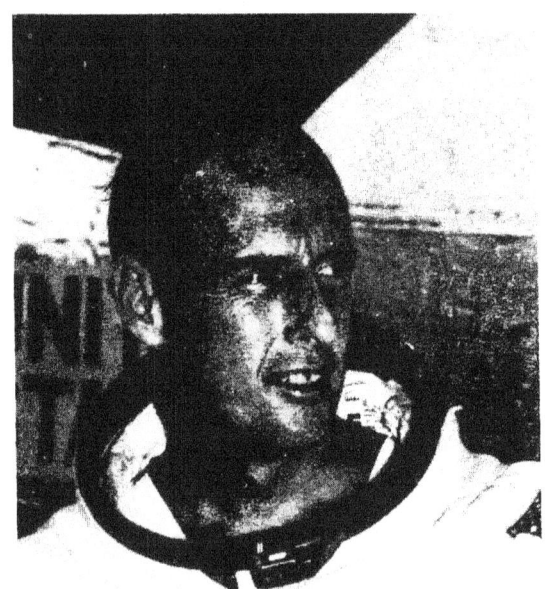

Name: Charles Conrad, Jr. (Captain, USN)

Birthplace and Date: Born on June 2, 1930, in Philadelphia, Pa.

Physical Description: Blond hair; blue eyes; height: 5 feet 6 ½ inches; weight: 138 pounds.

Education: Attended primary and secondary schools in Haverford, Pa., and New Lebanon, N.Y.; received a Bachelor of Science degree in Aeronautical Engineering from Princeton University in 1953, an Honorary Master of Arts degree from Princeton in 1966, an Honorary Doctorate of Laws degree from Lincoln-Weslyan University in 1970, and an Honorary Doctorate of Science from Kings College, Wilkes-Barre, Pa., in 1971.

Marital Status: Married to the former Jane DuBose of Uvalde, Tex.; her parents, Mr. and Mrs. W. O. DuBose, reside in Uvalde.

Children: Peter, December 24, 1954; Thomas, May 3, 1957; Andrew, April 30, 1959; Christopher, November 26, 1960.

Recreational Interests: His hobbies include golf, swimming and water skiing.

Organizations: Fellow of the American Astronautical Society, and Associate Fellow of the American Institute of Aeronautics and Astronautics and the Society of Experimental Test Pilots.

Special Honors: Awarded the NASA Distinguished Service Medal, two NASA Exceptional Service Medals, the Navy Astronaut Wings, the Navy Distinguished Service Medal, and two Distinguished Flying Crosses; recipient of Princeton's Distinguished Alumnus Award for 1965, the U.S. Jaycee's 10 Outstanding Young Men Award in 1965, the American Astronautical Society Flight Achievement Award for 1966, Pennsylvania's Award for Excellence in Science and Technology in 1967 and 1969, the Rear Admiral William S. Parsons Award for Scientific and Technical Progress in 1970, the Godfrey L. Cabot Award in 1970, the Silver Medal of the Union League of Philadelphia in 1970, and the National Academy of Television Arts and Sciences Special Trustees Award in 1970.

Experience: Conrad entered the Navy following graduation from Princeton University and became a naval aviator. He attended the Navy Test Pilot School at Patuxent River, Md., and upon completing that course was assigned as a project test pilot in the armaments test division there. He also served at Patuxent as a flight instructor and performance engineer at the Test Pilot School.

Current Assignment: Captain Conrad was selected as an astronaut by NASA in September 1962. In August 1965, he served as pilot on the 8-day Gemini 5 flight. He and command pilot Gordon Cooper, launched into Earth orbit on August 21, established a space endurance record of 190 hours and 56 minutes. The flight, which lasted 120 revolutions and covered a total distance of 3,312,993 statute miles, ended on August 29, 1965. It was also on this flight that the United States took over the lead in manhours in space.

On September 12, 1966, Conrad occupied the command pilot seat for the 3-day Gemini 11 mission. He executed orbital maneuvers to rendezvous and dock in less than one orbit with a previously launched Agena and piloted Gemini 11 through two periods of extravehicular activity performed by pilot Richard Gordon. Other highlights of the flight included the retrieval of a nuclear emulsion experiment package during the first EVA; the establishment of a new world space altitude record of 850 statute miles; the completion of the first tethered stationkeeping exercise, in which artificial gravity was produced; and the completion of the first fully automatic controlled reentry. Gemini 11 ended September 15, 1966, with the spacecraft landing in the Atlantic–2 ½ miles from the prime recovery ship USS Guam.

He was subsequently assigned as the backup spacecraft commander for the Apollo IX flight.

Conrad was spacecraft commander of Apollo XII, November 14-24, 1969. With him on man's second lunar landing mission were Richard F. Gordon (command module pilot) and Alan L. Bean (lunar module pilot). In accomplishing all of the mission's objectives, the Apollo XII crew executed the first precision lunar landing, bringing their lunar module, "Intrepid," to a safe touchdown on the Moon's Ocean of

Storms; and performed the first lunar traverse deploying the Apollo Lunar Surface Experiment Package (ALSEP), installing a nuclear power generator station which would provide the power source for long-term scientific experiments, gathering geologic samples of the lunar surface for return to Earth, and completing a close-up inspection of the Surveyor III spacecraft. Apollo XII lasted 244 hours and 36 minutes and was concluded with a Pacific splashdown and subsequent recovery operations by the USS HORNET.

Captain Conrad has completed three space flights, logging a total of 506 hours and 48 minutes in space—of which 7 hours and 45 minutes were spent in EVA on the lunar surface.

Current duties involve training for future manned space flights and, as Chief of Skylab Operations for the Astronaut Office, responsibility for monitoring astronaut activities in support of the Skylab Program.

Name: Joseph P. Kerwin (Commander, MC, USN)

Birthplace and Date: Born February 19, 1932, in Oak Park, Ill. His parents, Mr. and Mrs. Edward M. Kerwin, are residents of Chicago.

Physical Description: Brown hair; blue eyes; height: 6 feet; weight: 170 pounds.

Education: Graduated from Fenwick High School, Oak Park, in 1949; received a Bachelor of Arts degree in Philosophy from College of the Holy Cross, Worcester, Mass., in 1953; a Doctor of Medicine degree from Northwestern University Medical School, Chicago, in 1957; completed internship at the District of Columbia General Hospital in Washington, D.C.; and attended the U.S. Navy School of Aviation Medicine, Pensacola, Fla.

Marital Status: Married to the former Shirley Ann Good of Danville, Pa. Her parents, Mr. and Mrs. George D. Good, reside in Danville.

Children: Sharon, September 14, 1963; Joanna, January 5, 1966; Kristina, May 4, 1968.

Recreational Interests: His hobbies are reading and classical music.

Organizations: Fellow of the Aerospace Medical Association; and member of the Aircraft Owners and Pilots Association, and Phi Beta Pi.

Special Honors: Awarded the MSC Certificate of Commendation (1970).

Experience: Kerwin, a Commander, has been in the Navy Medical Corps since July 1958. Prior to becoming a naval aviator, he served two years as flight surgeon with Marine Air Group 14 at Cherry Point, N.C. He earned his pilot's wings at Beeville, Tex., in 1962.

He then became flight surgeon for Fighter Squadron 101 at Oceana Naval Air Station, Virginia Beach, Va., and was subsequently assigned as staff flight surgeon for Air Wing Four at the Naval Air Station, Cecil Field, Fla.

He has logged 2,000 hours flying time—1,800 hours in jet aircraft.

Current Assignment: Commander Kerwin was selected as a scientist-astronaut by NASA in June 1965. He is currently involved in training for future manned space flights.

Name: Paul Joseph Weitz (Commander, USN)

Birthplace and Date: Born in Erie, Pa., on July 25, 1932. His mother, Mrs. Violet Martin, now resides in Norfolk, Va.

Physical Description: Blond Hair; blue eyes; height: 5 feet 10 inches; weight: 180 pounds.

Education: Attended McKinley Elementary School in Erie, and Harborcreek High School in Harborcreek, Pa.; received a Bachelor of Science degree in Aeronautical Engineering from Pennsylvania State University in 1954 and a Master's degree in Aeronautical Engineering from the U.S. Naval Postgraduate School in Monterey, Calif., in 1964.

Marital Status: Married to the former Suzanne M. Berry of Harborcreek, Pa.; her father is John H. Berry.

Children: Mathew J., September 23, 1958; Cynthia A., September 25, 1961.

Recreational Interests: Hunting and fishing are among his hobbies.

Special Honors: Recipient of the Air Medal (5 awards) and the Navy Commendation Medal for combat flights in the Viet Nam area; presented the Secretary of Navy Commendation for Achievement as a result of leading a flak-damaged airplane to a safe landing; and for his performance during the period June through November 1965 as a member of CVW-7 embarked in USS INDEPENDENCE (CVA-62), he received the Navy Unit Commendation.

Experience: Weitz received his commission as an Ensign through the NROTC program at Pennsylvania State University and, upon graduation in 1954, was assigned to USS JOHN A. BOLE (DD-755) as CIC Officer. After having served in this capacity for one year, he completed his flight training at Corpus Christi, Tex., in September 1956.

He was an A-4 Tactics Instructor with VA-44 at the Naval Air Station in Jacksonville, Fla., from 1956 to 1960 and a project officer for various air-to-ground delivery tactics projects while on duty with VX-5 at China Lake, Calif., from September 1960 to June 1962. He completed the next two years at the U.S. Naval Postgraduate School and was then assigned to VAH-4 at the Naval Air Station in Whidbey, Wash., in June 1964. It was during this tour of duty, while serving as a detachment officer-in-charge, that announcement was made in April 1966 of his selection to the astronaut training program.

He has logged more than 3,700 hours flying time—3,200 hours in jet aircraft.

Current Assignment: Commander Weitz is one of the 19 astronauts selected by NASA in April 1966. He served as a member of the astronaut support crew for Apollo XII, and his present duties involve training for future manned space flights.

Name: Alan L. Bean (Captain, USN)

Birthplace and Date: Born in Wheeler, Tex., on March 15, 1932. His parents, Mr. and Mrs. Arnold H. Bean, reside in his hometown Fort Worth, Tex.

Physical Description: Brown hair; hazel eyes; height: 5 feet 9 ½ inches; weight: 155 pounds.

Education: Graduated from Paschal High School in Fort Worth; received a Bachelor of Science degree in Aeronautical Engineering from the University of Texas in 1955.

Marital Status: Married to the former Sue Ragsdale of Dallas; her parents, Mr. and Mrs. Edward B. Ragsdale, are residents of that city.

Children: Clay A., December 18, 1955; Amy Sue, January 21, 1963.

Recreational Interests: He enjoys being with his two children, and his hobbies include surfing, painting and handball. He also enjoys swimming, diving and gymnastics.

Organizations: Fellow of the American Astronautical Society; member of the Society of Experimental Test Pilots, and Delta Kappa Epsilon.

Special Honors: Awarded the NASA Distinguished Service Medal, the Navy Astronaut Wings and Navy Distinguished Service Medal; recipient of the Texas Press Association's Man of the Year Award (1969), the Rear Admiral William S. Parsons Award for Scientific and Technical Progress (1970), the University of Texas Distinguished Graduate Award (1970) and Distinguished Alumnus Award (1970), the Godfrey L. Cabot Award (1970), and the National Academy of Television Arts and Sciences Special Trustees Award (1970).

Experience: Bean, a Navy ROTC student at Texas, was commissioned upon graduation in 1955. Upon completing his flight training, he was assigned to Attack Squadron 44 at the Naval Air Station in Jacksonville, Fla., for four years. He then attended the Navy Test Pilot School at Patuxent River, Md. Upon graduation he was assigned as a test pilot at the Naval Air Test Center, Patuxent River, where he flew all types of naval aircraft (jet, propeller, and helicopter models) to evaluate them for Navy fleet use. Bean participated in the initial Navy evaluation of both the A5A and A4E jet attack airplanes. He attended the School of Aviation Safety at the University of Southern Califonria and was next assigned to Attack Squadron 172 at Cecil Field, Fla., as an A-4 light jet attack pilot.

During his career, he has flown 27 types of military aircraft as well as many civilian airplanes. He has logged more than 4,410 hours flying time—including 3,674 hours in jet aircraft.

Current Assignment: Captain Bean was one of the third group of astronauts named by NASA in October 1963. He served as backup command pilot for the Gemini 10 mission and backup lunar module pilot for the Apollo IX mission and was lunar module pilot for the flight of Apollo XII, November 14-24, 1969.

With him on man's second lunar landing mission were Charles Conrad (spacecraft commander) and Richard F. Gordon (command module pilot). The Apollo XII crew accomplished all mission objectives. Captain Conrad and Captain Bean executed a precision lunar landing, bringing their lunar module, "Intrepid," to a safe touchdown in the moon's Ocean of Storms—within 300 feet of their targeted landing point. On the surface, they performed a lunar geology traverse of one mile and deployed the Apollo Lunar Surface Experiment Package (ALSEP), installing the first nuclear power generator station on the Moon to provide the power source for these long-term scientific experiments which continue in operation today. Conrad and Bean then completed a close-up inspection of the Surveyor III spacecraft, returning several parts of the Surveyor to Earth.

Captain Bean has logged a total of 244 hours and 36 minutes of space flight, 31 hours and 31 minutes on the moon—of which 7 hours and 45 minutes were spent outside the spacecraft on the lunar surface.

Name: Owen K. Garriott (PhD)

Birthplace and Date: Born November 22, 1930, in Enid, Okla. His parents, Mr. and Mrs. Owen Garriott, reside in Enid.

Physical Description: Brown hair; blue eyes; height: 5 feet 9 inches; weight: 140 pounds.

Education: Graduated from Enid High School; received a Bachelor of Science degree in Electrical Engineering from the University of Oklahoma in 1953, a Master of Science degree and a Doctorate in Electrical Engineering from Stanford University in 1957 and 1960, respectively.

Marital Status: Married to the former Helen Mary Walker of Enid, Oklahoma. Her parents, Mr. and Mrs. Glenn A. Walker, reside in Enid.

Children: Randall O., March 29, 1955; Robert K., December 7, 1956; Richard A., July 4, 1961; Linda S., September 7, 1966.

Recreational Interests: his hobbies include amateur radio, sailing and scuba diving.

Organizations: Member of the American Geophysical Union, the Institute of Electrical and Electronic Engineers, Tau Beta Pi, Sigma Xi, the International Scientific Radio Union (URSI), and the American Astronomical Society.

Special Honors: National Science Foundation Fellowship at Cambridge University and at the Radio Research Station at Slough, England, 1960-1961.

Experience: From 1961 until 1965, he taught electronics, electromagnetic theory, and ionospheric physics in the Department of Electrical Engineering at Stanford University. He has performed research in ionospheric physics since obtaining his doctorate, and he has authored and co-authored more than 25 scientific papers and one book in this area.

Dr. Garriott has logged more than 1,600 hours flying time—including more than 1,200 in jet aircraft and the remainder in light aircraft and helicopters. In addition to NASA ratings, he maintains FAA commercial pilot and flight instructor certification.

Current Assignment: Dr. Garriott was selected as a scientist-astronaut by NASA in June 1965. He has since completed a 53-week course in flight training at Williams Air Force Base, Ariz. He is presently involved in training for future manned space flights.

Name: Jack Robert Lousma (Major, USMC)

Birthplace and Date: Born February 29, 1936, in Grand Rapids, Mich. His father, Jacob Lousma, resides in Ann Arbor, Mich.

Physical Description: Blond hair; blue eyes; height: 6 feet; weight: 185 pounds.

Education: Attended Tappan Junior High School and Ann Arbor High School; received a Bachelor of Science degree in Aeronautical Engineering from the University of Michigan in 1959 and the degree of Aeronautical Engineer from the U.S. Naval Postgraduate School in 1965.

Marital Status: Married to the former Gratia Kay Smeltzer of Ann Arbor. Her parents, Mr. and Mrs. Chester Smeltzer, reside in Ann Arbor.

Children: Timothy J., December 23, 1963; Matthew O., July 14, 1966; Mary T., September 22, 1968.

Recreational Interests: He is an avid golfing enthusiast and enjoys hunting and fishing.

Organizations: Member of the Society of the Sigma Xi, the University of Michigan "M" Club, and the Officers' Christian Union.

Special Honors: Awarded the MSC Certificate of Commendation (1970).

Experience: Lousma was assigned as a reconnaissance pilot with VMCJ-2, 2nd MAW, at Cherry Point, N.C., before coming to Houston and the Manned Spacecraft Center.

He has been a Marine Corps officer since 1959. He received his wings in 1960 after completing his training at the U.S. Naval Air Training Command. He was then assigned to VMA-224, 2nd MAW, as an attack pilot. He later served with VMA-224, 1st MAW, at Iwakuni, Japan.

He has logged 2,600 hours of flight time—2,400 hours in jet aircraft and 200 hours in helicopters.

Current Assignment: Major Lousma is one of the 19 astronauts selected by NASA in April 1966. He served as a member of the astronaut support crews for the Apollo IX, X, and XIII missions.

Name: Gerald Paul Carr (Lieutenant Colonel, USMC)

Birthplace and Date: Born in Denver, Colo., on August 22, 1932, but raised in Santa Ana, Calif., which he calls his home town. His mother, Mrs. Freda L. Carr, resides in Santa Ana.

Physical Description: Brown hair; blue eyes, height: 5 feet 9 inches; weight: 155 pounds.

Education: Graduated from Santa Ana High School; received a Bachelor of Science degree in Mechanical Engineering from the University of Southern California, a Bachelor of Science degree in Aeronautical Engineering from the U.S. Naval Postgraduate School in 1961, and a Master of Science degree in Aeronautical Engineering from Princeton University in 1962.

Marital Status: Married to the former JoAnn Ruth Petrie of Santa Ana.

Children: Jennifer Anne, July 31, 1955; Jamee Adele and Jeffrey Ernest, July 3, 1958; John Christian, April 4, 1962; Jessica Louise and Joshua Lee, March 12, 1964.

Recreational Interests: He enjoys sailing and playing golf, tennis, badminton and handball. His hobbies include woodworking and restoration of an old automobile.

Organizations: Member of the Marine Corps Association, the University of Southern California Alumni Association, and Tau Kappa Epsilon. President of the Houston Trojan Alumni Club, and Texas Chairman of the American Cancer Society's I.Q. Society.

Special Honors: Awarded the National Defense Service Medal, the Armed Forces Expeditionary Medal and the Marine Corps Expeditionary Medal, and a Letter of Commendation from the Commander of Carrier Division II.

Experience: When informed by NASA of his selection for astronaut training, he was assigned to the Test Directors Section, Marine Air Control Squadron Three, a unit responsible for carrying out testing and evaluation of Marine Tactical Data Systems.

Carr began his military career in 1949 with the Navy, and in 1950 he was appointed a midshipman (NROTC) and enrolled in the University of Southern California. Upon graduation in 1954, he received his commission and subsequently reported to the U.S. Marine Corps Officers' Basic School at Quantico, Va. He received his flight training at Pensacola, Fla., and Kingsville, Tex., and was then assigned to Marine All-Weather-Fighter-Squadron 114 where he gained experience in the F-9 and the F-6A Skyray. After postgraduate training, he served with Marine All-Weather Fighter-Squadron 122 from 1962 to 1965 piloting the F-8 Crusader in the United States and Far East. Other aircraft he has flown include the F-4, T-1, T-33, T-28, T-38, H-13, and ground effect machines.

He has logged more than 3,100 flying hours, 2,440 hours of which is jet time.

Current Assignment: Lt. Colonel Carr is one of the 19 astronauts selected by NASA in April 1966. He served as a member of the astronaut support crews for the Apollo VIII and XII flights, and was involved in the development and testing of the lunar roving vehicle which was used on the lunar surface by Apollo flight crews. His present duties involve crew training for Project Skylab.

Name: Edward G. Gibson (PhD)

Birthplace and Date: Born November 8, 1936, in Buffalo, N.Y. His parents, Mr. and Mrs. Calder A. Gibson, reside in Kenmore, N.Y.

Physical Description: Brown hair; brown eyes; height: 5 feet 9 inches; weight: 160 pounds.

Education: Graduated from Kenmore Senior High School; received a Bachelor of Science degree in Engineering from the University of Rochester in June 1959; a Master of Science degree in Engineering (Jet Propulsion Option) from the California Institute of Technology in June 1960; and a Doctorate in Engineering with a minor in Physics from the California Institute of Technology in June 1964.

Marital Status: Married to the former Julia Ann Volk of Township of Tonawanda, N.Y. Her father, Mr. John E. Volk, resides in Township of Tonawanda.

Children: Jannet, November 9, 1960; John E., May 2, 1964; Julie, October 12, 1968.

Recreational Interests: His hobbies include athletics, scuba diving and solar observations.

Organizations: Member of the American Institute of Aeronautics and Astronautics, Tau Beta Pi, and Sigma Xi. He was an R. C. Baker Fellow and a Fellow of the National Science Foundation at the California Institute of Technology.

Experience: While studying at Caltech, Gibson was a research assistant studying in the fields of jet propulsion and classical physics. His technical publications were in the field of plasma physics. He was a senior research scientist with the Applied Research Laboratories of Philco Corporation at Newport Beach, Calif., from June 1964 until coming to NASA. While at Philco, he did research on lasers and the optical breakdown of gases.

He has logged 1,500 hours flying time—1,400 hours in jet aircraft.

Current Assignment: Dr. Gibson was selected as a scientist-astronaut by NASA in June 1965. He completed a 53-week course in flight training at Williams Air Force Base, Ariz., and earned his Air Force wings. He served as a member of the astronaut support crew for the Apollo XII flight.

Name: William Reid Pogue (Lieutenant Colonel, USAF)

Birthplace and Date: Born January 23, 1930, in Okemah, Okla., the son of Mr. and Mrs. Alex W. Pogue, who live in Sand Springs, Okla.

Physical Description: Brown hair; blue eyes; height: 5 feet 9 inches; weight: 163 pounds.

Education: Attended primary and secondary schools in Oklahoma; received a Bachelor of Science degree in Education from Oklahoma Baptist University in 1951 and a Master of Science degree in Mathematics from Oklahoma State University in 1960.

Martial Status: Married to the former Helen J. Dittmar of Shawnee, Okla., whose parents, Mr. and Mrs. Franklin L. Dittmar, reside in Hugo, Okla.

Children: William R., September 5, 1953; Layna S., June 9, 1955; Thomas R., September 12, 1957.

Recreational Interests: He enjoys playing squash and handball, and his hobbies are handcrafts and stereo systems.

Organizations: Member of the Air Force Association; the Research Society of America; the Society of Experimental Test Pilots; and Sigma Xi.

Special Honors: Presented the MSC Superior Achievement Award (1970); winner of the Air Medal, the Air Force Commendation Medal, and the National Defense Service Medal; and, as a member of the USAF THUNDERBIRDS, is the recipient of an Outstanding Unit Citation.

Experience: Pogue came to the Manned Spacecraft Center from an assignment at Edwards Air Force Base, Calif., where he had been an instructor at the Air Force Aerospace Research Pilot School since October 1965.

He enlisted in the Air Force in 1951 and received his commission in 1952. While serving with the Fifth Air Force during the Korean conflict, from 1953 to 1954, he flew 43 combat missions; and from 1955 to 1957, he was a member of the USAF THUNDERBIRDS.

He has gained proficiency in more than 50 types and models of American and British aircraft and is qualified as a civilian flight instructor. Pogue served as a mathematics instructor at the USAF Academy in Colorado Springs from 1960 to 1963. In September 1965, he completed a two year tour as test pilot with the British Ministry of Aviation under the USAF/RAF Exchange Program, after graduating from the Empire Test Pilots' School in Farnborough, England.

He has logged 4,400 hours flight time—3,400 hours in jet aircraft.

Current Assignment: Lt. Colonel Pogue is one of the 19 astronauts selected by NASA in April 1966. He served as a member of the astronaut support crews for the Apollo VII, XI, and XIV missions.

Name: Russell L. Schweickart

Birthplace and Date: Born October 25, 1935, in Neptune, N.J. His parents, Mr. and Mrs. George Schweickart, reside in Sea Girt, N.J.

Physical Description: Red hair; blue eyes; height: 6 feet; weight: 161 pounds.

Education: Graduated from Manasquan High School, N. J.; received a Bachelor of Science degree in Aeronautical Engineering and a Master of Science degree in Aeronautics and Astronautics from Massachusetts Institute of Technology.

Marital Status: Married to the former Clare G. Whitfield of Atlanta, Ga. Her parents are the Randolph Whitfields of Atlanta.

IV-19

Children: Vicki, September 12, 1959; Randolph and Russell, September 8, 1960; Elin, October 19, 1961; Diana, July 26, 1964.

Recreational Interests: His hobbies are amateur astronomy, photography and electronics.

Organizations: Fellow of the American Astronautical Society; and member of the Society of Experimental Test Pilots, the Explorers Club, the American Institute of Aeronautics and Astronautics, and Sigma Xi.

Special Honors: Awarded the NASA Distinguished Service Medal, the FAI De La Vaulx Medal in 1970, and the National Academy of Television Arts and Sciences Special Trustees Award for 1969.

Experience: Schweickart served as a pilot in the U.S. Air Force and Air National Guard from 1956 to 1963.

He was a research scientist at the Experimental Astronomy Laboratory at MIT, and his work there involved research in upper atmospheric physics, star tracking, and stabilization of stellar images. His thesis for a Master's degree at MIT concerned stratospheric radiance.

He has logged 3,250 hours flight time—2,850 hours in jet aircraft.

Current Assignment: Schweickart was one of the third group of astronauts named by NASA in October 1963.

He served as lunar module pilot for Apollo IX, March 3-13, 1969. This was the third manned flight in the Apollo series, the second to be launched by a Saturn V, and the first manned flight of the lunar module.

Following a Saturn V launch into a near circular 102.3 x 103.9 nautical mile orbit, Apollo IX successfully accomplished command/service module separation, transposition, and docking maneuvers with the lunar module. The crew then separated their docked spacecraft from the S-IVB third stage and commenced an intensive five days of checkout operations with the lunar module. Highlight of this evaluation was completion of a critical lunar-orbit rendezvous simulation and subsequent docking, initiated by James McDivitt and Russell Schweickart from within the lunar module at a separation distance which exceeded 100 miles from the command/service module piloted by Dave Scott.

Schweickart and McDivitt also demonstrated and confirmed the operational feasibility of crew transfer and extravehicular activity techniques and equipment. During a 46-minute EVA, Schweickart evaluated external transfer capability, made photographs, and retrieved thermal samples from the lunar module exterior. This EVA marked the first space test of the Apollo portable life support system backpack which provided breathing oxygen and suit pressurization and cooling independent of the spacecraft life support system.

Apollo IX splashed down less than four miles from the helicopter carrier USS GUADALCANAL. Completing his first flight, Schweickart has logged 241 hours in space.

Schweickart continues training for future manned space flights and is assigned additional responsibility within the Astronaut Office for coordinating the ATM (Apollo Telescope Mount) and EVA (extravehicular activities) to be used in Project Skylab.

Name: Story Musgrave (MD)

Birthplace and Date: Born August 19, 1935, in Boston, Mass., but considers Lexington, Ky., to be his hometown. His mother, Mrs. Marguerite Swann Musgrave, resides in Upper Saddle River, N.J.

Physical Description: Blond hair; blue eyes; height: 5 feet 10 inches; weight: 155 pounds.

Education: Graduated from St. Mark's High School in Southborough, Mass.; received a Bachelor of Science degree in Statistics from Syracuse University in 1958, a Master of Business Administration degree in Operations Analysis and Computer Programming from the University of California at Los Angeles in 1959, a Bachelor of Arts degree in Chemistry from Marietta College in 1960, a Doctorate in Medicine from Columbia University in 1964, and a Master of Science in Biophysics from the University of Kentucky in 1966; and expects to receive a Doctorate in Physiology from the University of Kentucky in 1972.

Marital Status: Married to the former Patricia Marguerite Van Kirk of Patterson, N.J. Her mother, Mrs. Neil Van Kirk, resides in Wayne, N.J.

Children: Lorelei Lisa, March 27, 1961; Bradley Scott, July 3, 1962; Holly Kay, December 13, 1963; Christopher Todd, May 12, 1965; Jeffrey Paul, June 19, 1967.

Recreational Interests: His hobbies are playing chess, bicycling, parachuting, photography, flying, water skiing, running, scuba diving, handball, gardening and motorcycling.

Organizations: Member of the Aerospace Medical Association, the Aircraft Owners and Pilots Association, the Air Force Association, Alpha Kappa Psi, the American Association for the Advancement of Science, the American Institute of Aeronautics and Astronautics, the American Medical Association, the Civil Aviation Medical Association, the Flying Physicians Association, the National Aeronautics Association, the National Aerospace Education Council, the National Geographic Society, the New York Academy of Sciences, Phi Delta Theta, the Soaring Society of America, and the United States Parachute Association.

Special Honors: Awarded the National Defense Service Medal and an Outstanding Unit Citation as a member of U.S. Marine Corps squadron VMA-212; recipient of a U.S. Air Force Post-doctoral Fellowship (1965-1966) and a National Heart Institute Post-doctoral Fellowship (1966-1967).

Experience: Following graduation from high school in 1953, Musgrave entered the U.S. Marine Corps and completed basic training at Parris Island, S.C. He then reported to the Naval Air Technical Training Center at the Naval Air Station in Jacksonville, Fla., where he attended classes at the U.S. Naval Airman Preparatory School and the U.S. Naval Aviation Electrician and Instrument Technician School. He served as an electrician and instrument technician and as an aircraft crew chief while completing duty assignments in Korea, Japan and Hawaii and aboard the carrier USS WASP in the far east.

He has flown 40 different types of single- and multi-engine civilian and military aircraft, logging over 4,400 hours flying time, and he holds instructor, instrument instructor and airline transport ratings. An accomplished parachutist, he has made more than 200 free falls—including 70 experimental free fall descents involved with human aerodynamics. He holds an International Jumpmaster Class C License. He was President and Jumpmaster of the Bluegrass Sport Parachuting Association in Lexington, Ky., from 1964 to 1967.

Dr. Musgrave served a surgical internship from 1964 to 1965 at the University of Kentucky Medical Center in Lexington. He remained there until the summer of 1967 as a U.S. Air Force post-doctoral fellow in Aerospace Physiology and Medicine and as a National Heart Institute post-doctoral fellow.

Musgrave is presently a part-time resident in general surgery at the Denver General Hospital and a part-time instructor in the Department of Physiology and Biophysics at the University of Kentucky Medical Center.

Current Assignment: Dr. Musgrave was selected as a scientist-astronaut by NASA in August 1967. He has since completed the initial academic and flight training and is currently involved in further training for future manned space flights.

Name: Bruce McCandless II (Lieutenant Commander, USN)

Birthplace and Date: Born June 8, 1937, in Boston, Mass.

Physical Description: Brown hair; blue eyes; height: 5 feet 10 inches; weight: 150 pounds.

Education: Graduate of Woodrow Wilson Senior High School, Long Beach, Calif.; received a Bachelor of Science degree in Naval Sciences from the United States Naval Academy in 1958, and a Master of Science degree in Electrical Engineering from Stanford University in 1965. He is working towards a Ph.D. in Electrical Engineering at Stanford University.

Marital Status: Married to the former Bernice Doyle of Rahway, N.J. Her mother, Mrs. Charles Doyle, resides in Yuma, Ariz.

Children: Bruce III, August 15, 1961; Tracy, July 13, 1963.

Recreational Interests: His hobbies are electronics, photography, sailing and scuba diving; and he also enjoys swimming and playing volleyball.

Organizations: Member of the U.S. Naval Academy Alumni Association (Class of 1958); the National Geographic Society; the U.S. Naval Institute; the Institute of Electrical and Electronic Engineers; and the National Audubon Society.

Special Honors: Awarded the Expert Rifle and Pistol Shot Medals and holder of the National Defense Service Medal and the American Expeditionary Service Medal.

Experience: McCandless was graduated second in a class of 899 from Annapolis. He subsequently received flight training from the Naval Aviation Training Command at bases in Pensacola, Fla., and Kingsville, Tex. He was designated a Naval Aviator in March 1960. He proceeded to Key West, Fla., for weapons system and carrier landing training in the F-6A Skyray. He was assigned to Fighter Squadron ONE HUNDRED TWO (VF-102) from December 1960 to February 1964, flying the Skyray and the F-4B Phantom II. He saw duty aboard the USS FORRESTAL (CVA-59) and the USS ENTERPRISE (CVA(N)-65), including the latter's participation in the Cuban Blockade. For three months in early 1964, he was an Instrument Flight Instructor in Attack Squadron FORTY-THREE (VA-43) at the Naval Air Station Apollo Soucek Field, Oceana, Va., and then reported to the Naval Reserve Officers' Training Corps Unit at Stanford University for graduate studies in electrical engineering.

He has gained flying proficiency in the T-33B Shootingstar, T-38A Talon, F-4B Phantom II, F-6A Skyray, F-11 Tiger, TF-9J Cougar, T-1 Seastar, and T-34B Mentor aircraft, and the Bell 47G helicopter. He has logged more than 2,500 hours of flying time—2,250 in jet aircraft.

Current Assignment: Lt. Commander McCandless is one of the 19 astronauts selected by NASA in April 1966. He served as a member of the astronaut support crew for the Apollo XIV mission and is a co-investigator on the M-509 astronaut maneuvering unit experiment in the Skylab Program.

Name: Vance DeVoe Brand

Birthplace and Date: Born in Longmont, Colo., May 9, 1931. His parents, Dr. and Mrs. Rudolph W. Brand, reside in Longmont.

Physical Description: Blond hair; gray eyes; height: 5 feet 11 inches; weight: 175 pounds.

Education: Graduated from Longmont High School; received a Bachelor of Science degree in Business from the University of Colorado in 1953, a Bachelor of Science degree in Aeronautical Engineering from the University of Colorado in 1960, and a Master's degree in Business Administration from the University of California at Los Angeles in 1964.

Marital Status: Married to the former Joan Virginia Weninger of Chicago. Her parents, Mr. and Mrs. Ralph D. Weninger, reside in Chicago.

Children: Susan N., April 30, 1954; Stephanie, August 6, 1955; Patrick R., March 22, 1958; Kevin S., December 1, 1963.

Recreational Interests: Skin diving, skiing, handball and jogging.

Organizations: Member of the Society of Experimental Test Pilots, the American Institute of Aeronautics and Astronautics, Sigma Nu and Beta Gamma Sigma.

Special Honors: Presented the MSC Certificate of Commendation (1970).

Experience: Brand served as a commissioned officer and naval aviator with the U.S. Marine Corps from 1953 to 1957. His Marine Corps assignments included a 15-month tour in Japan as a jet fighter pilot. Following his release from active duty, he continued flying fighter aircraft in the Marine Corps Reserve and the Air National Guard until 1964, and he still retains a commission in the Air Force Reserve.

From 1960 to 1966, Brand was employed as a civilian by the Lockheed Aircraft Corporation. He first worked as a flight test engineer on the P3A "Orion" aircraft and later transferred to the experimental test pilot ranks. In 1963, he graduated from the U.S. Naval Test Pilot School and was assigned to Palmdale, Calif., as a experimental test pilot on Canadian and German F-104 development programs. Immediately prior to his selection to the astronaut program, Brand was assigned to the West German F-104G Flight Test Center at Istres, France, as an experimental test pilot and leader of a Lockheed flight test advisory group.

He has logged 4,030 hours of flying time, which include 3,260 in jets and 330 hours in helicopters.

Current Assignment: Brand is one of the 19 astronauts selected by NASA in April 1966. He served as a crew member for the thermal vacuum test of 2TV-1, the prototype command module; and he was a member of the astronaut support crews for the Apollo VIII and XIII missions. Brand served as backup command module pilot for Apollo XV.

Name: William B. Lenoir (PhD)

Birthplace and Date: Born on March 14, 1939, in Miami, Fla. His parents, Mr. and Mrs. Samuel S. Lenoir, reside in Miami.

Physical Description: Brown hair; brown eyes; height: 5 feet 10 inches; weight: 165 pounds.

Education: Attended primary and secondary schools in Coral Gables, Fla.; is a graduate of the Massachusetts Institute of Technology where he received a Bachelor of Science degree in Electrical Engineering in 1961, a Master of Science degree in 1962, and a Doctorate in 1965.

Marital Status: Married to the former Elizabeth May Frost, daughter of Mr. and Mrs. Thomas F. Frost, who reside in Brookline, Mass.

Children: William B., Jr., April 6, 1965; Samantha E., March 20, 1968.

Recreational Interests: His hobbies include sailing and woodworking, and he also enjoys ice hockey.

Organizations: Member of the American Geophysical Union the American Association for the Advancement of Science; the American Astronomical Society; the American Institute of Physics; Eta Kappa Nu; and the Society of Sigma Xi.

Special Honors: Sloan Scholar at MIT and winner of the Carleton E. Tucker Award for Teaching Excellence at MIT.

Experience: From 1964 to 1965, Lenoir was an instructor at MIT; and in 1965, he was named Assistant Professor of Electrical Engineering. His work at MIT included teaching electromagnetic theory and systems theory and performing research in the remote sensing of planetary atmospheres and surfaces and the theory of radiative transfer in anisotropic media.

He is participating as an investigator in several satellite experiments and continues his research in this area while completing astronaut training.

He has logged 900 hours flying time in jet aircraft.

Current Assignment: Dr. Lenoir was selected as scientist-astronaut by NASA in August 1967. He completed the initial academic training and a 53-week course in flight training at Laughlin Air Force Base, Tex. He is currently involved in training for future manned space flights.

Name: Don Leslie Lind (PhD)

Birthplace and Date: Born in Midvale, Utah, on May 18, 1930, His parents, Mr. and Mrs. Leslie A. Lind, reside in Midvale, Utah.

Physical Description: Brown hair; hazel eyes; height: 5 feet 11 3/4 inches; weight: 170 pounds.

Education: Attended Midvale Elementary School and is a graduate of Jordan High School, Sandy, Utah; received a Bachelor of Science degree with high honors in Physics from the University of Utah in 1953 and a Doctor of Philosophy degree in High Energy Nuclear Physics in 1964 from the University of California, Berkeley.

Marital Status: Married to the former Kathleen Maughan of Logan, Utah; her parents, Mr. and Mrs. J. Howard Maughan, reside in Logan.

Children: Carol Ann, January 24, 1956; David M., November 29, 1956; Dawna, September 1, 1958; Douglas M., October 26, 1960; Kimberly, April 30, 1963; Lisa Christine, March 26, 1970.

Recreational Interests: Hobbies are amateur theatricals, play writing and sculpturing; and he is an avid swimmer and skier.

Organizations: Member of the American Geophysical Union, the American Association for Advancement of Science, the Society of Sigma Xi, and Phi Kappa Phi.

Experience: Before his selection as an astronaut, he worked at the NASA Goddard Space Flight Center as a space physicist. He had been with Goddard since 1964 and was involved in experiments to determine the nature and properties of low energy particles within the Earth's magnetosphere and interplanetary space. Previous to this, he worked at the Lawrence Radiation Laboratory, Berkeley, Calif., doing research in ion-nucleon scattering, a type of basic high energy particle interaction.

Lind holds the rank of a Commander in the U.S. Naval Reserve and has held his Reserve status since September 1957. He served four years on active duty with the Navy and began his military career in January 1954 at the U.S. Navy Officer Candidate School in Newport, R.I. He received his wings in September 1955 after completing training with the Flight Training Command and was subsequently assigned to VF-143 then deployed at San Diego and later aboard the carrier USS HANCOCK.

He has logged more than 2,900 hours flying time—2,300 hours in jet aircraft.

<u>Current Assignment</u>: Dr. Lind is one of the 19 astronauts selected by NASA in April 1966. He is currently involved in training for future manned space flights.

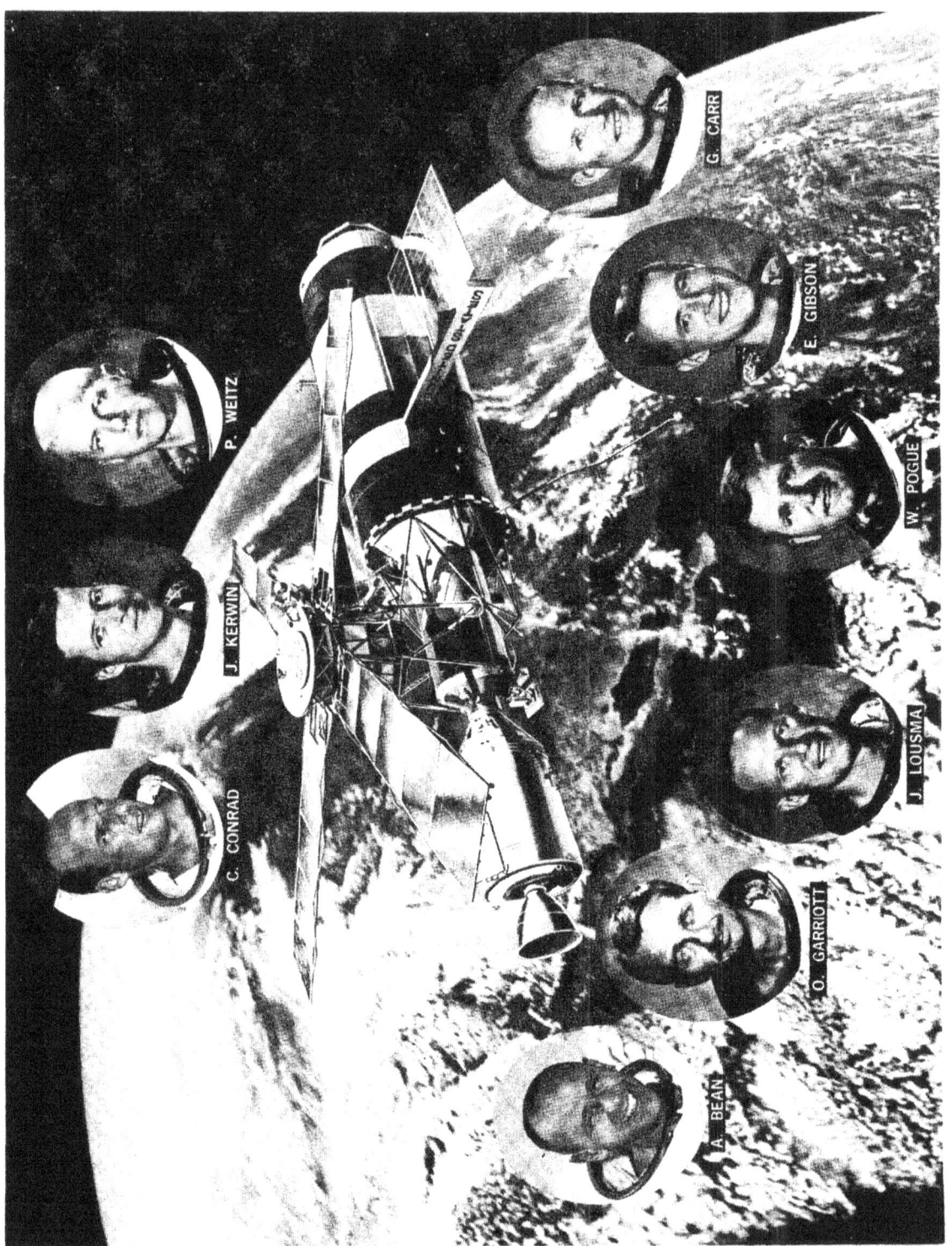

SECTION V

LAUNCH FACILITIES AND OPERATIONS

LAUNCH COMPLEX 39

Launch Complex 39 facilities at the Kennedy Space Center (see Saturn IB and Saturn V News Reference books) were planned and built specifically for the Apollo Saturn V space vehicle used to transport astronauts to the Moon and return them safely to the Earth.

Complex 39 uses the mobile concept of launch operations, a departure from the fixed launch pad techniques used previously at Cape Kennedy and other launch sites. The fixed site method called for assembly, checkout and launch at one site, tying up a pad for long periods and exposing flight hardware to the outside environment for extended intervals.

Using the mobile concept, the space vehicle is thoroughly checked out in an enclosed building before it is moved to the launch pad for final preparations. This affords greater protection, a more systematic checkout process using computer techniques, and pad time is minimal.

The major components of Complex 39 include the Vehicle Assembly Building (VAB) where space vehicles are assembled and prepared; the Launch Control Center (LCC) where the launch team conducts the preliminary checkout and final countdown; the Mobile Launchers (ML) upon which space vehicles are erected for checkout and from where they are launched; the Mobile Service Structure (MSS) which provides external access to the space vehicle at the pad; the Transporter, which carries the space vehicle, ML and MSS to the pads; the crawlerway over which the space vehicle travels from VAB to the launch pads; and Complex 39's two launch pads, Pad A for launching SL-1 and Pad B for launching SL-2, SL-3 and SL-4.

Vehicle Assembly Building

The VAB is the heart of Complex 39. Covering 3.25 hectares (8 acres), it is where the Saturn Skylab (SL-1) and Saturn IB/CSMs (SL-2, SL-3 and SL-4) are assembled and tested. The VAB has a volume of 3,624,000 cubic meters (129,428,000 cubic feet). It is 160 meters (525 feet) tall, 218 meters (716 feet) long and 158 meters (518 feet) wide.

The building is divided into a high bay area 160 meters (525 feet) tall and a low bay area with a height of 64 meters (210 feet), with both areas serviced by a transfer aisle for movement of vehicle stages.

The low bay work area contains eight stage preparation and checkout cells equipped with systems to simulate stage interface with other stages and the instrument unit.

The high bay provides facilities for assembly and checkout of both launch vehicle and spacecraft. It contains four separate bays for vertical assembly and checkout.

Work platforms in the high bays provide access by surrounding the vehicle at varying levels. Each platform consists of two bi-parting sections that move in from opposite sides and mate, providing a 360-degree access to the section of the space vehicle being checked.

The MLs, carried by transporters, move in and out of the VAB through four doors in the high bay area, one in each of the bays.

Launch Control Center

The four-story LCC is immediately adjacent to the VAB. The electronic "brain" of Complex 39, the LCC is used for checkout and test operations while Skylab hardware is being assembled and prepared in the VAB for movement to the pads. The LCC contains display, monitoring and control equipment used for both checkout and launch operations.

The LCC has telemeter checkout stations on its second floor and four firing rooms, one for each high bay of the VAB, on its third floor. Three firing rooms contain identical sets of control and monitoring equipment so that launch of a vehicle and checkout of others take place simultaneously.

A ground computer facility is associated with each firing room and a high speed computer data link is provided between the LCC and the ML for checkout of the launch vehicle. This link can be connected to the ML at either the VAB or at the pad.

The three firing rooms have some 450 consoles which contain controls and displays required for the checkout process. The digital data links connecting with the high bay areas of the VAB and the launch pads carry the data required during checkout and launch.

Mobile Launcher

The ML is a transportable launch base and umbilical tower for the space vehicle. Three MLs are available for use at Complex 39.

The 5,716 metric ton (12,600,000 pound) Mobile Launcher stands 135.9 meters (446 feet) tall when resting on its pedestals. The launcher base is a two-story steel structure 7.6 meters (25 feet) high, 48.8 meters (160 feet) long and 41.1 meters (135 feet) wide.

The launch vehicle sits over a 13.7 meter (45 foot) square opening which allows an outlet for engine exhausts into the launch pad trench containing a flame deflector. This opening is lined with a replaceable steel blast shield, independent of the structure, which is cooled by a water curtain.

Hydraulically-operated service arms on the launch tower support lines for the vehicle umbilical systems and provide access for personnel to the stages and the spacecraft.

These service and access arms are retracted at various intervals during the final stages of the countdown. They are equipped with a backup retraction system for use if the primary mode fails.

Transporters

Complex 39's facilities include two 2,721 metric ton (6,000,000 pound) transporters which move MLs into the VAB and MLs with assembled flight hardware to the launch pad. They are also used to transfer the MSS to and from the launch pads.

The transporters are 39.9 meters (131 feet) long and 34.7 meters (114 feet) wide. They move on four double-tracked crawlers, each 3 meters (10 feet) high and 12.5 meters (41 feet) long. Each shoe on the crawler track weighs 0.9 metric ton (2,000 pounds). Maximum speed unloaded is 3.2 kilometers (2 miles) per hour. Loaded speed is 1.6 kilometers (1 mile) per hour.

The transporters have a leveling system designed to keep the top of the space vehicles vertical within plus-or-minus 10 minutes of arc — about the dimensions of a basketball. This system also provides leveling operations required to negotiate the five percent ramp leading to the launch pads and to keep the load level when it is raised and lowered on pedestals at the pad and within the VAB.

The overall height of the transporter is 6 meters (20 feet) from ground level to the top deck on which the Mobile Launcher is mated for transportation. The deck is flat and about the size of a baseball diamond — 27 meters (90 feet) square.

Crawlerway

The transporter moves on a roadway 39.6 meters (130 feet) wide. This is almost as broad as an eight-lane turnpike. The roadway is built in three layers with an average depth of 2.13 meters (7 feet).

Mobile Service Structure

A 125 meter (410 foot) tall, 4,768 metric ton (10.5 million pound) tower is used to service Skylab hardware at the launch pads. This 40-story structure provides 360-degree platform access to the space vehicles. The service structure has five platforms — two self-propelled and three fixed but movable. Two elevators carry personnel and equipment between work platforms. The platforms can open and close around the space vehicles.

The Saturn V/Skylab and Saturn IB/CSM missions mark the first dual launch from Complex 39, made possible by the scheduling of the single MSS to service both space vehicles. The MSS is returned to its park site along the crawlerway prior to launch.

Launch Pads

Both Pad A and Pad B are roughly octagonal in shape. Each covers about 647,500 square meters (one fourth of a square mile) of terrain.

The center of the pad is a hardstand constructed of heavily reinforced concrete. In addition to supporting the weight of the ML and Skylab space vehicles, they must also support the MSS and transporters. The pad tops stand some 14.6 meters (48 feet) above sea level.

Saturn V and Saturn IB propellants are stored near the pad perimeters. Stainless steel, vacuum-jacketed pipes carry the liquid oxygen and liquid hydrogen from the storage tanks to the pads, up the ML and finally into the launch vehicle propellant tanks. The facilities include (for each pad) one liquid oxygen, one liquid hydrogen and three RP-1 tanks.

A total of 12 Saturn V launches were conducted during the Apollo program. All of these launches with the exception of Apollo 10 were conducted from Pad A. Apollo 10 was launched from Pad B.

KSC/Complex 39 Modifications

Operational modes for Skylab are based on precedence and experience gained during the processing of Apollo CSMs and LMs through the various phases of checkout. The resources of KSC have been effectively focused on the completion of the Skylab mission with minimum changes and maximum effectiveness. This same flexibility will make it possible to adapt Launch Complex 39 to future program requirements.

The major modification to Complex 39 facilities involved the construction of a 39-meter (127-foot) tall pedestal on an Apollo Saturn V mobile launcher to adapt the 68-meter (223-foot) tall Saturn IB to a tower built for the 110.6-meter (363-foot) tall Apollo Saturn V. (See Section I-8.)

Other Complex 39 modifications include several swing arm changes on both MLs to adapt them to the peculiar need of Skylab, modifications to VAB high bay work platforms, and construction of "clean rooms" for the ATM in the Manned Spacecraft Operations Building (MSOB) and High Bay 2 in the VAB. Pad B and Firing Room 3 of the LCC also required changes in the propellant servicing, control and monitoring areas to make them compatible with a two-stage launch vehicle. Pad A and Firing Room 2 required changes in the propellant, environmental control, instrumentation and control systems to make them compatible with a two-stage Saturn V with the OWS, ATM, AM/MDA payload.

Minor modifications to the MSS were also needed to provide access to the smaller diameter Saturn IB first stage.

Platform No. 1 on the MSS required extensive changes to provide access to the Saturn V second stage (S-II) with the MSS parked 2.73 meters (9 feet) south of its normal service position. This access is required for second stage insulation inspection after cryogenic loading during the Countdown Demonstration Test (CDDT).

Saturn IB Launch Pedestal — The new structure mounted on the platform of ML-1 allows the second stage, instrument unit and spacecraft to interface with the ML at the same vehicle stations as the Apollo/Saturn V. The steel pedestal stands 39 meters (127 feet) above the zero level of the 135.9-meter (446-foot) tall ML. The pedestal tapers upward from 14.6 meters (48 feet) square at the bottom to 6.6 meters (21 feet, 11 inches) square at the top. The columns are of 61 centimeter (24 inch) diameter, 2.54 centimeter (1 inch) thick pipe and the horizontal members are of 45.7 centimeter (18 inch) diameter, 2.54 centimeter (1 inch) thick pipe. Diagonal members are of 45.7 centimeter (18 inch) diameter, 1.9 centimeter (3/4 inch) and 1.27 centimeter (1/2 inch) thick pipe.

At the top of the tower is a launcher table with an outside diameter of 13.37 meters (44 feet), an inside diameter of 8.5 meters (28 feet) and a depth of 3 meters (10 feet) centered over the existing flame hole. Work platforms along the top deck increase the outside diameter to 15.2 meters (50 feet). The top deck, for alignment of holddown arms, fuel and cable mast, must be parallel to the horizontal within 1.5 millimeters (1/16 inch). The top plates are 5 centimeters (2 inches) thick and the bottom plates, walls and gussets are 2.54 centimeters (1 inch) thick.

The pedestal is designed as a free standing tower with no horizontal shear connection between the access bridges and the ML tower. This simplifies equal load distribution into each tower leg and assures that vehicle loads will be dumped through the desired support columns on the pad. Weight of the pedestal is about 226,800 kilograms (500,000 pounds). Associated equipment weighs an additional 226,800 kilograms (500,000 pounds). Weight of the pedestal was a constraint since it must be lifted and transported by the crawler transporter.

The pedestal is designed to launch a Saturn IB under maximum winds of 59.4 kilometers (37 miles) per hour at 9.1 meters (30 feet) elevation and to withstand a 200 kilometer (125 mile) per hour hurricane without the launch vehicle.

Installation of the pedestal required the removal of the Saturn V holddown arms, service arms 1, 2 and 3 and four tail service masts.

Design of the systems on the pedestal was by KSC in-house Civil Service engineers. These systems and their requirements are:

Liquid Oxygen (LOX) Transfer System — Provides vacuum jacketed piping, pumps and instrumentation to load oxidizer aboard the Saturn IB. Flow rates are 1,890 liters (500 gallons) per minute for slow fill and 4,730 liters (1,250 gallons) per minute for fast fill.

RP-1 System — Provides piping, pumps and instrumentation to load the kerosene-like fuel aboard the vehicle's first stage. Flow rates are 760 liters (200 gallons) per minute slow fill and 5,300 liters (1,400 gallons) per minute fast fill.

Environmental Control System — Provides cool or warm air to the boattail of the vehicle, via the pedestal, and to the pedestal for equipment and personnel. It is capable of transferring to gaseous nitrogen purge prior to fueling operations to inert the vehicle boattail cavity and pedestal.

Water System — Provides fire water for the boattail of the vehicle, a washdown capability for the pedestal for fuel spill and cooling water for launch. Flow rates are 15,140 liters (4,000 gallons) per minute for the boattail and 20,820 liters (5,500 gallons) per minute for washdown and launch. Pressure is 556 Newtons (125 pounds) per square inch.

<u>Power</u> — This system provides AC and DC power for lights and ground support equipment.

<u>Pneumatics</u> — Gaseous nitrogen and helium for vehicle and ground support equipment operating and charging are supplied at 13,344 Newtons (3,000 pounds) per square inch and various lower pressures.

<u>Intercom/Television System</u> — This system provides both voice and picture communications, giving the "eyes and ears" for checkout and launch operations.

<u>Emergency Access Platform</u> — The EAP provides limited access to the Saturn IB booster's eight engines for troubleshooting.

<u>Engine Service Platform</u> — This removable platform fills the engine exhaust hole and provides personnel access to the entire boattail engine area. It also provides the capability to change any one of the booster's eight H-1 engines.

The pedestal structure is of standard steel structural design. One unique requirement was that the pedestal withstand engine exhaust temperatures of approximately 2,760°C (5,000°F). Another was the problem of holding the vehicle to the pedestal during thrust build-up. This is accomplished by eight holddown arms mounted on top of the pedestal. If all engines are running smoothly and at proper thrust, the holddown arms release the vehicle for flight. If not, they support the vehicle during engine cut-off, thereby saving it for failure analysis, repair and launch at another time.

The holddown period (nominally four seconds) and subsequent release can set up detrimental oscillations unless proper resonant frequencies exist in each structure. This required vertical damping.

Also, the free standing vehicle mounted on the pedestal combination has a resonant frequency at a wind condition and direction that possibly can occur at KSC. This required horizontal damping at the top of the vehicle.

<u>"Clean Rooms"</u> — Two new clean rooms will support flight preparation of the ATM. The larger of the two is in the MSOB in the KSC Industrial Area. The second is on Platform A of the VAB's High Bay 2.

The MSOB clean room is 11.28 meters (37 feet) square and 14 meters (46 feet) high. Temperature and humidity are controlled carefully and the air in the clean room is circulated and filtered constantly to avoid contamination of the delicate lenses and other critical components. Contamination could distort spectrum measurements or, in the worst case, block out the entire optical system of one or more experiments. Air enters the clean room through specially designed filters which cover the entire ceiling. Inside the room, an elevated grated floor serves as a work platform for the checkout team and permits conditioned air to be collected and re-circulated.

Associated with the clean room is a personnel air lock and air shower, sticky mats, shoe cleaners and clean room clothing storage and distribution area. The clean room represents the major change in the MSOB required in support of the Skylab Program.

The second clean room — on Platform A of High Bay 2 — is an octagon measuring 10.8 meters (35 feet) across the flats. It is 7.9 meters (26 feet) high. Because the level of ATM checkout activity is less here than in the MSOB, the room is smaller and does not contain as many features. This clean room permits access to the ATM during integrated checkout operations.

Skylab-Related Modifications to MSS — The second (S-II) stage of the Skylab Saturn V launch vehicle requires access for insulation inspection after detanking of the cryogenic propellants loaded during the Countdown Demonstration Test (CDDT). The Payload Shroud of the Skylab Saturn V has a larger diameter than the Apollo service module. As a result, the MSS is prevented access to its normal position at Pad A and must be parked 2.7 meters (9 feet) to the south. To provide access to the second stage, Platform No. 1 has been modified to shift the vehicle clearance hole 2.7 meters (9 feet) in the northerly direction. Installation of this modification was completed shortly after launch of the Apollo 17 space vehicle.

Skylab-Related Modifications to Pad B, Complex 39 — Only minor changes were required to prepare Pad B for Saturn IB launchers. These include:

    RP-1 Fuel Transfer System — Changes in flow rates were required to meet the fill criteria of the Saturn IB. These changes were in controls, electrical components and routing of a new RP-1 transfer line from the Saturn V tail service mast interface at the 38.7 meter (127 foot) level of the ML. Patching changes were also required in the Pad Terminal Connection Room (PTCR) to prevent starting fast fill until the 15 percent level is reached.

    Liquid Oxygen (LOX) Transfer System — Modifications were required to the control and monitoring equipment in the PTCR for the LOX pumping system to meet the reduced flow rate requirements of the Saturn IB.

Skylab-Related Modifications to Launch Control Center, Firing Room 3 — Launch Control Center facilities were designed to accommodate launches of the three-stage Saturn V. Relatively minor changes were required to adapt it for launches of the two-stage Saturn IB. These include the addition of increased timing and countdown instrumentation to handle the precision requirements of a dual launch, eliminating S-IC and S-II measurements and replacing them with measurement racks to handle the Saturn IB booster. The work included removal of S-IC and S-II support panels and equipment which would have interfered with the installation of required Saturn IB booster and IU equipment.

Skylab-Related Modifications to High Bay 2, VAB, for OWS — Extensible work platforms in the VAB are designed to provide 360-degree access to the space vehicle while in the VAB. Platforms in the upper levels of the VAB normally used to service the 3.9-meter (12.8-foot) diameter Apollo spacecraft have been modified to accept the 6.7-meter (22-foot) diameter OWS package and its payload shroud.

Skylab-Related Modifications to Mobile Launcher 2 for Saturn V/OWS — Modifications to Mobile Launcher 2 have adapted this structure to the needs of a two-stage launch vehicle and a modified third stage housing the OWS. These changes include removal of propellant loading services to the S-IVB third stage and facilities required to service the Apollo spacecraft. ML changes include the addition of a 20.9 meter (68.5 foot) long service arm (designated 6A) at the 73.1 meter (240 foot) level. The new service arm includes an environmental shelter or "white room" which will provide access to Skylab during launch preparations in the VAB and on Pad A. The service arm was the spacecraft access arm over which the Apollo 11 crew passed on July 16, 1969, to begin man's first journey to the surface of the Moon. It will be used continuously while Skylab is in the VAB. It is designed for contingency use once the Mobile Launcher/space vehicle have been moved to the pad. Other changes include addition of pneumatic services (helium, nitrogen, oxygen), environmental control system services, power and instrumentation and a new umbilical to the Fixed Airlock Shroud that replaces the Lunar Module umbilical on Service Arm No. 7.

The overall cost of Skylab facility modifications was approximately $9 million with the single largest item the $1 million required to construct the pedestal.

V-2

SKYLAB LAUNCH OPERATIONS

NASA's John F. Kennedy Space Center performs preflight checkout, test and launch of the Saturn V/Skylab and the three Saturn IB launch vehicles which will carry manned crews to and from the nation's first space station.

Skylab and all its components will be launched by a two-stage Saturn V launch vehicle from Pad A at Complex 39. The first of the three manned spacecraft will be launched from Complex 39's Pad B by a Saturn IB about 24 hours later. Two additional Saturn IBs carrying three-man crews will be launched at intervals of about 90 days.

Earlier plans called for the launch of the Apollo/Saturn IB space vehicles from Complex 34 at KSC. The decision to conduct Saturn IB launches at Complex 39 rather than from Complex 34 was reached in May 1970 after a comprehensive study of the capabilities and costs of both locations. The decision was made to take advantage of the more modern facilities at Complex 39 and to save money by consolidating manpower, spacecraft support and checkout equipment requirements and to reduce transportation costs.

Advantages include operating in the VAB with its controlled environment as compared with long-term preparation at Launch Complex 34, which was exposed to the weather. Pad time for both vehicles will be minimal.

Hardware Preparation

Active preparations for the dual launches of SL-1 and SL-2 began at KSC in the late summer and early fall of 1972 with the arrival and "stacking" of Skylab hardware atop MLs in the VAB.

Saturn V/Skylab (SL-1) — Erection of the Skylab space vehicle on ML-2 in High Bay 2 of the VAB began on September 2 with placement of the S-IC stage on the zero level of the Mobile Launcher. The S-II stage was erected on September 20. The ATM arrived by aircraft at the Cape Kennedy Skid Strip from MSC on September 22. It was moved into its "clean room" in the MSOB for extensive testing and prelaunch checkout.

The Skylab OWS arrived at Port Canaveral aboard the Point Barrow on September 22. It was moved up the Banana River to the VAB Turn Basin by a MSC barge on the following day. Skylab OWS was mated with the S-II stage on September 28 (Figure V-1). The AM/MDA arrived by aircraft on October 6. Both were moved into the MSOB for prelaunch checkout. The Saturn V Instrument Unit was received at KSC on October 27 and mated with the launch vehicle on October 31 (Figure V-2).

The OWS solar array panels underwent a series of tests in the MSOB and were moved to the VAB where they were mechanically mated with the Workshop on November 15-16. The ATM solar arrays were installed on the ATM in the MSOB early in January 1973.

The AM/MDA was removed from its workstand in the MSOB (Figure V-3) early in December and moved into the high bay. The CSM was moved from an MSOB support stand shortly after the Apollo 17 launch and placed in an adjacent workstand in preparation for a docking test with the AM/MDA (Figure V-4). This docking test was successfully conducted during a four-day period in med-December.

The spacecraft and AM/MDA were undocked shortly before Christmas. The CSM was then moved into an altitude chamber in preparation for unmanned and manned vacuum tests. The AM/MDA was moved back into an integrated test stand and brazed with its FAS.

The ATM, AM/MDA/ATMDA, PS and other SL-1 components were moved to the VAB in late January to complete the stacking of the Skylab space vehicle.

Figure V-1 — OWS/S-II Mating

Figure V-2 — SL-1/IU Mating

Saturn IB/CSM (SL-2) — Stacking of the Saturn IB/CSM for the first manned visit to Skylab bagan with erection of the Saturn IB booster on its launch pedestal on ML-1 in the VAB's High Bay 1 on August 31, 1972. The S-IVB stage was mated with the booster on September 5 (Figure V-5) and the IU was erected on September 7. The CSM arrived by aircraft in July. It was moved to the MSOB for pre-flight checkout, including leak tests and a combined systems test. These were completed in late August.

A spacecraft mockup or "boilerplate" was mated with the Saturn IB launch vehicle and the space vehicle was moved to Complex 39's Pad B on January 9, 1973, for pad fit checks and launch vehicle fueling tests. The first flight of the Saturn IB/CSM from Complex 39 will permit testing of the compatibility of the space vehicle and its Mobile Launcher with pad facilities.

Fueling and systems tests were conducted in January and the space vehicle with its "boilerplate" spacecraft was returned to the VAB. The boilerplate spacecraft was removed and the flight spacecraft erected atop the Saturn IB in February.

Rollout

At the beginning of February, the SL-1 and SL-2 vehicles were in the VAB being readied for a series of tests to prepare them for the move to the pads. The overall philosophy is to conduct all possible tests within the protected environment of the VAB and move the space vehicles about a month before launch to the launch pads for limited testing and servicing.

Figure V-3 — AM/MDA In MSOB

SL-1 will be virtually in flight condition when moved to the pad. Completion of most pre-flight operations in the VAB has permitted diversion of KSC's single MSS to Pad B to service SL-2. The MSS will be used at Pad A during the dual launch operation to inspect second stage insulation of the Saturn V following the fueled or "wet" CDDT and in the event of a "scrub" of the SL-1 launch after cryogenic fueling has begun.

The MSS will be located at Pad B in support of SL-2 during the CDDT's early phases. The SL-1 countdown makes it necessary to return to Pad A to inspect the insulation panels on the S-II stage after draining of its cryogenic propellants. The move will get underway at the T minus 19 hour, 15 minute mark in the Saturn IB "wet" CDDT. After completion of the panel inspection and making any necessary repairs, the MSS will be moved back to Pad B. It will arrive there at the T plus 4 hour mark in the SL-2 "dry" CDDT.

The MSS will then remain at Pad B through the dual launch to follow unless a "scrub" turnaround during the Skylab countdown makes it necessary to return it to Pad A to inspect the S-II insulation panels on the Saturn V launch vehicle.

V-9

Figure V-4 — CSM/AM/MDA Docking Test

Figure V-5 — S-IB/S-IVB Mating

Countdown, Launch and Initial Activation

The Saturn V launch of the Skylab is scheduled for 1:30 p.m. EDT from Pad A at LC-39. This launch window extends for 1.5 hours through 3 p.m. EDT. The Skylab is to be launched on a flight azimuth of 40.88 degrees east of north and is to be inserted into a 435-kilometer (235 nautical miles) circular orbit 9 minutes, 47 seconds after launch.

First stage burnout is to occur at an altitude of 80 kilometers (43 nautical miles). The booster will enter a ballistic trajectory and impact in the Atlantic Ocean about 848 kilometers (458 nautical miles) downrange 10 minutes, 55 seconds after liftoff. Impact is to occur at approximately 34.3 degrees North and 74.45 degrees West. This is about 257 kilometers (139 nautical miles) due east of Wilmington, N.C. The S-II second stage will enter orbit along with the Skylab.

After insertion, the second stage is separated by retro-rockets. The Skylab is pitched down by the TACS and as the Workshop passes through the 90-degree nose-down attitude, the PS is jettisoned. The ATM and ATM solar arrays are deployed as the TACS orients the OWS to a solar inertial attitude. The OWS solar array is deployed and activation of the control moment gyro subsystem begins.

The Saturn IB launch of the manned spacecraft to rendezvous with Skylab is scheduled approximately 23.5 hours following SL-1 launch. The launch of SL-2 is scheduled for no earlier than 1 p.m. EDT during a window of ten munutes duration. The flight azimuth for the nominal mission is 51.3 degrees easth of north at the opening of the window and 37 degrees east of north at the closing of the window. Another launch window of seven munutes duration opens at 12:30 p.m. EDT the following day.

Based on the nominal liftoff time and flight azimuth, first stage burnout is to occur 2 minutes, 20 seconds after liftoff at an altitude of 55.6 kilometers (30 nautical miles). The expended rocket stage will impact 524 kilometers (283 nautical miles) downrange 6 minutes and 51 seconds after liftoff. Impact is to occur at 31.33 degrees North and 76.33 degrees West. The S-IVB stage is to enter orbit along with the spacecraft. It will be de-orbited later to impact in an ocean.

Launch window opening and closing times are the nominal times based upon an on-time Skylab launch and the planned Skylab trajectory. Actual times may vary depending upon the actual observed trajectory. Adjustments in the time of SL-2 liftoff — if required — will be accomplished during built-in holds prior to launch.

Launch opportunities for the first manned launch occur on the first two days out of every five-day period, that is, on the first and second days, the sixth and seventh days, the eleventh and twelfth days (etc.) following the Skylab launch.

A number of criteria concerning the Skylab must be met before the Apollo spacecraft with its three-man crew is committed to launch. These include:

> The Payload Shroud must have been jettisoned and be at a safe distance from the Skylab trajectory and in a position to preclude contact with the CSM during its ascent and de-orbit trajectories.
>
> The second stage must be separated and at a safe distance from the Skylab trajectory and in a position to preclude contact with the CSM during its ascent and de-orbit trajectories.
>
> The ATM must be deployed.
>
> Spaceflight Tracking and Data Network (STDN)/Skylab instrumentation and communications must be established.
>
> Pressure integrity of Skylab must be verified.
>
> The Attitude and Point Control System (APCS) must be activated and operating and solar inertial attitude maintained.
>
> The Skylab and ATM solar arrays must be deployed and operating.
>
> The orbital parameters must be adequate.
>
> Skylab attitude rates must be within tolerances to permit docking.
>
> The Skylab thermal control, electrical power, attitude control, communications and data systems must be operating within the limits required for successful manned habitation.
>
> The space radiation environment must be acceptable for manned orbital activity.

A government-industry team of about 400 will conduct the SL-1 countdown from Firing Room 2 of the LCC. A similar team in adjacent Firing Room 3 will conduct the parallel countdown of SL-2.

The preparations for SL-1 launch will begin at T-4.5 days with pickup of the countdown. The Workshop-related payload systems will be serviced and secured and the launch vehicle will be loaded with its cryogenic propellants. Final systems checks will be completed just prior to launch.

The SL-2 countdown will begin at T-6 days. Space vehicle pyrotechnics installation and electrical connections will be completed, mechanical buildup of the spacecraft will be accomplished and various gases and cryogenic fluids will be loaded aboard the spacecraft. Spacecraft batteries will be installed and fuel cells activated.

Certain functions will be staggered between the two space vehicles during the countdowns to support the dual launch preparations. Other functions will be "stretched out."

The countdown for the SL-2 launch will enter a built-in hold at the T-22 hours, 15 minute mark a short time before the SL-1 launch and be resumed at SL-1 liftoff. In the event of an SL-1 "scrub," the SL-2 count will be held at T-19 hours.

If SL-1 goes as planned, the trajectory will be followed closely and tracking data used to provide real-time data to the SL-2 Launch Vehicle Digital Computer (LVDC) which controls guidance functions. The final target update will be sent from MSC by radio frequency to the LVDC using the launch vehicle digital command system. A backup target updating capability is available via hard line using an RCA 110A computer at KSC.

SECTION VI

FLIGHT OPERATIONS

VI-1

OPERATIONS – MISSION CONTROL CENTER

Operations for Skylab at the Mission Control Center (MCC) call for significantly different ground-manning arrangements than used during Apollo. The eight-month-long mission will have low activity periods during the unmanned portions and high activity periods during the three manned portions of the mission.

In addition to increased manpower requirements, extensive facilities within the MCC will be needed. Analysis and evaluation of data transmitted on vehicles, systems and experiments will be conducted by engineers and scientists in special rooms established in MCC.

Mission control will be accomplished by providing in-flight analysis of the mission (trajectory, vehicle systems, experiment systems, scientific data and flight plan) and by controlling the progress of the in-flight phase of the mission through the use of voice, telemetry, tracking and updata capabilities. The general mission control functions which are heavily oriented to experiment performance support are:

Monitor and evaluate, in real and delayed time, the vehicle systems, experiment systems, scientific data and trajectory data. Based upon these data, decisions will be made concerning the progress of the mission toward satisfying mission objectives and mandatory, detailed test objectives and the need for proceeding to alternate flight plans or contingency plans.

Monitor and evaluate the condition of the flight crew.

Perform ephemeris and maneuver updating.

Monitor, evaluate and update flight plan activity, including experimental tasks, work/rest cycles, equipment checks, etc.

Advise the flight crew of update mission instructions, anomalies in spacecraft systems found during ground monitoring, ground evaluation and recommendations to solve or circumvent any spacecraft anomalies, and recovery-area weather conditions.

In addition to these general functions, the specific flight control functions which will be carried out at MCC are: Prelaunch test, countdown, unmanned launch, initial SWS activation, manned launch, rendezvous and activation, manned orbital operations, deactivation and entry, and unmanned operations.

Four teams of flight controllers, each working a 40-hour week, will support the unmanned launch and checkout of the Workshop and the first manned mission from the Mission Operations Control Room (MOCR) (Figure VI-1) and associated staff support rooms in MCC. A five-team schedule will be followed during the next two flights with each team working a 5-day-on/2-day-off schedule.

Launch Phase

During the unmanned launch phase, the controllers at MCC will monitor booster systems and trajectory in order to advise the Range Safety Officer of vehicle performance. The flight dynamics officer will monitor vehicle trajectory and will provide status information to the MCC team throughout powered flight and report insertion orbit conditions.

Flight controllers at MCC will monitor all cluster systems, track the SWS for orbit determination, and verify that the cluster systems are capable of supporting the mission prior to committing SL-2 to launch.

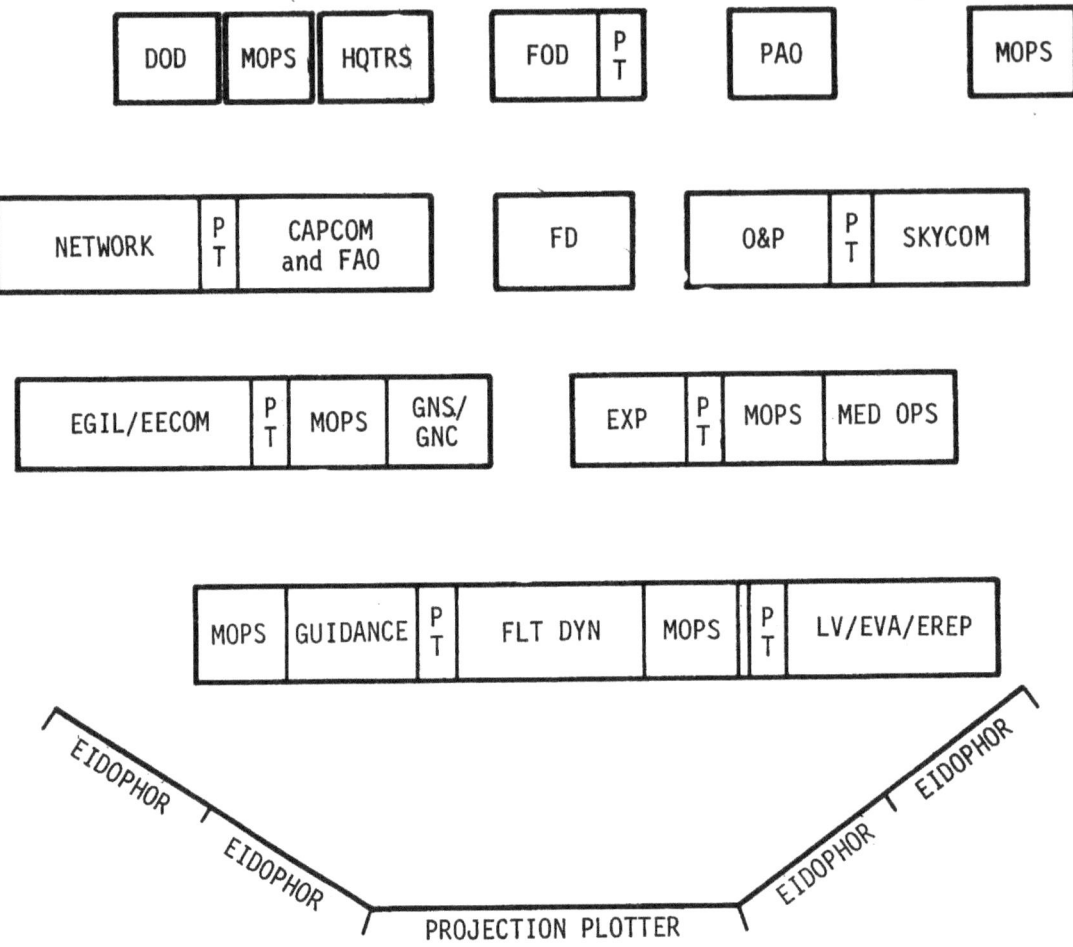

Figure VI-1 — Skylab MOCR Layout

Flight control procedures for the manned launch phase will consist of the evaluation of CSM systems, launch vehicle systems, flight crew conditions and space vehicle dynamics and trajectory. Specific flight control procedures which are accomplished between liftoff and orbital insertion are: Monitor and evaluate launch vehicle and CSM systems, the condition of the flight crew and space vehicle dynamics and trajectory; call marks on abort modes and send abort request if mission rules are violated; provide contingency maneuvers; monitor the flight plan and recommend alternate procedures to the flight crew in contingency situations; and advise the flight crew of launch vehicle, CSM and trajectory status.

The Skylab Flight Director will make the decision to start rendezvous precedures after CSM insertion. This decision will be based on satisfactory operation of all systems. After rendezvous and docking, MCC flight controllers will monitor activation and be prepared to provide assistance as required.

Manned Operations

During the manned periods, flight operations at MCC will be active in the following primary areas.

Mission Management — Although this task will necessarily resemble present-day mission activity, a very large interface with the "outside world" will exist. Representatives from numerous program and operations elements from other NASA centers and NASA Headquarters will all be involved in the decision-making process.

Systems Management — Although the onboard systems design for the OA incorporates a caution and warning system in each vehicle, ground controllers will remain responsible for system analysis, trend projection, consumables management and definition of crew malfunction procedures when required.

Special Systems Management — Aside from the type of systems monitoring and analysis required for Apollo (leak detection, temperature monitoring, electrical load sharing, attitudes and rates), design features unique to the SWS require that other tasks be accomplished. Systems flight controllers will be responsible for control moment gyro momentum management and electrical power system management.

Extravehicular Activity (EVA) Management — During EVA the monitoring of EVA systems data will be managed by flight controllers, and medical personnel will monitor medical data.

Thermal Management — Thermal monitoring of the OA will be a prime concern of the flight control team. The additional blanket heaters added to the Skylab CSM are automatic but require monitoring similar to Apollo thermal monitoring.

Quiescent CSM System Management — During the long duration phases of Skylab, many of the CSM systems will be powered down. Periodic testing of these powered-down systems will be conducted — some on a daily, others on a weekly basis — by the flight controllers (e.g., battery voltage checks, fuel cell checks on a daily basis, CM computer update, and topping off of the oxygen supply are conducted on a weekly schedule).

Medical Operations — The surgeon's basic responsibilities for crew health and life systems remains unchanged. Most of the crew health information will be obtained from the biomedical experiments and voice communications with the crew, and analysis of operational biomedical system data which is used primarily for EVA.

Experiment Support — Flight controllers will have the prime responsibility to ensure that the required amount of useful experiment data is retrieved, to advise crew-members of experiment operation malfunctions, and to provide operational interface between the principal investigators (PIs) and the flight control teams. A large portion of the Skylab experiments require the PIs to be in the MCC during operation of their experiments.

Trajectory Determination and Entry Planning — Although this function should remain at a low activity level during the manned orbital periods, routine ground tracking will be performed using the CSM unified S-Band (USB) Doppler data. Onboard computers will be updated periodically using vectors derived from ground tracking information.

Contamination and Radiation Monitoring — Data from operational contamination and radiation sensors (T025 and T027) will be telemetered to the ground, and this data, along with other sources of information, will provide data to the ground during both manned and unmanned phases.

Data Management — Communications and data management will continue to be a prime flight control function for the Skylab missions. Ground controllers will be responsible for maintaining optimum communications and data flow between the OA and the ground. This requires frequent monitoring and control of the onboard communications systems, the STDN and MCC. In addition, ground controllers will be responsible for the correlation of all onboard timers with GMT.

Deactivation and Entry — At the conclusion of each manned visit to the SWS, the flight controllers will assist the crew in verifying the detailed OWS, MDA/AM and ATM deactivation checklists and that the ATM C&D console is enabled to permit unmanned operations of the ATM. The flight controllers also provide the support to verify everything is nominal for vehicle closeout, undocking and entry

Unmanned Operations

After deactivation, deorbit and entry, the MCC flight control support teams will provide a minimum of ground monitoring of the SWS. Between SL-2 splashdown and SL-3 launch (and SL-3 splashdown to SL-4 launch) there are "active" monitoring periods and "reduced" monitoring periods.

Active Monitoring — The "active" flight control support will consist of about 15 people who will man the MOCR during an eight-hour period. These MCC personnel will do in-depth SWS status analysis and perform routine maintenance functions and science support.

Two remote tracking site contacts are planned for each revolution of the SWS whenever possible. Nominal ground functions for the "active" unmanned period are:

Navigation and Timing Updates — Once-daily, a three-hour period of collection and processing of C-band skin tracking information from the remote sites will be completed for use in ephemeris determination and navigation updating.

Systems Monitoring — Active monitoring of the SWS systems will include checking consumable quantities, providing systems support for operation of the ATM experiments, evaluating systems management and forecasting systems status for the ensuing 24 hours.

Experiments Operations — Several of the ATM experiments will be operated remotely from the ground during the unmanned period. Ground tasks for ATM operation fall into the "active" as well as "reduced" monitoring periods.

Before the crew leaves the SWS, it will deploy through the SAL the experiment extension mechanism to which the S149 micrometeorite collection experiment is attached. After crew departure, ground controllers will command the activation of the S149 for the remainder of the unmanned period. It will be remotely deactivated prior to docking of SL-3 in order to avoid contamination of the experiment during docking operations.

Communications, Tracking and Data Management — Routine MCC communications management commands of the ATM and AM tape recorder dumps will be made throughout the unmanned periods. The ATM will be dumped at least once per revolution if station contact permits. Updates of the AM will not be made on a regularly scheduled basis.

Contamination Monitoring — Real-time telemetered monitoring of the SWS contamination environment will be made and used as a guide by the ATM personnel in planning the operation of their experiments during the unmanned periods.

Reduced Monitoring — A team of three to six people will man the MOCR during the "reduced" monitoring periods. They will be responsible for collecting data for monitoring systems. This team will implement a callup procedure if an out-of-limits or serious problem is detected.

Prior to entering the 16-hour period of reduced monitoring, the SWS will be placed in the desired configuration (ATM canister offset as required for experiment S055). The operation will then be handed over to the reduced monitoring team who will monitor the MCC data retrieval system and be responsible for implementing contingency plans in the event of anomalies.

Throughout the unmanned period, a day-to-day assessment will be made of the SWS to assure its suitability for the next mission. Periodic checks will be made of the following: Remaining consumables (oxygen, nitrogen and water within limits for manned habitation); adequate thermal control; adequate electrical power (OWS and ATM solar arrays operating); Attitude Pointing and Control Subsystem operating and solar inertial attitude maintained; adequate communications and instrumentation; and orbital parameters are such that the next manned mission can be completed satisfactorily.

Critical analysis will be made of the SWS starting at about 24 to 36 hours prior to the launch of the next crew. The SWS will then be pressurized and monitored for several hours to verify pressure integrity.

Several SWS events will be supported or controlled from the ground during the pre-docking period. Some of these are: turn-on and control of the OWS heaters, activation of the VHF ranging system in the SWS, acquiring Z-LV (Z-axis is parallel to local vertical), attitude mode for rendezvous, and turn-on and control of the SWS activation lighting and running lights.

MCC Computer Functions

Any computer processing necessary in real-time will be accomplished by the Communication, Command and Terminal System computers — CCATS (Univac 494) and Mission Operations Computers MOC (IBM 360). These computers will perform functions for Skylab basically similar to functions they performed in Apollo.

The primary function of CCATS is to interface the MCC with the Goddard Space Flight Center and to route telemetry, tracking and command data to the proper areas for processing.

The MOC is required to process telemetry, tracking and command data for various operational purposes. A data flow and transfer capability is required between the MOC and both CCATS and the Mission Operation Planning System (MOPS).

The MOPS computers will satisfy the bulk of the near-real-time monitoring requirements and also contain the majority of the planning tools.

Due to the complexities of the Skylab mission and the fact that there are far more systems and experiments on board than in Apollo, the operations at MCC will have additional personnel and SSR's. Some of the new flight control positions will include the Earth Resources Experiment Package (EREP) and the ATM experiment officers.

Additions to the list of SSR's and Science Rooms which will provide support to the MOCR will include the ATM experiments, ATM science room, EREP experiments SSR, EREP science room and corollary science room.

OPERATIONS SUPPORT — MARSHALL SPACE FLIGHT CENTER

MSFC, as the Skylab integration center, is required to provide consultation and assistance to KSC during the checkout and launch operations and to MSC throughout the Skylab missions for the Saturn Workshop modules and experiments which were MSFC's development responsibility. This includes in-depth technical support to the launch and flight operations teams concerning system operational status, system and experiment operating trends and problem resolution. It also includes contingency review, adjustment to experiment priority and planned mission activities and the MSFC contributions to the Mission Evaluation reports. To fulfill this responsibility, operations support organizations have been established, existing operational support facilities have been modified, the Apollo data management system has been expanded, and a system of automatically scanning operational data has been devised.

Huntsville Operations Support Center

During the total mission support period, the Huntsville Operations Support Center (HOSC) organization and its facilities provide support to KSC prelaunch activities, MSC flight operations and support to both Saturn and Skylab program organizations. Figure VI-2 illustrates the major elements of the Huntsville Support Organization. The HOSC is staffed with personnel from the MSFC program and engineering line organizations. The HOSC Operations Support Manager is directly assisted by an operations support staff and a Facility and Data Support Organization. The operations staff is responsible for overall coordination and direction of all operations support activities. The Data Support Organization is responsible for acquisition, handling and dissemination of all data for the Skylab mission.

The level of support activities will vary significantly as the total mission activities change. The mission can be divided into five characteristic support phases, some of which are repeated over the course of the mission.

Premission phase support covers that period from approximately T-6 months until SL-1 launch. During this phase both SL-1 and SL-2 are in prelaunch testing at KSC, and both the Saturn and Skylab interfaces with KSC will be active.

The SL-1 launch phase support consists of the SL-1 launch and Saturn Workshop (SWS) deployment and activation. Also supported during this phase are the continued prelaunch activities of SL-2. This support phase represents a peak activity period for the support organization, and full continuous manning of the HOSC with staff and support engineers is expected for Skylab support to the Flight Operations Management Room (FOMR). The initial deployment and activation of the Saturn Workshop will involve all Skylab elements in evaluating the initial hardware performance and verifying its readiness for manning. During this phase, both the FOMR and Booster System Engineer (BSE) interfaces to the MCC will be active, as well as both the Saturn and Skylab interfaces to KSC. This phase will last about 24 hours.

The Saturn IB launch phase support for SL-2 will be repeated during the launches of SL-3 and SL-4, beginning at liftoff and extending through IU active lifetime. During these phases the Saturn Workshop will be in an unmanned mode. The Saturn interface to KSC and both the FOMR and BSE interfaces to the MCC will be active. The HOSC will be fully manned at a peak level at the beginning of this phase, with both Saturn and Skylab engineering support personnel. The Saturn support will diminish with CSM separation, but orbital support will be required until the experiments carried in the IU and S-IVB/IU launch functions are completed. During the SL-2 launch, for the first manned entry of the Orbital Assembly, all Skylab elements will be required for immediate support.

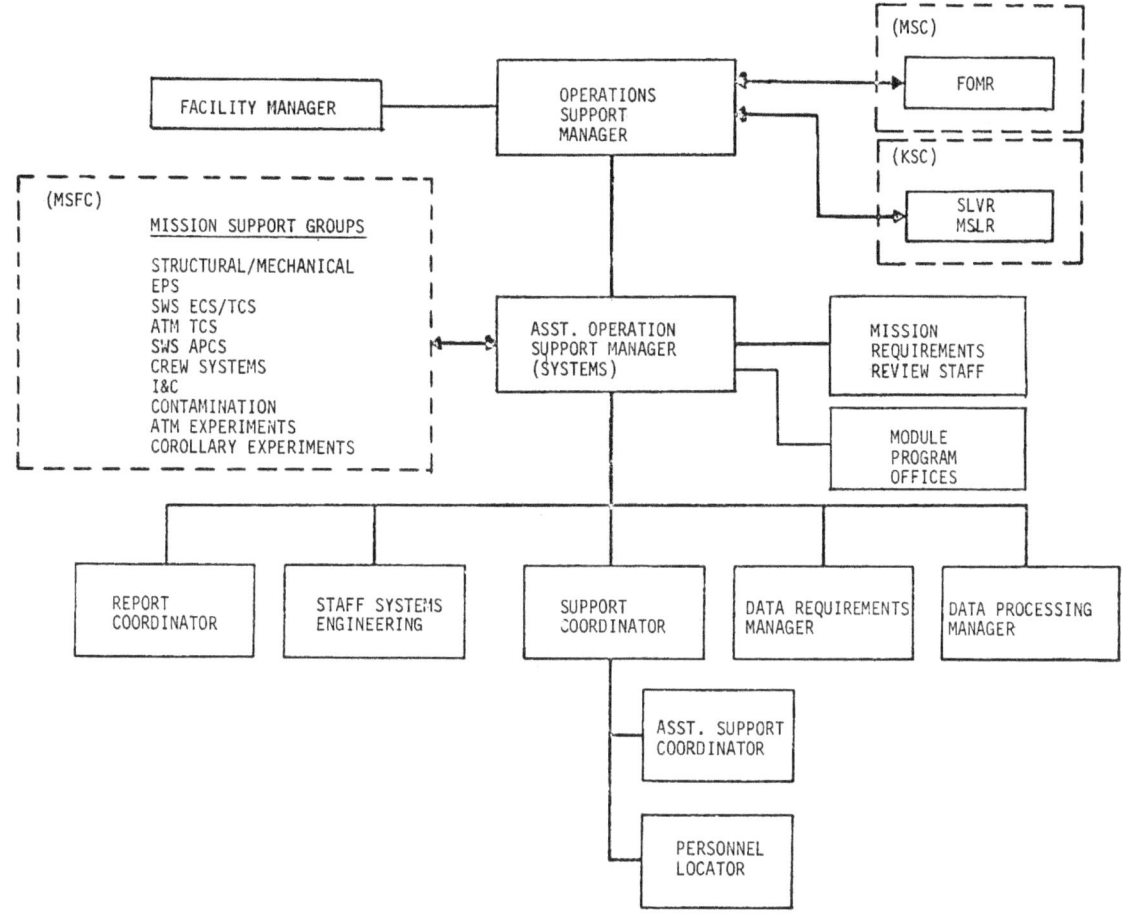

Figure VI-2 — MSFC Mission Support Organization

Manned operations phase support covers the period during which the Orbital Assembly is manned for SL-2, SL-3 and SL-4 missions. Only the FOMR interface will be active to the MCC; however, the Saturn interfaces to KSC will be active in support of prelaunch activities in preparation for the SL-3 and SL-4 launches as well as the contingency rescue mission. HOSC manning and support activities will vary, depending upon time in the mission, crew activity and the behavior of the Saturn Workshop hardware. The support coordination staff will be fully manned during crew activity periods and partially manned during crew rest periods, but minimum staff functions will be performed on an around-the-clock basis. Mission Support Groups (MSG) engineers will be required in the HOSC after initial crew entry and until the OA is fully activated, checked out and routinely operating. After this initial phase, engineering evaluation and non-critical support activities will be carried out as much as possible during normal work-week hours. MSG engineers will be called into the HOSC in the event of contingencies, or as scheduled for special events. Full Skylab manning of the HOSC is required during OA deactivation prior to a crew's return to Earth.

The orbital storage phase begins at the time the CSM leaves the Orbital Assembly to return to Earth and continues until launch of SL-3; it is repeated between the SL-3 and SL-4 missions. During the orbital storage phase the Saturn Workshop is unmanned, but its primary systems will continue to operate, and a significant amount of ATM experiment operation is planned. HOSC manning will be reduced to one shift per day, but all Saturn Workshop support functions will be maintained on stand-by, and routine engineering evaluation will be continued. As the SL-3 and SL-4 launches approach, support activities and HOSC manning will be increased to provide GO/NO-GO evaluation for re-manning of the Saturn Workshop.

The HOSC organization provides technical support to KSC through the MSFC Launch Vehicle Representative (SLVR) in the areas of Saturn System operation and through the Marshall Skylab Representative (MSLR) for SWS systems by:

Resolving significant technical problems which occur during prelaunch testing.

Accomplishing ground and flight wind monitoring during the terminal countdown and providing GO/NO-GO recommendation to the Launch Director.

Performing simulations and troubleshooting at the Saturn System Development Facility (Breadboard) for known or potential launch vehicle problems.

Generating target update math model analyses to confirm the Saturn IB performance capability to achieve the rendezvous target parameters.

The HOSC organization also provides support to MSC flight operations for both the Saturn launch vehicle and the Saturn Workshop through the BSE and the FOMR. The launch vehicle flight control responsibility is assigned to the MCC Booster Systems Engineer and his supporting staff. The launch vehicle flight controllers are provided from the permanent MSC based personnel of the MSFC Flight Control Office. Launch vehicle support from HOSC is normally provided directly to the BSE; however, for SL-1 all post-insertion HOSC support will be provided through the FOMR.

Launch vehicle flight controller tasks during the SL-1 flight include:

Flight control of the S-IC and S-II powered flights and Saturn Workshop separation from the spent S-II Stage.

Monitoring and confirmation of initiation of orbital safing of the spent S-II Stage.

Primary flight control of the Saturn Workshop deployment and activation functions controlled by the IU, in conjunction with the MSC Orbital Assembly systems flight controllers.

For the Saturn IB launches (SL-2, SL-3 and SL-4), the Launch Vehicle Flight Controllers have the following responsibilities:

Exercising flight control to protect crew safety and enhance mission success during the Saturn powered ascent to orbit until CSM separation, similar to Apollo.

Exercising flight control of S-IVB/IU orbital operations, including the S150 (Thermal Control Coating) and M415 (Galactic X-ray Mapping) experiments.

In conjunction with the MCC Flight Dynamics Officer, update prelaunch target parameters of the Saturn IB to permit CSM rendezvous with the Saturn Workshop.

MSFC support to the Booster System Engineers include:

Providing technical information and advice upon BSE request on any vehicle systems.

Providing predicted vehicle flight dynamics based on measured prelaunch winds.

Verification of Saturn IB performance capability to achieve the rendezvous target parameters computed by the MCC.

MSFC support to OA flight operations is provided through the MCC Flight Operations Management Room (FOMR). These support functions include:

    Responding to MCC support requests to analyze off-nominal Skylab conditions identified by the MCC, and to provide additional engineering and test data required by MCC.

    Reviewing and/or proposing mission changes to the daily MSC Flight Planning inputs based on systems assessment results and assessing and recommending changes to mission requirements, experiment priorities and detailed test objectives.

    Providing systems simulations required for systems performance or Contingency analysis using the Attitude and Pointing Control System Simulator, the Neutral Buoyancy Simulator, the Thermal Control System math model and the Power Systems Breadboard.

    Evaluating systems and experiment hardware performance.

    Providing MSFC input to Flight Operations Management Room Daily Mission Report.

MSFC FOMR Team

The FOMR provides a single management interface between MCC flight operations and the MSFC and MSC Program Offices with their respective engineering support organizations. The MSFC Skylab Program Manager will be represented by the MSFC Senior Program Office Representative and his staff in the FOMR. Support requests upon MSFC originated by the flight control organizations will be forwarded through the FOMR to the HOSC. Responses to such requests and other inputs to the mission will flow back through the FOMR to the flight control organization. MSFC will provide continuous FOMR support during the manned mission periods. The level of MSFC FOMR manning will vary as a function of mission period. Full FOMR support is planned for the high activity periods of countdown, launch, rendezvous and activation.

Mission Support Groups

The Skylab mission technical support is provided by the responsible program, engineering and contractor organizations in their respective technical discipline areas. Designated personnel from these organizations are available to provide engineering data and contingency problem solving capability. They will also perform systems assessment and engineering evaluation activities for continuing hardware certification and capability to complete the Skylab mission. In order to provide the most direct and timely response to mission problems, Mission Support Groups have been identified for each primary Orbital Assembly system area. A responsible leader and assistant leader(s) are identified for each system group, plus the necessary representatives from other line organizations to provide complete cognizance of the system area. MSG leaders are the recognized authority in the applicable systems area and are able to identify from the group membership those specialists necessary to work a given problem. The MSG leaders are responsible for coordinating problem disposition and preparing the required FOMR response. Sufficient contractor personnel are located on-site at MSFC to assist in performing systems assessment activities and react to real time critical problems. Additional support personnel are on-call at contractor plants.

Mission Evaluation

In addition to the direct FOMR support functions, engineering evaluation of the OA will be performed to maintain MSFC cognizance of systems and experiment behavior through current assessment of systems performance, provide an input to the SL-2 launch commit, provide an input for the SL-3 and SL-4 Flight Readiness Reviews and prepare final post-mission program and technology reports. Due to the nature of these functions and the common work force involved in both engineering evaluation and response to MCC Mission Action Requests (MAR), the engineering activities during the mission are considered an integral part of the overall MSFC operations support effort.

HOSC Facility

The HOSC facility consists of two floors and a basement, totaling over 929 square meters (10,000 square feet), in the MSFC Computation Laboratory. Modification of the existing facilities has been accomplished to allow the dual support of the launch and orbital activities of Skylab, whereas previously the HOSC was configured to support only the Apollo launch vehicle.

Figures VI-3 and VI-4 illustrate the operational areas within HOSC. Equipment allocation within the HOSC is discussed below.

Operations Managers Room provides an office for the Operations Support Manager and Support Coordination Staff. It is used as a coordination center for operations support activities during the mission.

Operations Support Room is used as a data display area for evaluating real time mission data for technical support to KSC and MSC during launch phase. During the orbital phase it is used as a work area for special orbital problem analysis and data review.

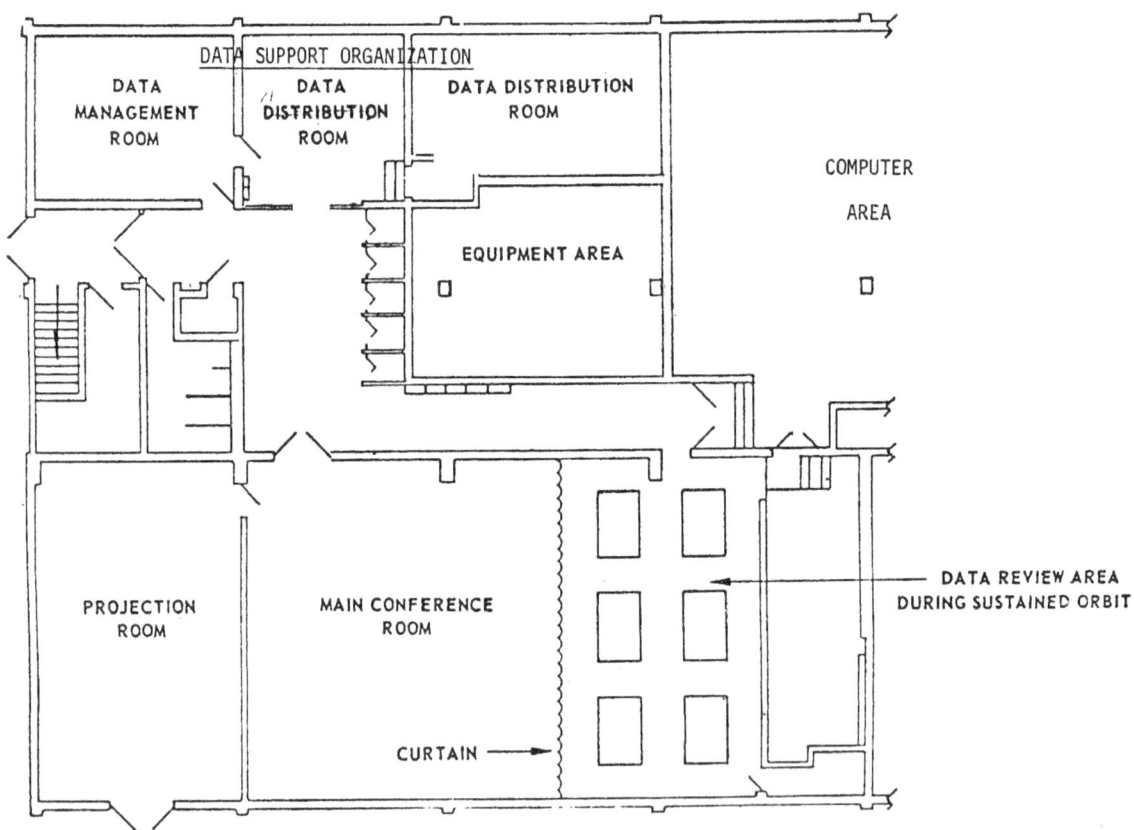

Figure VI-3 — HOSC First Floor

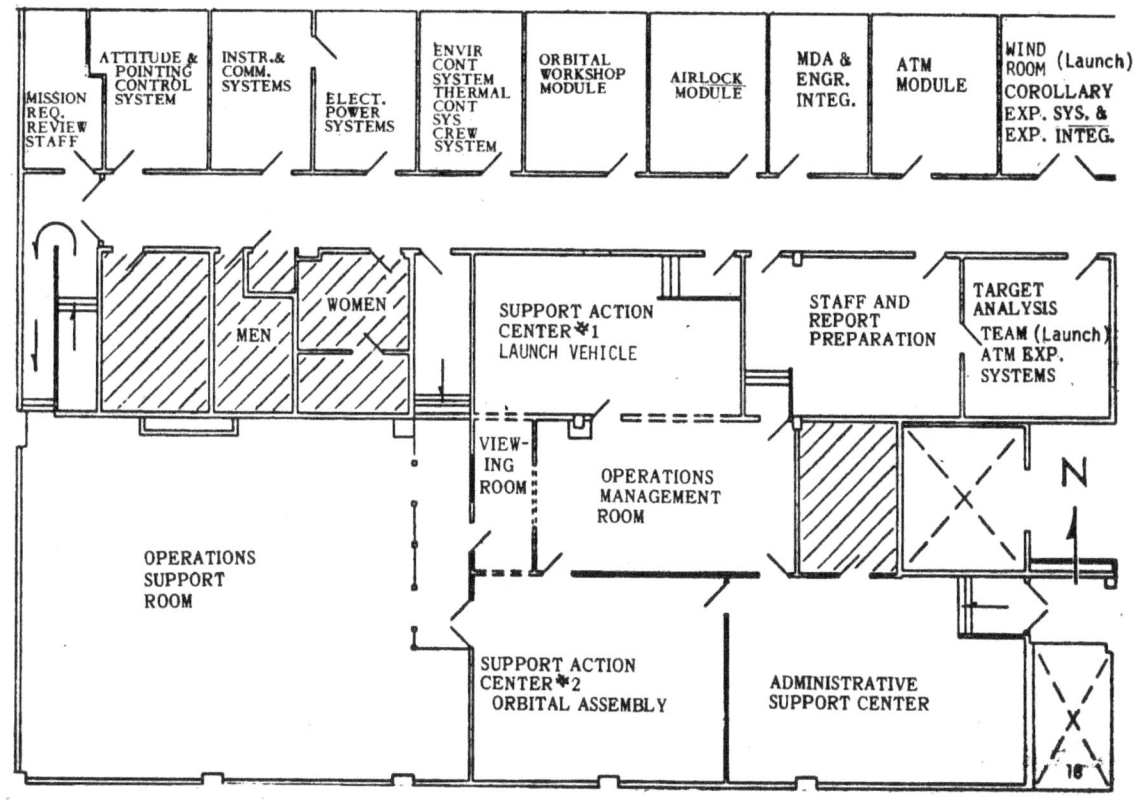

Figure VI-4 — HOSC Second Floor

Support action Center #2 is used by the OA Technical Support Team as a primary teleconference area for communication to KSC in the identification and resolution of OA problems requiring development center support during the prelaunch tests, countdown and launch phases. During the orbital phase it is used by the Operations Staff for meetings, conferences and teleconferences. Support Action Center #1 is used by the Launch Vehicle Technical Support Team for the same purposes as the OA TST uses #2.

Administrative Support Center

The extensive support operations taking place during the Skylab mission require a general administrative base of support. The Administrative Support Center, on the second floor of the HOSC, will provide this support.

The Administrative Support Center will be staffed by NASA and contractor personnel. Services supplied by this staff include secretarial, reproduction, library and Long Distance Xerography (LDX) service.

The Main Conference Room is used as a location for backup engineering support personnel and observers during launch phase. During orbital phase it is used as a Flight Evaluation Working Group/Mission Evaluation Working Group (FEWG/MEWG) meeting room and as a general conference room.

The Wind Room is used to house the wind monitoring team during prelaunch and launch phase. Primary use is monitoring ground and flight winds at KSC and providing advisory support to launch operations personnel in the event of marginal wind conditions. During the orbital phase the room is used by Corollary Experiment Systems and Experiment Integration personnel as a conference work area.

The Target Analysis Room is used by the Target Update Team for analysis of capability of launch vehicle performance to accomplish required rendezvous during launch of SL-1/SL-2. After SL-2 launch, it will be used by ATM Experiment Systems personnel as a conference work area.

Data Support Organization is located on the first floor of the HOSC adjacent to the computer complex. These personnel process data requirements and coordinate the implementation of these requirements with appropriate agencies during launch and orbital phases, maintain current status of data acquisition and direct the dissemination of acquired data to requesting agencies.

Conference Work Areas are designated for NASA Skylab Project Office personnel and their respective contractor supporting personnel. The rooms will be utilized by different disciplines during the launch and orbital mission phases to provide telecommunications interfaces among Principal Investigators, contractor plans, module offices and Mission Support Groups.

Simulator Facilities

MSFC has various simulator facilities which can be used to troubleshoot systems on board the spacecraft. These facilities use existing engineering development hardware and math models not provided at the Mission Control Center at Houston and provide support as requested by the FOMR.

Attitude and Pointing Control Simulator Facilities — The APCS will be activated during periods of anomalous conditions on board Skylab, and during certain critical mission phases. Four different APCS simulators available for callup by the flight operations team through the FOMR are:

> IBM 360-75 Simulator — This is an all-digital simulation of Apollo Telescope Mount Digital Computer (ATMDC) and hardware. It is used to verify program calculations, timing, order of calculations. It has the ability to ignore false inputs.
>
> IBM 360-44/ATMDC Simulator — This simulator exercises the flight program in the ATMDC. It is used to simulate the vehicle reactions to the program.
>
> Hardware Simulation Laboratory (HSL) — This is the only simulator that runs real time. It has the capability to test computer switchover, with hardware and software simulations of hardware.
>
> Hybrid Simulator — The hybrid simulation model of the ATMDC/vehicle system is used to test the redundancy management system with perturbations.

Power Systems Breadboard — The MSFC power systems breadboard is available during the mission for callup by the flight operations team through the FOMR. This breadboard uses flight-type hardware and can be used for modeling the operation and interaction of the total Skylab electrical power system. This breadboard is used during the mission for troubleshooting electrical power system problems and for verifying emergency operating procedures.

Saturn Workshop Thermal Control System Math Model — The SWS Thermal Control System analysis computer program at MSFC is available for mission support to determine the thermal behavior of the SWS under various conditions. The model will enable mission control personnel to enhance trend and performance predictions and to refine procedures for operating the spacecraft equipment.

Neutral Buoyancy Facility — The neutral buoyancy facility was used for EVA training of the Skylab crews. The facility is available during the mission for any required testing and verification of EVA procedure changes.

Voice Communications — The communications circuits connecting MSC and KSC to the HOSC are voice, high-speed data and television circuits, all designed to meet the requirements of the Skylab mission. The voice system is a special telephone system with facilities to allow the Mission Support Groups to communicate with KSC, MSC and contractors' plants (Figure VI-5).

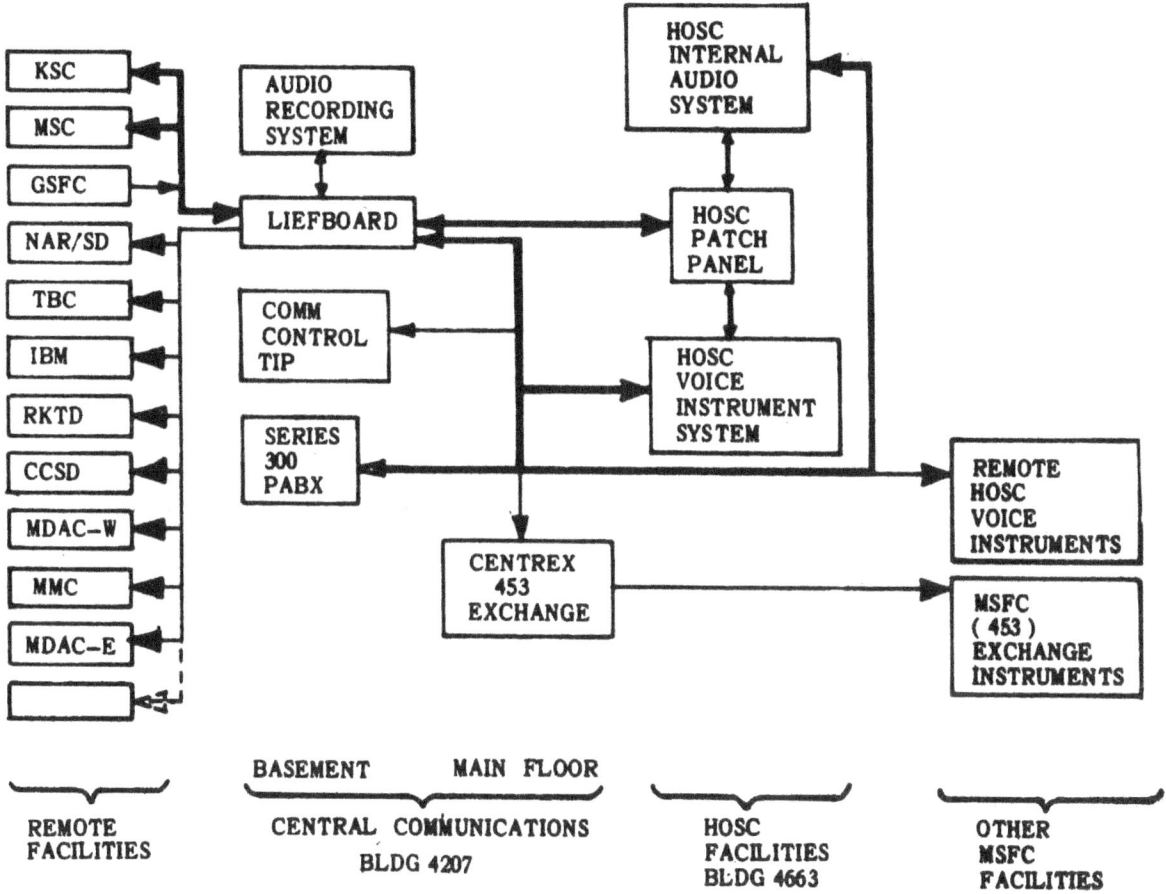

Figure VI-5 — Mission Support / Voice Distribution

Data Lines — High-speed data circuits provide the HOSC a means of obtaining digital data from MSC and KSC at the rate of 50 kilobits per second. This is the mission data that is put into computers for further processing and distribution.

Mission Operations Planning System (MOPS) — Within HOSC are special computer controlled terminals, MOPS, equipped with a video screen and a printing device, allowing mission support personnel to search the memories of computers at MSC. The user views on the screen the current flight and experiment data available at MSC, and with a push of a button obtains a paper copy of what he sees on the screen.

Television – Television is transmitted from KSC and MSC to the HOSC where it is input to the HOSC video distribution system which services 80 television monitors in the HOSC and Mission Support Group activity centers external to the HOSC. Mission Support Groups using the closed circuit TV system have available at their request 360 formats of data including digital TV from computers and TV inputs from document readers, KSC and MSC.

Computer Program and Facilities

The computer complex at MSFC supporting Skylab consists of two large digital computer systems with peripheral equipment. This complex is augmented by several computers at Slidell, La. The computers are used for processing data collected and received from Skylab for distribution to data users. The computer processing of data includes controlling display consoles, shipping data to 20 channels of closed circuit television and converting electronic data into lists of information called data books. Data books, along with other computer created information in the form of graphs and charts, are sent to nearby working areas for immediate examination, or are shipped to data users elsewhere for a more in-depth analysis.

In the Operations Support Room a number of computer controlled data display consoles provide real-time information during the launch, and, when necessary, during the flight. Some 300 meters, 700 lights, six graph-making machines and 20 channels of televised data make it possible for MSFC to keep abreast of the mission activities.

The large number of experiments and systems generating data made it necessary to develop some means of reducing the quantity of information transmitted from the remote tracking sites. A data compression technique has been developed where only changes in generally constant data (cabin temperature, for example) are transmitted, instead of continuous readings which are repetitive.

Experimenters, engineers and research and development agencies who require any data from systems, subsystems and experiments on board the spacecraft submit a standard form which details the information needed and the format in which it is required.

This data is in many different forms, such as: television displays of the event as it takes place; a light which illuminates to show an expected event has occured; a magnetic tape which contains electronic signals; tape recordings of the astronaut's comments during an experiment; or charts, graphs and tables generated by a computer. Data may also be of the type which is returned with each returning space capsule, such as still and motion picture film, hand written experiment logs and return samples of many kinds.

Up to 5,000 of these requests for data are expected by the time the mission begins. In order that any one of these requests be available at any time (for scheduling, examination of its contents, etc.), a computerized filing system has been initiated at MSFC. This system allows the personnel who plan for the handling of data on a day-to-day basis the flexibility to change with changes in the Mission Flight Plan.

Auto Scan

The large amount of data generated by Skylab (2,000 measurements repeated approximately once every 2 seconds or 86,400,000 measurements per day) made it necessary to generate a computer program called Auto Scan for analyzing the data. Auto Scan automatically scans the predefined telemetered measurements. When a particular data measurement exceeds predetermined limits, Auto Scan senses it and identifies the areas of system degradation or high interest. The system engineer can then perform an in-depth analysis to determine if any corrective actions are required.

RECOVERY

Recovery of the Skylab flight crew and CM would appear to be simply a continuation of Apollo recovery planning and procedures, but there are factors unique to the Skylab Program which required major changes in the recovery concept. These factors include greater mission lengths, expanded contingency landing areas and special medical requirements.

The short time intervals between CM landings and subsequent launches in essence requires a commitment of support forces for the duration of the Skylab program. Since probabilities of abort cannot be localized in terms of geographic area, the use of aircraft becomes increasingly important due to their flexibility and rapid movement.

Skylab recovery support consists of a minimum use of dedicated forces, especially ships, and places a high reliance on aircraft search and rescue capabilities for contingency operations. Since Skylab will be flying over a very large area from 50 degrees North to 50 degrees South, a highly flexible contingency support capability, including the ability to call upon both foreign and domestic, military and commercial planes and ships, search, rescue and retrieval capabilities will be maintained.

One of the prime objectives of the Skylab program is to determine the effect of long duration weightlessness has upon man. As a result, upon recovery it is mandatory to minimize the time required for flight crews to begin their post-medical examinations.

A group of six mobile laboratory units will be aboard the Prime Recovery Ship (PRS). Each of the units will house equipment and personnel in various disciplines and will serve as a base of operations for the initiation of post flight evaluation of the physiological condition of Skylab crews.

Because of the nature and time criticality of this post-recovery phase, the following stringent requirements have been established:

> The time interval between CM landing and flight crew delivery to the Skylab Mobile Laboratory (SML) should be as short as possible, not more than about one hour.
>
> The capability to accomplish this one-hour transfer must exist once per day throughout the mission.
>
> The flight crew must be retrieved with a minimum of exertion on their part and a capability must exist to maintain the crew in a supine position until delivery to the SML.

The Skylab launch site area will be oriented more toward the north than during Apollo. Launch abort support will consist of two helicopters with the capability to transport the CM, a landing craft utility (LCU) and a salvage ship.

Launch Abort Area

The northerly launch azimuth results in a launch groundtrack passing near the east coast of the U.S. and Canada, crossing the North Atlantic, and then dipping south of Great Britain. The launch abort area is defined by this groundtrack and by the launch abort modes (Figure VI-6).

Mode I Abort — This is an abort which occurs after launch escape system (LES) arming and before launch escape tower (LET) jettison, which occurs about three minutes into flight. A Mode I abort would result in a CM landing between the vicinity of the launch pad and approximately 741 kilometers (400 nautical miles) downrange.

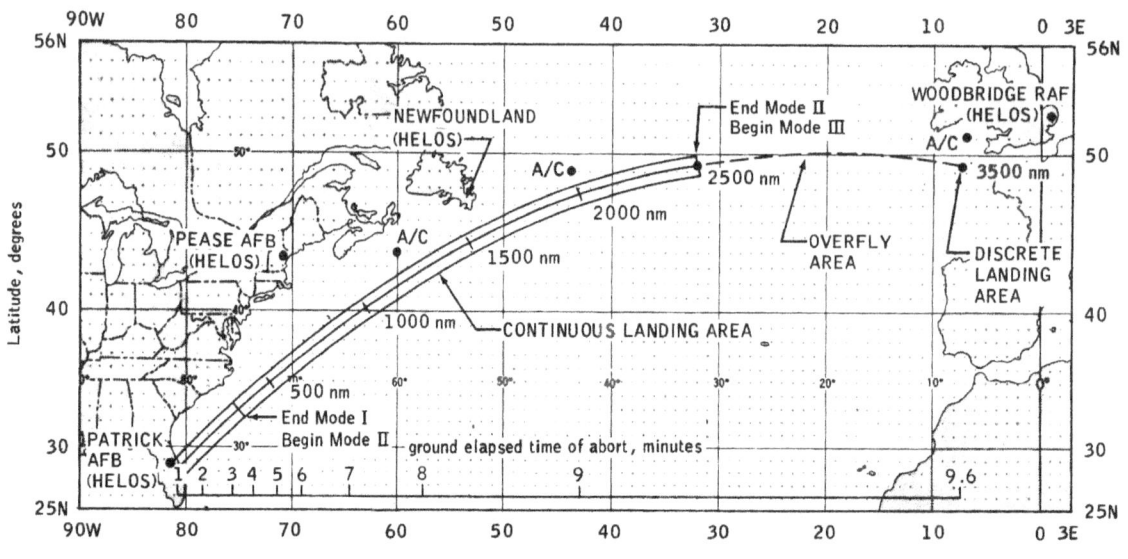

Figure VI-6 — Launch Abort Area

<u>Mode II Abort</u> — This is an abort which occurs after LET jettison, but which is non-propulsive, utilizing a full-lift CM entry profile. The resulting CM landing point would be between approximately 741 and 4,630 kilometers (400 and 2,500 nautical miles) downrange and would correspond to about three to nine minutes elapsed time of flight. Although a Mode II abort could result in a CM landing further downrange than 4,630 kilometers (2,500 nautical miles), a Mode III abort would be attempted in this region of flight.

<u>Mode III Abort</u> — This is an abort which utilizes either a posigrade or retrograde service propulsion system (SPS) burn to constrain the CM landing to a discrete area. This capability will exist after about nine minutes of flight and will be used, if possible, to overfly the area between 4,630 and 6,482 kilometers (2,500 and 3,500 nautical miles) downrange. A Mode III abort would be targeted so that a landing would occur in a discrete area south of Great Britain. This mode would also be used to prevent a landing on the continent of Europe.

The launch abort area, therefore, consists of a continuous band, with the groundtrack at the center, from the vicinity of the launch site out to 4,630 kilometers (2,500 nautical miles), and a discrete area 6,482 kilometers (3,500 nautical miles) downrange and south of Great Britain. The area from 4,630 to 6,482 kilometers (2,500 to 3,500 nautical miles) downrange is an overflight area with relatively low probability of CM landing.

Three fixed-wing search and rescue aircraft (HC-130's) will be positioned along the groundtrack. One aircraft will support the target area south of Great Britain and two will support the continuous landing area. These aircraft will provide CM location and flotation collar installation capability throughout the launch abort area. They are position biased toward the area of adverse weather. Portions of the launch abort area immediately downrange of the launch site will be supported by refuelable helicopters from that area.

Flight crew retrieval capability will be provided in the launch abort area with refuelable helicopters at Loring AFB, Me., at Gander International Airport in Newfoundland and at Woodbridge RAF, England. Helicopters at these locations will be on alert for immediate flight to the landing area if needed.

It is planned to retrieve the CM with a ship-of-opportunity, military or commercial. Figure VI-7 depicts the statistical merchant ship density in the North Atlantic. Still under consideration is the pre-positioning of one or two secondary recovery ships (SRS) which might conduct CM retrieval.

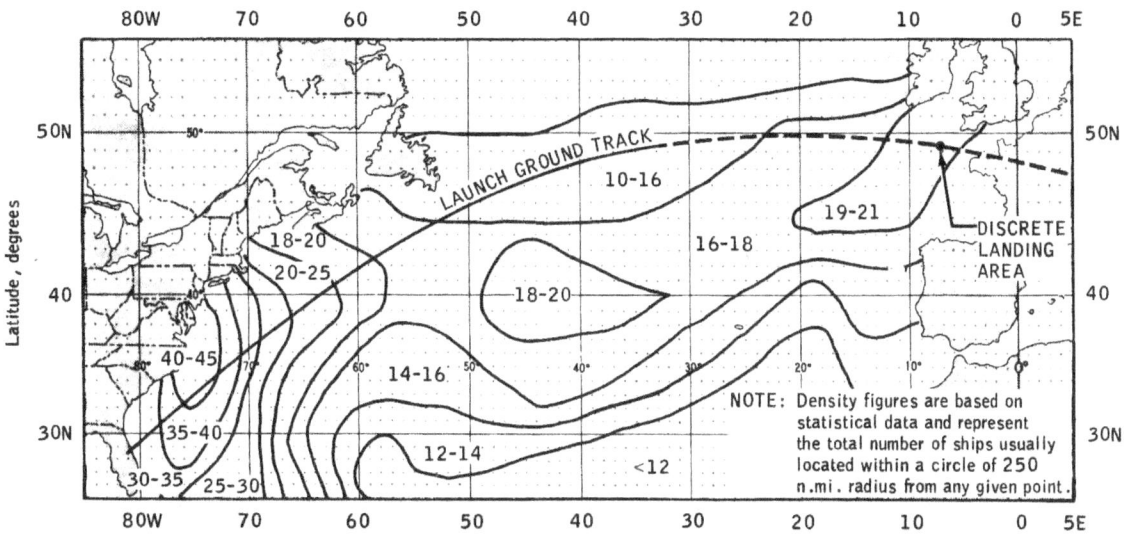

Figure VI-7 — Merchant Ship Density

Primary Recovery Area

Current plans are to locate the primary landing area in the Pacific Ocean. SL-2 and SL-3 primary recovery areas will be off the coast of California, west of San Diego, and SL-4 primary recovery area will be the mid-Pacific.

A PRS (USS Ticonderoga) with a complement of fixed wing aircraft and helicopters will be used in the primary recovery area. The Skylab PRS will be on station for approximately the last week of each flight, unlike Apollo recovery operations during which the PRS was on station from before launch until recovery.

Secondary Recovery Area

The secondary landing area, a location to which the CM would be targeted if necessary to terminate the mission after orbital insertion and before the planned end-of-mission, will be supported by home-based refuelable helicopters for the flight crew retrieval and an SRS to recover the CM.

To assure adequate recovery force support a decision for mission termination must be made at least 24 hours prior to landing. This 24-hour lead time provides sufficient time for transfer, set-up and calibration of Skylab Mobile Laboratories (SML). The SML will be flown from Ellington Air Force Base near Houston, Tex., (via a C5A) to the island nearest the target point in the Pacific.

Fence Concept

The ground track of Skylab shifts to the east each day, varying the distance from a fixed recovery base to a changing target point. This makes it difficult to maintain fixed wing aircraft and home-based helicopters for a given area. As a result, a series of land bases perpendicular to the groundtracks will be used to match the easterly shift of the groundtracks.

VI-17

Since the groundtrack repeats about every five days, this sequence of base support will also be repetitive and, in essence, will form a fence arrangement when related to the groundtracks.

Contingency Landing Area

All of the Earth's surface overflown by Skylab, between 50 degrees North and 50 degrees South can be considered the contingency landing area. Recovery forces will not be dedicated to support a contingency landing; instead, various sources which could provide aid would be called upon as required. This would include not only the Aerospace Rescue and Recovery Service aircraft, continually on alert, but also search and rescue capabilities of other DOD elements and of foreign nations.

Contingency operations will be determined and directed in the Recovery Operations Control Room (ROCR) in the MCC. The ROCR will maintain a data base of information on the locations and capabilities of the world's airfields, aircraft, ports and communications facilities.

Target points will be selected, based on predicted ship locations, aircraft availability, port availability, lighting conditions, weather forecasts and land mass clearance. This can be done in advance as part of the routine of daily target point determinations for each revolution.

If an emergency dictates a landing away from a selected target point, the ROCR will determine what recovery facilities are located in the area of the landing and how they can be contacted. The ships and aircraft in the landing area will then be contacted through appropriate channels and requested to assist in the location and retrieval efforts.

Recovery Lighting Constraints

The time required for recovery operations and the safety of these operations is influenced by the existence of daylight in the landing area. Visual location of the CM, swimmers and equipment deployment from aircraft, and retrieval operations are all more difficult at night.

It is highly desirable that any landing occur with enough daylight left for recovery operations to be completed before dark. Time limits imposed by medical requirements and the potential need for flight crew retrieval in a supine position add to the desire for daylight operations.

The time and day of the SL-1 launch will set the pattern for lighting conditions at the launch site and landing areas for the three manned missions. Because of this pattern, it is not possible to have daylight for recovery operations in all situations. Therefore, lighting constraints have been determined which are categorized as mandatory, highly desirable or desirable, depending on landing probabilities and medical requirements.

Mandatory:
Daylight in the planned end-of-mission landing area.
    CM landing no earlier than the beginning of morning civil twilight.
    CM landing no later than two hours prior to the end of evening civil twilight.

Highly Desirable:
Daylight in the Mode I/II launch orbit landing area (Pad to 4,630 kilometers or 2,500 nautical miles downrange).
    Launch no earlier than the beginning of morning civil twilight at the launch site.
    CM landing no later than three hours prior to the end of evening civil twilight at the end of the Mode II abort landing area (4,630 kilometers or 2,500 nautical miles downrange).

Daylight in the secondary landing areas during the first 48 hours of the three manned missions.

CM landing no earlier than the beginning of morning civil twilight.

CM landing no later than two hours prior to the end of evening civil twilight.

Maximize recovery lighting during the latter portion of each mission.

CM landing no earlier than the beginning of morning civil twilight.

CM landing no later than two hours prior to the end of evening civil twilight.

<u>Desirable:</u>

Daylight in the Mode III discrete landing area.

CM landing no earlier than the beginning of morning twilight.

CM landing no later than three hours prior to the end of evening civil twilight.

Skylab Mobile Laboratories (SML)

Post-flight medical examinations of the crew, and completion of all post-flight Skylab medical experiments required to be carried out in the recovery area, will be accomplished in mobile laboratories aboard the primary recovery ship or at the nearest land base.

The Skylab Mobile Laboratory (SML) consists of six basic U.S. Army Medical Unit Self-Contained Transportables (MUST) modified to meet the post-mission medical requirements of Skylab. The units are outfitted to support the following medical disciplines: blood, cardiovascular, metabolic studies, microbiology nutrition and endocrinology and operational medicine.

The SML system will be maintained in an operational ready configuration at MSC and, at a predetermined time, will be transferred to Ellington Air Force Base (near MSC) and airlifted via USAF C-5A (Figure VI-8) to an airfield near the location of the PRS. The design of the SML permits it to be deployed on the PRS and made operational within 36 hours from the time the SML is received at dockside.

If splashdown becomes necessary in a secondary recovery zone, the SML will be airlifted (via C-5A) to the land base nearest the splash site. The SML will not be unloaded at the secondary recovery area. All procedures related to post-mission medical testing will be accomplished while the SML is in the C-5A on the ground.

A minimum of 24 hours mission termination notification is required for the SML to be fully operational and ready to receive the astronauts when they arrive from any secondary recovery area. A maximum of 17 hours is necessary to transport the SMLs to a "worst case" secondary recovery area. The SML can then be made operationally ready within 7 hours after arrival and thus comply with the 24 hour notification.

The Blood Laboratory will support the M100 series of medical experiments (see Life Sciences Experiments). This lab will have five scientists and technicians operating the various centrifuges, microscopes and other laboratory equipment.

The Cardiovascular Laboratory will support post-mission testing in the M090 series of medical experiments. Five persons will be responsible for the operation of the equipment in support of the post-flight testing of the flight crews.

The Metabolic Laboratory will be operated by five people responsible for the operation of equipment in the collection of post-flight crew data in the M171 experiment.

Four persons will operate inside the Microbiology Laboratory which is equipped to analyze immediate post-flight microbiological samples from each crewman. In addition to performing these time critical analyses, personnel in the lab will conduct prerequisite operations on those samples which shall be returned to MSC for subsequent analysis.

Figure VI-8 — Skylab Mobile Laboartory

The Nutrition and Endocrinology Laboratory is equipped with the personnel (eight) and the equipment to perform post-flight tests in relation to the M070 medical experiments. The food for the returned astronauts will be prepared in this lab, which will also serve as a dining area for the crewmen.

Standard physical examinations, including audiometric and visual examinations, will be performed on the Skylab crews in the Operational Medicine Laboratory. Two people will be assigned to this lab. (See Life Sciences Experiments.)

SECTION VII

CREW TRAINING

The Skylab crew training program has incorporated all the requirements to prepare flight crews to effectively accomplish the mission objectives and goals assigned to them. Each member of the Skylab prime and backup crews will have completed by their respective launch time a total of approximately 2,150 hours of training (Table VII-1).

The introduction of specialization has had the net effect of reducing each crewmember's overall requirements. The Skylab commander's specialty is the CSM launch, rendezvous and docking, EVA and deorbit and entry; the scientist pilot has the specialty of the ATM, EVA and the medical experiments; and the pilot is the specialist in areas of the AM and MDA, the OWS and Earth resources experiments. Although each of the crewmen receive specialized training, there has been a sufficient amount of cross training to assure successful completion of the more critical and important mission tasks and experiments. All training is accomplished at the Manned Spacecraft Center with the exception of EVA training using the Marshall Space Flight Center's neutral buoyancy simulator and spacecraft test at the contractors' plants and Kennedy Space Center.

The primary elements which form the basis of the Skylab crew training program are:
Background training (scientific/academic).
Systems, experiments and operations briefings and reviews.
Operation procedures training (simulators and mockup trainers).

Briefings and Reviews — 450 hours

CSM briefings take up 95 hours of crew training time. Emphasis of these sessions is on major subsystems and operations. A final six-hour system briefing is scheduled four weeks prior to launch, concentrating on system anomalies and late modifications.

The OA briefings account for 112 training hours. These sessions provide the crew with a thorough working knowledge of the OWS, AM, MDA and ATM. Launch vehicle briefings total eight hours.

Other briefings and reviews and the hours of crew time spent are: Solar Physics; Flight Plan and Checklists, 75; and Mission Rules and Techniques, 50.

Systems Training — 350 Hours

The flight crews participated in selected spacecraft and experiments tests conducted to confirm the operational acceptability of the crew station and crew equipment (150 hours). Tests in the CSM are repeated for each mission vehicle, all other tests (OWS, MDA, AM, etc.) have been conducted one time. Prime flight crews with their backups have participated in spacecraft testing at the prime contractor plants at North American Rockwell, Downey, Calif., CSM; Martin-Marietta Aerospace, Denver, Colo., MDA; McDonnell Douglas Astronautics Co., St. Louis, Mo., MDA/AM; McDonnell Douglas Astronautics Co., Huntington Beach, Calif., OWS; and Marshall Space Flight Center, Huntsville, Ala., ATM.

Periodically the flight crews inspected the equipment stowed aboard the CSM and the OWS. This activity, which totals 68 hours, lets the crew examine minutely their flight gear and check it for function, operation and suitability.

Training for a possible fire or decompression and end-of-mission water recovery takes up 40 hours.

Other significant areas of systems training time are: briefings on crew systems (food, waste, EMU, hygiene), 46 hours; TV and photo, 26 hours; and activation and deactivation of the OWS (employing mockup trainers), 20 hours.

Table VII-1 – Skylab Crew Training (Hours)

ACTIVITY	SL-2	SL-3	SL-4
Briefings and Reviews			
CSM	95	95	95
OWS	112	112	112
Launch Vehicle	8	8	8
Solar Physics	110	110	110
Flight Plan, Checklist	75	75	75
Mission Techniques & Rules	50	50	50
Systems Training			
Crew Systems	46	46	46
TV & Photo	26	26	26
SWS Act/Deact	20	20	20
Stowage & Bench Checks	68	68	68
Fire & Egress	40	40	40
Spacecraft Tests	150	50	50
EVA/IVA			
One-g	108	127	119
NBS	48	57	42
Medical	98	98	98
Simulators			
CMS	300	300	300
SLS	300	300	300
CMPS	80	80	80
DCPS	15	15	15
Experiments			
Medical	134	134	98
EREP	72	72	72
ATM	46	46	46
Corollary	178	209	165
Rescue	8	16	24
TOTAL HOURS	2187	2154	2059

EVA/IVA – 156 Hours

Full-scale mockups of the spacecraft modules, referred to as trainers (Figure VII-1), have been employed extensively for experiment training, procedural and timeline development, conducting stoware exercises, performing one-g procedures, and particularly for performing training associated with EVA and IVA in-flight tasks. Major training hardware elements used are the CM Trainer, MDA Trainer, AM Trainer and the OWS Trainer in the 108 hours of one-g EVA and IVA training.

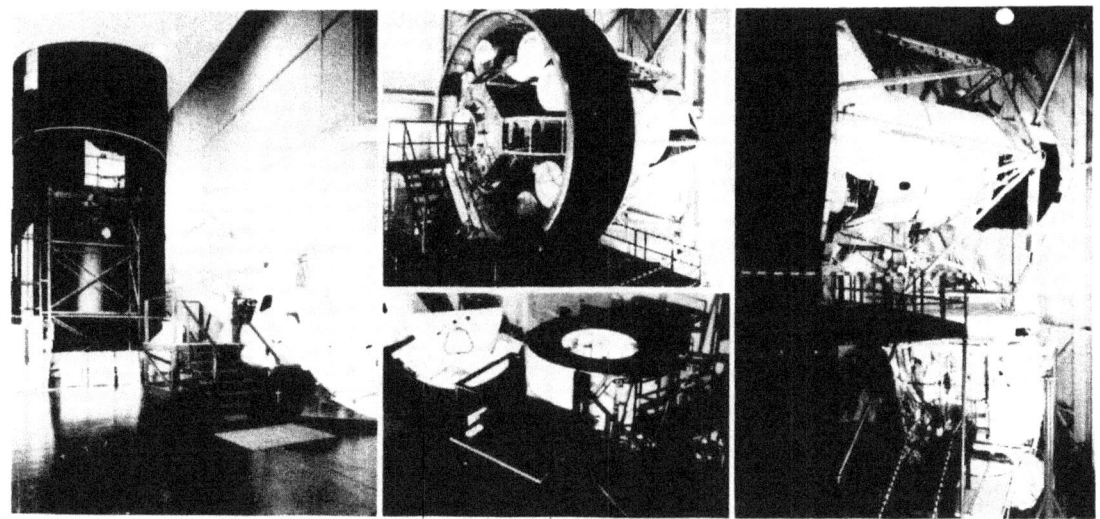

Figure VII-1 — Skylab 1-G Trainers

The MSFC neutral buoyancy simulator (NBS) was used for the 48 hours of zero-g training. The training was to prepare the crewmen for performing assigned EVA tasks, such as installation and retrieval of film magazines associated with ATM solar experiments and the recovery of thermal control coatings sample panels. Vacuum intravehicular transfer, a contingency activation of the MDA or OWS using the Astronaut Life Support Assembly (ALSA), was also practiced.

Medical — 98 Hours

The crewmen received practical training in the diagnosis of illnesses of an outpatient nature. This included the employment of diagnostic equipment (thermometer, stethoscope, sphygmomanometer, opthalmoscope/otoscope, etc.); the use of the various laboratory parts of the Inflight Medical Support System (IMSS) (microbiology, hematology, urinalysis); and the use of the incubator for growing microbial cultures.

Therapeutic procedures for treatment of crew illness or accidents was practiced by the crew in the areas of minor injury, dental care and chemotheraphy, using items such as the tracheotomy, dental and medication kits. Additional training stressed first aid, resuscitation and supportive measures, in case of major illness or injury, sufficient to stabilize the patient for transport to definitive medical care, and the use and care of the Operational Bioinstrumentation System.

Simulators — 695 Hours

Full mission simulators, part task simulators, experiment task simulators, and various engineering development simulators (Figure VII-2) have all been utilized by the flight crews to rehearse and practice the procedures. This activity totals approximately one-third of all crew training.

Command Module Simulator (CMS) — 300 Hours — The CMS provides comprehensive training in all mission segments of the CSM-launch, orbital insertion, rendezvous, docking, orbital assembly, pointing, deorbit and entry. Training in the CMS can be performed independently or integrated with the Skylab Simulator (SLS) and Mission Control Center.

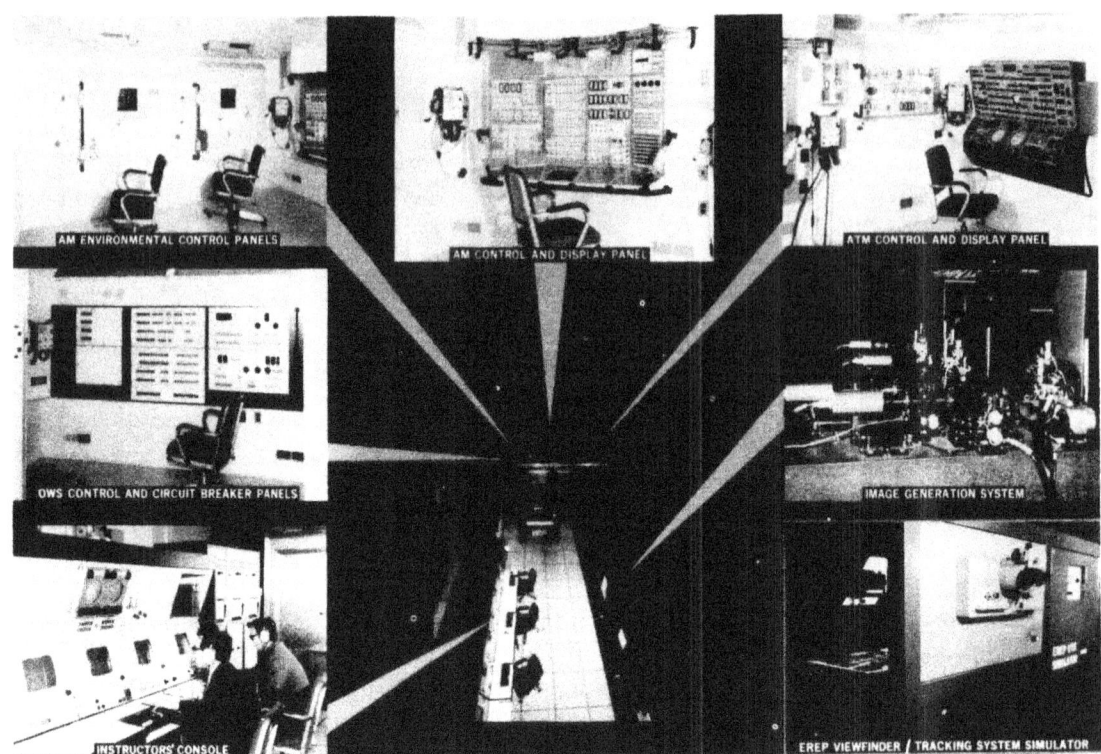

Figure VII-2 — Skylab Simulators

Skylab Simulator (SLS) — 300 Hours — The SLS is an integrated SWS procedures trainer composed of a functional ATM C&D console, STS C&D panel, Oxygen/Nitrogen control panel, Aft Compartment panel, Lock Compartment control panel, and OWS Electrical Display and C&W control panel. Major mission segments practiced are the MDA/AM, OWS and ATM systems activation and deactivation, and orbital operations—150 hours. Although operational training for ATM solar experiments (S052, S054, S055A, S056 and S082A and B) is conducted with the SLS, the highly specific nature of this training lends itself to the establishment of a training syllabus separate from other SLS training efforts. The training is accomplished on the SLS ATM console by the integrated operation of the experiments in compatibility groups arranged to maximize concurrent experiment operations—150 hours.

CM Procedures Simulator (CMPS) — 80 Hours — The CMPS has been utilized primarily to develop crew proficiency in rendezvous and entry procedures, both nominal and contingency. Rendezvous training commences at CSM orbital insertion and terminates upon docking with the SWS.

Dynamic Crew Procedures Simulator (DCPS) — 150 Hours — The DCPS provided flight crews with additional practice for launch and launch aborts. Training consisted of familiarization of the nominal launch timeline, launch vehicle failures and recovery modes. Follow-on sessions expanded upon this, developing crew proficiency in the recognition of the various abort situations and launch vehicle contingency mode operations.

Experiments — 430 Hours

Crew operational training for all experiments consisted of about 160 hours of briefings and 270 hours of operational training primarily involving mockups and special simulators.

Training was accomplished with mockups, flight-type gear or with specialized simulators, such as the Six-Degree-of-Freedom Simulator (Experiment M509), the Air Bearing Simulator (M509 and T020 Experiments), and the EREP Simulator (Experiments S190 through S194).

Experiment briefings were applied to developing knowledge and proficiency in the conduct of scientific, technological, engineering, medical and Department of Defense experiments. This training included sessions with the crews and the principal investigators.

Rescue — 8 to 24 Hours

Either as a rescued crew or a crew required to carry out a rescue mission in case of a disabled CSM, Skylab astronauts received from 8 to 24 hours of rescue training involving briefings, reviews, mockups and simulators as a part of their training for a nominal mission. If a rescue is required, additional training for the rescue crew will be given.

Dr. Joseph Kerwin, scientist pilot for the SL-2 mission, handles a film cassette during a training exercise in the Neutral Buoyancy Simulator at the NASA-Marshall Space Flight Center. Here Dr. Kerwin's position is evidenced by bubbles rising. Astronauts training for missions are submerged in the giant water tank and weighted to a neutrally buoyant state, thus simulating weightlessness.

SECTION VIII

CREW OPERATIONS

The major responsibilities for crew members are divided as follows: <u>Commander (CDR)</u> – Overall mission, CSM systems and flight planning. <u>Science Pilot (SPT)</u> – ATM system, medical experiments and systems. <u>Pilot (PLT)</u> – Earth observations, experiment systems and cluster systems (AM, MDA, OWS).

All crewmen will be able to perform normal operations of all systems; however, each crewman is an expert in certain systems.

VIII-1

CREW SCHEDULE

Skylab crews will operate on a 24-hour day with an 8-hour sleep period, assigned work schedules and have a day off each week, which is referred to as "rest and relaxation."

The typical crew duty day is between 6 a.m. and 10 p.m., Houston time. MCC operations in support of crew activities, including experiment performance, will not be constrained by the Spaceflight Tracking and Data Network. Mission time will be Greenwich Mean Time (GMT), Mission Day and Day of Year (MD/DOY).

Crew manhour allocation in Skylab is divided among operations, experiment performance, eating, personal hygiene, rest and relaxation and sleeping. The sleep and rest periods take the major portion of the allotted time with 33.6 percent. Eating is assigned 12.3 percent.

Each crewman is scheduled about one-half hour daily for personal hygiene and about 45 minutes for systems housekeeping. "R&R" days are the 7th, 14th and 23rd days of the first manned mission.

The first Skylab crew will spend about 511 manhours during the 28-day mission performing experiments and another 66 manhours in systems housekeeping, daily chores vital to crew comfort and to vehicle safety and operation.

The major portion of the experiment time—177 manhours—is allotted to medical experiments. About 166 manhours are scheduled for operation and maintenance at the C&D Panel of the ATM.

Operation of the Earth Resources Experiment Package (EREP) takes up to 88 manhours. The other experiments — scientific, technological and operational — are scheduled for 80 manhours of mission time. (See Figure VIII-1 for SL-2 manhour allocation.)

A hypothetical day in Skylab would begin at 6 a.m. (Houston time) with the crew awakening.

6 a.m. (1100 hours GMT) Crew awakens, personal hygiene, breakfast.

7:30 a.m. All three astronauts perform M071 medical experiment which includes measuring left-over food, recording their water intake and reporting their status to the ground.

8 a.m. CDR and PLT move out to the MDA to prepare for EREP pass. The SPT remains in the OWS and prepares one of the scientific experiments using the SAL.

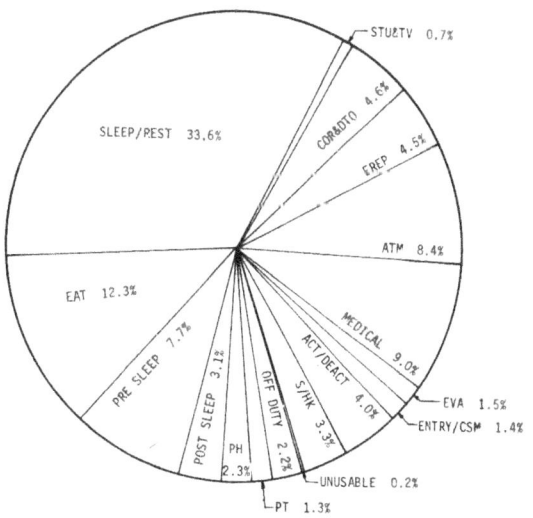

Figure VIII-1 — Manhour Allocations, SL-2

10 a.m. While the PLT moves to the ATM C&D panel in the MDA to monitor the activity of the Sun, the CDR and SPT move back to the experiment area in the OWS to participate in the M092/M093 experiment, which is designed to evaluate the status of the CDR's cardiovascular system. (See Life Sciences Experiments.)

12 noon. Lunch time for the SPT and PLT in the OWS Wardroom. The CDR eats a snack while monitoring the ATM console in the MDA. After lunch, the M071 experiment is conducted again.

1:30 p.m. With the PLT serving as an observer, the SPT performs the cardiovascular experiments. The two change places after one and one-half hours with the PLT performing the tests and the SPT as the observer. The CDR, meanwhile, is at the ATM console.

6 p.m. Dinner time for the Skylab crew. After dinner, the M071 experiment, weighing, measuring and recording food and water intake is done.

7 p.m. The CDR and SPT review the day's work, report the accomplishments to MCC and look over the next day's flight plan. The PLT moves down to the ATM C&D panel for an hour and reports film usage and ATM data.

About the time most householders in the U.S. are cleaning up the dinner dishes and taking out the garbage, the Skylab crew is assembling the garbage to be placed in the trash airlock.

10 p.m. Sleep preparations get underway and the SPT checks out the equipment he will use for the M133 Sleep Experiment. He will don the special sleep cap. His sleep patterns will be recorded on tape for later analysis on the ground. The crew begins an eight-hour sleep period shortly after 10 p.m.

Each crewman is allowed 30 minutes per day for exercise, normally scheduled prior to dinner.

Crew days off will be scheduled to avoid conflict with EREP passes of primary interest. No crew duty activities will be scheduled on days off with the exception of routine system housekeeping, realtime monitoring of solar flares, M071 mineral balance, re-entry simulations, debriefings, and crew activities required for passive experiments.

Every day is a medical day, with a particular crewman performing one series of M092/M093 or M092/M171 every third day. If this conflicts with an EREP pass of primary interest, the medical series will be moved 24 hours, before or after the EREP pass.

EVA is scheduled for a maximum of three hours from the start of egress to completion of ingress. One EVA is scheduled on the 26th day of the first manned mission for retrieval of ATM film and D024 plates; three EVAs are planned for the second manned mission (the 4th, 28th, and 55th days of the mission), and two EVAs are scheduled for the final manned mission.

Two crewmen will be fully suited for the EVAs. The third man, who will be located forward of the airlock, will be "soft suited" (pressure suit without helmet and gloves) and will perform required systems monitoring and read the procedures. EVAs will be scheduled over the continental U.S. to achieve maximum network coverage.

At the end of each day the crew will make a status report which will include the status of consumables, unscheduled maintenance, housekeeping checks, anomalies and stowage.

Re-entry simulations are scheduled toward the closing days of each mission.

Crew sleep time occupies 33.3 percent of the mission time during SL-3. Eating requires 11.7 percent of their time with experiments, off duty and other duties taking up the remainder of the mission. (See Figures VIII-2 and VIII-3 for SL-3 and SL-4 manhours allocations, respectively.)

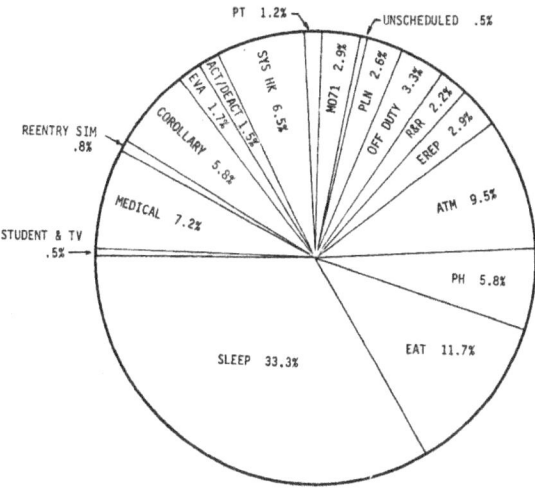

Figure VIII-2 — Manhour Allocations, SL-3

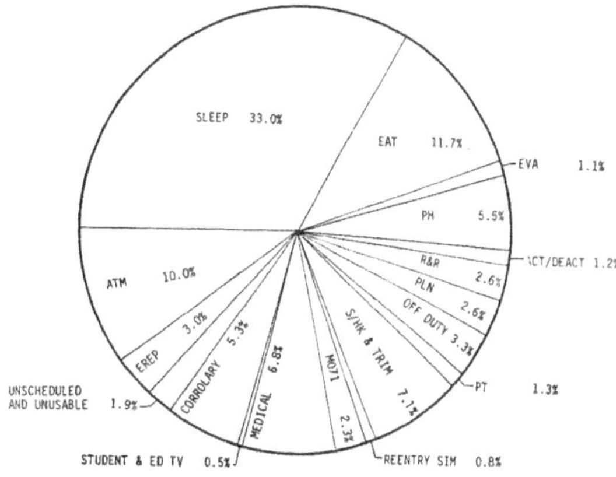

Figure VIII-3 — Manhour Allocations, SL-4

VIII-2

EXTRAVEHICULAR ACTIVITIES

A total of six EVAs are planned for Skylab — one on the 26th day of SL-2, three during SL-3 and two during SL-4. A maximum duration of three hours will be scheduled for each EVA.

The EVA crew will recover exposed photographic film from ATM experiments (S052, S054, S056, S082A and H-1) and load fresh film into the proper experiment cameras. Sample panels of the D024, a thermal-coating experiment, located on a truss of the AM, and S230 on the second and third missions, will also be retrieved.

EVAs will be performed through a hatch (a modified Gemini spacecraft hatch) located in the lock compartment of the AM. Skylab EVA crewmen will be referred to as EV1, EV2 and EV3, as opposed to commander, science-pilot and pilot (Table VIII-1).

Prior to the start of EVA preparations, which includes suiting up in their extravehicular mobility units (See Crew Equipment section), EV-3 deploys the television camera on the T027 extendible boom (through the SAL) so that live and taped TV can be made of the EVAs. While EV3 is performing this task, EV1 and EV2 are in the OWS unstowing their extravehicular mobility units (pressure suits) for the EVA.

All three Skylab crewmembers will be suited up for the EVA. EV1 and EV2, who actually perform the EVAs through the AM hatch, will wear their EMUs. EV3, who will remain in the MDA, will be "soft suited," wearing a pressure suit without helmet or gloves. EV3 will be a different crewman for each EVA.

EV3 will remain in the MDA to monitor systems and perform experiment operations. He will start the EVA clock and advise EV1 and EV2 of EVA elapsed time.

EV1 will stand in the FAS throughout the EVA, assisting the EV2, who is at the outside workstations. EV1 will pass out film, tend the life support umbilical lines, and assist in bringing the exposed film back into the AM.

In addition to retrieving the exposed film from the ATM cameras, EV2 will retrieve a thermal control panel from Experiment D024 which is attached to the exterior of the AM. While EV2 retrieves the panel, EV1 will film the activity. (See D024 Experiment.)

The other panel of D024 will be retrieved during EVA on SL-3. The first sample will have been exposed to the space environment about one month, depending on the exact launch time and flight schedule, and the second sample will have been exposed about five months.

Table VIII-1 — Crew Scheduling, EVA

CREWMAN POSITION DESIGNATIONS FOR EVA						
CREWMAN	SL-2 EVA 1		SL-3 EVA 1		SL-3 EVA 2	
	NOMINAL	ALTERNATE	NOMINAL	ALTERNATE	NOMINAL	ALTERNATE
CDR	EV1	EV2	EV1/EV2		EV1	
SPT	EV2		EV1		EV2	
PLT		EV1	EV2			EV1/EV2
CREWMAN	SL-3 EVA 3		SL-4 EVA 1		SL-4 EVA 2	
	NOMINAL	ALTERNATE	NOMINAL	ALTERNATE	NOMINAL	ALTERNATE
CDR	EV2		EV1/EV2		EV2	
SPT		EV1/EV2	EV2		EV1	EV2
PLT	EV1		EV1			EV1

EV1 - Crewman who egresses first and performs operations at the FAS Work Station (VF)

EV2 - Crewman who egresses second, transfers along handrails and performs operations at the Center Work Station (VC) and Sun-end Work Station (VS)

At the end of the EVA the AM hatch will be closed and sealed and the AM repressurized. The ATM film will be stowed in the CM for return to Earth.

Crew members will then doff their EMUs and place them in the suit drying facility where they will remain for a period of 10 hours and then be stowed in the MDA. The crew will also log the readings of the Passive Radiation Dosimeters, which recorded the amount of radiation the crew received during the EVA period.

The crew members will also wear their pressure suits during the run of the M509 Astronaut Maneuvering Unit and T020 Foot Controlled Maneuvering Unit experiments. The same donning and doffing and drying procedures will be followed. The T020 experiment will be used only during SL-3 and SL-4.

The EVAs are performed at five workstations on the AM exterior and on the ATM (Figure VIII-4). The EVA bay on the AM contains a workstation on the FAS and a replacement workstation located by the EVA hatch. The ATM workstation consists of the center and Sun end workstations for film retrieval and replacement and the transfer workstation used to transfer the Sun end cameras.

Each workstation consists of a foot restraint to hold a crew member while he performs assigned tasks (Figure VIII-5).

The workstation on the FAS serves as the main operations area in support of the nominal EVA activities.

The replacement workstation is within this same bay area and is utilized in the event of a contingency or hardware failure when equipment must be changed.

The center workstation, transfer workstation and Sun end workstation on the ATM serve as the work area. One crewman will conduct the work tasks at each of the ATM workstations.

The workstations include numerous aids, such as lighting (Figure VIII-4), handrails, foot restraints, clamps, hooks, film tree receptacle, extendible booms and a clothesline transporter.

The minimum time required to depressurize the AM lock compartment from 5.0 to 0.15 psia is 50 seconds and to repressurize takes approximately 25 seconds. Slower depressurization or pressurization is achievable by partially opening the appropriate equalization valve.

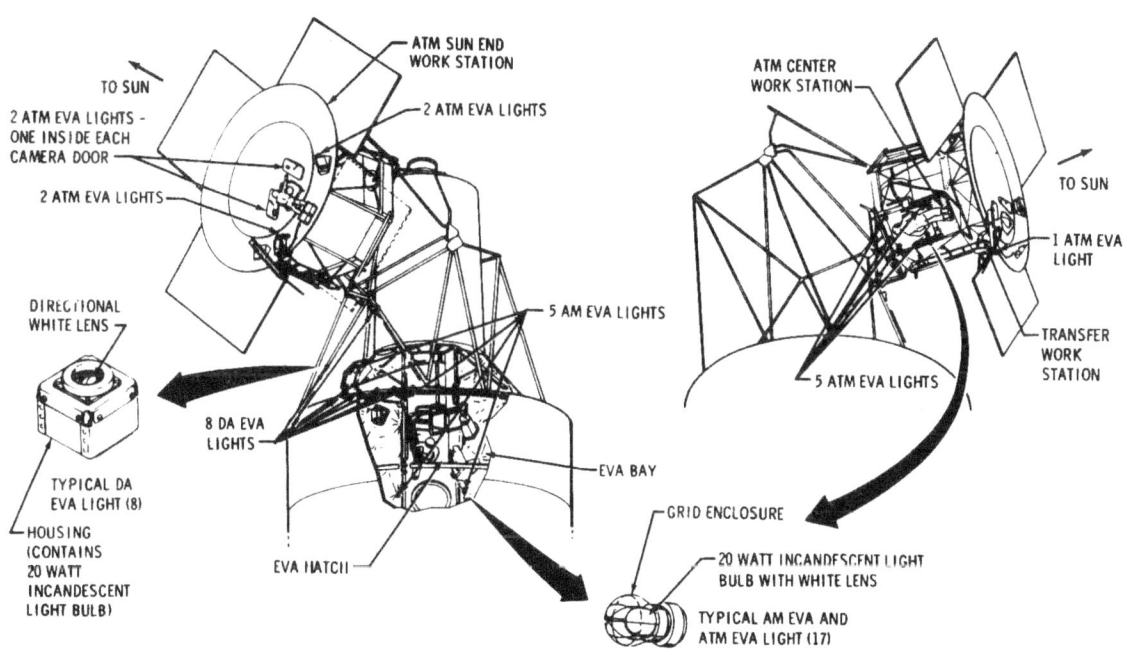

Figure VIII-4 — EVA Workstations and Lighting

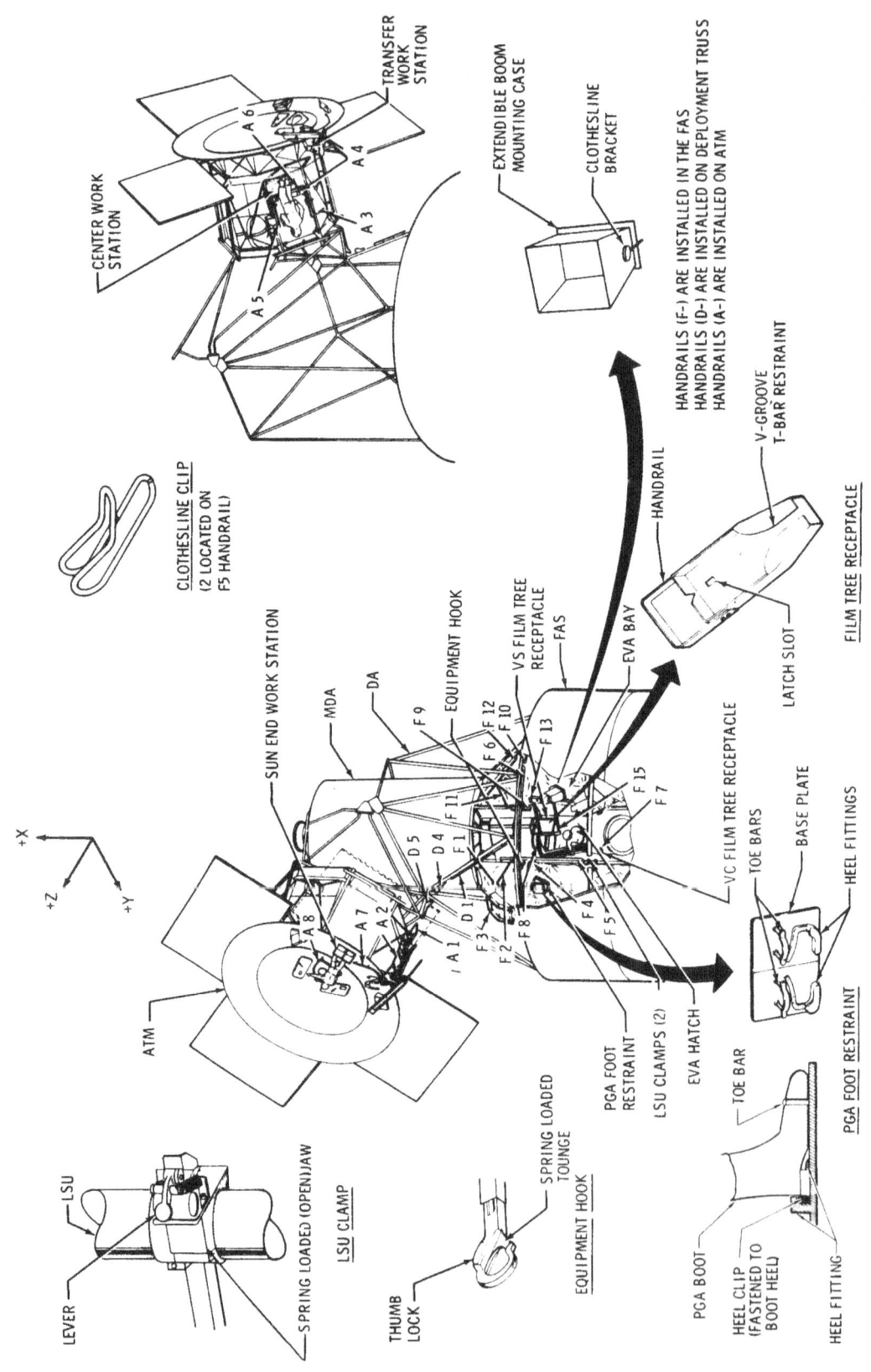

Figure VIII-5 — External Restraints and Mobility Aids

VIII-6

SCHEDULED MAINTENANCE

Scheduled maintenance consists of those in-flight housekeeping and components replacement tasks to be accomplished on a regular basis to insure optimum life of equipment and systems. One-hundred manhours are scheduled for these tasks during SL-2 and more than 200 hours for SL-3 and SL-4.

Housekeeping for the crew will follow the typical down-to-Earth household pattern of cleaning up the dishes, putting out the trash, checking the air conditioning/heating systems and making the beds. Skylab housekeeping tasks will consist of: periodic systems verification, changing teleprinter paper, dumping the condensate control system, disposing of trash, vacuuming air filter screens and picking up loose floating particles, and conducting pre and post-sleep checklists.

A few of the chores crew members will perform before turning in for the night are: stow any loose objects, turn down the intercom systems and adjust the ventilation valves and set heat control at desired crew comfort level, configure the bunks for sleep, and set-up the breakfast trays and set the warmer switch. Next morning, some of the tasks they must perform are: Make-up the bunks, return the communication intercom system to normal operations, readjust the ventilators and heat controls, check status of experiments which operated during the sleep period, and check the teleprinter for messages which may have been passed up from MCC during the night.

Some 317.5 kilograms (700 pounds) of spare parts and more than 27.2 kilograms (60 pounds) of maintenance tools are launched aboard the SWS. Skylab tools include numerous wrenches, screw drivers, pliers and other work tools in the portable maintenance equipment kits. A tool caddy which the crew can wear around their waist contains the tools required to perform any scheduled inflight maintenance or servicing.

Maintenance activities and housekeeping chores are performed on a daily basis, others every 7, 10, 14 or 28 days. Time to complete these tasks range anywhere from a two minute periodic verification of the SWS environmental control system to 30 minutes to clean and disinfect the waste management center.

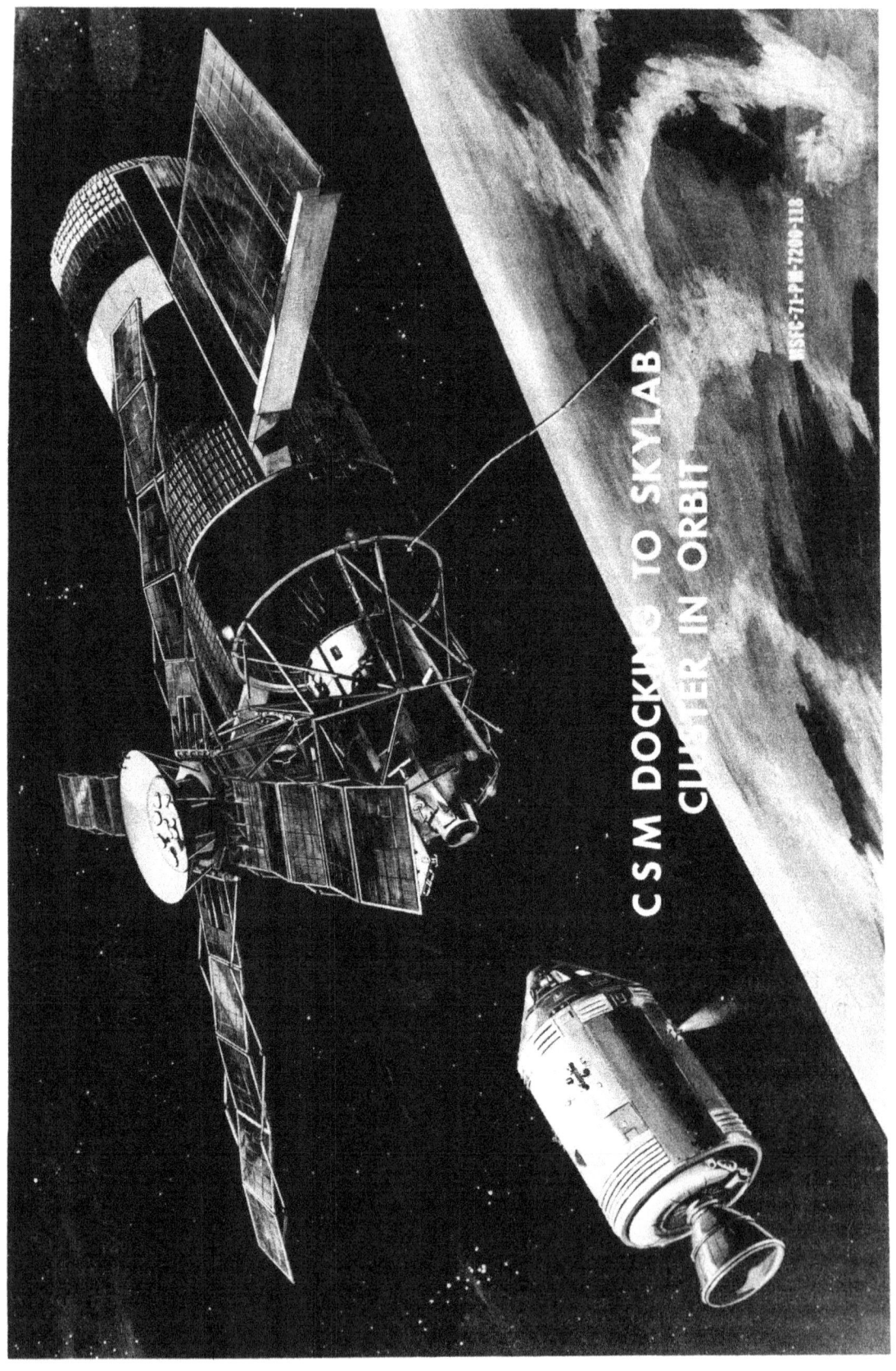
CSM DOCKING TO SKYLAB CLUSTER IN ORBIT

SECTION IX

THE SPACEFLIGHT TRACKING AND DATA NETWORK

The Spaceflight Tracking and Data Network (STDN) is a world-wide complex of tracking facilities which provides support to all NASA Earth-orbiting spacecraft. The network consists of land stations at 19 locations, Advanced Range Instrumented Aircraft (ARIA), and an instrumented ship, the USNS Vanguard (Figure IX-1). The northernmost station in the network is in Fairbanks, Alaska (65 degrees north latitude) with the southernmost station at Orroral Valley, Australia (35 degrees south latitude).

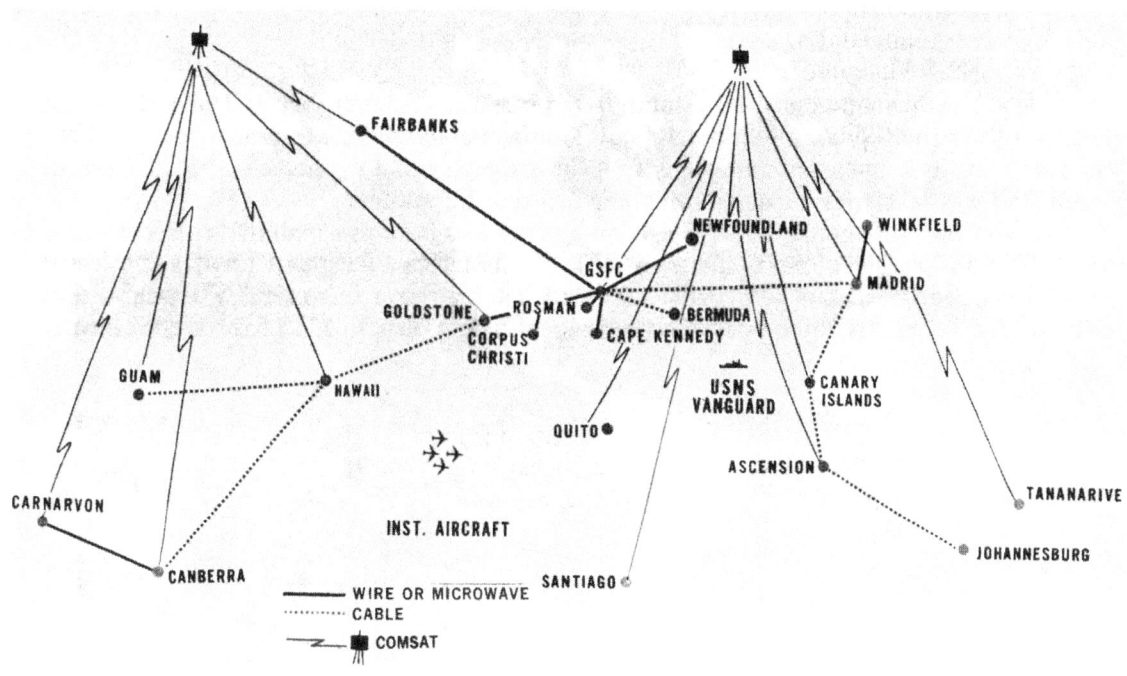

Figure IX-1 — Spaceflight Tracking and Data Network

Real-time network operational control and scheduling is provided by the Goddard Space Flight Center, Greenbelt, Md. A central computing center at Goddard, provides computational capability required for network operation and data analysis. Spacecraft operations for unmanned spacecraft are conducted through the STDN stations by the Project Operations Control Center at Goddard, and, for manned flights, like Skylab, by the Mission Control Center (MCC) at the Manned Spacecraft Center.

The STDN stations are linked and supported by the facilities of the National Aeronautics and Space Administration Communications Network (NASCOM), a global communications network established by NASA to provide operational ground communications for support of all spaceflight operations. The STDN equipment encompasses telemetry, command, tracking, communications and data handling systems. A variety of equipment configurations provides extreme flexibility in supporting the varied spaceflight missions.

Thirteen STDN stations, equipped with either 9 meter (30-foot) or 26 meter (85-foot) antennas, will be supporting the Skylab program. They are:

- Merritt Island, Florida
- Bermuda
- Ascension Island
- Grand Canary Island
- Carnarvon, Australia
- Guam
- Hawaii
- Corpus Christi, Texas
- Goldstone, California
- Canberra, Australia
- Madrid, Spain
- Newfoundland, Canada
- USNS Vanguard

The USNS Vanguard will be temporarily positioned in the port of Mar Del Plata in Argentina during the Skylab mission to obtain additional network coverage in the southern hemisphere. This is being accomplished with the assistance and cooperation of the Comision Nacional De Investigaciones Espaciales of the Argentine Republic.

All stations supporting the Skylab missions will have the capability to receive and record television transmissions. Three stations — Goldstone, Corpus Christi and Merritt Island — have been designated "prime" stations for television reception. TV transmissions received by these stations will be provided on a "real-time" basis to MSC for special events.

SECTION X

MANUFACTURING AND TESTING FACILITIES, TRANSPORTATION

The Skylab Program uses facilities that were in existence at the initiation of the program. With minor modifications, a vast network of shops, laboratories, engineering offices, special test fixtures and transporting equipment support the development, testing and transportation of Skylab hardware.

CONTRACTOR FACILITIES

McDonnell Douglas — Western Division

The McDonnell Douglas Astronautics Company-Western Division (MDAC-WD) outfitted and tested the Orbital Workshop and fabricated the Payload Shroud at its facility in Huntington Beach, Calif. (Initial fabrication of the S-IVB which was converted to the OWS was completed at this same facility.) The Huntington Beach facility provided administrative, engineering, manufacturing and testing floor space. Proof pressure and leak tests were performed at the MDAC-WD Seal Beach, Calif., facility.

The OWS (Figure X-1) was modified and outfitted with qualified subsystem components, subassemblies and experiments at Huntington Beach and then moved to their vertical checkout laboratory (VCL) for subsystem acceptance testing. At the VCL all subsystems and experiments were functionally tested, experiment and crew interfaces verified and crew equipment and supplies storage activities accomplished. At the completion of this complex and comprehensive acceptance test program, the OWS was shipped to KSC.

McDonnell Douglas — Eastern Division

The McDonnell Douglas Astronautics Company-Eastern Division (MDAC-ED) manufactured and tested the Airlock Module, the Fixed Airlock Shroud and the ATM Deployment Assembly at its facility located near the Lambert Field Airport in St. Louis, Mo. The Eastern Division facility provided manufacturing and testing floor space and a vacuum chamber where the AM and MDA were tested under simulated altitude conditions of 45,720 meters (150,000 feet).

At this facility, qualified and accepted components were brought together for acceptance testing at the subsystem and system level after installation in the AM. Final tests and factory checkout were performed before delivery, as a prelude to final module acceptance. The major test activities and sequences performed at the St. Louis facility are depicted in Figure X-2.

At the completion of post-manufacturing acceptance testing, the AM was mated to the MDA and further module/system integration testing was performed.

The AM/MDA, without the FAS or DA, was then installed in the MDAC-ED altitude chamber where further integrated spacecraft system testing was performed at simulated flight altitudes and with flight crew personnel participation. This test verified the AM/MDA system's capability to operate properly at orbital altitudes. Crew capability and functional performance was also evaluated.

The AM/MDA, Fixed Airlock Shroud and the ATM Deployment Assembly were then shipped to KSC.

Figure X-1 — OWS At MDAC-WD

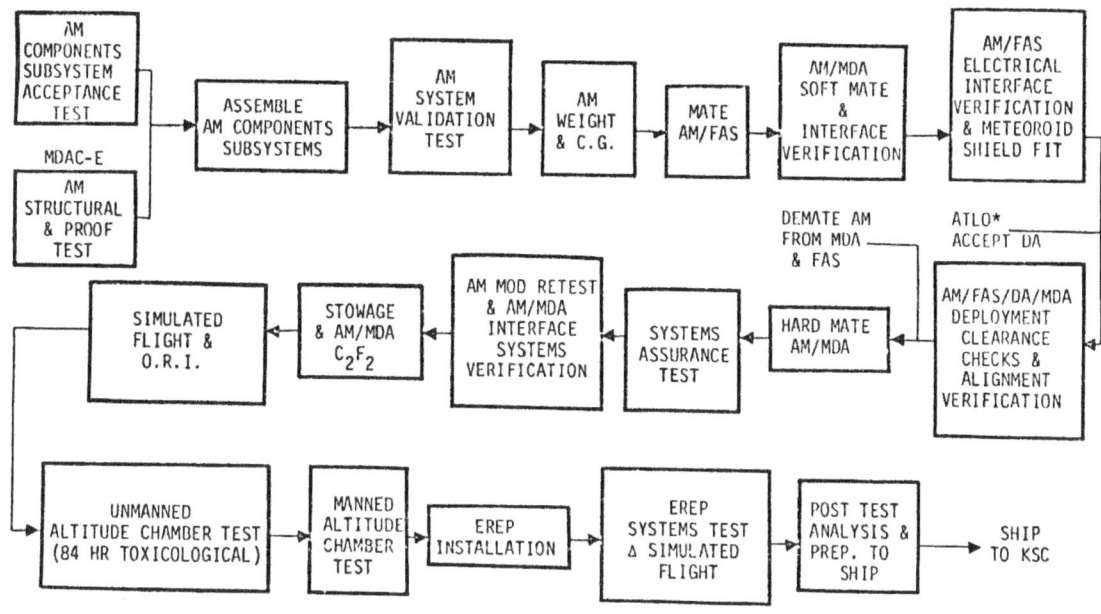

Figure X-2 — MDAC-ED Test Flow

Martin Marietta Aerospace

The Martin Marietta Aerospace Denver Division outfitted and tested the Multiple Docking Adapter, and developed, manufactured and tested Skylab experiments at their facility about 25 miles southwest of Denver, Co. The Denver Division Facility consists of administrative and engineering office space and manufacturing and testing floor space in 46 Martin Marietta-owned buildings. The Engineering and Administration buildings, factory, general purpose laboratories, Space Simulation Complex and supporting service facilities are closely located.

It was from these Denver facilities that Martin Marietta engineers and management personnel performed their assigned Skylab Program Cluster Systems engineering analysis and payload integration tasks and managed sub-contractor work related to Skylab.

The MDA shell was fabricated and proof-pressure tested at the Marshall Space Flight Center in Huntsville, Ala. It was then shipped to Denver where its outfitting with qualified experiments and subsystems was completed (Figure X-3). Further structural and pressure tests, and electrical-electronic, electromechanical and mechanical system acceptance tests were performed. Scientific experiments were integrated with the MDA systems and functionally operated.

The MDA was then shipped via airplane to the McDonnell Douglas Astronautics Company in St. Louis, Mo., where it was mated with the Airlock Module and further integration testing was performed.

Bendix Corporation

The Bendix Corporation Navigation & Control Division designed, manufactured and tested several major components and systems at their facility in Teterboro, N.J. The Navigation & Control Division facility consists of administrative and engineering office space and manufacturing and testing floor space allocated to Skylab related hardware. In addition, Bendix provided an engineering facility in Denver to provide close communication with Martin Marietta Aerospace, their prime contractor, on a portion of their engineering and manufacturing activity.

Figure X-3 — MDA In Vertical Test

One such activity was the design and manufacture of the ATM Control and Display console. The C&D console was fabricated and tested in Teterboro then shipped to MDAC-ED where it was fitted to the MDA and tested further.

Bendix also designed and manufactured major components of the Attitude and Pointing Control System Teterboro. Control Moment Gyros (CMGs), the Star Tracker and the Experiment Pointing Electronics Assembly (EPEA) were manufactured and tested in Teterboro then shipped to Marshall Space Flight Center for mounting on the ATM.

North American Rockwell Corporation*

The North American Rockwell Corporation (NR) was prime contractor for the second stage (S-II) of the Saturn V launch vehicle, the H-1, J-2 and F-1 engines that power all Saturn stages, the Apollo Command and Service Modules, the Launch Escape System and the Spacecraft Lunar Module Adapter. (See Apollo News Reference.)

NR's Rocketdyne Division at Canoga Park, Calif., produced the H-1 engines for the Saturn IB vehicle's first stage, the F-1 engines for the Saturn V first stage, and the J-2 engines for the Saturn V second (S-II) and third (S-IVB) stages.

---

*Now known as Rockwell International Corporation.

The Rocketdyne facility at Neosho, Missouri, was involved in producing and performing preliminary tests of H-1 engines. The engines were further tested at the Marshall Space Flight Center. F-1 engines were tested at the Rocket Engine Test Site, Edwards, Calif., and J-2 engines were tested at Rocketdyne's facility at Santa Susana, Calif., as were the S-II stages.

The Space Division of NR at Downey, Calif., was responsible for development of the Command and Service Modules and related spacecraft hardware. (See Saturn V News Reference and Saturn IB News Reference books.)

Chrysler Corporation

The Chrysler Corporation, Space Division, was prime contractor for the first stage (S-IB) of the Saturn IB launch vehicle. The S-IB is a redesigned and more powerful version of the Saturn I first stage developed by the Marshall Space Flight Center.

Chrysler, at the Marshall Center's Michoud Assembly Facility, New Orleans, built 12 of the S-IB vehicles. The S-IB's were shipped by barge to MSFC for test firings and then back to Michoud for storage awaiting use. Five have been launched. The sixth, seventh and eighth are to launch the crews for SL-2, SL-3 and SL-4, respectively. (See Saturn IB News Reference.)

TRW Systems

TRW Systems of Redondo Beach, Calif., served as a subcontractor to McDonnell Douglas Astronautics Company, producer of the Orbital Workshop. The task assigned to TRW Systems was to design, manufacture and test a solar array system to supply power for the OWS. Tests of the resulting SAS indicated that it would sustain predicted operational loads and environments and would function reliably under adverse conditions. TRW devised mechanisms for retaining the array during flight from Earth to orbit, releasing the array upon command, and deploying it into the fully open position.

The Boeing Company

The Boeing Company was prime contractor for the first stage (S-IC) of the Saturn V launch vehicle. The S-IC was developed jointly by Boeing and the Marshall Space Flight Center. Four S-IC stages were assembled at MSFC, two test stages and the first two flight boosters. Boeing manufactured a dynamic test stage, a facilities vehicle and 13 flight stages at the Marshall Center's Michoud Assembly Facility, New Orleans. The first flight stage built by Boeing was tested at MSFC, the other 12 at MSFC's Mississippi Test Facility. (See Saturn V News Reference.)

International Business Machines, Inc.

International Business Machines, Inc., is prime contractor for the instrument unit (IU) for Saturn IB and Saturn V launch vehicles. The IUs for these two vehicles are almost identical. IBM had limited responsibilities in the integration of Saturn I instrument units before assuming full integration responsibilities in the Saturn IB program.

The IU's for Saturn vehicles were manufactured at IBM's plant at Huntsville, Ala. Major suppliers of IU components were: Electronic Communications, Inc., St. Petersburg, Fla., Control Computer; Bendix Corporation, Teterboro, N. J., ST-124 inertial platform; and IBM Federal Systems Division, Owego, N. Y., Launch Vehicle Digital Computer and Launch Vehicle Data Adapter. (See Saturn IB News Reference and Saturn V News Reference Books.)

NASA FACILITIES

Marshall Space Flight Center

Marshall Space Flight Center (MSFC) in Huntsville, Ala., was responsible for the design, manufacture and assembly of the Apollo Telescope Mount (ATM). Contractors such as Ball Brothers Research Corporation, Bendix, Honeywell, and Perkin-Elmer designed and manufactured the experiments and blackboxes to be used on the ATM and supplied them to MSFC where they were received, tested and stored until required for installation.

The ATM structure which supports the experiments and blackboxes was designed, manufactured and assembled by NASA at its MSFC facilities. The complete assembly of the ATM was also performed at MSFC. Skilled specialists, working in clean room facilities where stringent cleanliness levels were maintained, installed the instrumentation, experiments, blackboxes, cables, etc., and performed critical alignment procedures to set the experiments before checkout of the ATM began.

Post-manufacturing checkout (PMC) of the complete ATM and vibration testing was also accomplished at MSFC but required moving to different facilities. For each move the ATM was sealed in large nylon bags to maintain the cleanliness at the required levels (Figure X-4). During PMC, tests were performed on every blackbox and system such as the electrical power and network, instrumentation and communications, attitude and pointing control, thermal control, etc., to verify correct operation within preset requirements. The vibration tests performed on the ATM were simulating loads which will be present during flight. Following vibration testing a system checkout was again performed to reverify the correct performance of the ATM before preparing it for shipment to the Manned Spacecraft Center for thermal vacuum testing.

Figure X-4 — ATM Movement

Manned Spacecraft Center

At MSC, thermal vacuum testing was performed in the space environmental simulator laboratory. The vacuum chamber is 19.8 meters (65 feet) in diameter, 36.6 meters (120 feet) high. Major structural and vibroacoustics design verification tests were performed on payload hardware in the Structural Vibration and Acoustical Laboratory. This facility provided a test area complete with supporting vibration "shaker" and acoustic generators, with supporting control and instrumentation systems.

Michoud Assembly Facility and Mississippi Test Facility

The NASA-Michoud Assembly Facility (MAF) at New Orleans has a main plant with 16.2 hectares (40 acres) under one roof. S-IC and S-IB stages were assembled in MAF's vehicle assembly building. At the Mississippi Test Facility (MTF), S-IC stages built at MAF and S-II stages from the West Coast underwent preflight static tests. NASA's Computer Operations Office at Slidell, La., operates digital and analog computers and other data reduction equipment in support of MAF and MTF.

The Skylab Program launch vehicles were fabricated and tested as part of the Apollo/Saturn program. Since completion of that test program, they have been in storage. Refurbishment has been accomplished. Prelaunch checkout is the responsibility of KSC.

Kennedy Space Center

The KSC facility was developed by NASA to support the complex operations required to prepare launch and space vehicles for their space missions. No new major facilities have been constructed in support of the Skylab mission. Three of the many laboratories and launch complexes are required for the Skylab hardware, the Operations and Checkout (O&C) building, the Vehicle Assembly Building (VAB) and Launch Complex 39.

At the completion of the Skylab contractors' flight hardware acceptance test program, the modules were shipped to KSC where they were prepared for launch. The O&C building provided engineering, test and checkout support floor space. In addition to computerized automatic checkout equipment and supporting ground support equipment, the O&C building provided an altitude chamber where final CSM checkout was done.

Upon completion of the final checkout, the modules were moved to Launch Complex 39 where they were mated with the Saturn launch vehicle and prepared for launch.

TRANSPORTATION

The major elements of the Skylab space station and launch vehicles arrived for final assembly at NASA's Florida launch site from such distant points as Denver, Colo., St. Louis, Mo., Huntington Beach, Calif., and Huntsville, Ala. Many had made other stops before reaching the Kennedy Space Center.

Three methods were used for transporting the Skylab hardware between the manufacturing, test and assembly sites. Land routes were used for short hauls with specially designed transporters (Figure X-5); water routes were used mostly for long hauls of large cargo such as Saturn launch vehicle stages; and air routes also were used for long hauls of certain hardware.

Water Transportation

From the beginning of the Apollo program, it was recognized that water transportation would be required for the larger articles. Consequently, Saturn manufacturing, test and launch sites were all located on major waterways.

The Marshall Space Flight Center has several vessels to transport large rocket stages and components between manufacturing, test and launch sites.

Saturn IB barges "Promise" and "Palaemon" have been in operation for several years. The smaller of the two, "Palaemon," 54 meters (180 feet) long, was used primarily to transport Saturn IB boosters between the Michoud Assembly Facility and the Marshall Center, where they were static fired.

The "Promise," 78 meters (260 feet) long, carries Saturn IB boosters from Huntsville and Michoud to the Kennedy Space Center where they are launched.

Figure X-5 — AM/MDA Transporter

The U.S. Navy turned over to NASA five additional sea-going barges of the Promise type, one of whick was sent to the Kennedy Space Center. They are 78 meters (260 feet) long and 14.4 meters (48 feet) wide, with an operating draft of 2.1 meters (seven feet). All have been modified for Saturn roles. Two of them, like the Promise, have deck-mounted cargo hangars for enclosing rocket stages. The other three, used for short hauls, have open decks.

One of the vessels has been outfitted for movement of the S-IC stage between Michoud, MSFC and KSC. This craft, the Poseidon, has a large deckhouse 15 meters (50 feet) wide, 12.9 meters (43 feet) maximum height, and 60 meters (200 feet) long to enclose the S-IC and S-II stages.

The Military Sealift Command provided a self-propelled Landing Ship Dock (LSD) for transporting the flight S-II stages, the OWS and Payload Shroud from the point of manufacture to the launch site. The ship, the USNS Point Barrow, was modified to provide desired characteristics (Figure X-6). These modifications included construction of a covered hangar and installation of a stabilization system. The ship is 135.5 meters (465 feet) long, 22.2 meters (74 feet) wide, with 5.7-meter (19-foot) maximum draft. Its speed is about 15 knots.

Air Transportation

The "Super Guppy" (Figure X-7) aircraft was manufactured by Aero Spacelines, Inc., Van Nuys, Calif., primarily to carry the S-IVB third stage of the Saturn V launch vehicle, as no other aircraft could accommodate that size cargo.

The aircraft has an inside diameter of 7.5 meters (25 feet) and a total length of 42.3 meters (141 feet). Tail height is about 14 meters (46.5 feet) or almost five stories above the ground. Usable volume is 1,400 cubic meters (49,790 cubic feet).

The Super Guppy was used to transport: the MDA from Martin Marietta Aerospace facilities in Denver to McDonnell Douglas facilities in St. Louis, Mo.; the AM/MDA from St. Louis to KSC; and the ATM from MSFC in Huntsville, Ala., to KSC.

Figure X-6 — USNS Point Barrow

Figure X-7 — Super Guppy Aircraft

SECTION XI

MANAGEMENT

Management of the Skylab Program is shared by officials of NASA Headquarters, NASA centers involved in manned space flight, and aerospace industries. Overall management is the responsibility of Dale D. Myers, NASA Associate Administrator for Manned Space Flight. Other top management officials for Skylab are:

NASA Headquarters:
    William C. Schneider, Director, Skylab Program.
    John C. Disher, Deputy Director, Skylab Program.
    Robert O. Allen, Director, Skylab Operations.
    Thomas E. Hanes, Director, Skylab Experiments.
    Melvin Savage, Director, Skylab Engineering.
    Haggai Cohen, Director, Skylab Reliability, Quality and Safety.
    J. Pemble Field, Jr., Director, Skylab Budget and Control.

Marshall Space Flight Center:
    Dr. Rocco A. Petrone, Director, Marshall Space Flight Center.
    Dr. William R. Lucas, Deputy Director, Technical, MSFC.
    Richard W. Cook, Deputy Director, Management, MSFC.
    Leland F. Belew, Manager, Skylab Program, MSFC.
    Stanley R. Reinartz, Deputy Manager, Skylab Program, MSFC.
    George B. Hardy, Chief, Program Engineering and Integration Project.
    Porter Dunlap, Chief, Ground Support Equipment Project.
    Jack H. Waite, Chief, Experiment Development and Payload Evaluation Project.
    Floyd M. Drummond, Chief, Airlock Module/Multiple Docking Adapter Project.
    William K. Simmons, Chief, Saturn Workshop Project.
    Rein Ise, Chief, Apollo Telescope Mount Project.
    Richard G. Smith, Manager, Saturn Program.
    H. Fletcher Kurtz, Manager, Mission Operations.
    Dr. James B. Dozier, ATM Project Scientist.
    Eugene H. Cagle, ATM Engineering Manager

Manned Spacecraft Center:
    Dr. Christopher C. Kraft, Jr., Director, Manned Spacecraft Center.
    Kenneth S. Kleinknecht, Manager, Skylab Program Office, MSC.
    Arnold D. Aldrich, Deputy Manager, Skylab Program Office, MSC.
    Alfred A. Bishop, Manager, Missions Office, Skylab Program Office, MSC.
    Willis B. Mitchell, Jr., Manager, Engineering Office, Skylab Program Office, MSC.
    W. Harry Douglas, Manager, Manufacturing and Test Office, Skylab Program Office, MSC.
    Reginald M. Machell, Manager, Orbital Assembly Project Office, Skylab Program Office, MSC.
    Clifford E. Charlesworth, Manager, Earth Resources Program Office, MSC.
    Donald K. Slayton, Director, Flight Crew Operations, MSC.
    Richard S. Johnston, Director, Life Sciences, MSC.
    Howard W. Tindall, Director, Flight Operations, MSC.
    Eugene F. Kranz, Chief, Flight Control Division, Flight Operations, MSC.
    Glynn S. Lunney, Skylab CSM Manager

Kennedy Space Center:

    Dr. Kurt H. Debus, Director, Kennedy Space Center.
    Miles J. Ross, Deputy Director, Kennedy Space Center.
    Robert C. Hock, Manager, Apollo/Skylab Program, KSC.
    W. L. Halcomb, Manager, Skylab Ground Systems Office, KSC.
    J. C. Leeds, Manager, Skylab Program Control Office, KSC.
    O.L. Duggan, Manager, Skylab Reliability, Quality Assurance and Systems Safety Office, KSC.
    A. R. Raffaelli, Manager, Skylab Space Vehicle Office, KSC.
    P. A. Minderman, Director of Technical Support, KSC.
    Raymond L. Clark, Director of Design Engineering, KSC.
    Grady Williams, Deputy Director of Design Engineering, KSC.
    Walter J. Kapryan, Director of Launch Operations, KSC.
    Dr. Robert H. Gray, Deputy Director of Launch Operations, KSC.
    Dr. Hans F. Gruene, Director, Launch Vehicle Operations, KSC.
    John J. Williams, Director, Spacecraft Operations, KSC.
    Paul C. Donnelly, Associate Director, Launch Operations, KSC.
    Isom A. Rigell, Deputy Director for Engineering, KSC.

Industry:

    John J. Eckle, Manager, S-IC Project, The Boeing Company.
    Richard Schwartz, Program Vice President, Saturn Launch Vehicles, North American Rockwell Corporation.
    J. D. Shields, Manager, S-IB Project, McDonnell Douglas Astronautics Company.
    H. D. Lowrey, Manager, S-IVB Project, Chrysler Corporation.
    James E. Howard, Manager, IU Project, International Business Machines, Inc.
    George Merrick, Vice President and CSM Programs, Manager, North American Rockwell Corporation.
    R. M. Davis, Skylab-MSFC Systems Engineering and Integration, and MDA Project, Martin Marietta Aerospace.
    Normand Ruest, Manager, Skylab Program, International Business Machines, Inc.
    F. J. Sanders, Manager, OWS Project, McDonnell Douglas Astronautics Company.
    E. T. Kisselburg, Manager, Am Project, McDonnell Douglas Astronautics Company.
    Miles R. Robinson, Manager, Skylab Computers System, International Business Machines, Inc.
    Milton Brown, Program Manger, Denver Facility, Bendix Corporation, Navigation and Control Division, Teterboro, New Jersey.
    Larry Sullivan, Manager, Ground Support Equipment, General Electric Company.
    R. L. Bruckwick, Manager, Solar Array Project, TRW Systems.
    Sherman Schrock, Experiments Integration, Martin Marietta Aerospace.
    John Kolvek, Manager, CMG, Star Tracker and EPEA Projects, Bendix Corporation, Navigation and Control Division, Teterboro, New Jersey.
    Tim Russo, Manager, ATM Control & Display Project, Bendix Corporation, Navigation and Control Division, Teterboro, New Jersey.
    Eugene C. Wood, Skylab-MSC, Martin Marietta Aerospace.
    Kenneth P. Timmons, Skylab-KSC, Martin Marietta Aerospace.

SECTION XII

CONTRACTS

Hundreds of private firms, educational institutions, research organizations and federal agencies have been involved in creating Skylab. Some firms supplied small items, such as nuts and bolts; others supplied complete systems, such as modules of the OA or launch vehicle stages. Smaller items were acquired through direct purchase, some items through subcontracts, and larger items through prime contracts. Various institutions and government agencies, including the Departments of Defense and Transportation, contributed to the effort through research and development.

The following table lists major contractors, suppliers or subcontractors and the units and services provided by each. The table is divided into three sections, the first showing procurements under the cognizance of the Marshall Space Flight Center, the second section those of the Manned Spacecraft Center, and the third section those of the Kennedy Space Center.

Organization	Location	Items
MSFC		
American Science Engineering, Inc.	Cambridge, Mass.	Experiment S054, X-ray Spectrographic Telescope. Field Support for S054.
The Boeing Company	New Orleans, La.	S-IC Stage of Saturn V.
Chrysler Corporation Space Division	New Orleans, La.	S-IB Stage of Saturn IB.
Department of Transportation	Washington, D.C.	Experiment T003, Inflight Aerosol Analysis.
General Electric Co.	Huntsville, Ala.	Electrical and Ground Support Equipment.
IBM Corporation	Huntsville, Ala.	ATM Digital Computer System, Instrument Units for Saturn V and Saturn IB.
Marshall Space Flight Center	Huntsville, Ala.	ATM. Experiment M415, Thermal Control Coating. Experiment M479, Zero Gravity Flammability. Experiment M512, Materials Processing in Space.

Organization	Location	Items
Martin Marietta Aerospace	Denver, Colo.	Experiment T027, Contamination Measurement. Payload Integration, Multiple Docking Adapter, ATM C&D Console. (The Bendix Corporation under Contract to Martin Marietta.)
Massachusetts Institute of Technology	Cambridge, Mass.	Experiment M562, Indium Antimonide Crystals.
McDonnell Douglas Astronautics Co.	Huntington Beach, Calif.	S-IVB Stages for Saturn IB and Saturn V. Orbital Workshop. Payload Shroud.
McDonnell Douglas Astronautics Co.	St. Louis, Mo.	Airlock Module.
North American Rockwell Corp.	Downey, Calif.	S-II Stage, Saturn V.
Naval Research Laboratory	Washington, D.C.	Experiment S082, Spectroheliograph. Experiment S009, Nuclear Emulsion.
Perkin Elmer Corp.	Norwalk, Conn.	ATM H-Alpha Telescopes. Field Support for H-Alpha Telescopes.
President and Fellows of Harvard College	Cambridge, Mass.	Experiment S055, Ultraviolet Scanning Polychromator Spectroheliometer.
Radio Corporation of America	Camden, N. J.	Skylab Video Tape Recorders.
Regents of University of California	Los Angeles, Calif.	Experiment M564, Metal & Halides Eutectics.
Rensselaer Polytechnic Institute	Troy, N. Y.	Experiment M556, Vapor Growth of II-IV Compounds.

Organization	Location	Items
U. S. Atomic Energy Commission	Oak Ridge, Tenn.	Experiment M558, Radioactive Tracer Diffusion.
University of Alabama	Huntsville, Ala.	Experiment M560, Growth of Spherical Crystals.
University of California at Berkeley	Berkeley, Calif.	Experiment S228, Trans-Uranic Cosmic Rays.
University Corporation for Atmospheric Research (UCAR)	Boulder, Colo.	Experiment S052, White Light Coronograph.
University of Southern California	Los Angeles, Calif.	Experiment M563, Mixed III-IV Crystal Growth.
University of Wisconsin	Madison, Wis.	Experiment S150, Galactic X-ray Mapping.
Westinghouse Electric Corporation	Pittsburgh, Pa.	Experiment M518, Multipurpose Electric Furnace.

MSC

Organization	Location	Items
Actron Industries	Monrovia, Calif.	Earth Terrain Cameras.
AiResearch Division of Garrett Corp.	Torrence, Calif.	Astronaut Life Support Assembly.
AIL/Cutler-Hamner, Inc.	Melville, N. Y.	Experiment S194, L-Band Microwave Radiometer.
Block Engineering, Inc.	Cambridge, Mass.	Experiment S191, IR Spectrometer (EREP).
David Clark Co.	Worcester, Mass.	Skylab Communication Carriers.
Delco Electronics	Santa Barbara, Calif.	Guidance & Navigation Systems.
Dallas County Hospital	Dallas, Tex.	Experiment S015, Effects of Zero Gravity on Single Living Human Cells.

Organization	Location	Items
Dudley Observatory	Albany, N. Y.	Experiment T025, Coronagraph Particulate Measurements Study. Experiment S149, Particle Collection.
Ehrenreich Photo	Garden City, N. Y.	35 mm Cameras.
General Electric Co.	Houston, Tex.	Flight Garments and Crew Provisions. Experiment S193, Radiometer Scatterometer Altimeter.
ILC Industries	Dover, Del.	APO LLO/Skylab Spacesuits.
Honeywell Radiation Center	Lexington, Mass.	Experiment S192, Multispectral Scanner.
Itek Corporation	Lexington, Mass.	Experiment S190, Multispectral Photographic Facility. Airborne Multispectral Photographic Facility.
Lockheed Electronics Company, Inc.	Plainfield, N. J.	Electron/Proton Spectrometer.
Martin Marietta Aerospace	Denver, Colo.	Skylab Payload Integration, EREP Integration, T025 Coronagraph, Contamination Measurement, M092 Lower Body Negative Pressure Experiment, M093 Vectorcardiogram, M133 Sleep Monitoring Experiment, S063 UV Airglow Horizon Photography, Crew Operations and Training, Flight Operations.
J. A. Maurer, Inc.	Long Island City, N. Y.	16mm Cameras. Film Magazines.
McDonnell Douglas Astronautics Co.	St. Louis, Mo.	Skylab Mission Operational Requirements.

Organization	Location	Items
Massachusetts Institute of Technology	Cambridge, Mass.	Guidance & Navigation Systems.
Naval Research Laboratory	Washington, D.C.	Experiment S020, Solar Photography.
North American Rockwell Corp., Space Division	Downey, Calif.	Command and Service Modules.
Northwestern University	Evanston, Ill.	Experiment S019, UV Stellar Astronomy and Articulated Mirror.
Pacific Platronics, Inc.	Santa Cruz, Calif.	Lightweight Headsets.
Paillard, Inc.	Linden, N. J.	70mm Cameras.
University of Kansas	Lawrence, Kan.	Support for Experiment S193, Radiometer Scatterometer Altimeter.
University of Texas	Austin, Tex.	Experiment S019, UV Stellar Astronomy.
Whirlpool Corp.	St. Joseph, Mich.	Skylab Food System.
Westinghouse Electric Corporation	Baltimore, Md.	Color TV Camera Systems. Color TV Remote Control.
KSC		
The Boeing Company	Kent, Wash.	S-IC Stage Support.
International Business Machines Corp.	Huntsville, Ala.	Instrument Unit Support.
North American Rockwell Corp.	Seal Beach, Calif.	S-II Stage Support.*
	Downey, Calif.	Command-Service Module Support.
McDonnell Douglas Astronautics Co.	Huntington Beach, Calif.	S-IVB Stage and Orbital Workshop Support. Airlock Module Support.

*Funded under Apollo.

Organization	Location	Items
Chrysler Corporation, Space Division	New Orleans, La.	S-IB Stage Support.
General Electric	Huntsville, Ala.	Automatic Checkout Equipment Maintenance and Checkout.*
Delco Electronics Div., General Motors Corp.	Milwaukee, Wis.	CSM Guidance and Navigation Checkout.
Martin Marietta Aerospace	Denver, Colo.	Multiple Docking Adapter Support. Experiment Facilities and Support.

*Funded under Apollo.

APPENDIX A

ACRONYMS AND ABBREVIATIONS

Å	Angstrom
AAP	Apollo Applications Program
AC	Alternating Current
A/C	Audio Center
ACN	Ascension Island, STDN Station
ACQ SS	Acquisition Sun Sensors
ACS	Attitude Control System
ACTH	Adrenocortecotropic Hormone
ADH	Antidiurectic Hormone
AES	Apollo Extension Systems
Ag	Silver
AGAVE	Automatic Gimballed Antenna Vectoring Equipment
A-H	Ampere Hour
Al	Aluminum
ALC	Audio Load Compensator
ALDS	Apollo Launch Data System
ALSA	Astronaut Life Support Assembly
AM	Airlock Module
AMS	Articulated Mirror System
AMU	Astronaut Maneuvering Unit
APCS	Attitude and Pointing Control System
APP	Antenna Position Programmer
ARC	Ames Research Center
ARIA	Apollo Range Instrumented Aircraft
AS&E	American Science and Engineering
ASMU	Automatically Stabilized Maneuvering Unit
A7LB	Pressure Suit
ATM	Apollo Telescope Mount
ATMDC	Apollo Telescope Mount Digital Computer
B&W	Black and White
BDA	Bermuda, United Kingdom, STDN Station
BMMD	Body Mass Measurement Device
BSE	Booster System Engineer
C	Celsuis
C&D	Control and Display
CBRM	Charger/Battery/Regulator Module
CCATS	Communication, Command and Terminal System
C^2F^2	Crew Compartment Fit & Function
CCS	Command Communication System
CCSD	Chrysler Corporation-Space Division
CCU	Crewman Communication Umbilical
CDDT	Countdown Demonstration Test

CDP	Central Data Processing
CDR	Commander
CEDAR	Contingency Engineering Data for Analog Recording
CG	Center of Gravity
CIF	Central Instrumentation Facility
cm	Centimeter
CM	Command Module
CMD	Command
CMG	Control Moment Gyro
CMGEA	Control Moment Gyro Electronics Assembly
CMGIA	Control Moment Gyro Inverter Assembly
CMGS	Control Moment Gyro Subsystem
CMS	Command Module Simulator
CMPS	Command Module Procedures Simulator
CO_2	Carbon Dioxide
CONIE	National Commission for Space Science (Spanish)
CRO	Carnarvon, Australia, STDN Station
CSM	Command and Service Module
CTM	Cardiotachometer
CWG	Constant Wear Garment
CWS	Center Work Station
Cu	Copper
CYI	Grand Canary Island, Spain, STDN Station
DA	Deployment Assembly
DAC	Data Acquisition Camera
DAS	Digital Address System
DCPS	Dynamic Crew Procedures Simulator
DCS	Digital Command System
DDAS	Digital Data Acquisition System
DDC	DC to DC Power Converter
DMS	Data Management Summary
DNA	Desoxyribonucleic Acid
DOD	Department of Defense
DOS	Australian Department of Supply
DSE	Data Storage Equipment
D/T	Delayed Time
EAFB	Ellington Air Force Base
EAP	Emergency Access Platform
EBW	Exploding Bridge Wire
ECF	Extracellular Fluid
ECG	Electrocardiogram
ECS	Environmental Control System
EDS	Emergency Detection System
EEG	Electroencephalographic
EHS	Electrical Harness Assembly
EMU	Extravehicluar Mobility Unit
EOG	Electro-oculographic

EPCS	Experiment Pointing Control Subsystem
EPEA	Experiment Pointing Electronics Assembly
EPS	Electrical Power System
EREP	Earth Resources Experiment Package
ERTS	Earth Resources Technology Satellite
ESE	Experiment Support Equipment
ESS	Experiment Support System
ETC	Earth Terrain Camera
ETR	Eastern Test Range
EV1	EVA 1
EV2	EVA 2
EV3	EVA 3
EVA	Extravehicular Activity
EXP	Experiment
F	Fahrenheit
FAS	Fixed Airlock Shroud
FCC	Flight Control Computer
FCMU	Foot Controlled Maneuvering Unit
FDF	Flight Data File
FEWG	Flight Evaluation Working Group
FM	Frequency Modulation
FMC	Forward Motion Compensation
FMS	Force Measuring System
FMU	Force Measuring Unit
FOMR	Flight Operations Management Room
fps	Frames per Second
FSS	Fine Sun Sensor
g	Gravity, Gram
GDS	Goldstone, Calif., STDN Station
GET	Ground Elapsed Time
GHz	Gigahertz
GMT	Greenwich Mean Time
GN_2	Gaseous Nitrogen
GOSS	Ground Operational Support System
GSE	Ground Support Equipment
GSFC	Goddard Space Flight Center
GWM	Guam, United States, STDN Station
Hα	Hydrogen Alpha
HAO	High Altitude Observatory
HAW	Kauai, Hawaii, STDN Station
HCO	Harvard College Observatory
HDC	Hasselblad Data Camera
Hg	Mercury
HHMU	Hand Held Manuevering Unit
HOSC	Huntsville Operations Support Center
HSK	Honeysuckle Creek, Australia, STDN Station
HSL	Harvard Simulation Laboratory

IBM	International Business Machines, Inc.
I&C	Instrumentation and Communications
I/C	Intercom
IMSS	Inflight Medical Support System
IMU	Inertial Measuring Unit
INTA	Spain's Instituto Nacional de Tecnica Aerospacial
I/O	Input/Output
IPHM	Individual Personal Hygiene Modules
IR	Infrared
IU	Instrument Unit
IVA	Intravehicular Activity
KeV	Thousand Electron Volts
Kg	Kilogram
Km	Kilometer
KSC	Kennedy Space Center
KV	Thousand Volts
LaRC	Langley Research Center
LBNP	Lower Body Negative Pressure
LBNPD	Lower Body Negative Pressure Device
LC	Launch Complex
LCC	Launch Control Center
LCG	Liquid Cooled Garment
LCU	Landing Craft Utility
LDX	Long Distance Xerography
LES	Launch Escape System
LET	Launch Escape Tower
LH_2	Liquid Hydrogen
LiOH	Lithium Hydroxide
LIMS	Limb Motion Sensing System
LM	Lunar Module
LOX	Liquid Oxygen
LSD	Landing Ship Dock
LSS	Life Support System
LSU	Life Support Umbilical
LVDC	Launch Vehicle Digital Computer
LVMS	Leg Volume Measuring System
m	Meter
mA	Milliampere
MAD	Madrid, Spain, STDN Station
MAF	Michoud Assembly Facility
MAK	Medical Accessories Kit
MAR	Mission Action Request
MCC	Mission Control Center
MDA	Multiple Docking Adapter
MDAC-E	McDonnell Douglas Astronautics Company-East, St. Louis, Mo.
MDAC-W	McDonnell Douglas Astronautics Company-West, Huntington Beach, Calif.

MD/DOY	Mission Day/Day of Year
MeV	Million Electron Volts
MEWG	Mission Evaluation Working Group
mg	Milligram
MHz	Megahertz
MIL	Merritt Island, Florida, STDN Station
ML	Mobile Launchers
MLA	Merritt Island, Florida. C Band Radar
MLU	Memory Load Unit
mm	Millimeter
MMC	Martin Marietta Corporation
MOC	Mission Operations Computer
MOCR	Mission Operations Control Room
mol	Molecular
MOPS	Mission Operation Planning System
MPC	Manual Pointing Controller
MSC	Manned Spacecraft Center (Now Lyndon B. Johnson Space Center)
MSFC	Marshall Space Flight Center
MSG	Mission Support Group
MSLR	Marshall Skylab Representative
MSOB	Manned Spacecraft Operations Building
MSS	Mobile Service Structure
MTF	Mississippi Test Facility
MULT	Multiplex
MUST	Medical Unit Self-Contained Transportables
N_2	Nitrogen
NAR/SD	North American Rockwell Corporation-Space Division, Downey, Calif. (Now Rockwell International Corp.)
NASCOM	NASA Communications Network
NASA	National Aeronautics and Space Administration
NBS	Neutral Buoyancy Simulator
N/cm^2	Newtons per Square Centimeter
NFL	St. Johns, Newfoundland, STDN Station
Ni	Nickel
NM	Nautical Mile
NOAA	National Oceanographic & Atsmopheric Administration
NOCC	Network Operations Control Center
NPV	Non-Propulsive Vent
NR	North American Rockwell Corporation (Now Rockwell International Corp.)
NRL	Naval Research Laboratory
O_2	Oxygen
OA	Orbital Assembly
OAM	Orbital Assembly Module
OAO	Orbiting Astronomical Observatory
OBS	Operational Bioinstrumentation System
O&C	Operations and Checkout
ODAE	Off Duty Activities Equipment
OM	Optical-Mechanical

OMSF	Office of Manned Space Flight
OWS	Orbital Workshop
PAD	Program Applications Document Update
PAT	Patrick AFB, Florida
PBI	Polybenzimidazoic
PCG	Power Conditioning Group
PCSA	Power and Control Switching Assembly
PCM	Pulse Code Modulation
PCU	Pressure Control Unit
PGA	Pressure Garment Assembly
PI	Principal Investigator
PLT	Pilot
PM	Phase Modulation
PMC	Post Manufacturing Checkout
ppm	Parts per Million
PPCO$_2$	Partial Pressure Carbon Dioxide
PPO$_2$	Partial Pressure Oxygen
PRS	Prime Recovery Ship
PS	Payload Shroud
psi	Pounds per Square Inch
psia	Pounds per Square Inch Absolute
PSK	Phase Shift Keyed
PTL	Photographic Technology Laboratory
RASM	Remote Analog Submultiplexer
R&R	Rest and Relaxation
RBC	Red Blood Cell
RCS	Reaction Control System
RDM	Remote Digital Multiplex
RF	Radio Frequency
RG	Rate Gyro
RGP	Rate Gyro Package
RLC	Rotating Litter Chair
RNA	Ribonucleic Acid
RNBM	Radio Noise Burst Monitor
ROCR	Recovery Operations Control Room
RPM	Roll Positioning Mechanism
RS	Refrigeration Subsystem
R/T	Real Time
RTTA	Range Tone Transfer Assembly
SAA	South Atlantic Anomaly
SAG	Solar Array Group
SAL	Scientific Airlock
SAS	Solar Array System
SCAMA	Switching, Conferencing
SEVA	Skylab Extravehicular Visor Assembly
SI	Solar Inertial
SIA	Speaker Intercom Assembly

SID	Subject Identification
S-IC	Saturn V, First Stage
S-IB	Saturn IB, First Stage
S-II	Saturn V, Second Stage
S-IVB	Saturn IB, Second Stage
SL	Skylab
SLA	Spacecraft LM Adapter
SLVR	Saturn Launch Vehicle Representative
SML	Skylab Mobile Laboratory
SMMD	Specimen Mass Measurement Device
Sn	Tin
SOP	Secondary Oxygen Pack
SPS	Service Propulsion System
SPT	Scientist Pilot
SRS	Secondary Recovery Ship
SS	Sun Sensor
SSR	Staff Support Room
ST	Star Tracker
STDN	Spaceflight Tracking & Data Network
STE	Star Tracker Electronics
STS	Structural Transition Section
SWS	Saturn Workshop
TACS	Thruster Attitude Control Subsystem
TBC	The Boeing Company
TCS	Thermal Control System
TELTRAC	Telemetry Tracker
TEX	Corpus Christi, Texas, STDN Station
TM	Telemetry
TST	Technical Support Team
TV	Television
TVIS	Television Input Station
UCTA	Urine Collection & Transfer Assembly
UHF	Ultra High Frequency
USB	Unified Side Band
UV	Ultraviolet
Vac	Volts Alternating Current
VAB	Vehicle Assembly Building
VAN	USNS Vanguard (T-AGM 19) Instrumentation Ship
VC	Center Work Station (EVA)
VCG	Vectorcardiographic
VCL	Vertical Checkout Laboratory
Vdc	Volts Direct Current
VF	FAS Work Station (EVA)
VHF	Very High Frequency
VS	Sun-End ATM Work Station (EVA)
VSS	Video Selector Switch

VTR	Video Tape Recorder
V/TS	Viewfinder Tracking System
VX	Voice
W	Watt
WCIU	Workshop Computer Interface Unit
WM	Waste Management
W/m^2	Watts per Square Meter
WMC	Waste Management Compartment
WR	Ward Room
WT	Water Tank
X-REA	X-Ray Event Analyzer
X-RT	X-Ray Telescope
Z-LV	Z Local Vertical
ZPN	Impedance Pneumogram

APPENDIX B

GLOSSARY

Agar suspension — A nutritional medium for growing microbial cultures
Airlock — An airtight compartment with adjustable air pressure and located between places of unequal pressure
Analog signal — A signal whose voltage variance is proportional to the variance of the parameter being measured
Angstrom Unit — A unit of length equal to one ten millionth of a millimeter and used to express electromagnetic wavelengths
Apogee — The point of an orbit that is farthest from the Earth
Arc-min — One sixtieth of a degree of arc
Arc-sec — One sixtieth of an arc-min
Astrophysics — Physics pertaining to the outer atmosphere of the Earth and of interplanetary space
Atmosphere — The envelope of gases which surrounds a heavenly body. The environment within a spacecraft
Attitude — The position of a spacecraft as determined by the inclination of its axes to some frame of reference
Axis — Any of three straight lines: the first running through the center of the vehicle lengthwise; the second at right angles to this; and the third perpendicular to the first two at their point of intersection
Azimuth — The angle between true north and the direction of travel
Beacon — A device that transmits a signal used in tracking
Biocide — A chemical substance that can kill living organisms; disinfectant
Biosensors — Biomedical sensors attached to the crew to sense such conditions as heart rate and respiration
Booster — A propulsive stage of a launch vehicle
Bungee — An elastic restraint
Bus — A main circuit for transfer of electrical power
C-Band — A radio frequency band of 3.9 to 6.2 gigahertz used in tracking radar
Calfax fastener — A hand operated fastener that is locked with a twisting action
Canard — A short stubby wing-like element affixed to the spacecraft or launch vehicle to provide better stability
Cardiotachometer — A device for measuring heart rate
Centrifuge — A device using centrifugal force to separate particles of varying densities
Chromosphere — The reddish layer of incandescent gases around the Sun
Closed loop — A control system that uses feedback or self correction signals
Constant wear garment — Thermal knit underwear
Consumables — The food, water, gas, etc., that are used during the mission
Contamination — Molecular and particulate matter which interferes with transmission of light, e.g., film on a spacecraft window caused by spacecraft venting
Convection — Transmission of heat by the movement of heated particles
Coolant — A fluid in a thermal control system for transferring heat
Corona, solar — The outer atmospheric shell of the Sun
Cosmic ray flux — Stream of ionizing radiation of extraterrestrial origin, chiefly of protons, alpha particles and other atomic nuclei, but including some high energy electrons and photons

Cytoplasmic motion — The motion of the protoplasm, exclusive of the nucleus, of a cell

Deployment — The unfolding and releasing of the various Skylab mechanisms in preparation for manned visits, e.g., Solar array deployment

Desaturation (CMG) — Realignment of the CMG spin axes to avoid coalignment

Dew point — The temperature at which the atmosphere can hold no more water vapor

Dielectric — The insulation between the layers of a solar cell or the plates of a capacitor and having specific electrical properties

Discone antennas — The antennas on the extended booms of Skylab that permit improved spacecraft to ground transmission quality

Discrete signal — A signal that is either "ON" or "OFF" signifying an event

Dosimeter — A small device for measuring amounts of radiation absorbed

Downlink data — Data from the spacecraft to the ground

Downrange — Ahead of the spacecraft in the flight path

Drogue — The hollow part of the docking mechanism that captures the docking probe when proper alignment is attained

Electrocardiogram — A tracing showing the changes in electric potential produced by contractions of the heart

Electrolyte — A substance in which the conduction of electricity is accompanied by chemical reaction

Electromagnetic spectrum — Entire range of radiation. In order of decreasing frequency: cosmic ray photons, gamma rays, X-rays, ultraviolet, visible light, infrared, microwave, radio waves, heat and electric currents

Electrostatic — Having a static electric charge

Ephemeris — A publication giving the computed places of the celestial bodies for each day of the year, or for other regular intervals

Ergometer — A device for measuring energy expended or work done

Eutectic alloy — An alloy of two or more metals proportioned such that the lowest possible melting point is achieved

Expendables — Items used by the crew that are disposed of after use

Exothermic — Giving off heat

Exploding bridge wire — Wire which heats to a high temperature and burns, thus igniting a charge

Explosive bolts — Bolts surrounded with an explosive charge which can be detonated by an electrical impulse

Extravehicular — Outside the spacecraft

Footprint — Area of possible landing sites or ground photographic targets

Fuel cell — An electrochemical generator in which the chemical energy from the reaction of oxygen and a fuel is converted directly to electricity.

Gegenschein — The faint visible background light of the sky outside the Earth's atmosphere

Gimbal — Mechanical frame connecting two mutually perpendicular intersecting axes of rotation

Ground tracks — The trace on the ground of the orbiting spacecraft

Ground truth data — Data taken on the ground that is used for comparison with data taken from the spacecraft over the specific ground sites

Gyro — A spinning reference wheel

Heat exchanger — A device for transferring heat from one substance to another

Heat flux — Flow of heat

Heat sink — A device for absorption or transfer of heat away from a component or system

Hematology — The study of the blood and its diseases

Hertz — The unit for expressing frequency

Impedance pneumogram — A trace of respiration rate

Inclination, orbital — An angle between two lines or planes and the plane of the Earth's Equator, e.g., Skylab orbital inclination is 50 degrees

Infrared — Electromagnetic radiation of wavelengths from the red end of the visible color spectrum to the microwaves used in radar

Klaxon — A warning horn

Load compensator — A device for matching the CSM audio system to the SWS audio system

Magnetosphere — Asymmetric region surrounding the Earth extending from about 400 to several thousand miles above the Earth's surface in which charged particles are trapped and their behavior dominated by the Earth's magnetic field

Mate — The joining of two or more major portions or modules of a launch vehicle or spacecraft

Math model — A method of expressing the performance of a system mathematically for use in performance simulations and analysis. This simulation is performed with the aid of a computer

Max Q — Maximum dynamic pressure during launch

Meteoroid — A solid particle of matter traveling in space at a considerable speed

Molecular sieve — A bed of adsorbing material used to remove one or more constituents in a gas stream

Momentum — Mass times velocity; as used in this reference, it strictly refers to the angular momentum of the spacecraft or of the spinning CMG wheel

Momentum dump — An orbital maneuver or TACS firing to desaturate the CMGs

Multiplex — Simultaneous transmission of two or more messages or units of information on a single channel

Multispectral — Covering several portions of the electromagnetic spectrum

Neutral buoyancy — Neither rising nor sinking when submerged in water; in effect, weightless

Non-propulsive vents — Outlets for venting spacecraft gases that, in themselves, cause no resultant thrust.

Occult — To block, e.g., to block the solar disk for coronal observations

Open loop — A control system in which there is no feedback or self corrective action

Opthalmascope — An instrument for examining the interior of the eye

Orbit — Spacecraft path around the Earth beginning and ending at a fixed point in space and requiring 360 degrees of travel

Orbital insertion — Entering the desired orbit

Ordnance — Explosives for separating joined surfaces

Ozone layer — Layer in the Earth's atmosphere about 20 miles above sea level which strongly absorbs ultraviolet radiation

Partial pressure — A portion of the total pressure that one of the constituent gases exert

Payload — The hardware to be placed into orbit

Photometer — An instrument for measuring the intensity of light

Plasma physics — The science dealing with fully ionized gases

Pneumatics — Branch of physics that deals with the properties of gases, such as pressure or density

Posigrade — Orbital motion in the same direction that the spacecraft is moving

Probe — A device used in docking as a guide and latching mechanism. It engages with a hollow drogue to facilitate mating of two spacecraft

Precipitin reaction — A chemical reaction that yields a precipitate as a product

Pyrotechnic device — A device using ordnance

Quick disconnect — A quick mating and release connector for hoses

Radiant heater — A heater, generally an electrical heater, that transfers heat by radiation rather than conduction or convection

Rate gyro — A gyroscope whose precession rate is provided to a computer to determine spacecraft maneuver rate information

Real time — Transmitting data simultaneous with the occurrence

Rehydratable food — Dehydrated food which is prepared for use by adding water

Rendezvous — Two objects brought together, as CSM to Skylab

Retrograde — Orbital motion opposite to the direction in which the spacecraft is moving. Retrograde firing decreases spacecraft velocity

S Band — A radio frequency band of 1550 to 5200 megahertz

Simulator — A mockup or device that simulates the operational spacecraft hardware. Either used for training or for ground simulations of spacecraft performance

Single point ground — A common point where all electrical returns are tied to prevent ground loop currents

Solar array — Solar cells wired and mounted into an array for providing electrical power to the spacecraft

Solar cell — A small "sandwich" unit used for converting solar energy to electrical energy

Solar disk — The view of the Sun as seen from the spacecraft

Solar emission — Electromagnetic emission from the Sun, e.g., X-rays, ultraviolet, etc.

Solar flare — A solar storm which is accompanied by a high emission of electromagnetic energy

Solar limb — The outer edge of the Sun

Spectrograph — A device for displaying radiant energy with the waves arranged in the order of their frequency

Spectroheliogram — A picture at specific wavelengths showing emission features of the Sun

Stadimeter — An optical device for determining spacecraft altitude

Strapdown computation — Mathematical calculation of an inertial attitude reference based upon sensed vehicle rates

Sublime — To change from a solid to a gas, bypassing the liquid state

Tape dump — Transmission of data which has been stored; the data is dumped when the spacecraft is over a ground station

Telemetry — A system for taking measurements in the spacecraft and transmitting them to ground stations

Tether device — A device for restraining objects

Thermal coatings — A passive thermal control system applied to a component to either absorb or radiate heat

Thermal curtain — Protective shield to block thermal radiation onto certain components of the spacecraft

Thermonuclear energy — Energy derived from a thermonuclear reaction, as in the energy of the Sun

Thermostabilized food — Food which has been thermally conditioned to retard spoilage, such as refrigerated

Transducer — A device for converting physical measurements to electrical signals

Transponder — A radio or radar transceiver that automatically and promptly transmits a reply upon reception of an interrogation signal

Two gas atmosphere — In Skylab the atmosphere is composed of oxygen and nitrogen

Ultra High Frequency (UHF) — The radio frequency band of 300 to 3000 Megahertz

Ultraviolet — Beyond violet in the spectrum; light having wavelengths shorter than 4000 Angstrom units

Unified S band — See S Band. The primary communications frequencies for CSM to the ground
Uplink data — Ground commands and data to the spacecraft
Van Allen Radiation Belt — Two doughnut shaped belts of high energy particles trapped in the Earth's magnetic field
Velcro — A hook and pile fastener commonly used in garments
Velometer — A device for measuring air flow velocities
Very High Frequency (VHF) — The radio frequency band of 30 to 300 Megahertz

INDEX

	Page
Abort	VI-15
Areas	VI-15
Mode I	VI-15
Mode II	VI-16
Mode III	VI-16
Acquisition Sun Sensor	II-59
Activation	IV-4
Administrative Support Center	VI-11
Aft Skirt	I-43
Airlock	I-22
Module	I-22
Scientific	I-9
Trash disposal	I-2, II-18
Airlock Module	I-22
Airlock Tunnel Assembly	I-24
Fixed Airlock Shroud	I-22, I-26
Stowage	II-93, II-94
Structural Transition Section	I-23
Truss Assemblies	I-26
Air transportation	X-9
Aperture doors	I-28
Apollo Applications Program	iii-3
Apollo Telescope Mount	I-27
Control Moment Gyro	I-29
Deployment Assembly	I-32
Experiment canister	I-28
Rack	I-27
Solar arrays	I-32
Apparel, wearing	II-80
ATM Digital Computer/WCIU	I-29, II-60
Attitude and Pointing Control	II-54
Attitude hold	II-55
CMGS	II-55, II-57
Experiment pointing	II-55
Solar inertial mode	II-55
Standby mode	II-55
Z local vertical mode	II-55
TACS	II-55, II-61
Audio	II-62
Auto Scan	VI-14
A7LB Suit	II-83
Backup hardware	I-46
Bandage Kit	II-30
Bendix Corporation, Teterboro, N.J.	X-3
Biographies, crew	IV-8
Bioinstrumentation, operational	II-31
Boeing Company, New Orleans, La.	X-5
Breadboard, power systems	VI-12

INDEX (Continued)

	Page
Briefing and reviews, training	VII-1
Cameras	II-78
Data acquisition, 16 mm	II-78
Earth Terrain	III-33, II-79
Hasselblad	II-79
Multispectral	III-33
Nikon, 35mm	II-78
Twilight airglow	III-42
Canister	I-28
Catherization kit	II-30
Caution and Warning system	II-68
Caution and warning	II-68
Emergency	II-74
Chrysler Corporation Space Division, New Orleans, La.	X-5
Clothing module	II-80
Cold plate	II-38
Command and Service Module	I-34
Command Module	I-35
Commands, ground	II-61
Communications	II-61
Air to ground	II-61, IX-1
Commands	II-61
Caution and warning	II-68
Internal	II-62
Ranging	II-67
Teleprinter	II-67
Television	II-64
Compartment, OWS	I-10
Crew	I-10
Experiment	I-13
Forward	I-7
Sleep	I-12
Waste management	I-11
Configuration, launch	I-5
Contracts	XII-1
Control Moment Gyro	II-55, II-57
Corona	III-22
Countdown, launch	V-10
Crawlerway	V-2
Crew compartment	I-10
Crew equipment	II-80
Crew operations	VIII-1
Crew schedule	VIII-1
Crew training	VII-1
Deactivation	IV-5
Dental kit	II-30
Deorbit	IV-5
Deployment Assembly	I-32

INDEX (Continued)

	Page
Deployment reel	I-33
Docking	IV-4, I-18
Docking ring	I-18
Docking target	I-16
Dosimeter	I-14
Drogue	I-18
Earth observations experiments	III-29
Earth Resources Experiments Package	III-29
Earth terrain camera	III-33
Electrical power system	II-40
Engineering and technology experiments	III-52
Environmental control	II-1
Ergometer	III-19, III-14
Experiment compartment	I-13
Experiment canister, ATM	I-28
Experiment Package Caging and Gimbal Assembly	II-60
Experiment pointing	I-31, II-56
Control system	II-56
Electronics assembly	II-56, II-60
Mode	II-55
Experiments	III-1
Astrophysics	III-39
Earth observation	III-29
Engineering and Technology	III-52
Materials Science and Manufacturing	III-47
Solar physics	III-21
Student	III-63
Extravehicular activities	VIII-3
Extravehicular Mobility Unit	II-83
Fence concept	VI-17
Film	II-77, III-32
Boom	VIII-5
Clothesline	VIII-5
Types	II-77, III-32
Film vault	I-7, II-94
Fine Sun Sensor	II-58
Fire sensor control panel	II-74
Fire sensors	II-74
Firing room	V-2
Flight data file	II-94, I-10
Flight Operations	VI-1
Flight Operations Management Room (FOMR)	VI-9
Fixed Airlock Shroud	I-22, I-26
Food management	II-9
Food management table	I-11, II-8
Food trays	II-8
Foot Controlled Maneuvering Unit	III-59
Forward Compartment	I-7

INDEX (Continued)

	Page
Ground track	VI-17
H-Alpha telescopes	III-27
Handwasher	I-12, II-20
Hardware	I-1
Hematology kit	II-27
History, Skylab	iii-3
Huntsville Operations Support Center (HOSC)	VI-10
Inflight Medical Support Systems	II-22
Infrared spectrometer	III-34
Intercom boxes	II-64
International Business Machines, Huntsville, Al.	X-5
Kennedy Space Center, Florida	X-7
L-Band radiometer	III-37
Launch configurations	I-5
Launch Escape System	I-37
Launch facilities	V-1
Complex 39	V-1
Crawlerway	V-2
Launch Control Center	V-1
Mobile launcher	V-2
Mobile Service Structure	V-3
Saturn IV launch pedestal	V-4
Transporter	V-2
Vehicle Assembly Building	V-1
Launch operations	V-7
Countdown	V-10
Rollout	V-8
Vehicle Assembly	V-7
Launch vehicles	I-38, I-41
Saturn IB	I-40, I-41
Saturn V	I-37, I-38
Launch windows	IV-3
Leak detector	II-96
Life Sciences experiments	III-1
Life Support Systems	II-1
Environmental control	II-1
Food management	II-9
Inflight Medical Support System	II-22
Microbial control	II-26
Operational bioinstrumentation	II-31
Personal hygiene	II-20
Portable CO_2/Dew point sensor	II-36
Sleep provisions	II-19
Waste management	II-12
Water management	II-5
Lighting	II-51
Liquid Cooled Garments	II-83
Magnetic soap holders	I-12, II-20

INDEX (Continued)

	Page
Maintenance, scheduled	VIII-7
Management	XI-1
Manned Spacecraft Center	X-6
(Now Lyndon B. Johnson Space Center) Houston, Tex.	
Manual pointing controller	II-60
Manufacturing and testing facilities	X-1
Contractor facilities	X-1
NASA facilities	X-6
Marshall Space Flight Center, Huntsville, Al.	X-6
Martin Marietta Aerospace, Denver, Colo.	X-3
McDonnell Douglas Astronautics Co.	
Eastern Division, St. Louis, Mo.	X-1
Western Division, Huntington Beach, Calif.	X-1
MDA stowage	II-89
Medical Accessories Kit	II-31
Medical kit	II-22
Bandage	II-30, II-34
Catheterization	II-30
Dental	II-30, II-35
Hematology	II-27
Microbiology	II-27
Therapeutic	II-27
Memory Load Unit	II-61
Metabolic analyzer	III-14
Meteoroid shield	I-2, I-16
Michoud Assembly Facility, New Orleans, La.	X-7
Microbial control	II-20
Mission Control Center	VI-1
Mission objectives	IV-2
Mission Operations Planning Systems (MOPS)	VI-13
Mission profile, Skylab	iii-1
Mission sequence	iii-5
Mission Support Groups	VI-9
Mississippi Test Facility, Bay St. Louis, Miss.	X-7
Mobile Launcher	V-2
Mobile Service Structure	V-3
Molecular sieve	II-2
Multiple Docking Adapter	I-15
Multispectral photographic facility	III-33
Multispectral scanner	III-34
Neutral Buoyancy Facility	VI-13
North American Rockwell Corp.	X-4
(Now Rockwell International Corp.) Downey, Calif.	
Ophthalmoscope	II-24
Orbital Workshop	I-1
Crew compartment	I-10
Experiment compartment	I-13
Forward compartment	I-7

INDEX (Continued)

	Page
Sleep compartment	I-12
Solar Arrays	I-2, II-45, II-46
Stowage	II-89
Wardroom	I-10
Waste Management Compartment	I-11
Waste tank	I-15
Passive thermal control	II-39
Payload Shroud	I-32
Personal hygiene	II-20
Personal hygiene kit	II-23, II-85
Photography	II-74
POGO	I-42
Portable CO_2/Dew Point Sensor	II-36
Power, electrical	II-40
Circuit protection	II-51
Conditioning	II-44
Control	II-50
Generation	II-44
Monitoring	II-51
Pressure equalization valve	I-18
Probe	I-18
Provisions and Stowage	II-89
AM stowage	II-89
Food stowage	II-93
Film vault	II-94
Flight data file	II-94
MDA stowage	II-89
OWS stowage	II-89
Tools	II-95
Urine freezer	II-93
Pyrotechnic devices	I-32
Rack, ATM	I-27
Radio noise burst monitor	I-19
Ranging	II-67
Rate gyro processor	II-60
Reactivation	IV-6
Recovery	VI-15
Contingency landing areas	VI-18
Lighting	VI-18
Primary	VI-17
Secondary	VI-17
Rendezvous	IV-3, IV-6
Repeater panel, Caution and Warning	II-74
Rescue missions	IV-6
Restraints	I-43, VIII-5
Crew	I-44
Equipment	I-45
Foot	I-44, I-45

INDEX (Continued)

	Page
Thigh	I-44, I-45
Sleep	I-44, I-45
Restraints and Mobility Aids	I-43
Rollout	VI-15
Rotating litter chair	I-13, III-12
Sample return containers	II-15
Saturn IB	I-40
Saturn V	I-37
Schedule, crew	VIII-1
Service Module	I-34, I-37
Simulator facilities	VI-12
Skylab History	iii-3
Skylab Mission profile	iii-1
Skylab Mobile Laboratories (SML)	VI-15, VI-19
Skylab modules	
Airlock Module	I-22
Apollo Telescope Mount	I-27
Command Module	I-34, I-35
Instrument Unit	I-1, I-39, I-43
Multiple Docking Adapter	I-15
Orbital Workshop	I-1
Service Module	I-34, I-37
Sleep compartment	I-12
Sleep provisions	I-2, I-12
Snoopy emblem	I-12
Solar	
Arrays	I-1, I-2, I-27, I-32
Corona	III-22
Flares	III-22
Limb	III-23
Physics	III-21
Solar inertial	II-55
Solar Physics experiments	III-21
Spacecraft LM Adapter	I-37
Spaceflight Tracking and Data Network (STDN)	IX-1
Specimen Mass Measurement Device	III-6
Star Tracker	II-58
Stowage	II-89
MDA	II-89
AM	II-89
OWS	II-89
Food	II-9, II-93
Film	II-94
Structural Transition Section	I-23
Student experiments	III-63
Sun Sensor	I-27, I-28, I-29, II-58
Survival equipment	II-87

INDEX (Concluded)

	Page
System	
Attitude and Pointing Control	II-54
Communication	II-61
Crew Equipment	II-80
Electrical Power	II-40
Life Support	II-1
Provisions and Stowage	II-89
Thermal Control	II-37
Teleprinter	II-67
Television	II-64
Therapeutic kit	II-27
Thermal Control System	II-37
Thruster Attitude Control System	I-2, II-54, II-61
Thrusters	I-2, I-7, II-54, II-61
Tool caddy	
Tools	I-14, I-22, II-95
Tracking lights	II-51
Training crew	VII-1
Transporters	Fig. X-5, P-X-7
Transportation	X-7
Air	X-9
Water	X-7
Truss assemblies	I-22
TRW System, Redondo Beach, Calif.	XI-2, X-5
TV Input Station (TVIS)	II-66
Umbilical	II-96
Crew communication	II-64
Launch vehicle	V-2
Life Support	VIII-3
Urine freezer	I-12, II-93
Vehicle Assembly Building	V-1
Velcro	I-44, I-45
Wardroom	I-2, I-10
Washcloth squeezer	I-12
Waste management	I-11, I-12
Waste Management Compartment	I-2, I-11
Water dump	I-15
Water management	II-5
Wearing apparel	II-80
Window heaters	I-2
Z local vertical	II-55

NATIONAL AERONAUTICS AND SPACE ADMINISTRATION
Washington, D. C. 20546
202-755-8370

FOR RELEASE:
May 1, 1973

PROJECT: SKYLAB 1 and 2

contents

GENERAL RELEASE	1-5
OBJECTIVES OF THE FIRST SKYLAB MISSION	6-7
SKYLAB EXPERIMENTS	8-10
MISSION PROFILE: LAUNCHES, DOCKING AND DEORBIT	11-13
MISSION PROFILE IV-2: Real-time Flight Planning	14
MISSION PROFILE IV-3: Crew Work Day	15
MISSION PROFILE IV-4: The Workshop Between Visits	16
COMMUNICATIONS AND DATA	17-27
COUNTDOWN AND LIFTOFF	28-36
SATURN WORKSHOP	37
SATURN LAUNCH VEHICLES	38

NOTE: Details of Skylab spacecraft elements, systems, crew equipment and experimental hardware are contained in the Skylab News Reference distributed to the news media. The document also defines the scientific and technical objectives of Skylab activities. This press kit confines its scope to the first manned visit to Skylab and briefly describes features of the mission.

NATIONAL AERONAUTICS AND SPACE ADMINISTRATION
Washington, D. C. 20546
AC 202/755-8370

William Pomeroy
(Phone 202/755-3114)

FOR RELEASE:
Monday, May 1,

RELEASE NO: 73-80

SKYLAB AIMS AT BEING USEFUL

An inward look at man's home planet as well as an outward look toward his life-giving Sun will begin May 14 when the Skylab spacecraft is launched into an Earth-hugging orbit. Man himself and an examination of how well he fares during long periods in space will also be a major objective of this first United States experimental space station.

A three-man astronaut crew in a modified Apollo command/service module (CSM) will be launched into a rendezvous orbit the following day, and will spend almost a month aboard Skylab conducting solar astronomy, Earth resources, medical and other scientific and technical investigations. Two additional three-man crews will spend two months each living and working aboard Skylab later in the year.

-more-

Ranging as far north as the U.S.-Canadian border and as far south as the tip of Argentina, Skylab's instruments will scan a major portion of the inhabited regions of the Earth. The large, 90,600-kilogram (100-ton) space station is expected to be visible at times to people on the ground as it glides overhead.

Skylab is aimed at gaining in space new knowledge for the improvement of life on Earth. Its investigations and experiments will help develop new methods of learning about the Earth's environment and resources and new ways to evaluat programs directed at preserving or enhancing those resources throughout the world.

The astronauts aboard will perform medical experiments aimed at a better knowledge of man's own physiology and they will seek new knowledge about our star, the Sun, and its interaction with our earthly environment.

Among other experiments will be ones directed at developing new industrial processes utilizing the unique advantages of weightlessness.

-more-

The Skylab Program is predominantly utilitarian in nature, capitalizing on the vehicles and know-how developed in the Apollo Program to service and advance a wide range of interests while greatly increasing the opportunities for men to function in space. Skylab will be operational the better part of a year, permitting the economy of extended usage and laying the groundwork for future long-duration missions.

America's young people have a stake in Skylab also, for carried aboard the space station are experiments selected from a nationwide competition among secondary school students.

Crewmen for the first manned visit to Skylab are Charles Conrad, Jr., commander; Dr. Joseph P. Kerwin, science pilot; and Paul J. Weitz, pilot. In the all-Navy crew, Conrad holds the naval rank of captain and Kerwin and Weitz are commanders. Conrad flew as pilot on Gemini 5, command pilot on Gemini 11 and was commander of the second manned lunar landing, Apollo 12. Kerwin and Weitz have not been in space before.

Skylab will be boosted from the NASA Kennedy Space Center, Fla., into a 433.4-kilometer (268.7 mile) circular Earth orbit by a Saturn V launch vehicle. Approximately 24 hours after Skylab reaches orbit, the CSM with the crew aboard will be launched atop a Saturn-1B into a 150x222.2-kilometer (93x137.8 mile) elliptical orbit from which they will follow a rendezvous maneuver sequence using the CSM service propulsion system.

-more-

The launch of the Saturn V with its unmanned spacecraft is referred to as Skylab 1, SL-1, or sometimes, the workshop launch. The launching of the first astronaut crew on a Saturn 1B is referred to as Skylab 2, SL-2 or the first manned visit. The first Skylab mission begins with the Skylab 1 launch and ends with recovery of the crew.

After docking with Skylab, the crew will remain aboard the command module until the following morning, when they will enter and activate the space station for the 28-day mission. Crew activity days will be on Houston local time, starting at 6 a.m. CDT and ending at 10 p.m. CDT. Mission Control is at the Johnson Space Center, Houston, Texas, home of the astronauts.

While real-time flight planning was usually forced by contingencies in Gemini and Apollo, real-time flight planning will be the standard method for accomplishing the most in Skylab. The operational flight plan will serve mainly as a basic guide into which the performance of experiments will be scheduled on a daily basis by flight planners in the Mission Control Center. For example, Earth resources Experiment package (EREP) experiments depend upon clear skies over ground sites to be scanned and photographed. Scheduling of EREP passes will depend upon weather forecasts and actual conduct of the experiments may be deferred if the cloud cover turns out to be excessive.

-more-

Near the end of the 28-day first manned mission, two crew members will don pressure suits and go outside where one will retrieve solar telescope film canisters for return to Earth. These film canisters, data recordings from other Skylab experiments and other forms of information gathered during the month in space will be stowed aboard the command module for the return home.

Ground controllers will keep an electronic eye upon Skylab, its experiments and systems, after the crew completes their stay.

After undocking, the crew will perform two deorbit burns to bring the command module to splashdown in the eastern Pacific about 1,280 km (800 miles) southwest of San Diego, Calif. Extensive medical examinations in Skylab mobile laboratories aboard the prime recovery vessel, the USS Ticonderoga, will be conducted before the crew is flown back to Houston.

The second manned visit to Skylab is planned for early August.

END OF GENERAL RELEASE

II - OBJECTIVES OF THE FIRST SKYLAB MISSION

The Skylab Program was established for four explicit purposes: to determine man's ability to live and work in space for extended periods; to extend the science of solar astronomy beyond the limits of Earth-based observations; to develop improved techniques for surveying Earth resources from space; and to increase man's knowledge in a variety of other scientific and technological regimes.

Skylab, the first space system launched by the United States specifically as a manned orbital research facility, will provide a laboratory with features which cannot be found anywhere on Earth. These include: a constant zero gravity environment, Sun and space observation from above the Earth's atmosphere, and a broad view of the Earth's surface.

Dedicated to the use of space for the increase of knowledge and for the practical human benefits that space operations can bring, Skylab will pursue the following:

Physical Science - Increase man's knowledge of the Sun, its influence on Earth and man's existence, and its role in the universe. Evaluate from outside Earth's atmospheric filter, the radiation and particle environment of near-Earth space and the radiations emnating from the Milky Way and remote regions of the universe.

Life Science - Increase man's knowledge of the physiologica and biological functions of living organisms - human, other animal, and tissues - by making observations under conditions not obtainable on Earth.

Earth Applications - Develop techniques for observing Earth phenomena from space in the areas of agriculture, forestry, geology, geography, air and water pollution, land use and meteorology.

Space Applications - Augment the technology base for future space activities in the areas of crew/vehicle interactions structures and materials, equipment and induced environments.

The First Skylab Mission has three specific objectives as follows:

1. Establish the Skylab orbital assembly in Earth orbit.

    (a) Operate the spacecraft cluster (including CSM) as a habitable space structure for up to 28 days after the SL-2 launch.

- (b) Obtain data for evaluating the total spacecraft performance.
- (c) Obtain data for evaluating crew mobility and work capability in both intravehicular and extravehicular activity.

2. <u>Obtain medical data on the crew for use in extending the duration of manned space flights.</u>

- (a) Obtain medical data for determining the effects on the crew which result from a space flight of up to 28 days duration.
- (b) Obtain medical data for determining if a subsequent Skylab mission of up to 56 days duration is feasible and advisable.

3. <u>Perform in-flight experiments.</u>

- (a) Obtain ATM solar astronomy data for continuing and extending solar studies beyond the limits of Earth from low Earth orbit.
- (b) Obtain Earth resources data for continuing and extending multisensor observation of the Earth from low Earth orbit.
- (c) Perform the assigned scientific, engineering and technology experiments.

The Gemini 7 mission demonstrated that man can readily adapt to space flight for up to two weeks without ill effects. Six Apollo lunar landings proved that man can go into space a quarter million miles away from his mother planet, adapt to a lower gravity field and do useful work in the hostile environment of a hard vacuum.

Skylab will push forward the threshold of human adaptability to spaceflight first by doubling Gemini 7's time in space with the first Skylab crew, then doubling that experience in the next two manned visits.

In the total of 140 manned days of operation, the nine Skylab astronauts will amass medical, scientific and engineering data that will influence the design and operation of future generations of space vehicle systems.

III - SKYLAB EXPERIMENTS

The Skylab space station carries the largest array of experimental scientific and technical instruments ever flown in space. They total 58 and fall into four broad categories: medical, Earth Resources Experiments Package (EREP), Apollo Telescope Mount (ATM), and corollary. This experimental equipment will permit more than 200 ground-based principal investigators to supervise 271 scientific and technical investigations.

Skylab medical experiments are aimed toward measuring man's ability to live and work in space for extended periods of time, his responses and aptitudes in zero gravity, and his ability to readapt to Earth gravity once he returns to a one-g field.

EREP experiments will use six devices to advance the technology of Earth remote sensing and at the same time gather data that may be applied to research in agriculture, forestry, ecology, geology, geography, meteorology, hydrology, hydrography, oceanography and such representative tasks as: mapping snow cover and assessing water-runoff potentials; mapping water pollution; assessing crop conditions determining sea state; classifying land use; and determining land surface composition and structure.

ATM experiments utilize an array of telescopes and sensors to improve knowledge of the Sun and its influence on the Earth.

A wide range of experiments fall into the corollary category, ranging from stellar astronomy and materials processing in zero-g to the evaluation of astronaut maneuvering devices for future extravehicular operations.

Seven experiments selected through a national secondary school competition in the Skylab Student Project also are assigned to the first manned mission.

Experiments assigned to the First Skylab Missions are listed below.

In-flight medical experiments (on all missions):

M071 Mineral Balance
M073 Bioassay of Body Fluids
M074 Specimen Mass Measurement
M092 Lower Body Negative Pressure
M093 Vectorcardiogram
M110 ⎫
M113 ⎬ Series, Hematology and Immunology
M114 ⎪
M115 ⎭
M131 Human Vestibular Function
M133 Sleep Monitoring

-more-

M151   Time and Motion Study
M171   Metabolic Activity
M172   Body Mass Measurement
    (These are three ground-based medical experiments –
    M078, M11 and M112 involving pre- and post-flight data

Earth Resources Experiments Package (EREP) experiments (on all missions):

S190   Multispectral Photographic Facility comprised of:
S190A Multispectral Photographic Cameras
S190B Earth Terrain Camera
S191   Infrared Spectrometer
S192   Multispectral Scanner
S193   Microwave Radiometer/Scatterometer and Altimeter
S194   L-Band Radiometer

The ATM experiments (on all missions):

S052   White Light Coronagraph
S054   X-Ray Spectrographic Telescope
S055A Ultraviolet Scanning Polychromator-Spectroheliometer
S056   Extreme Ultraviolet and X-Ray Telescope
S082A Coronal Extreme Ultraviolet Spectroheliograph
S082B Chromospheric Extreme Ultraviolet
    (Two hydrogen-alpha telescopes are used to point the
    ATM instruments and to provide TV and photographs of
    the solar disk.)

The corollary experiments:

D008   Radiation in Spacecraft
D024   Thermal Control Coatings
M415   Thermal Control Coatings
M487   Habitability/Crew Quarters
M509   Astronaut Maneuvering Equipment
M516   Crew Activities/Maintenance Study
M551   Metals Melting
M552   Exothermic Brazing
M553   Sphere Forming
M555   Gallium Arsenide Crystal Growth
M556   Single Crystals Growth
M566   Copper Aluminum Eutectic
S009   Nuclear Emulsion
S015   Zero Gravity Single Human Cells
S019   Ultraviolet Stellar Astronomy
S020   X-Ray/Ultraviolet Solar Photography
S149   Particle Collection
S183   Ultraviolet Panorama
S228   Trans-Uranic Cosmic Rays

T002 Manual Navigation Sightings
T003 In-flight Aerosol Analysis
T025 Coronagraph Contamination Measurements
T027 Contamination Measurement (Sample Array System)
T027, Contamination Measurement
S073 Gegenschein Zodiacal Light

The student investigations:

ED11 Atmospheric Absorption of Heat
ED12 Volcanic Study
ED22 Objects within Mercury's Orbit
ED23 UV from Quasars
ED26 UV from Pulsars
ED31 Bacteria and Spores
ED76 Neutron Analysis

(Details of the above experiments may be found in Skylab Experiments Overview, available from the Government Printing Office, at $1.75 a copy. Stock number is 3300-0461.)

IV - MISSION PROFILE: LAUNCHES, DOCKING AND DEORBIT

Two launches approximately 24 hours apart will place into Earth orbits the Skylab Saturn Workshop and the Command/Service Module with the first crew who will work and live in the space station for up to 28 days. The crew's docking will take place in the fifth CSM orbit.

The Saturn Workshop (the unmanned spacecraft cluster) will be launched atop a Saturn V launch vehicle from Pad A of the NASA Kennedy Space Center Launch Complex 39 at 1:30 pm EDT, May 14, 1973. At orbital insertion, Skylab will be in a 433.4-km (268.7-mile) circular orbit with an inclination of 50 degrees.

The Skylab 2 CSM will be launched into an initial 150 x 222.: (93 x 137.8-mile) orbit by a Saturn 1B launch vehicle from Pad B of Complex 39 with liftoff at 1 pm EDT, May 15, 1973. Both launches will go northerly from the Florida site.

A five-step rendezvous maneuver sequence will be followed to bring the CSM into Skylab's orbit --- two phasing maneuvers, a corrective combination maneuver, a coelliptic maneuver, terminal phase initiation and braking. The CSM will dock with Skylab's multiple docking adapter at about seven hours, 40 minutes GET.

Timekeeping will be on a ground-elapsed-time (GET) basis until Skylab 2 GET of eight hours, after which timing will switch over to day of year (DOY), or mission day (MD), and Greenwich Mean Time (GMT) within each day. Mission Day 1 will be the day the crew is lauched.

After docking, the Skylab crew will verify that all docking latches are secured, then relax with a meal period and eight hours of sleep. The crew will enter and begin activating Skylab following morning.

At the completion of the 28-day manned operation period, the crew will board the CSM undock and perform two deorbit burns the first of which will lower CSM perigee to 166 km (103 miles) and the second burn will again lower perigee to an atmospheric entry flight path. Splashdown will be in the eastern Pacific about 1280 km (800 miles) southwest of San Diego, Calif. Splashdown coordinates are 25° 20' N, 127° 04' W.

Following is the preliminary timeline of certain Skylab 1 and 2 key events:

	Date	Time
Launch	May 14	1:30 p.m. EDT
	(launch window closes at 5:00 p.m.)	
S-IC/S-II Separation		1:32:40
S-II Ignition		1:32:42
Payload separation		1:40
Orbit insertion		1:40
Jettison payload shroud		1:45
Rotate ATM 90°		1:46
Deploy ATM solar array system		1:55
Deploy OWS solar array system		2:11
Deploy meteoroid shield		3:06

SKYLAB 2 (First manned launch)

	Date	Time
Launch	May 15	1:00 p.m. EDT*
S-IB/S-IVB Separation		1:02:22
S-IVB Ignition		1:02:23
S-IVB Engine Cutoff		1:10
Orbit Insertion		1:10
CSM/S-IVB Separation		1:16
Phasing burns		3:20 to 6:59
Station keeping		7:49 to 8:22
Docking		8:40
Pressurize tunnel	May 16	8:30 a.m.
MDA hatch open		9:00 a.m.
EVA Egress (EVA 2 hours 25 minutes)	June 10	1:00 p.m.
Undock	June 12	8:46 a.m.
Separation		9:35 a.m.
Deorbit		1:03 p.m.
Entry interface		1:27 p.m.
Splashdown		1:44 p.m.

*Launch window can vary from 7 to 15 minutes depending on the orbital parameters of the space space station.

IV-2 MISSION PROFILE: Real-time Flight Planning

In pre-Skylab United States manned space flight programs, the pre-mission flight plans were followed "by the numbers". Such will not be the case in Skylab flight planning, for the pre-mission printed flight plan will serve mainly as a guideline for planners in the Mission Control Center who each day will be developing the upcoming day's activity to yield the highest return of experiment data.

The daily flight plan, radioed to the crew for on-board teleprinter readout before the astronauts waken, will be designed to take advantage of unique opportunities such as cloud-free forecasts for desired EREP observations and solar event viewing tasks that will accomplish the greatest gain for worldwide ATM joint observing programs (JOPS).

Flight planners will have their hands full. The Skylab flight planning cycle begins at midnight Houston time, or CDT, with a team of flight planners in Mission Control Center developing a "summary flight plan" for the following crew work day. This first team will be relieved by the so-called "execution" team of flight controllers who will carry out the existing detailed flight plan for that day and leave the planning for the next work shift. Flight planners on the next, or "swing", shift will take the summary plan and develop a "detailed flight plan" for the following day, locking up the operational details first developed in the early morning hours ----- and so on, in leapfrog fashion.

Considerations that go into planning each day's flight plan include the different requirements of various experiments which have to be resolved, the optimum use of crew time, and objectives still to be met. A process of review of summary flight plans proposed by the planners takes into account the viewpoints of Skylab systems engineers, experiment principal investigators, flight surgeons, mission management, the flight crew and the weather outlook for potential EREP survey sites.

In planning the crew's work day, precedence is given ATM, EREP and medical experiments, with other experiments scheduled in the remaining time.

Daily flight plans sent up to the Skylab teleprinter will be reproduced and distributed to newsmen at the JSC News Room.

-more-

IV-3 MISSION PROFILE: Crew Work Day

Space days for the Skylab crew will not be a whole lot different from Earth days, for the normal activity day will start at 6 a.m. and run until 10 p.m. CDT. Days off, however, will be fewer and farther between.

All three crewmen will eat breakfast at 7 a.m., lunch at noon and dinner at 6 p.m. CDT---except for the man on duty at the ATM console during lunch, who will shift his meal time so that he can be relieved at the console. A standard eight hours sleep will be scheduled each day.

Crew days off will fall about every seventh day, depending upon experiment scheduling conflicts. For example, if an opportunity for a fruitful EREP pass over an unclouded portion of the Earth arises, the day off will be delayed to allow the EREP pass to be made.

Two 15-minute personal hygiene periods will be scheduled each day for each crewman and 30 minutes each day for physical exercise. Additionally, an hour a day will be set aside for "R&R"---rest and relaxation.

Another regularly-scheduled activity is two and a half hours each day for systems housekeeping.

The remaining eight hours in the crew day will be filled with experiment operation planned in real-time by flight planning teams in the Mission Control Center.

IV-4 MISSION PROFILE: The Workshop Between Visits

Ground controllers will become absentee landlords of Skylab during the periods between manned visits. Housekeeping and experiment status monitoring will be handled remotely by information telemetered to Earth, and required commands can be sent up to activate or deactivate many systems.

As the Skylab crew prepares to undock and return to Earth, it will leave the cluster in a "solar inertial attitude" with the ATM instruments pointed at the Sun. The Attitude and pointing control system will keep the vehicle in this solar attitude throughout the two-month unmanned period. Fresh film loaded by Skylab crewmen before undocking will allow ATM S052 White Light Coronagraph and S054 X-ray Spectrographic Telescope experiments to record solar activity in their respective spectra during the unmanned interval.

Immediately after the crew has undocked, the ground will command Skylab to vent down to a pressure of about two pounds per square inch. The pressure will then be allowed to gradually decay to a minimum of one-half pound.

Skylab's attitude pointing and control system and both major electrical systems will remain fully "up" during unmanned operations periods. The telemetry and command systems also will stay "live" to relay systems information to ground controllers and to accept commands for housekeeping functions and data retrieval. The environmental control system will be inactive, except for the refrigeration system and some thermal control components.

-more-

V COMMUNICATIONS AND DATA

The magnitude of the support requirements for Skylab in tracking and data acquisition has been summed up as follows: "One day's coverage is equal to an entire Apollo mission."

What this means to the people manning the far flung global network of tracking stations is that many innovations in data acquisition, communications and command functions have occurred since Apollo. Skylab transmits so much data that only 10 percent of the data collected by each station can be sent to Houston while the spacecraft passes over the station. The other 90 percent will be stored by computers and sent later. The supplying of all the vital information being generated by Skylab to ground controllers at Houston instantaneously will be done by tracking network facilities which were configured to handle about half the amount of data during the Apollo missions.

Flight control personnel will maintain contact with the Skylab spacecraft through the Spaceflight Tracking and Data Network (STDN). This network is a complex of fixed ground stations, portable ground stations, specially equipped aircraft and an instrumented ship used for transmitting signals to and receiving and processing data from the spacecraft during the Skylab mission from launch to Earth return. Each station includes tracking telemetry, television and command systems; the communications systems and switching systems.

Under the overall supervision of NASA Headquarters Office of Tracking and Data Acquisition (OTDA), the Goddard Space Flight Center, Greenbelt, Maryland is responsible for the operation and maintenance of the world-wide network.

Thirteen of the 22 STDN stations will be supporting the Skylab mission. They are:

Merritt Island, Fla.	Carnarvon, Australia
Grand Canary Island, Spain	Corpus Christi, Texas
Hawaii	Madrid, Spain
Honeysuckle Creek, Australia	Ascension Island
USNS Vanguard (tracking ship)	Guam
Bermuda	Goldstone, Calif.
Newfoundland	

To obtain the required coverage for Skylab it was necessary that a station be established in the Southern Hemisphere between 35 and 40 degrees South Latitude and 55 and 60 degrees West Longitude. Economics and time being the prime considerations, it was not feasible to install a permanent facility for short term programs such as Skylab, therefore, the Vanguard Tracking Ship will be positioned at Mar Del Plata, Argentina for this purpose.

Since the close of the Apollo program, the network has been engaged in augmenting station equipment and personnel to support Skylab.

The Skylab will be in an Earth orbit with at least one station pass approximately every 90 minutes; therefore requiring a total effort on a 24-hour basis. To insure adequate support of the long duration missions (28/56 days) all stations have been equipped with dual channel receivers, additional decommutation equipment and special gear to handle Skylab voice communications. In addition to increased equipment, the staffing at each site was augmented to provide the capability for 24-hour operations.

The entire network is linked by the facilities of the NASA Communications Network (NASCOM), a global communications network established by NASA to provide operational ground communications for support of all spaceflight operations.

Communications

The NASA Communications Network, one of the most extensive and sophisticated communications networks in existence, links all the STDN stations and NASA installations together. Over two million circuit miles covered by the network includes data and voice channels, medium and high speed message circuits. The majority of these circuits connecting and servicing these centers are leased from common carriers such as AT&T, Western Union, ITT, and various local telephone companies throughout the world. The circuits are specially engineered and maintained for NASA.

Control Center for the NASCOM Network is the NASA Goddard Space Flight Center, Greenbelt, Md. Special computers are used in the system to act as traffic policemen. The computers are programmed to recognize specific types of information and automatically direct or switch it to the proper destination. Switching centers located in London, Madrid, Honolulu, and Australia are used to augment the network, receive data from the tracking stations and route it to Goddard.

-more-

The complexity of programs such as Skylab has required the network to continuously revise and sophisticate its total system in order to handle the voluminous amount of information the network handles on a daily basis. As an example: during Project Mercury, the amount of information handled per second was the equivalent of a single 8 1/2 x 11 printed page per minute -- with the Gemini flights, this increased to the equivalent of 10 pages -- the Apollo series saw this increased by a factor of 50 to 1 over Gemini and estimates for the Skylab indicate growth of 10 to 1 over Apollo, or 5000 times from Mercury.

Network Operations

Network stations supporting the Skylab will use the "S" Band systems developed and employed during the Apollo flights. The "S" Band system is not only more powerful for longer reach and better coverage during near Earth activities, but also simplifies the ground task by combining all tracking and communications functions into a single unit.

During the mission, stations will view the spacecraft for periods of 6 to 10 minutes. Not unlike the Gemini mission one of the major differences people will be quick to recognize is the changed quality of voice and TV transmissions when compared to Apollo.

The orderly flow of mission information, command and data between the station actively tracking the spacecraft and Mission Control Center in Houston is the prime considerations during manned missions. Prior to each pass over a particular station, ground controllers at MCC transmit information to the station to update the flight plan. At the station, high-speed computers compare the information to preprogrammed parameters for validity before transmitting it to the spacecraft.

The 13 STDN stations supporting the Skylab mission are equipped with unified "S" band systems (USB).

The "unified" concept of the unified "S"-band system permits the multiple functions -- command, telemetry, tracking and two-way voice communications -- to be accomplished simultaneously using only two carrier frequencies: an uplink frequency between 2090 and 2120 MHz and a downlink frequency between 2200 and 2300 MHz. The system will also receive television from Skylab.

-more-

As used in the Apollo program, the USB uplink, voice and updata (command information) frequency modulates subcarriers; these subcarriers are combined with ranging data and the composite signal comprises the uplink carrier frequency. A subcarrier is also used for uplinking voice information. Subcarrier use is required only when multiple uplink functions are required; for example, uplink command data is phase modulated onto the main carrier frequency for transmissions to the workshop. All USB systems can transmit two uplink frequencies simultaneously.

The USB downlink system includes four main receivers and is capable of receiving four downlink frequencies simultaneously in the 2200-2300 MHz frequency range. Normally the downlink carrier will be modulated with a composite signal consisting of ranging data and modulated subcarriers, but as with the uplink, other data can be modulated directly onto the main carrier. Two Signal Data Demodulate or Systems (SDDS) are in each USB system to demodulate the various downlink signals. Television signals are taken directly from the carrier and filtered to remove subcarrier information, and then remoted directly to JSC, over wideband lines. Astronaut voice is normally sent over regular communications lines of the NASA Communications Network (NASCOM).

During the Skylab mission the CSM will act as an interface between the workshop and the ground for all voice and television communications. Command and voice will be uplinked to the CSM on USB frequency 2106.4 MHz while "realtime" telemetry and voice will be downlinked to the stations on a frequency of 2287.5 MHz. Recorded telemetry, voice and real-time television will be downlinked on a frequency of 2272.5 MHz. A VHF system with frequencies of 296.8-259.7 will be used to provide backup two-way voice communications with the CSM.

The ATM and Orbital Workshop equipment will use different systems and frequencies for transmitting and receiving data and voice communications. A UHF uplink will be employed for transmitting data to the spacecraft with VHF used to downlink data from the spacecraft. The ATM and Workshop have two VHF transmitters for downlinking real-time and recorded data. One transmitter will be used for real-time and recorded data. One transmitter will be used for real-time "dump" of data during each station pass and one transmitter will be used to "dump" recorded data.

-more-

Two recorders are aboard the Workshop to record data obtained between station passage. Data is recorded at a speed of 4 ips and is "dumped" over each station at a speed of 72 ips. Real-time and recorded telemetry is transmitted at 72.0 kb/sec. There will normally be no "dump" data from the CSM as the nominal CSM configuration calls for up to 80 percent of the systems to be powered down except during launch and reentry.

Data Management

Due to the immense proportion of usable data being recovered during the Skylab mission, a "data-compression" system will be employed in order that 100 percent of the data can be provided to Mission Control.

Data compression is simply a method of reducing the amount of information received during a mission by extracting only that portion of the data which is meaningful prior to sending it on to Mission Control Center. In this system, each parameter in the telemetry downlink is represented by a succession of samples. Data compression uses a mathematical standard to judge which of those samples contain redundant information and deletes those samples. Thus only the meaningful information is transmitted by the data compression computer. As an example, the computer compares the current value of a particular sample to the value of the last sent sample and if it is the same, that sample of information will not be transmitted. If the sample is less than the last sample, or more than the value selected, the information will be transmitted.

All data received will be recorded at each station during a pass and either sent to JSC real-time or post pass. As an example of the requirements for data handling at each station the following is a typical schedule for transmitting data during and after a station pass:

SUPPORT PHASE	FUNCTIONS	
	TELEMETRY PROGRAM	COMMAND PROGRAM
PASS	COMPRESS & TRANSMIT PCM DATA OVER THREE 7.2 KBPS LINES	COMMANDING
		TRANSMIT FIXED FORMATTED BIOMED DATA ON 7.2 KBPS LINE
	LOG PCM DATA ON DIGITAL TAPES (ADDT)	

-more-

POST PASS PHASE I	COMPRESS AND TRANSMIT REAL-TIME ADDT DATA OVER THREE 7.2 KBPS LINES	LOG DUMP PCM DATA ON DIGITAL TAPE (ADDT)
POST PASS PHASE II	COMPRESS AND TRANSMIT DUMP ADDT DATA OVER THREE 72. KBPS LINES	TRANSMIT CMD HISTORY DATA OVER TTY CIRCUITS

Range Instrumented Aircraft

Four instrumented aircraft will be used to support the Skylab mission, operating from Spanish, Australian and Indian Ocean airfields. The instrumented aircraft are used primarily to fill the voids between land and ship stations during the launch and early orbital phases of the flight.

One aircraft will operate out of Madrid, Spain and support the mission at a location 100 miles off the coast of Greece, in the Mediterranean Ocean for the purpose of monitoring the ATM deployment phase of the mission. Upon completion of the ATM deployment maneuver, the aircraft will reposition to a new area in the North Atlantic at 48 degrees North - 38 degrees West to monitor the CSM/SIVB separation.

One aircraft will stage out of Mahe, Seychelles Islands and support at a location 100 miles East of Mahe during the solar array beam deployment. Upon completion of the deployment maneuver, the aircraft will move to Capetown, South Africa to monitor the SIVB deorbit maneuver during SL-2.

Two aircraft will stage at Perth, Australia and be positioned 1500 miles South of Perth in the Indian Ocean to provide voice communications with the CSM and to monitor the SIVB deorbit maneuver during SL-2.

On-board Television Distribution

Television coverage during the mission will be both real-time and recorded. All stations in the STDN network are capable of receiving and recording video; however, only Goldstone, Calif. (GDS), Corpus Christi, Texas (TEX), and Merritt Island, Fla. (MILA) have been designated as "prime" for live television and will transmit video to the Johnson Space Center, Houston, in real-time.

-more-

"Live" television will be transmitted to Houston via hardline, color-converted and released to the news media under the direction of the Public Affairs Office, Johnson Space Center.

Recorded television will be stored aboard the Skylab and dumped daily to selected stations in the network. Video recorded at US stations will be transmitted daily to JSC where it will be edited, color-converted and released. Other stations will record video as directed.

The stations at Madrid, Spain and Honeysuckle Creek, Australia will have a "real-time" receive, record and transmit capability; however, they will record only, unless otherwise directed. The Guam STDN site will record video of the CSM/OA rendezvous and station-keeping maneuver and transmit this video via satellite to JSC within 30 or 40 minutes after completion of the event.

Still photographs of the video signal will be obtained and released to the news media at the MILA STDN station, Merritt Island, Fla. and JSC.

Color television from Skylab will be fed to ground stations by a portable TV camera. The camera, attached to a 9.1-meter (30-foot) cable, can be connected to six TV locations throughout the cluster: Multiple Docking Adapter, Airlock Module, Workshop forward dome, forward compartment, experiment compartment and the CSM. A 3.7-meter (12-foot) cable is supplied for use in the CSM. Additionally, black and white TV from the ATM solar telescopes can be relayed to Earth. Both color and black and white TV signals are relayed by the CSM FM transmitter.

An on-board videotape recorder permits delayed relay of up to 30 minutes of TV from either the color camera or the ATM equipment.

Mission events planned for TV relay include rendezvous and station keeping, experiment operations, a tour of the Workshop and other spacecraft elements, systems housekeeping, ATM console operations, EVAs for ATM film canister loading and retrieval, and undocking at the end of the mission.

Data collected and relayed in real-time to STDN stations from the Skylab cluster by the instrumentation system includes vehicle systems conditions such as pressures, voltages and temperatures; crew medical status such as respiration and heart rates; scientific information from experiments, and confirmation of mission events triggered by on-board sequencers or by ground command. Additionally, the instrumentation system furnishes data to on-board crew displays and to an array of data recorders for delayed transmission to the ground.

The Skylab intercom system has speaker boxes in 13 locations: two in the MDA forward compartment and one in the aft compartment; one in the Workshop dome; two in the Workshop forward compartment; two in the experiment compartment; one in the wardroom; one in the waste management compartment; and three in the sleep compartments.

-more-

SKYLAB I & II

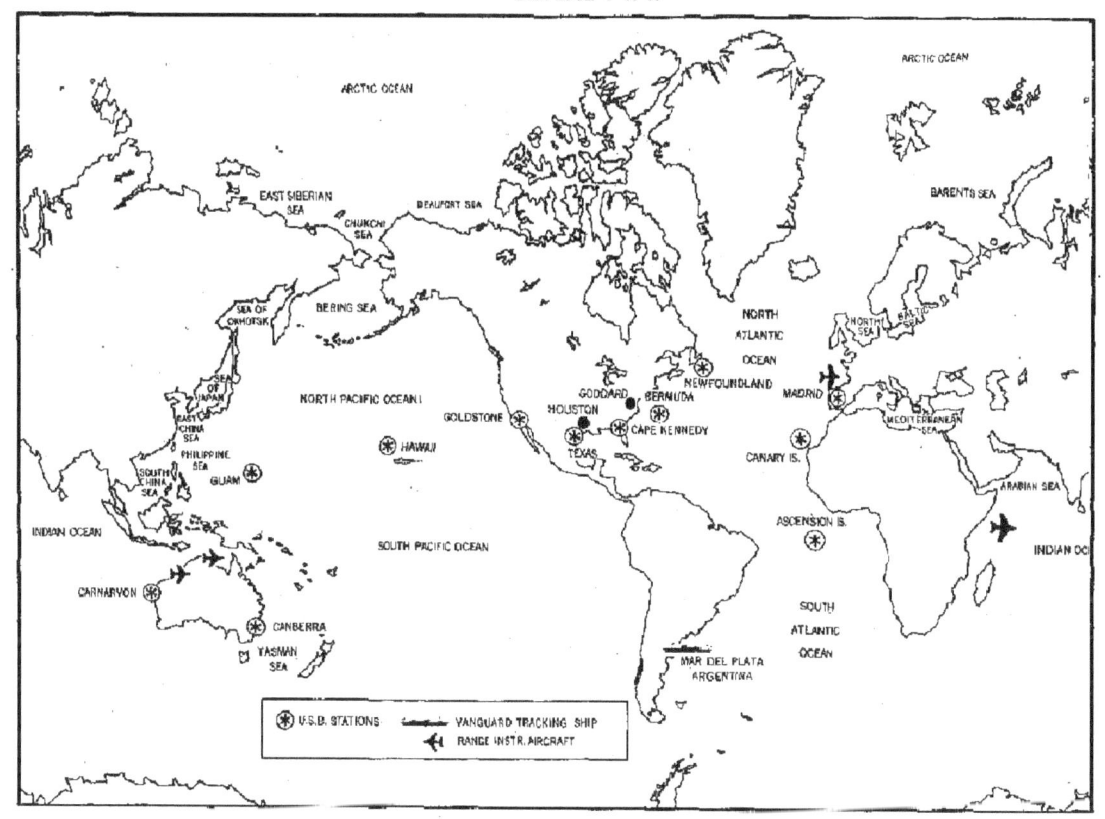

SPACEFLIGHT TRACKING AND DATA NETWORK

TYPICAL PRIME STATION TV CONFIGURATION FOR SKYLAB:

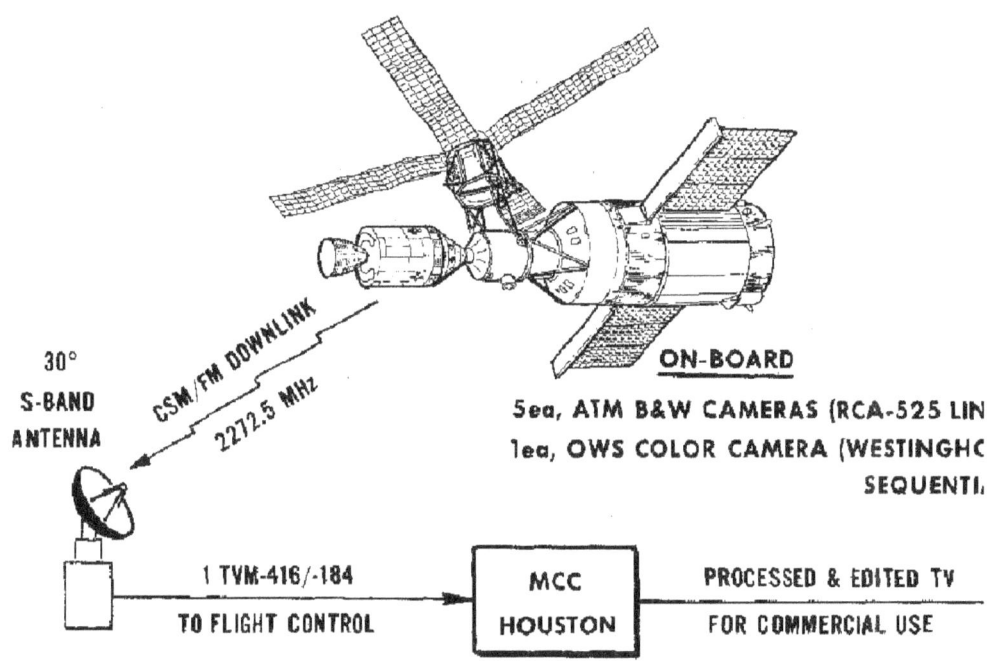

30° S-BAND ANTENNA

CSM/FM DOWNLINK 2272.5 MHz

ON-BOARD
5 ea, ATM B&W CAMERAS (RCA-525 LIN
1 ea, OWS COLOR CAMERA (WESTINGHO
SEQUENTI

1 TVM-416/-184 TO FLIGHT CONTROL → MCC HOUSTON → PROCESSED & EDITED TV FOR COMMERCIAL USE

5 ORBITAL PASSES DAILY, 10 MIN DURATION EACH. REAL-TIME INTERFACE TO MCC 1 HOUR DAILY, AT TIME OF FINAL PASS. AFTER FINAL PASS, ALL TV RECORDED WILL BE PLAYED BACK TO HOUSTON.

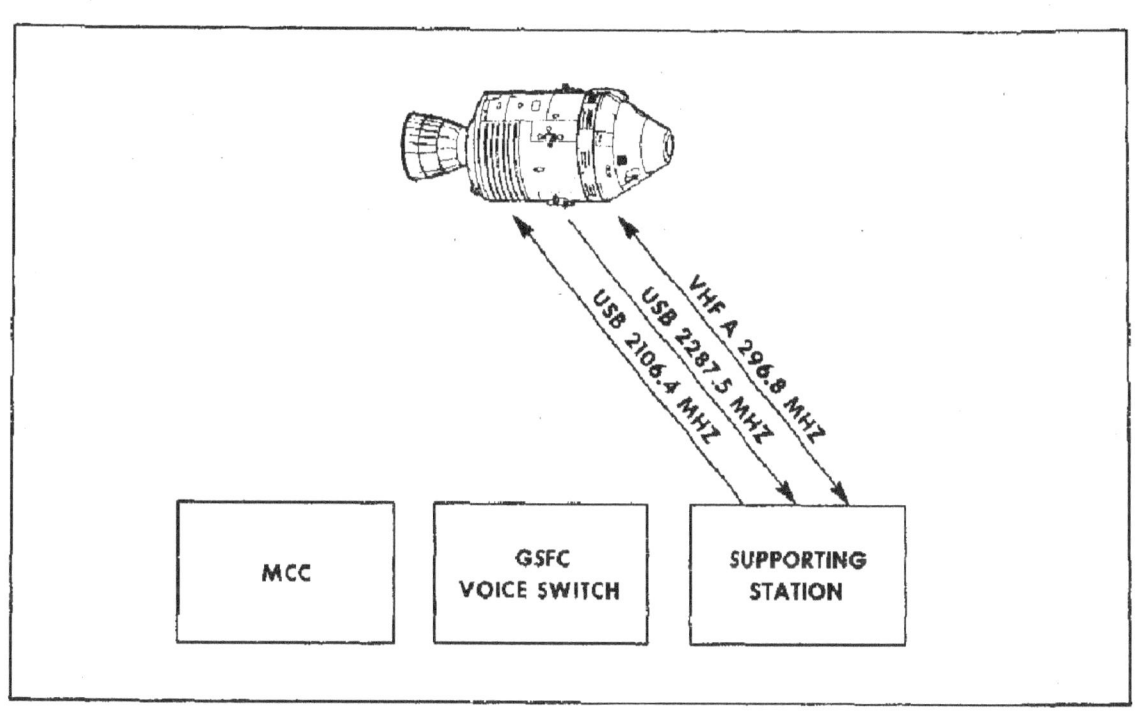

NORMAL COMMUNICATIONS MODE

VI - COUNTDOWN AND LIFTOFF

A government-industry team of about 1,300 at the Kennedy Space Center, will conduct the dual countdown of Skylab 1 and Skylab 2, the first time a parallel launch operation involving two complex spacecraft and two Saturn-class launch vehicles has been performed at Complex 39.

A team of about 500 will conduct the launch of Skylab 1 (Saturn V/Orbital Workshop) from the Launch Control Center's (LCC) Firing Room 2. Launch is to be from Pad A. Another team of approximately 500 will conduct the launch of Skylab 2 (Saturn 1B/Apollo) from the LCC's Firing Room 3. Skylab 2 will be launched from Pad B.

Approximately 300 team members will control the Orbital Workshop and Apollo spacecraft aspects of the launch from the Manned Spacecraft Operations Building in the KSC Industrial Area five miles south of Complex 39. Of these, about 200 will be involved in the Orbital Workshop launch and 100 in that of the Apollo spacecraft.

Final precount activities for Skylab 1 will begin 4.5 days before launch and those for Skylab 2 will get underway six days prior to launch.

During the early portion of the SL-1 precount, space vehicle pyrotechnics and electrical connections are completed. The Orbital Workshop and its related payload systems will be closed out and final systems checks conducted just prior to launch.

Precount activities for SL-2 includes mechanical buildup of spacecraft components and servicing the spacecraft with various gases and cryogenics (liquid oxygen and liquid hydrogen). Space vehicle pyrotechnics and electrical connections are also completed.

The final countdown for SL-1 will begin at T-minus 7 hours; that for SL-2 will begin at T-minus 9 hours.

The intricate dual precount and countdown will be conducted in parallel, with the SL-2 precount entering a built-in hold at T-22 hours, 15 minutes. This is 15 minutes prior to liftoff of Skylab 1.

-more-

It is a distance of 1.3 kilometers (.8 statute miles) from the center of Pad A to the center of Pad B and for safety reasons the latter pad will be cleared of personnel two hours prior to Skylab 1 liftoff.

In the event of a "scrub" in the launch of SL-1, the SL-2 count will be held at the T-minus 19 hour mark.

The RP-1 fuel used in both the Saturn V and Saturn 1B boosters was loaded preceeding the Countdown Demonstration Tests (CDDT). Cryogenic propellant loading liquid oxygen and liquid hydrogen, takes place during the terminal portions of both countdowns Liquid oxygen is the oxidizer in each propulsion stage and liquid hydrogen the fuel for the upper stages of both rockets.

Movement of the Mobile Service Structure, normally associated with a Saturn V operation, will not take place during the SL-1 launch. The MSS was used to inspect the second stage insulation of the Saturn V during the CDDT and then moved from Pad A to Pad B to support the SL-2 launch. It is to be returned to Pad A only if a "scrub" is called after cryogenic loading of the Saturn V begins and another inspection of the second stage insulation panels is necessary. The MSS will be moved from Pad B to its park site - in a nominal launch sequence - after the T-20 hour mark in the SL-2 countdown.

Key events late in the respective countdowns include:

SKYLAB - 1

T-7 hours	Clearing of blast danger area for launch vehicle propellant loading begins.
T-5 hours, 30 minutes	Launch vehicle propellant loading begins. Liquid oxygen for first stage. Liquid oxygen and liquid hydrogen for second stage. Continue through T-2 hours, 15 minutes
T-5 hours, 15 minutes	Open Multiple Docking Adapter vent valves
T-2 hours, 30 minutes	Thruster Actuated Control System (TACS) covers removed.
T-2 hours, 15 minutes	Retract primary damper.
T-2 hours	One-hour built-in hold begins.

-more-

T-40 minutes	Final launch vehicle range safety checks (to 35 minutes).
T-33 minutes, 30 seconds	Arm destruct system.
T-30 minutes	Launch vehicle power transfer test. Turn on AM transmitter and Digital Command System receiver.
T-6 minutes	Space vehicle final status schecks.
T-3 minutes, 7 seconds	Firing command (automatic sequence).
T-50 seconds	Launch vehicle transfer to internal power.
T-8.9 seconds	Ignition sequence start.
T-2 seconds	All engines running.
T-0	Liftoff.

SKYLAB - 2

T-9 hours	Begin clearing of blast danger area for launch vehicle propellant loading
T-8 hours, 8 minutes	Initial target update to the Launch Vehicle Digital Computer (LVDC) for rendezvous with OWS.
T-6 hours, 45 minutes	Launch vehicle propellant loading. Liquid oxygen in first stage and liquid oxygen and liquid hydrogen in second stage. Continues through 4 hours, 15 minutes.
T-4 hours, 17 minutes	Flight crew alerted.
T-4 hours, 2 minutes	Crew medical examination.
T-3 hours, 32 minutes	Brunch for crew.
T-3 hours, 30 minutes	One-hour, 13 minute, built-in hold. The lift-off time will be adjusted at the pickup of the count following this hold based on OWS target update information received at T-8 hours, 8 minutes.
T-3 hours, 7 minutes	Crew leaves Manned Spacecraft Operati Building for LC-39 via transfer van.

-more-

T-2 hours, 55 minutes	Crew arrives at Pad B.
T-2 hours, 40 minutes	Start flight crew ingress.
T-1 hour, 51 minutes	Space Vehicle Emergency Detection System (EDS) test.
T-58 minutes	Launch vehicle power transfer test.
T-45 minutes	Retract Apollo access arm to stand-by position (12 degrees).
T-44 minutes	Arm launch escape system.
T-42 minutes	Final launch vehicle range safety checks (to 35 minutes).
T-35 minutes to T-15 minutes	Last target update of the LVDC for rendezvous with the OWS.
T-33 minutes	Arm destruct system.
T-15 minutes	Maximum 2 minute hold for adjusting liftoff time.
T-15 minutes	Spacecraft to full internal power.
T-6 minutes	Space vehicle final status checks.
T-3 minutes, 7 seconds	Firing command (automatic sequence)
T-50 seconds	Launch vehicle transfer to internal power.
T-3 seconds	Ignition sequence start.
T-1 second	All engines running
T-0	Liftoff.

NOTE: Some changes in the counts are possible as a result of the experience gained in the countdown demonstration test which is held before launch.

SL-1 (SATURN V) LAUNCH EVENTS

Time Hrs Min Sec	Event	Vehicle Wt Kilograms (Pounds)*	Altitude Meters (Feet)*	Velocity Mtrs/Sec (Ft/Sec)*	Range Kilometers (Naut Mi)*
00 00 00	First Motion	2,818,085 (6,212,815)	59 (194)	0 (0)	0 (0)
00 01 15	Maximum Dynamic Pressure	1,824,200 (4,021,673)	12,298 (40,348)	460 (1,511)	4 (2)
00 02 20	S-IC Center Engine Cutoff	942,067 (2,076,903)	61,533 (201,880)	1,951 (6,400)	53 (29)
00 02 38	S-IC Outboard Engine Cutoff	748,876 (1,650,989)	84,670 (277,788)	2,537 (8,324)	85 (46)
00 02 40	S-IC/S-II Separation	581,417 (1,281,805)	87,301 (286,422)	2,543 (8,342)	89 (48)
00 02 42	S-II Ignition	580,816 (1,280,479)	89,593 (293,940)	2,534 (8,314)	92 (50)
00 03 10	S-II Aft Interstage Jettison	548,758 (1,209,805)	127,462 (418,181)	2,631 (8,631)	153 (83)
00 05 14	S-II Center Engine Cutoff	389,184 (858,004)	271,923 (892,137)	3,566 (11,699)	491 (265)
00 09 50	Payload Separation	89,439 (197,180)	442,027 (1,450,221)	7,332 (24,056)	1,818 (982)
00 09 58	Orbit Insertion	89,439 (197,180)	442,128 (1,450,552)	7,333 (24,057)	1,871 (1,010)

*English measurements given in parentheses

SL-2 (SATURN IB) LAUNCH EVENTS

Time Hrs Min Sec	Event	Vehicle Wt Kilograms (Pounds)*	Altitude Meters (Feet)*	Velocity Mtrs/Sec (Ft/Sec)*	Range Kilometers (Naut Mi)*
00 00 00	First Motion	586,421 (1,292,836)	0 (0)	0 (0)	0 (0)
00 01 13	Maximum Dynamic Pressure	374,532 (825,702)	12,438 (40,807)	468 (1,536)	4 (2)
00 02 18	Inboard Engine Cutoff	189,435 (417,632)	55,418 (181,817)	1,981 (6,498)	59 (32)
00 02 21	Outboard Engine Cutoff	184,281 (406,270)	58,310 (191,306)	2,037 (6,684)	64 (35)
00 02 22	S-IB/S-IVB Separation	183,517 (404,586)	59,636 (195,655)	2,037 (6,682)	66 (36)
00 02 23	S-IVB Ignition	137,362 (302,830)	60,893 (199,781)	2,031 (6,663)	69 (37)
00 02 51	Launch Escape Tower Jettison	131,377 (289,636)	84,379 (276,834)	2,104 (6,903)	119 (64)
00 08 57	S-IB Stage Impact	45,495 (100,300)	0 (0)	90 (295)	498 (269)
00 09 40	S-IVB Engine Cutoff	30,878 (68,075)	158,368 (519,581)	7,625 (24,690)	1,760 (950)
00 09 50	Orbit Insertion	30,803 (67,910)	158,510 (520,047)	7,532 (24,711)	1,834 (990)

*English measurements given in parentheses

The solar cell arrays are folded against the sides of the OWS for launch and deployed in orbit. The OWS is protected by meteoroid shield which is deployed by swinglinks to stand five inches from the OWS wall is space.

The liquid oxygen tank of the S-IVB was converted to serve as a receptacle for liquid and solid wastes. Trash is placed in the tank through an airlock in the floor of the Crew Quarters. Liquid is fed to the tank via inlet lines and, in some cases, collected in receiving bags and introduced like trash through the airlock.

(See Skylab News Reference book for detailed information on the Orbital Workshop and other elements of the Skylab cluster.)

Apollo Telescope Mount

The Apollo Telescope Mount is the first manned astronomical observatory for performing solar research from Earth orbit. It weighs 11,181 kilograms (24,650 pounds), is 4.4 meters (14.7 feet) long and almost 6 meters (20 feet) in diameter with solar arrays folded, or 31 meters (102 feet) wide with arrays extended.

The ATM consists of five major hardware elements: experiment canister, attitude and pointing control system, solar array wings, control and display console (in the MDA) and the rack assembly.

The experiments canister consists of the spar, the MDA and the Sun end canister halves and the canister girth ring.

The rack is made of two large octagonally-shaped rings separated by eight vertical beams. Equipment-mounting panels are provided in seven of the bays between beams. One bay is left open for an astronaut work station. The Sun end mounts the solar shield assembly and acquisition Sun sensor.

A girth ring around the center of the spar is the structural interface between the experiments canister and the rack-mounted experiment pointing control-roll positioning mechanism.

The MDA end canister includes four film retrieval doors. The Sun end canister half has two film retrieval doors and ten aperture doors on the Sun end bulkhead.

Mounted on the ATM are major elements of Skylab's Attitude and Pointing Contol System that provides three-axis stabilization and maneuvering capability for the orbiting vehicle.

-more-

The ATM solar array consists of four wing assemblies which are stowed in a folded position for launch and deployed upon reaching orbit. The wings expose 112 square meters (1,200 square feet) of solar cells to the Sun.

Airlock Module

The Airlock Module is between the MDA and the OWS. It is 5.3 meters (17.5 feet) long, weighs 22,226 kilograms (49,000 pounds) and has 17.4 cubic meters (622 cubic feet) of habitable volume. It consists of a Structural Transition Section (STS), tunnel assembly, four truss assemblies, the lower truss of the Deployment Assembly, a flexible tunnel extension and a Fixed Airlock Shroud (FAS).

The STS connects the tunnel assembly to the MDA. The tunnel has an airlock and hatch to permit astronauts to perform extravehicular activities without depressurizing the entire spacecraft. The FAS provides a shroud around the aft portion of the AM and structural mounting for the AM and MDA, the Deployment Assembly and the Skylab oxygen supply tanks. It supports the Payload Shroud, ATM, AM and MDA during boost.

The four truss assemblies attach the AM to the FAS and provide exterior mounting for battery, electronic, thermal and experiment equipment.

The STS contains the AM data file, control panels, lights, circuit breaker panels, ducts, stowage containers, the molecular (mol) sieve, and cabin heat exchanger, ATM tank, water tank, condensate and carbon dioxide sensor modules.

Although relatively small, the AM tunnel contains dozens of items of equipment, including lights, ATM film tree support, ducts, vent valves, stowage, spare mol sieve fan and replacement liquid/gas separators, tape recorder module, portable timers, spare batteries, light bulbs and teleprinter head and numerous other items.

The FAS protects the AM aft compartment and serves as mounting structure for two discone antennas.

Multiple Docking Adapter

The Multiple Docking Adapter is a cylindrical structure with a forward conical bulkhead. It is 5.2 meters (17.3 feet long and 3 meters (10 feet) in diameter. It weighs 6,260 kilograms (13,800 pounds and contains about 32 cubic meters (1,140 cubic feet) of space. It has an axial docking port at the forward end, to which the Apollo CSM will normally dock, and a radial port which could be used as a backup if necessary.

The MDA serves as the docking interface for the Command Module and permits the transfer of personnel, equipment, power and electrical signals between the docked CSM and the Airlock and Orbiting Workshop. In orbit the MDA functions as a major experiment control center for solar observations, metals and materials processing and Earth resources experiments. Major experiment items in the MDA are the ATM Control and Dispaly console, Earth Resources Experiment Package, and the M512 and M518 materials processing facilities.

Major items on the outside include the S192 10-band multispectral scanner, S191 infrared spectrometer, S194 L-Band antenna, proton spectrometer, inverter lighting control assembly, orientation lights, docking lights and docking targets.

The MDA also contains special tool kits and spare parts for selected types of orbital maintenance and activation/ deactivation sequences to be performed by the astronauts.

Payload Shroud

The Payload Shroud is a smooth structure which surrounds and protects the ATM, MDA, AM and associated equipment during launch and climb to orbit. Once in orbit, the PS is split into four quarters and jettisoned.

The PS is 6.5 meters (21.7 feet) in diameter at the aft end, 17.1 meters (56 feet) long and weighs 11,794 kilograms (26,000 pounds). It has a nose cap, a forward cone which tapers at a 25-degree angle, and an aft cone which tapers at a 12.5-degree angle. The aft cone connects to a cylindrical section which attaches to the Fixed Airlock Shroud.

VII - SATURN WORKSHOP

The Skylab cluster, without the CSM attached is called the Saturn Workshop. The following describes the major elements.

The Orbital Workshop (OWS) portion of Skylab is S-IVB-212, the second stage of Saturn IB vehicle SA-212, which has been converted into living and working quarters for three astronauts. It is divided into two main sections, the forward compartment and crew quarters. The living area, formerly the liquid hydrogen tank, affords 295 cubic meters (10,414 cubic feet) of space. The OWS weighs 35,380 kilograms (78,800 pounds).

Mounting on opposing sides of the OWS are solar array "wings" which provide electrical power. At the aft end, the engine was replaced by cold gas storage bottles and a refrigeration radiator Thrusters for attitude control are mounted on the circumference at the aft end.

The crew quarters section, in the aft end of the former hydrogen tank, is divided into a wardroom with about 9.3 square meters (100 square feet) of floor space, a waste management compartment of 2.8 square meters (30 square feet), a sleep compartment of about 6.5 square meters (70 square feet) and an experiment area of about 16.7 square meters (180 square feet).

The forward compartment is separated from the crew quarters by an eight-inch beam structure with a grid on each side, serving as floor and ceiling. In the forward compartment are lockers for storing food, clothing, film and other items and water tanks holding enough for the entire mission.

On the water tank mounting ring are 25 lockers holding supplies such as bundles of urine bags, portable lights, electrical cables, hoses, umbilicals, pressure suits, tape recorder, charcoal filters, fans, lamps and intercom boxes. Major items on the floor and around the wall include the food lockers and freezers, several major items of experiment equipment, astronaut maneuvering equipment, EVA suits and various scientific instrumen

The thermal control and ventilation system will provide the astronauts with a habitable environment with temperatures ranging from 15.6 to 32.2 degrees Centigrade (60 to 90 degrees Fahrenheit) and an oxygen-nitrogen atmosphere with internal pressure of 3.45 Newtons per square centimeter (5 pounds per square inch).

VIII - SATURN LAUNCH VEHICLES

The launch vehicles for the Skylab program are Saturn multi-stage rockets developed by the NASA-Marshall Space Flight Center for the Apollo Program.

A two-stage Saturn V will send the unmanned Skylab cluster into Earth orbit. This will be the 13th flight of a Saturn V. Ten of the previous 12 vehicles have been manned, the rocket havin been proven safe for manned flight after only two launches.

For its Skylab role, the Saturn V does not carry a "live" third stage. In its place will be the Orbital Workshop and mounte atop the Workshop, enclosed in a shroud, will be the Airlock Module, Multiple Docking Adapter and Apollo Telescope Mount. The Saturn V will place this unmanned payload into an Earth orbit at an altitude of 433.4 kilometers (268.7 miles).

The smaller Saturn IB vehicles will carry Skylab crews into orbit to rendezvous and dock with the orbiting space station. Each of these vehicles consists of the S-IB (first) stage and S-IVB (second) stage and the Instrument Unit with the manned Apoll Command-Service Module above.

Twelve Saturn IB vehicles were manufactured. Five have been launched successfully. The sixth (Sa-206) will carry the first crew (Skylab 2) into orbit, the next (SA-207) will transport the Skylab 3 crew, and the eighth (SA-208) will take the Skylab 4 crew to the space station.

In case of emergencies, the next vehicle in line will be used for a rescue mission. SA-209 would be used if rescue of the Skylab 4 crew was required.

NOTE: Robert O. Aller of NASA Headquarters, is Director of Skylab Operations. His name was spelled improperly in the Skylab News Reference.

-end-

NATIONAL AERONAUTICS AND
SPACE ADMINISTRATION
Washington, D. C. 20546
202-755-8370

FOR RELEASE:
July 23, 1973

PROJECT: SKYLAB 3
Second Manned Missi

contents

GENERAL RELEASE	1-5
OBJECTIVES OF THE SKYLAB PROGRAM	8-10
OBJECTIVES OF THE SECOND MANNED SKYLAB MISSION	11-12
MISSION PROFILE: LAUNCH, DOCKING AND DEORBIT	13-18
COUNTDOWN AND LIFTOFF	19-21
SKYLAB EXPERIMENTS	22-24
REAL-TIME FLIGHT PLANNING	26-28
SKYLAB STATUS: WHAT HAPPENED	29-34
ACCOMPLISHMENTS	35-39
SKYLAB BETWEEN VISITS	40-42
SKYLAB AND RELATED OBJECTS VISIBLE	43-44

NOTE: Details of the Skylab spacecraft elements, systems, crew equipment and experimental hardware are contained in the Skylab News Reference distributed to the news media. The document also defines the scientific and technical objectives of Skylab activities. This press kit confines its scope to the second manned visit to Skylab and briefly describes features of the mission.

NATIONAL AERONAUTICS AND
SPACE ADMINISTRATION
Washington, D. C. 20546
AC 202/755-8370

William Pomeroy
(Phone 202/755-3114)

FOR RELEASE:
July 23, 1973

RELEASE NO: 73-131

NEXT SKYLAB CREW GOES UP JULY 28

Three American astronauts will begin a two-month stay in space July 28 when the second Skylab crew is launched into orbit to man the Skylab space station. The second crew will further extend the long-term quest for knowledge about man's home planet, his Sun and himself which was begun by the Skylab 2 mission lasting 28 days.

The Skylab 3 crew will live and work aboard the space station for up to 56 days while measuring the human adaptability to long-duration spaceflight, conducting solar astronomy experiments above the distorting effects of the atmosphere, and surveying conditions and resources down on the fragile spacecraft Earth. Launched May 14, the Skylab space station is in an orbit tilted 50 degrees to the equator and ranges over most of the Earth's populated regions -- from the Canadian Border to the tip of Argentina.

-more-

Early in the space station's launch, known as Skylab 1, an aluminum micrometeoroid shield tore loose, taking with it one of the large power-generating solar array panels on the Skylab Workshop, and causing higher-than-normal temperatures in the Workshop living space. The first Skylab crew launch was delayed for 10 days while sunshields were fabricated and the crew was trained in erecting the shields. Once the temperatures were brought down by the parasol-like device that was deployed and the remaining solar array was freed by an innovative EVA repair using tools aboard Skylab, the space station settled down to a more or less normal operation.

The contingency repairs performed in-flight by the Skylab 2 crew of Charles Conrad, Joseph Kerwin and Paul Weitz, yielded an unexpected return by demonstrating that man can indeed tackle difficult repair and construction tasks in space.

In spite of the adversities at the outset of the first manned Skylab mission, all planned operational objectives were met, and much of the expected experimental data were gathered.

Taking up where the first crew left off, the second Skylab crew will double the information gained from medical experiments that measure man's physical responses to long-term exposure to weightlessness and other aspects of the space environment. The Sun and its influence upon life on Earth will again come under scrutiny as the Skylab crew focuses the astronomical telescopes and instruments of the Skylab space station toward our star some 93 million miles across space.

Closer to home, Skylab's Earth Resources Experiment Package (EREP) will scan and photograph physical and environmental features of the Earth's surface and atmosphere in 26 planned EREP "passes." Additionally, a group of scientific and technological experiments will be conducted during the 56 days of flight, including seven investigations selected in a nationwide competition among high school students.

Skylab 3 crewmen are Alan L. Bean, commander; Dr. Owen K. Garriott, science pilot; and Jack R. Lousma, pilot. Bean is a US Navy captain, Garriott a civilian scientist-astronaut, and Lousma a US Marine Corps major. Bean was lunar module pilot on the second manned lunar landing, Apollo 12, and with Apollo 12 commander Charles "Pete" Conrad, explored the region around the Surveyor III landing site. Garriott and Lousma have not flown in space.

-more-

Liftoff for Skylab 3 is scheduled for 7:08 a.m. EDT, July 28 atop a Saturn IB from NASA Kennedy Space Center Launch Complex 39, Pad B. Rendezvous and docking will occur during the fifth command/service module orbit after a standard rendezvous maneuver sequence.

After docking with the space station, the Skylab crew will open the hatch, enter Skylab and begin to activate the station's systems. Skylab crew work days begin at 6 a.m. and end at 10 p.m. Houston time, Central Daylight.

Three EVAs are scheduled for the second crew: one to deploy a twin-boom sunshield to replace the parasol erected by the previous crew, and to the Sun end of the Apollo Telescope Mount (ATM) to retrieve and replenish film cannisters. The second and third EVAs will be for ATM retrieval and replacement.

On September 22 the crew will undock the CSM from Skylab to deorbit and land in the eastern Pacific, about 1,830 km (990 nm) southwest of San Diego. Command module splashdown will be at 8:38 p.m. EDT September 22. Prime recovery vessel will be the landing platform-helicopter (LPH) USS New Orleans.

The launch vehicles for the Skylab program are Saturn multi-stage rockets developed by the NASA-Marshall Space Flight Center for the Apollo Program. A two-stage Saturn V placed the unmanned Skylab cluster into Earth orbit. This was the 13th flight of a Saturn V. The smaller Saturn IB vehicles carry Skylab crews into orbit to rendezvous and dock with the orbiting space station. The seventh Saturn IB to be launched will transport the Skylab 3 crew.

(END OF GENERAL RELEASE; BACKGROUND INFORMATION FOLLOWS)

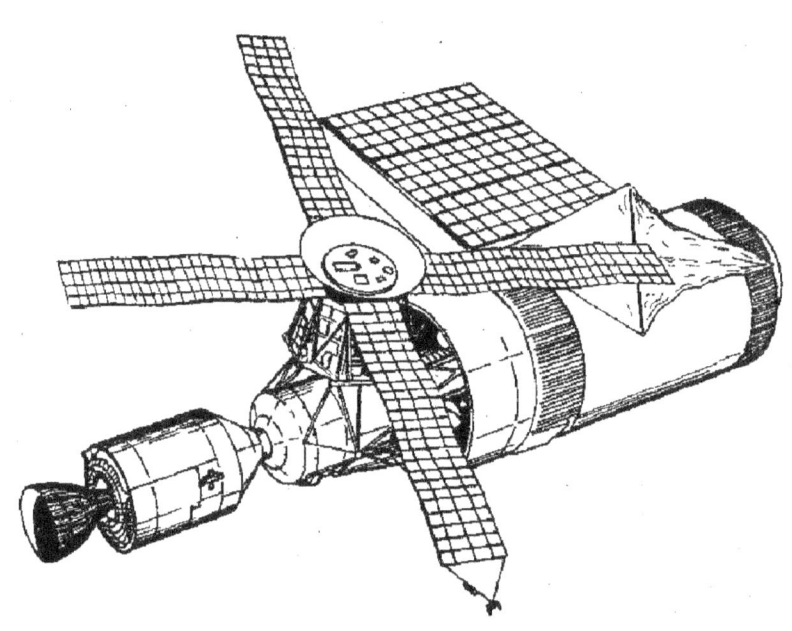

How Skylab appeared at end of the first manned visit.

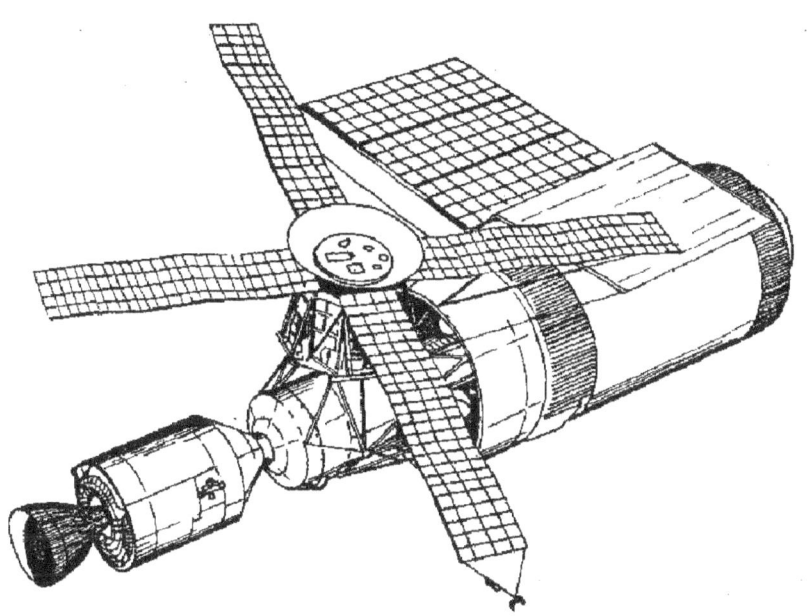

How Skylab would look with twin-boom sunshade installed.

OBJECTIVES OF THE SKYLAB PROGRAM

The Skylab Program was established for four explicit purposes: to determine man's ability to live and work in space for extended periods; to extend the science of solar astronomy beyond the limits of Earth-based observations; to develop improved techniques for surveying Earth resources from space; and to increase man's knowledge in a variety of other scientific and technological regimes.

Skylab, the first space system launched by the United States specifically as a manned orbital research facility, will provide a laboratory with features which cannot be found anywhere on Earth. These include: a constant zero gravity environment, Sun and space observation from above the Earth's atmosphere, and a broad view of the Earth's surface.

Dedicated to the use of space for the increase of knowledge and for the practical human benefits that space operations can bring, Skylab will pursue the following:

Physical Science - Increase man's knowledge of the Sun, its influence on Earth and man's existence, and its role in the universe. Evaluate from outside Earth's atmospheric filter, the radiation and particle environment of near-Earth space and the radiations emanating from the Milky Way and remote regions of the universe.

Life Science - Increase man's knowledge of the physiological and biological functions of living organisms - human, other animal, and tissues - by making observations under conditions not obtainable on Earth.

Earth Applications - Develop techniques for observing Earth phenomena from space in the areas of agriculture, forestry, geology, geography, air and water pollution, land use and meteorology.

Space Applications - Augment the technology base for future space activities in the areas of crew/vehicle interactions, structures and materials, equipment and induced environments.

-more-

The first Skylab mission achieved its three specific objectives. They were as follows:

1. <u>Establish the Skylab orbital assembly in Earth orbit.</u>

    a. Operate the spacecraft cluster (including CSM) as a habitable space structure for up to 28 days after the launch of the crew.

    b. Obtain data for evaluating the total spacecraft performance.

    c. Obtain data for evaluating crew mobility and work capability in both intravehicular and extravehicular activity.

2. <u>Obtain medical data on the crew for use in extending the duration of manned space flights.</u>

    a. Obtain medical data for determining the effects on the crew which result from a space flight of up to 28 days duration.

    b. Obtain medical data for determining if a subsequent Skylab mission of up to 56 days duration is feasible and advisable.

3. <u>Perform in-flight experiments.</u>

    a. Obtain ATM solar astronomy data for continuing and extending solar studies beyond the limits of Earth from low Earth orbit.

    b. Obtain Earth resources data for continuing and extending multisensor observation of the Earth from low Earth orbit.

    c. Perform the assigned scientific, engineering and technology experiments.

The Gemini 7 mission had demonstrated that man could readily adapt to space flight for up to two weeks without ill effects. Now Skylab has pushed forward the threshold of human adaptability to spaceflight by doubling Gemini 7's time in space with the first Skylab crew.

-more-

SKYLAB MAJOR EVENTS

(Central Daylight Time)

MISSION	LAUNCH	LANDING	DURATION DAY:HR:MIN
SL-1	MAY 14 - 12:30P CDT (134:17:30 GMT*)		
SL-2	MAY 25 - 8:00A CDT (145:13:00 GMT)	JUNE 22 - 8:50A CDT (173:13:50 GMT)	28:00:50
SL-3	JULY 28 - 6:08A CDT (209:11:08 GMT)	SEPT 22 - 7:38P CDT (266:00:38 GMT)	56:13:30
SL-4	TBD	TBD	56 DAYS

* **DAY OF YEAR: HR: MIN** in Greenwich Mean Time

OBJECTIVES OF THE SECOND MANNED SKYLAB MISSION

The second Skylab mission officially began June 22 when the first CSM and its crew separated from the space station just prior to reentry. The unmanned portion of this SL-3 mission will continue until the second crew is launched. After docking, the SL-3 crew will enter Skylab, reactivate its systems, and proceed to inhabit and operate the orbital assembly for up to 56 days. During this time the crew will perform systems and operational tests and the assigned experiments.

The four objectives of the second Skylab mission are as follows:

1. <u>Perform unmanned Saturn Workshop operations</u>

    a. Obtain data for evaluating the performance of the unmanned station.

    b. Obtain solar astronomy data by unmanned ATM observations.

2. <u>Reactivate and Man Skylab in Earth orbit</u>

    a. Operate the cluster (SWS plus CSM) as a habitable space structure for up to 56 days after the SL-3 launch.

    b. Obtain data for evaluating the performance of the space station.

    c. Obtain data for evaluating crew mobility and work capability in both intravehicular and extravehicular activity.

3. <u>Obtain medical data on the crew for use in extending the duration of manned space flights</u>

    a. Obtain medical data for determining the effects on the crew which result from a space flight of up to 56 days duration.

    b. Obtain medical data for determining if a subsequent Skylab mission of greater than 56 days duration is feasible and advisable.

4. Perform in-flight experiments

   a. Obtain ATM solar astronomy data for continuing and extending solar studies beyond the limits of Earth-based observations.

   b. Obtain Earth resources data for continuing and extending multisensor observations from Earth orbit.

   c. Perform the assigned scientific, engineering, technology and DOD experiments.

-more-

MISSION PROFILE: Launch, Docking and Deorbit

Skylab 3, the second manned visit to the Skylab space station, will be launched at 7:08 am EDT July 28 from NASA Kennedy Space Center Launch Complex 39 Pad B for a fifth-orbit rendezvous with the space station. The Skylab space station, designated Skylab 1, was launched into an initial 431x432.9km (233 by 234 nm) orbit which is expected to be 424.6 by 439.5km (229x237 nm) at Skylab 3 rendezvous.

The standard five-step rendezvous maneuver sequence will be followed to bring the Skylab 3 CSM into the space station's orbit---two phasing maneuvers, a corrective combination maneuver, a coelliptic maneuver, terminal phase initiation and braking. The CSM will dock with Skylab's axial docking port at about eight hours 20 minutes after launch.

After verifying that all docking latches are secured, the Skylab 3 crew will begin actication of the space station, but will sleep aboard the command module the first night.

Timekeeping will be on a ground-elapsed-time (GET) basis until Skylab 3 GET of eight hours, after which timing will switch over to day of year (DOY), or mission day (MD), and Greenwich Mean Time (GMT or "Zulu") within each day. Mission day 1 will be the day the crew is launched.

At the completion of the 56-day manned operation period, the crew will board the CSM, undock and perform two deorbit burns---the first of which will lower CSM perigee to 166.5 km (90 nm) and the second burn will again lower perigee to an atmospheric entry flight path. Splashdown will be in the eastern Pacific about 1830 km (990 nm) southwest of San Diego, Calif. after 874 CSM revolutions. Splashdown coordinates are 23°28' N, 129°26' W. Command module touchdown time will be 8:38 pm EDT September 22.

Skylab 3 (Second manned launch)

Event	Date	Time (EDT)
Launch	July 28	7:08:50 a.m.
Orbital insertion		7:18:53 a.m.
CSM/S-IVB separation, 3 fps RCS		7:33:50 a.m.
Phasing 1 (NC1), 221.1 fps SPS		9:26:19 a.m.
Phasing 2 (NC2), 158 fps SPS		11:42:12 a.m.
Corrective combination (NCC), 29.6 fps SPS		12:28:21 p.m.
Coelliptic (NSR), 19.2 fps SPS		1:05:21 p.m.
Terminal Phase initiate (TPI), 20.9 fps SPS		2:21:12 p.m.
Terminal phase finalize (TPF), 27.3 fps SPS		2:54:54 p.m.
Docking		3:38:50 p.m.
Orbit trim burn 1, 2.4 fps RCS	August 1	10:04:18 a.m.
Orbit trim burn 2, 1.4 fps RCS	August 26	10:36:11 p.m.
Orbit trim burn 3, 1.3 fps RCS	Sept. 17	9:26:12 p.m.
Undocking	Sept. 22	3:21:33 p.m.
Separation, 5 fps RCS		4:08:19 p.m.
Shaping burn, 258.5 fps SPS		4:55:33 p.m.
Deorbit burn, 191.9 fps SPS		7:57:11 p.m.
Entry interface (400,000 feet)		8:22:35 p.m.
Landing at 23°28' N x 129°26' W		8:38:29 p.m.

RENDEZVOUS SEQUENCE

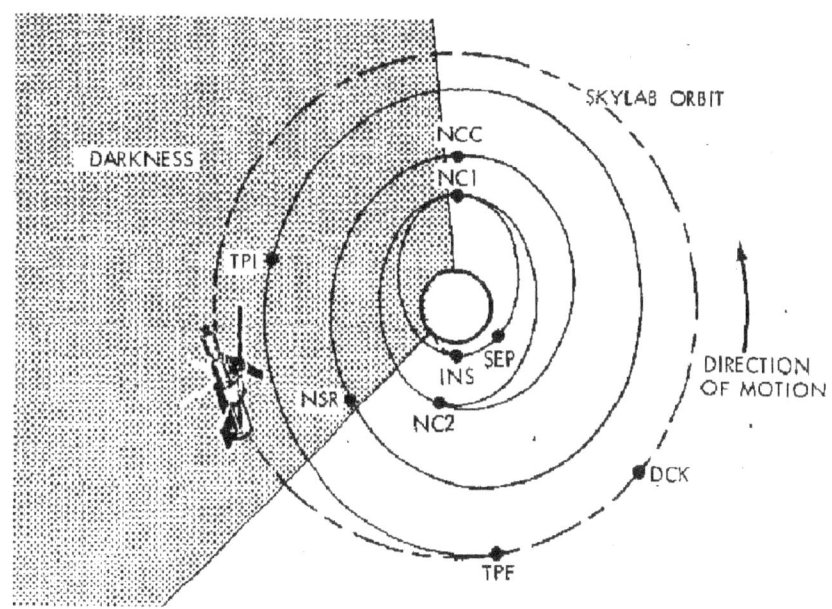

		TIME, G.E.T., HR:MIN:SEC	DELTA V, ADDED FT. PER SECOND	RESULTANT PERIGEE/APOGEE, N. MI.
INS	INSERTION	00:10:03.1	—	81/120
SEP	SEPARATION MANEUVER	00:25:00.0	3.0	81/121
NC1	PHASING 1	02:17:29.4	221.1	120/208
NPC	PLANE CHANGE	PLANE CHANGE, IF NECESSARY		
NC2	PHASING 2	04:33:22.8	158.0	202/215
NCC	CORRECTIVE COMBINATION	05:19:31.7	29.6	208/228
NSR	COELLIPTIC	05:56:31.7	19.2	219/227
TPI	TERMINAL PHASE INITIATION	07:12:22.0	20.9	223/234
TPF	TERMINAL PHASE FINALIZATION	07:46:04.0	27.3	230/238
DCK	DOCKING	08:30:00	—	—

ML73-2330

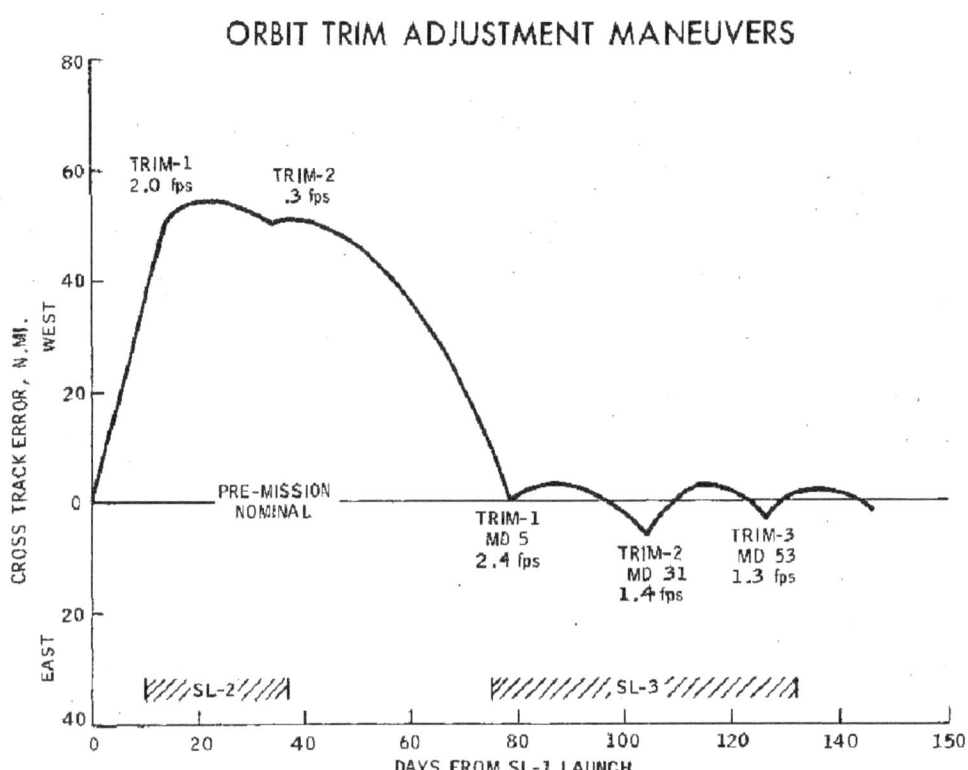

END OF MISSION SEQUENCE FOR SL-3 SPS DEORBIT

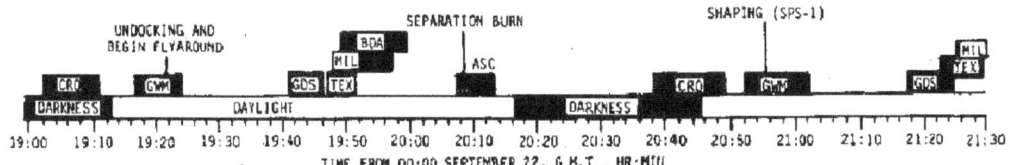
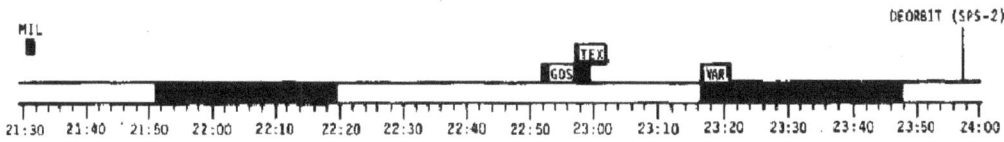
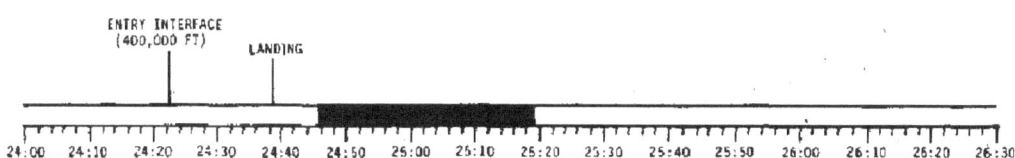

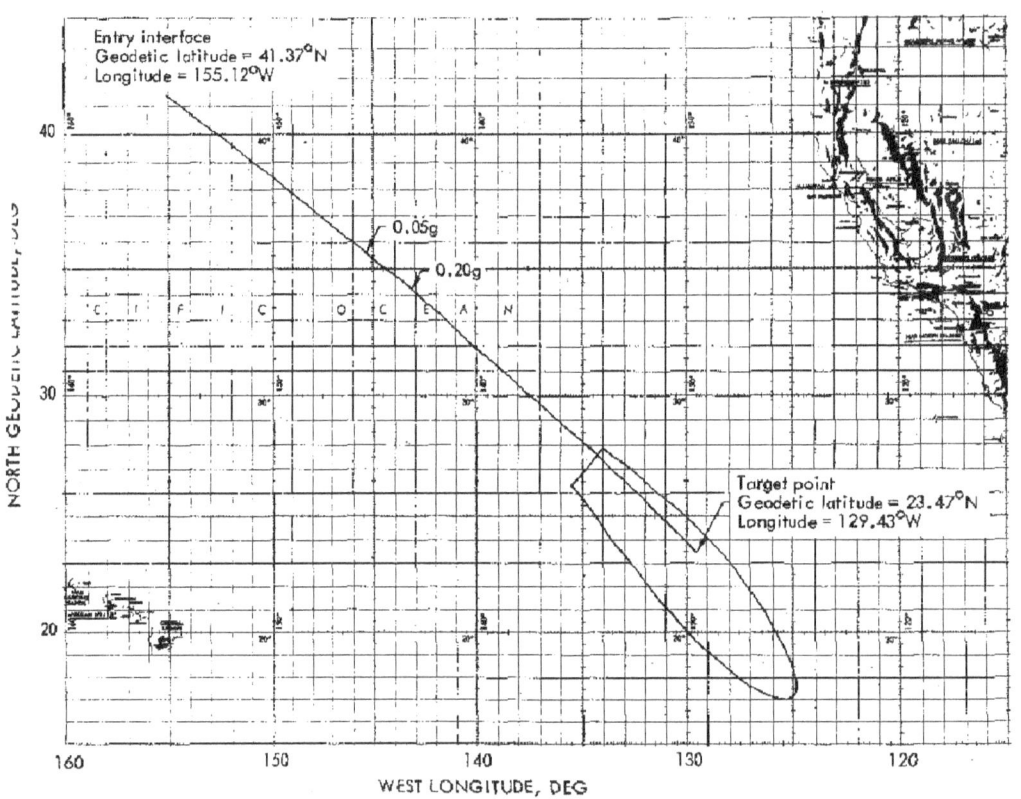

CM reentry track and maneuver envelope SW of NAS North Island, San Diego

COUNTDOWN AND LIFTOFF

After the May 25 launch of the first crew to man Skylab the mobile launcher was brought back to the Vehicle Assembly Building at the NASA Kennedy Space Center in Florida. The stages of the next Saturn IB launch vehicle and boilerplate spacecraft were erected on the mobile launcher, beginning May 28.

Integrated testing of the launch vehicle stages was conducted while the spacecraft underwent thorough testing, including simulated flights in the altitude chamber, in the Manned Spacecraft Operations Building at KSC's industrial area.

On June 8, the flight spacecraft was moved to the VAB and erected atop the launch vehicle three days later, the fully assembled space vehicle was moved to Launch Complex 39, Pad B for pad integration and final tests prior to the launch countdown.

The countdown for this third Skylab launch differs from previous ones in that the Countdown Demonstration test CDDT and the final countdown have been incorporated into a single launch countdown. The early portion of the count will include launch vehicle cryogenic fueling and final countdown activities without astronaut participation.

Following the simulated T-0, the count will be recycled to the T-47 hour mark instead of recycling for a dry test with crew participation, then going through the entire count again as had been done on earlier missions.

Key events in the final count, beginning at T-47 hours include:

T-45 hours 30 minutes	Install launch vehicle batteries
T-39 hours	Launch vehicle power transfer test
T-36 hours	Command service module cryogenic fueling. Takes approximately 6 hours
T-26 hours	Complete CSM mechanical buildup. Takes approximately 12 hours
T-9 hours	Begin clearing pad area
T-8 hours	Replenish RP-1 (first stage fuel)

-more-

T-6 hours 50 minutes	Begin launch vehicle cryogenic propellant load. (Loading takes approximately 3 hours - replenish continues through remainder of countdown)
T-4 hours	Primary damper retracted
T-3 hours 45 minutes	CSM closeout crew on station
T-2 hours 40 minutes	Flight crew enters spacecraft
T-1 hour 51 minutes	Emergency detection system tests (to T-1 hour, 21 minutes)
T-57 minutes	Clear closeout crew from pad area
T-45 minutes	Retract swing arm 9 to park position
T-44 minutes	Arm Launch Escape System
T-42 minutes	Final launch vehicle range safety check (to T-35 minutes)
T-35 minutes	Last target update of the LVDC for rendezvous with the OWS
T-15 minutes	Hold for liftoff adjustment - maximum 2 minutes
T-5 minutes	Swing arm 9 fully retracted
T-3 minutes 7 seconds	Start automatic sequence
T-50 seconds	Launch vehicle transfer to internal power
T-3 seconds	Ignition sequence starts
T-0	Liftoff

SL-3 (SATURN IB) LAUNCH EVENTS

Time Hrs Min Sec	Event	Vehicle Wt Kilograms (Pounds)*	Altitude Meters (Feet)*	Velocity Mtrs/Sec (Ft/Sec)*	Range Kilometers (Naut Mi)*
00 00 00	First Motion	586,647 (1,293,314)	90 (295)	1.8 (5.9)	0 (0)
00 01 13	Maximum Dynamic Pressure	375,026 (826,776)	12,599 (41,334)	473 (1,552)	4.2 (2.3)
00 02 16	Inboard Engine Cutoff	190,013 (418,900)	56,167 (184,275)	1,976 (6,483)	57 (31)
00 02 19	Outboard Engine Cutoff	184,822 (407,455)	59,152 (194,069)	2,033 (6,669)	62 (34)
00 02 21	S-IB/S-IVB Separation	184,059 (405,774)	60,522 (198,562)	2,032 (7,781)	65 (35)
00 02 22	S-IVB Ignition	138,028 (304,294)	61,821 (202,826)	2,064 (6,771)	67 (36)
00 02 49	Launch Escape Tower Jettison	132,141 (291,317)	86,160 (282,676)	2,119 (6,953)	117 (63)
00 09 03	S-IB Stage Impact	45,370 (100,021)	0 (0)	90 (295)	503 (271)
00 09 53	S-IVB Engine Cutoff	30,749 (67,789)	158,402 (519,692)	7,561 (24,807)	1,807 (975)
00 10 03	Orbit Insertion	30,694 (67,668)	158,544 (520,157)	7,568 (24,829)	1,881 (1,015)

*English measurements given in parentheses

SKYLAB EXPERIMENTS

The Skylab space station carries the largest array of experimental scientific and technical instruments the United States has ever flown in space, a total of 58. They fall into four general categories: life sciences, Earth resources, solar physics and corollary. Data received will permit 200 principal investigators to supervise 271 scientific and technical investigations. While most of the detailed experiment runs are planned pre-mission, there are occasions when specific observations are scheduled in real-time to take advantage of an unique opportunity, such as the solar flare and Hurricane Ava that developed during the first manned mission.

Skylab medical experiments are aimed toward measuring man's ability to live and work in space for extended periods of time, his responses and aptitudes in zero gravity, and his ability to readapt to Earth gravity once he returns to a one-g field.

Earth resources experiments (EREP) employ six devices to advance the technology of remote sensing and at the same time gather data applicable to research in agriculture, forestry, ecology, geology, geography, meteorology, hydrology, hydrography and oceanography through surveys of site/task combinations such as mapping snow cover and water runoff potentials; mapping water pollution; assessing crop conditions; determining sea state; classifying land use; and determining land surface composition and structure. On days that EREP passes are scheduled, the JSC News Center will publish site/task guides identifying principal investigators, specific locations or areas and scientific disciplines. The second manned mission has 26 EREP passes scheduled, including one pass over the Japanese island chain. Eleven EREP passes were run on the first manned visit out of 15 that had been scheduled.

ATM solar astronomy experiments utilize an array of eight telescopes and sensors to expand knowledge of our planet's Sun and its influence upon the Earth. Almost 82 hours, 80 percent of the premission scheduled ATM experiment time, were logged by the first Skylab crew while gathering some 17,000 frames of ATM film. Some 45,000 frames of ATM film will be available for the next manned mission.

A wide range of experiments falls into the corollary category, ranging from stellar astronomy and materials processing in zero-g to the evaluation of astronaut manuvering devices for future extravehicular operations.

Seven experiments selected through a national secondary school competition in the Skylab Student Project are also assigned to the second manned mission.

- more -

Experiments assigned to second Skylab mission are listed below

In-flight medical experiments (on all missions):

M071	Mineral Balance
M073	Bioassay of Body Fluids
M074	Specimen Mass Measurement
M092	Lower Body Negative Pressure
M093	Vectorcardiogram
M110	
M113	
M114	Series, Hematology and Immunology
M115	
M131	Human Vestibular Function
M133	Sleep Monitoring
M151	Time and Motion Study
M171	Metabolic Activity
M172	Body Mass Measurement

(These are three ground-based medical experiments - M078, M111 and M112 involving pre- and post-flight da

Earth Resources Experiment Package (EREP) experiments (on all missions):

S190	Multispectral Photographic Facility comprised of:
S190A	Multispectral Photographic Cameras
S190B	Earth Terrain Camera
S191	Infrared Spectrometer
S192	Multispectral Scanner
S193	Microwave Radiometer/Scatterometer and Altimeter
S194	L-Band Radiometer

The ATM experiments (on all missions):

S052	White Light Coronagraph
S054	X-Ray Spectrographic Telescope
S055A	Ultraviolet Scanning Polychromator-Spectroheliometer
S056	Extreme Ultraviolet and X-Ray Telescope
S082A	Coronal Extreme Ultraviolet Spectroheliograph
S082B	Chromospheric Extreme Ultraviolet

(Two hydrogen-alpha telescopes are used to point the ATM instruments and to provide TV and photographs of the solar disk.)

The corollary experiments:

M508	Astronaut Maneuvering Equipment
M512	Materials Processing Facility
M516	Crew Activities/Maintenance Study
* M555	Gallium Arsenide Crystal Growth
* S015	Zero-g Single Hunman Cells
S019	Ultraviolet Stellar Astronomy
S063	Ultraviolet Airglow Horizon Photography
# S071	Circadian Rhythm Pocket Mice
# S072	Circadian Rhythm Vinegar Gnats
S073	Gegenschein/Zodiacal Light
S149	Particle Collection
S150	Galactic X-Ray Mapping
S230	Magnetospheric Particle Collection
T003	Inflight Aerosal Analysis
T020	Foot-Controlled Maneuvering Unit

The student investigations:

# ED21	Libration Clouds
# ED25	X-Rays from Jupiter
# ED32	In-Vitro Immunology
# ED52	Web Formation
# ED 63	Cytoplasmic Streaming
# ED74	Mass Measurement
ED76	Neutron Analysis

* Deferred from Skylab 2

\# Unique to Skylab 3

(Details of the above experiments may be found in Skylab Experiments Overview, available from the Government Printing Office (Stock No. 3300-0461) $1.75/copy; or from experiment booklets and manuals in the KSC and JSC Newsrooms.)

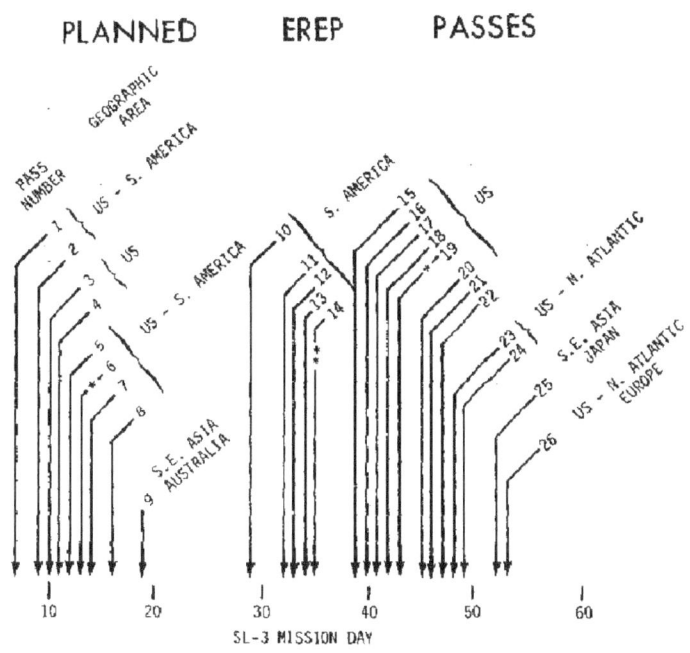

REAL-TIME FLIGHT PLANNING

Time was when pre-mission flight plans were followed "by the numbers" with few changes except those caused by systems malfunctions. Skylab flight planning, however, is almost wholly done in real-time, with the pre-mission flight plan serving mainly as a guide to Mission Control Center flight planners. Each day's flight plan is designed to yield the highest experiment data return.

Teleprinted to the Skylab space station early in the morning before the crew wakens, the daily flight plan takes advantage of unique opportunities that enhance data gathering for particular experiments. For example, forecasts of cloud-free EREP sites and ground observatory predictions of unusual solar activity have a bearing upon when EREP passes and ATM runs are scheduled in the flight plan.

The Skylab flight planning cycle begins at midnight Houston time (CDT) with a team of flight planners in Mission Control Center drafting a "summary flight plan" for the following crew work day that will start 32 hours later. This first team is relieved by the so-called "execution" team (day team) of flight controllers which carries out the existing detailed flight plan for the immediate day. Then the flight planners on the next, or "swing" shift develop from the summary flight plan a detailed flight plan for the following day, nailing down the activity details first summarized in the early morning hours --- and so on in leapfrog fashion.

Daily flight plans pivot around experiment requirements which have to be resolved, optimum crew time use, and mission objectives still have to be met. Proposed summary flight plans embrace the viewpoints of Skylab systems engineers, experiment principal investigators, flight surgeons, mission management, the flight crew and the weatherman's forecast for potential EREP survey sites. Precedence is given to mandatory operations, ATM, EREP and medical experiments, with other experiments and operations filling the remaining time.

Revised summary flight plans will be reproduced daily and distributed to newspersons at the JSC Newsroom, and the daily crew teleprinter "loads" will be available for review at the query desk.

DAILY CREW ACTIVITY

Skylab crew work days in space are not a whole lot different from work days on Earth. The normal day starts at 6 a.m. and runs until 10 p.m. CDT. Days off, however, are fewer and farther between.

Breakfast is at 7 a.m., lunch at noon and dinner at 6 p.m. CDT --- except for the man on duty at the ATM console during lunch, who shifts his meal time so that he can be relieved at the console. Eight hours of sleep are normally scheduled each day

During the mission the astronauts will be operating and monitoring about 60 items of experimental equipment and performing a wide variety of tasks associated with the several hundred Skylab scientific and technical investigations.

Depending upon experiment scheduling requirements, Skylab crews have a day off about every seventh day.

About two 15-minute personal hygiene periods are scheduled each day for each crewman and one hour and 30 minutes for physical exercise. Additionally, an hour a day maybe set aside for R&R -- rest and relaxation. Another regularly scheduled activity each day is two and a half hours of systems housekeeping, such as cleaning of environmental control system filters, trash disposal and wiping down the walls of the space station.

Mission Control Center flight planners fill the remaining eight hours of the crew work day with experiment operations.

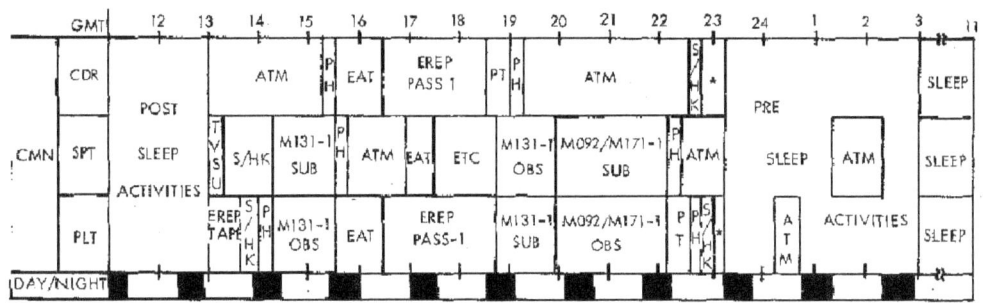

SKYLAB STATUS: WHAT HAPPENED

The unmanned Skylab space station was launched on May 14. Approximately one minute after liftoff, at the time of highest aerodynamic pressure, the meteoroid shield around the outside of the workshop was torn off and apparently caused one of two solar panels used to generate electricity for the laboratory also to be torn away and jammed the other in a way that prevented its full deployment.

The net result was that Skylab was in good orbit, but had only about half of its power-generating capability in operation and the spacecraft was overheating. The overheating occurred because the lost meteoroid shield also provided thermal balance. It was painted in a way to reflect enough sunlight so that the laboratory would stay cool.

A principle purpose of the meteoroid shield -- a thin aluminum skin .025 inch thick -- was to protect the Skylab vehicle from the possible impacts of tiny space particles by providing enough resistance to make them splatter and lose energy before striking the inner walls of the workshop.

Without the shield the workshop will be exposed to more potential direct hits which might result in some minor air leaks by the end of the 240-day mission. The pressurization system is adequate to meet such a contingency.

The Skylab team responded quickly to the situation. The first task was to stabilize conditions. Temperatures were increasing rapidly. External skin temperatures were estimated to be as high as 325 degrees Fahrenheit. There was concern that the unrefrigerated on-board food, medicines, and film might spoil.

The flight control team tried to find an attitude or position of Skylab which would minimize the heating and at the same time cast sufficient sunlight on the remaining solar cells, those attached to the ATM, to generate the electricity required to operate the space station. Ground controllers oriented the orbiting space station from one attitude to another to control temperatures and still obtain enough sunlight for power generation

After a great deal of calculation, analysis and some experimentation, inside temperatures were stabilized at approximately 125 degrees and power levels at about 2800 watts, which barely covered the unmanned housekeeping requirements. Although some food and medicines were assumed to have been spoiled, there remain sufficient unspoiled food on board for all three missions, and some of the medicines were replaced by the first crew to go aboard

While the laboratory was being stabilized, it became very clear that a fix would be required. The laboratory was too hot for normal habitation and the temperature was too high to carry out the medical experiments.

The temporary pitched-up attitude of the laboratory was determined by the need to balance solar heating and power generation, and was therefore not fully appropriate for either the solar experiments (which require precise pointing at the Sun) or the Earth resources experiments (which require equally precise pointing at the Earth). The best way to fix the Skylab was to provide quickly a sunshade which would once again reflect away the proper amount of sunlight so that the laboratory would remain cool and regain its pointing flexibility.

By the third day after launch, a number of approaches to thermal control had been well enough defined to develop a firm design, development, manufacturing, test and training schedule. The aerospace industry and NASA centers has responded fast and well to the call for help. The crew launch date was then reset for Friday, May 25, a delay of 10 days.

On the day before launch, three different sunshades were selected to go along with the crew because no one really knew what the astronauts would find when they rendezvoused with Skylab. Officials didn't know if the meteoroid shield was completely and cleanly severed or whether parts of it were obstructing areas where the sunshade might be installed. By carrying several different sunshades, the crew would at least have one suitable for the situation.

One sunshade, called a SEVA sail, was a trapezoidal awning to go on ropes that would stretch from the base of the Skylab work shop to a hand rail on the apollo telescope mount. (SEVA refers to standup extravehicular activity). One of the astronauts, standing up in the hatch of the undocked command/service module, would first attach ropes and hooks to the Skylab base. The CSM would then be maneuvered toward the ATM where the converging ropes would be attached at a single point, pulled taut and the 22-by-24-foot sail would be positioned over the workshop.

A different "twin-boom" sunshade was designed to be deployed from the ATM truss assembly during an EVA. Two of the astronauts would step out of the airlock in pressure suits, affix a special bracket on the ATM structure, and attach to the bracket two long poles that they had assembled from short sections. At the end of each pole is a pulley with a rope threaded through it. With poles forming an inverted vee extending back over the workshop, a sheet of reflecting material would be hooked on the ropes and pulled, like a sail, to a position over the workshop where the meteoroid shield should have been.

Actually used by the first Skylab crew was the simpler parasol concept that did not require an EVA. After docking and entering the Skylab, the crew enxtended a folded canopy through the scientific airlock on the Sun side of the workshop. Once outside the spacecraft, the nylon and aluminized Mylar material was deployed mechnically, like a parasol, to form a 22-by-24-foot rectangular thermal shield over the workshop's exposed area. This approach offered the least difficult means of quickly bringing the heating problem under control.

The astronauts had trained with all three concepts at the Johnson Space Center and in the zero-gravity simulator at the Marshall Space Flight Center.

Prior to launch, program officials approved a stand-up EVA from the undocked command module to remove any debris that covered the scientific airlock and to attempt, if feasible, to free the jammed solar array. The decision was made to carry bolt cutters, tin snips, and a bending tool to help with the tasks.

On launch day, Pete Conrad, Joe Kerwin and Paul Weitz went through launch and rendezvous, soft docked, prepared for stand-up EVA, undocked, and tackled the salvage problem. Weitz did the stand-up EVA as Kerwin steadied his legs and Conrad maneuvered the CSM.

The scientific airlock was clear of debris but the crew found a length of 3/4-inch angle aluminum bent up and over the solar array beam. The beam, deployed about five degrees, was firmly restrained by the metal strap.

The angle aluminum strap has a series of bolts, one of which apparently was driven into the .025 aluminum of the solar wing, securely fastening it. The slope of the strap along the side of the beam was such that the tools could not get a grip to pry it away.

The next day, the astronauts followed procedures written just two days earlier and deployed the parasol. By the 11th mission day the inside temperatures had dropped to 75 degrees.

Immediately after parasol deployment the crew started operating experiments. They found that one, the S019 ultraviolet stellar astronomy experiment had a mirror tilt gear drive mechanism that was jammed. They promptly disassembled and assembled it again. It's working fine.

Then, as temperatures dropped and flight planners began to see daylight, Skylab encountered a new problem on day five.

-more-

During the first full EREP pass, the space station left solar orientation and went to "local vertical" as planned. This moves the solar arrays out of the sunlight and the batteries go to discharge. On that first pass, four battery systems which had gotten hot in the unmanned "pitch-up" attitude showed they were taking less than one-half charge, and one battery system dropped off the line completely. The loss was serious even though there are 18 such battery packs in the ATM power supply system.

However, the backup astronaut crew, plus a small sleepless group of specialists had been continuing to work on procedures to remove the strap that held the solar wing undeployed.

The procedures were radioed up on day 12, the crew practiced in space (inside the workshop) on day 13, and went EVA on day 14. Kerwin and Conrad cut the strap, broke a restraining bolt, and erected the solar wing. Within hours the solar wing was supplying electricity. Skylab was in full working order to carry out its planned 270 scientific and technical investigations.

In addition, the crew performed a number of other actions that saved certain experiments which otherwise could not have functioned. And, on their EVA they solved the problem of a malfunctioning ATM battery relay by banging on it with a hammer, a repair technique warmly endorsed by appliance owners and machinery operators everywhere.

The following summarizes the status of Skylab as it awaits the next working crew:

1. ELECTRICAL POWER SYSTEM

The Skylab power system was operating well and no failures or degradations were experienced in the latter phase of the first manned visit. The average power generating capacity after the crew left was ranging between 6700 watts and 8500 watts, depending on the Sun angles. The average cluster loads were about 4700 watts without command module loads and will increase to about 5900 watts when command module loads are added. All eight airlock module battery regulator modules have been functioning well since the astronauts deployed solar wing no. 1. Electricity generated by the workshop wing is fed to the airlock module (AM) battery system.

Some degradations have occurred in the ATM Power System due to the thermal stresses induced early in the mission. One of the 18 ATM charger battery regulator modules (CBRM) is inoperative. Four CBRM's exposed to high temperatures had shown some degradation in battery storage capacity but have recovered most of their original capacity. One of the solar cell modules has had a 10 per cent degradation due to high temperatures and one regulator was operating below specification during the daylight passes reducing its integrated output to 80 per cent of capability. However, the total Skylab power system shows sufficient margin to accomplish the remaining two missions.

-more-

2. ENVIRONMENTAL CONTROL SYSTEMS

The workshop internal temperatures were stabilized originall at about 73 degrees Fahrenheit by the deployment of the parasol. During the unmanned operations, temperatures rose to the mid-90s due to increased time in the sunlight during each orbit. It is planned to deploy the twin-boom sunshade early in the next mission to improve the shade coverage and to counteract the effects of any ultraviolet degradation of the parasol. In addition, a parasol of improved material will be brought up by the crew to be available if needed.

The airlock cooling system has been operating well and effectively cooling the equipment. The system has supplied sufficient cooling during EVA and is effectively controlling the cluster humidity. An earlier malfunction of the thermal control valve has been eliminated by a thermal/pressure cycling procedure and the valves in both coolant loops are now modulating properly.

The ATM passive and active cooling systems also are operatin well.

In summary, the environmental control systems, except for the loss of the meteoroid shield, retains original redundancies and should satisfactorily complete the Skylab missions.

3. ATTITUDE CONTROL SYSTEM

In general, the attitude control system has functioned as planned. Gyro drift rates have required more ground management than was anticipated. The high drift has been attributed tentatively to bubbles in the gyro fluid during vacuum operation.

The high drift rates do not generally present a problem during solar inertial orientation since continuous gyro update is possible during the daylight portion of each orbit. However, during the early part of the mission, when off-nominal pointing modes were required to control the thermal environment, alternate means to verify the proper attitude were required.

The three control moment gyros (CMG) and digital computer system are operating satisfactorily. Gravity gradient dumping of angular momentum during the dark portion of the orbit has been satisfactory and has prevented momentum saturation or unnecessary usage of the thruster attitude control system (TACS).

Due to the early off nominal pointing modes, significantly more TACS propellants were used than anticipated. The amount remaining, about 44 per cent of the pre-mission total, is sufficie for nominal 3 CMG or 2 CMG operation for the last two missions. If problems develop similar to the initial SL-1 operation, TACS augmentation is possible by means of the CSM Reaction Control System during the manned phases.

-more-

4. HABITABILITY SUPPORT SYSTEMS

All elements of the Habitability Support Systems have been functioning as specified without any significant anomalies. The workshop waste management system operation has received very favorable comments from the crew. The system has functioned as planned and the crew has been pleased with the shower. Similarly, the Skylab food and operation of the systems for food preparation have satisfied the crew.

The food refrigeration system operated flawlessly throughout the first unmanned and manned phases. During Skylab deactivation, however, a malfunction in the radiator by-pass valve resulted in rising freezer temperatures. The redundant secondary refrigeration loop exhibited similar characteristics. Continuous on-off cycling of the by-pass valve in the primary system resulted in reversal of the warming trend and brought temperatures back to near normal. A trouble shooting procedure has been developed to insure proper operation of the secondary system in the event its use becomes necessary during the manned part of the next mission.

5. INSTRUMENTATION AND COMMUNICATIONS SYSTEM

Voice communication between the Skylab and Mission Control has been good during station passes and tape recorder dumps. TV quality, both in real time and through video tape recorder transmission, was excellent.

One of Skylab's two color TV cameras became inoperative but two new cameras will be resupplied on the SL-3 launch.

One of the three active airlock module tape recorders became inoperative after 843 hours of operation and was replaced by the crew. Later, this replacement recorder malfunctioned during the unmanned phase after 320 hours of operation. Four spare tape recorders were aboard Skylab originally.

As a result of these malfunctions, tape recorder operations during the second mission's unmanned phase has been reduced to three hours per day. Two new tape recoders will be brought up on SL-3 to fully restore the spares inventory.

One of the airlock's three 10-watt transmitters failed and was replaced by switching to the 2-watt transmitter without degradation of experiment or systems data transmissions. Additional transmitter failures, however, would degrade data transmission capability. Consequently, studies are underway at MSFC to determine the feasibility of transmitter replacement during the final mission.

ACCOMPLISHMENTS

The first Skylab manned mission made significant contributions to the basic purpose for which the space station program was established. All mission objectives of SL-1/2 were successfully accomplished.

Broadly summarized, the accomplishments were as follows:

1. Approximately 80% of the solar data planned has been obtained. Major scientific accomplishment was monitoring of solar flare on June 15.

2. Eleven of the fourteen Earth resources data runs planned were accomplished. (6 experiments/instruments were operated for 77 Principal Investigators)

3. All medical experiments (16) were conducted as require by the operational medical protocols. The time histor of man's adaptation to the zero-g environment obtained for the first time.

4. Data was taken on all experiments scheduled for SL-2 except those that could not be accomplished due to use of the solar airlock for parasol deployment and weight or power limitations.

5. Data was obtained on five student investigations. Two student investigations are rescheduled for SL-3 (ED12 Volanic Study, ED22 Objects in Mercury's Orbit, data could not be obtained because of orbit track or location of astronomical body).

Major support form the astronauts included:

Maintenance: Experiment door pinned; coronagraph occulating disk dusted off; faulty camera replaced; and battery package relay was released.

Scientifically: Through astronaut alertness the early portion, or development, of a solar flare was observed with all ATM instruments.

- more -

EXPERIMENTS SUMMARY:

Not all of the returned pictures and other data are expected to be completely useful for the scientific investigations. For example, cloud cover and procedural problems will have reduced the usefulness of some of the EREP pictures. Similarly, equipment problems, exposure settings and other difficulties may have reduced the scientific product to be expected form some ATM and other astronomy pictures. As data from the first manned mission are analyzed procedures are being developed to provide improved efficiency for obtaining scientific observations on the second mission.

ATM ACCOMPLISHMENTS SUMMARY

MANNED VIEWING TIME		81 hrs	
SOLAR VIEWING PERIODS (passes)		76 FULL	29 PARTIAL
FILM USAGE (frames)	USED	PLANNED	
S052	4519	8025	
S054	6739	6976	
S056	4296	6000	
S082A	219	201	
S082B	1608	1608	
TOTAL	17377	22810	76%

*S052, S054 and S055 CONTINUE TO OPERATE IN UNMANNED MODE

EREP ACCOMPLISHMENTS SUMMARY

- DATA COLLECTED

MULTISPECTRAL CAMERA (S190A)	6500 FRAMES
EARTH TERRAIN CAMERA (S190B)	960 FRAMES
INFRARED SPECTROMETER (S191 Data Acq. Camera)	5400 FRAMES
SCANNER (192), INFRARED SPECTROMETER (191) & MICROWAVE SENSORS (193, 194)	41,000 FT. MAGNETIC TAPE

- DATA COLLECTED OVER
    - X 31 STATES & PUERTO RICO
    - X 6 FOREIGN COUNTRIES, MEXICO, BRAZIL, BOLIVIA, NICARAGUA, COLUMBIA, AND CANADA.
    - X GULF OF MEXICO, CARIBBEAN SEA, PACIFIC/ATLANTIC OCEANS
- DATA OBTAINED FOR 75 PRINCIPAL INVESTIGATORS, (66 U.S. and 9 FOREIGN) AND FOR SENSOR PERFORMANCE EVALUATIONS
- DATA WAS COLLECTED FOR 186 INDIVIDUAL TASKS ON SL-2

	ACHIEVED	PLANNED	

ATM

MANNED VIEWING TIME	81 HRS	101 HRS.	81%
EXPERIMENT FILM	17,352 FRAMES	22,810 FRAMES	75%
H-ALPHA-1 FILM	13,000 FRAMES	16,000 FRAMES	

EREP

PASSES	11 (5 SHORT)	14	79%
ETC PASSES	6	10	60%
PHOTOS	7460	9000	83%
TAPE REELS	6	6	

186 TASK SITES COVERED, DATA TAKEN FOR 75 INVESTIGATIONS

MEDICAL

| PERFORMANCES | 137 | 147 | 93% |
| MAN HOURS | 148 | 158 | 94% |

ALL PLANNED URINE, BLOOD, & FECES SAMPLES TAKEN (EXCEPT FIRST 3 DAYS URINE IS UNKNOWN)

COROLLARIES

SCIENTIFIC AIRLOCK	32 MAN HRS.	38 MAN HRS.	84%
OTHER COROLLARIES	22 MAN HRS.	14 MAN HRS.	157%
UV ASTRONOMY PASSES	10	16	
MATERIALS SCIENCE OPERATIONS	9	10	

FOUR ASSIGNED EXPERIMENTS NOT PERFORMED - S020, T025, S015, M555

STUDENTS

| BACTERIA & SPORES, NEUTRON ANALYSIS | 4 MAN HRS. | 4½ MAN HRS. |

DATA FOR ATMOSPHERIC ABSORPTION OF RADIANT HEAT, U.V. FROM QUASARS, U.V. FROM PULSARS

SKYLAB BETWEEN VISITS

The second Skylab mission is in two parts: Unmanned and manned. The unmanned portion has been underway since, June 22 at 4:55 AM EDT when the Conrad/Kerwin/Weitz crew undocked from Skylab. The manned portion, a 56-day workout, will start when the Bean/Garriott/Lousma crew docks with the space station.

The ATM experiments which can operate in the unmanned configuration (S052, S054, and S055) are not only continuing their long range observational programs, but gathered unique data in support of numerous international ground based and rocket observations of the June 30 eclipse.

Highest priority was placed upon the eight to ten orbits bracketing the eclipse where the combined ATM and ground based observations were used to determine temporal evolution of solar features. Observations during the days before and after the eclipse permitted the study of the three-dimensional structure of various solar features and hence, increased the value of non-ATM observations during the eclipse.

Each of the ATM experiments also has more specific goals during this time.

The data obtained by the S052 White Light Coronagraph (High Altitude Observatory) provided a cross calibration with 30 collaborators on the polarization of the corona since ground-based observers must contend with an additional polarization contribution from the Earth's atmosphere.

The S054 X-Ray Spectrographic Telescope (American Science and Engineering) obtained a series of solar images with its thinnes filter (in the wavelength ranges 3.5-36 and 44-60 Angstroms) for collaboration with ground based observations and to identify transient features during the time of the eclipse.

-more-

The S055 Ultraviolet Scanning Polychromator-Spectroheliomete (Harvard College Observatory) studied specific features which occurred at the solar limb at the time of the eclipse. Additiona ultraviolet spectra of these features obtained over a longer time span will specifically augment the data taken by a rocket experiment launched in Mauritania.

The data taken by S055 were sent via telemetry to the ground to be processed by the experimenter for use by the ground observers of the eclipse. Experiments S052 and S054 photographed the eclipse events. Their film will be retrieved at the end of the Skylab 3 mission in late September.

Between crew visits to the Skylab space station, ground controllers become sort of absentee landlords for the station. Experiments and systems status monitoring and off/on commanding is handled remotely through data and command telemetry links from the Mission Control Center at Houston.

The Skylab cluster remained in the solar inertial attitude after the first Skylab crew undocked for return to Earth. The space station's attitude and pointing control system kept the ATM telescopes aligned with the Sun.

Skylab internal pressure is vented down from five to about two pounds per square inch after the Skylab crews depart.

Attitude pointing and control systems and both major electri systems in the space station remain fully "up" during unmanned periods. The telemetry and command systems also stay "live" to relay systems information to ground controllers and to accept commands for housekeeping functions and data retrieval. The environmental control system remains inactive, except for the refrigeration system and some thermal control components.

A number of passive Skylab experiments require long term exposure in space to acquire the desired scientific data. While the orbiting station has been unmanned the following experiments have been in operation:

1. S149 - particle collection - Four cassettes with polished surfaces are being exposed to collect micrometeorites and dust particles. The cassette holder is extended on a boom through a scientific airlock.

2. S228 - trans-uranic cosmic rays - An array of plastic modules comprised of 0.010 inch thick sheets will be exposed till the end of the third manned mission to attempt to determine the existence of high-Z cosmic rays. Unit is inside the workshop.

3. S230 - magnetospheric particle composition - A collection of foils (aluminum, platinum, aluminum oxide) are mounted on an exterior strut where they can be bombarded by rare gases (helium, neon, argon). Samples will be returned after each mission and the isotopic abundance of the gases collected in the foils will be compared with the abundance found on the lunar surface.

4. D024 - thermal control coatings - One set of paint and film samples (2 arrays) were returned by the first crew. A second set, with longer exposure to the space environment, will be returned by the next crew.

5. ED76 - neutron analysis - Ten detectors are measuring the ambient neutron flux at Skylab orbital altitudes. Four detectors were returned by the first crew. The remaining six will be returned on the last mission.

SKYLAB AND RELATED OBJECTS VISIBLE

NASA will continue to distribute information enabling people in most populated areas of the world to see the Skylab space station as it orbits the Earth.

Skylab is visible to the unaided eye only in clear skies during the two hours before dawn and after dusk -- when the viewer is in the Earth's shadow and the space station is in the Sunlight at its orbiting altitude of 435 kilometers (270 miles). Sunlight reflected off portions of the large spacecraft is what makes Skylab visible.

The space station under the best visual conditions, will appear approximately as bright as the brightest star in the sky. It will be moving easterly fast enough to be easily distinguishable from stars and may be visible for as long as seven minutes.

There has been some confusion on the part of Skylab watchers because of other visible objects traveling along the same track both ahead of and behind the space station.

These other objects were launched with Skylab and include four jettisoned, separate panels from the payload cover, the Saturn booster's S-II second stage, a jettisoned radiator shield and one unidentified object. Relative positions of the objects and Skylab keep changing from day to day. As in a 500-mile auto race, some of the objects go faster and overlap the slower ones.

Skylab is in a stable attitude and its brightness varies gradually as it moves across the star field. The other objects are tumbling in flight and seem to slowly blink on and off. An exception is the large S-II stage which is even brighter than Skylab. The stage is large, cylindrical and painted white. These features give it high and fairly steady visibility even though it tumbles as it orbits.

Sighting information for key cities is computed and issued every two weeks by the NASA Marshall Space Flight Center, Huntsville, Ala.

Ground areas that Skylab crosses include all of the U. S. except Alaska, a strip of Southern Canada, all of South America, China, Africa, Australia, India, most of Asia and southern portions of USSR.

The space laboratory flies over 89 per cent of the world's population and 65 per cent of the Earth's land areas as it orbits from 50 degrees north of the Equator to 50 degrees south.

Skylab is 36 meters (118 feet) long and its solar cell arrays are about 31 meters (100 feet) from tip to tip.

-end-

NASA NEWS

NATIONAL AERONAUTICS AN
SPACE ADMINISTRATION
Washington, D.C. 20546
202-755-8370

FOR RELEASE:
November 5, 1973

PROJECT: SKYLAB 4
Third Manned Mission

PRESS KIT

contents

GENERAL RELEASE	1-5
OBJECTIVES OF SKYLAB PROGRAM 1	6
OBJECTIVES OF THIRD MANNED MISSION	7-8
EXPERIMENTS	9-11
COMET KOHOUTEK AND SKYLAB	13-23
ACCOMPLISHMENTS	24
REAL-TIME FLIGHT PLANNING	25-27
MISSION PROFILE	28-29
COUNTDOWN	30-32
SKYLAB RESCUE VEHICLE	33

NOTE: Details of the Skylab spacecraft elements, systems, crew equipment and experimental hardware are contained in the Skylab News Reference distributed to the news media. The document also defines the scientific and technical objectives of Skylab activities. This press kit confines its scope to the third manned visit to Skylab and briefly describes features of the mission.

NATIONAL AERONAUTICS A SPACE ADMINISTRATION
Washington, D. C. 20546
AC 202/755-8370

FOR RELEASE:

November 5, 1973

Bill Pomeroy
Headquarters, Washington, D.C.
(Phone: 202/755-3114)

RELEASE NO: 73-229

SKYLAB PUTS OUT WELCOME MAT FOR COMET

Space Station Skylab's final tenants will move into their orbiting home 270 miles above the Earth on or about November 10 to complete a harvest of scientific information about our home planet and our life giving star, the Sun. Two earlier threesomes of tenants occupied the space station for 28 and 59 days before "leaving the key under the mat" for the final crew that will live aboard Skylab for up to two months.

Earth resources, solar astronomy, medical and other experiments will fill the waking hours of the Skylab crewmen, with the opportunity to view the comet Kohoutek as an added bonus in December or January.

Flying above the distorting layers of Earth atmosphere, Skylab's solar telescopes and astronomical cameras are expected to provide valuable data about the make-up of comets as well as continuing the surveillance of the flares, prominences and other dynamic events taking place on the face of the Sun.

The manned spacecraft's crew can provide fast reaction times in monitoring and recording the sudden emergence and rapid development of unpredicted events on the Sun or the Kohoutek comet. Unmanned satellites, losing valuable time in ground-based data analysis and in radio relay of information, cannot react as quickly.

For the final Skylab manned mission three categories of experiments -- solar physics, Earth resources and medical -- will follow patterns set during the first two missions, with additional tasks in gathering scientific data. The Apollo Telescope Mount (ATM) has been assigned comet Kohoutek observations in addition to its other chores. Although 30 Earth resources surveys are planned, options are open for up to 10 additional surveys. Medical investigations have been broadened to increase the knowledge of how the human body adapts to long periods of space flight.

Other corollary scientific and technological experiments -- including ten student experiments selected by the National Science Teachers Association in a nationwide competition among high school students -- will round out Skylab's final services as an orbiting scientific station.

-more-

A scientific highlight for the next Skylab crew will be the passage near the Earth-Sun region of the comet Kohoutek in December and January. The mission period will cover the comet's perihelion (closest approach to Sun) on December 28, allowing observations both before Kohoutek passes behind the Sun as well as the post-perihelion changes as the comet swings around and retreats into deep space.

Comet observations will be made with solar and astronomical instruments and cameras, including two mounted outside on the ATM truss for operation by the astronauts during an EVA. Skylab's vantage point above the atmosphere offers a unique opportunity to observe the changing composition of a comet in the ultraviolet spectrum. Little is known about the structure of comets, other that the popular theory that they are like giant snowballs hurtling through the solar system. Skylab consumables will be closely husbanded to keep open an option to extend the final mission to about 70 days, depending upon crew health and the condition of space station systems. An extension of the mission would provide an opportunity for northern hemisphere winter Earth resources surveys of ice and snow distribution and the start of the growing season.

Crewmen are Gerald P. Carr, commander; Dr. Edward G. Gibson, science pilot; and William R. Pogue, pilot. Carr is a U.S. Marine Corps lieutenant colonel, Gibson a civilian scientist-astronaut, and Pogue a U.S. Air Force lieutenant colonel. None has flown in space.

-more-

This liftoff, designated Skylab 4, is scheduled for 11:41 a.m. EST November 10 atop a Saturn 1B from NASA Kennedy Space Center's Launch Complex 39, Pad B. Rendezvous and docking will occur during the fifth command/service module orbit after a standard rendezvous maneuver sequence.

After docking with the space station, the crew will begin "turning on" Skylab in preparation for two months or more of working and living in space.

Undocking for a 56-day mission would be January 6 with a two-impulse service propulsion retrofire sequence bringing comma module splashdown in the north-central Pacific, about 509 km (310 statute miles) north-north-west of Honolulu, Hawaii. Splashdown would be at 5:44 p.m. EST on January 6. If the mission is extended to 69 days, splashdown would be on January 19 at about 4 p.m. EST in the central Pacific.

Like any new home, the experimental space station had several defects -- anomalies -- which had to be corrected before Skylab became habitable for the planned long missions. At launch, an aluminum micrometeoroid shield ripped off, taking with it one of the main solar cell arrays for generating electrical power and jamming the other to prevent its unfolding. Loss of the shield caused temperatures inside the space station to rise to an uncomfortably high level.

-more-

The first crew to visit Skylab, Charles Conrad, Joseph Kerwin and Paul Weitz, carried along a parasol-like device which they erected to help bring the temperatures down. The jammed solar cell array was freed during an EVA and the first manned visit turned out to be successful, not only from the standpoint of gathering scientific data, but also in demonstrating that men can take on difficult repair and construction jobs in space.

The second Skylab crew was not without its share of repair work around the space station. Alan Bean, Owen Garriott and Jack Lousma erected a second sunshield to supplement the first one. They installed a "six-pack" of gyros in the Skylab attitude control system to replace balky gyros that had become undependable.

Two repair chores are planned for the final Skylab crew. They include replacing the fluid in a primary coolant loop that was leaking during the second Skylab manned mission; and inspection of the antenna on the S193 Microwave Radiometer/Scatterometer Altimeter in the Earth Resources Experiment Package (EREP).

OBJECTIVES OF THE SKYLAB PROGRAM

The Skylab Program was established for four purposes: (a) to determine man's ability to live and work in space for extended periods; (b) to extend the science of solar astronomy beyond the limits of Earth-based observations; (c) to develop improved techniques for surveying Earth resources from space; and (d) to increase man's knowledge in a variety of other scientific and technological regimes.

Skylab, the first space system launched by the United States specifically as a manned orbital research facility, is providing a laboratory with features not available anywhere on Earth. These include: a constant zero gravity environment, Sun and space observation from above the Earth's atmosphere, and a broad view of the Earth's surface.

Dedicated to the use of space for the increase of knowledge and for the practical human benefits that space operations can bring, Skylab is pursuing the following objectives:

Physical Science - Increase man's knowledge of the Sun, its influence on Earth and man's existence, and its role in the universe. Evaluate from outside Earth's atmospheric filter, the radiation and particle environment of near-Earth space and the radiations emanating from the Milky Way and remote regions of the universe.

Life Science - Increase man's knowledge of the physiological and biological functions of living organisms by making observations under conditions not obtainable on Earth.

Earth Applications - Develop techniques for observing Earth phenomena from space in the areas of agriculture, forestry, geology, geography, air and water pollution, land use and meteorology.

Space Applications - Augment the technology base for future space activities in the areas of crew/vehicle interactions, structures and materials, equipment and induced environments.

-more-

OBJECTIVES OF THE THIRD MANNED SKYLAB MISSION

The third Skylab manned mission officially began September 25 when the second CSM and its crew separated from the space station just prior to reentry. The unmanned portion of this SL-4 mission will continue until the third crew is launched. After docking, the crew will enter Skylab, reactivate its systems, and proceed to inhabit and operate the orbital assembly for up to 56 days. During this time the crew will perform systems and operational tests and the assigned experiments.

The objectives of the third Skylab manned mission are as follows:

1. <u>Perform unmanned Saturn Workshop operations</u>

    a. Obtain data for evaluating the performance of the unmanned station.

    b. Obtain solar astronomy data by unmanned ATM observations.

2. <u>Reactivate and Man Skylab in Earth orbit</u>

    a. Operate the cluster (SWS plus CSM) as a habitable space structure for up to 56 days after the SL-4 launch.

    b. Obtain data for evaluating the performance of the space station.

    c. Obtain data for evaluating crew mobility and work capability in both intravehicular and extravehicular activity.

3. <u>Obtain medical data on the crew for use in extending the duration of manned space flights</u>

    a. Obtain medical data for determining the effects on the crew which result from a space flight of up to 56 days duration.

    b. Obtain medical data for determining if a subsequent Skylab mission of greater duration is feasible and advisable.

-more-

4. Perform in-flight experiments

   a. Obtain ATM solar astronomy data for continuing and extending solar studies beyond the limits of Earth-based observations.

   b. Obtain Earth resources data for continuing and extending multisensor observations from Earth orbit.

   c. Obtain data of the comet Kohoutek beyond the limits of Earth-based observations.

   d. Perform the assigned scientific, engineering, technology and DOD experiments.

-more-

SKYLAB EXPERIMENTS

The Skylab space station carries the largest array of experimental scientific and technical instruments the United States has ever flown in space, a total of 58. They fall into four general categories: life sciences, Earth resources, solar physics and corollary. Data received will permit 200 principal investigators to supervise 271 scientific and technical investigations. While most of the detailed experiment runs are planned pre-mission, there are occasions when specific observations are scheduled in real-time to take advantage of unique opportunities such as solar flares and hurricanes observed during the first and second mission.

Skylab medical experiments are designed to measure man's ability to live and work in space for extended periods, his responses and aptitudes in zero gravity, and his ability to readapt to Earth gravity once he returns to a one-g field.

Earth resources experiments (EREP) employ six devices to advance remote-sensing technology and at the same time gather data applicable to research in agriculture, forestry, ecology, geology, geography, meteorology, hydrology, hydrography and oceanography through surveys of site/task combinations such as mapping snow cover and water runoff potentials; mapping water pollution; assessing crop conditions; determining sea state; classifying land use; and determining land surface composition and structure. On days that EREP passes are scheduled, the JSC News Center will publish site/task guides identifying principal investigators, specific locations or areas and scientific disciplines. The third manned mission has 30 EREP passes scheduled with possible options for up to 10 more, including passes over the United States, South America, Europe, Africa, Australia/New Zealand, Malaysia and Japan. An extension to a 70-day mission duration would allow coverage of snow and ice distribution in the northern hemisphere and the start of the growing season in the United States.

ATM solar astronomy experiments utilize an array of eight telescopes and sensors to expand knowledge of our planet's Sun and its influence upon the Earth. Additionally, ATM instruments will be used in observations of the comet Kohoutek between December 14 and January 2 and other non-solar events such as the planet Mercury's transit and a solar eclipse.

-more-

A wide range of experiments falls into the corollary category, ranging from stellar astronomy and materials processing in zero-g to further evaluation of astronaut maneuvering devices for future extravehicular operations. Several instruments in the corollary category will also be used in observations of the comet Kohoutek, as will a new experiment, S201, a modified version of an instrument taken to the Moon on Apollo 16.

Ten experiments selected by the National Science Teachers Association through a national secondary school competition in the Skylab Student Project are assigned to the third manned mission.

Experiments assigned to the third manned Skylab mission are listed below:

In-flight medical experiments (on all missions):

M071	Mineral Balance
M073	Bioassay of Body Fluids
M074	Specimen Mass Measurement
M092	Lower Body Negative Pressure
M093	Vectorcardiogram
M112	
M113	
M114	Hematology and Immunology
M115	
M131	Human Vestibular Function
M133	Sleep Monitoring
M151	Time and Motion Study
M171	Metabolic Activity
M172	Body Mass Measurement

(These are two ground-based medical experiments – M078 and M111 – involving pre- and post-flight data.)

Earth Resources Experiment Package (EREP) experiments (on all missions):

S190A	Multispectral Photographic Cameras
S190B	Earth Terrain Camera
S191	Infrared Spectrometer
S192	Multispectral Scanner
S193	Microwave Radiometer/Scatterometer and Altimeter
S194	L-Band Radiometer

The ATM experiments (on all missions):

S052	White Light Coronagraph
S054	X-Ray Spectrographic Telescope
S055A	Ultraviolet Scanning Polychromator-Spectroheliometer
S056	Extreme Ultraviolet and X-Ray Telescope
S082A	Coronal Extreme Ultraviolet Spectroheliograph
S082B	Chromospheric Extreme Ultraviolet Spectrograph

(Two hydrogen-alpha telescopes are used to point the ATM instruments and to provide TV and photographs of the solar disk.)

The corollary experiments:

D024	Thermal Control Coatings
M479	Zero Gravity Flammability
M487	Habitability/Crew Quarters
M509	Astronaut Maneuvering Equipment
M516	Crew Activities/Maintenance Study
M556 thru M566	Multipurpose Electric Furnace Experiments
S009	Nuclear Emulsion Experiment
S019	UV Stellar Astronomy
S020	X-Ray/Ultraviolet Solar Photography
S063	UV Airglow Horizon Photography
S149	Particle Collection
S183	UV Panorama
S201B	Far UV Electronographic Camera
S228	Trans-Uranic Cosmic Rays
S230	Magnetospheric Particle Composition
S233	Hand-held Photography of Comet Kohoutek
T002	Manual Navigation Sightings
T003	Inflight Aerosol Analysis
T020	Foot-Controlled Maneuvering Unit
T025	Coronograph Contamination Measurements
T053	Earth Laser Beacon

The student investigations:

ED12	Volcanic Study
ED22	Objects within Mercury's Orbit
ED24	X-Ray Stellar Classes
ED25	X-Rays from Jupiter
ED31	Bacteria and Spores
ED41	Motor Sensory Performance
ED61/62	Plant Growth/Plant Phototropism
ED63	Cytoplasmic Streaming
ED72	Capillary Study
ED76	Neutron Analysis

(Details of most of the above experiments may be found in Skylab Experiments Overview, available from the Government Printing Office (Stock No. 3300-0461) $1.75/copy; or from experiment booklets and manuals in the KSC and JSC newsrooms.)

-more-

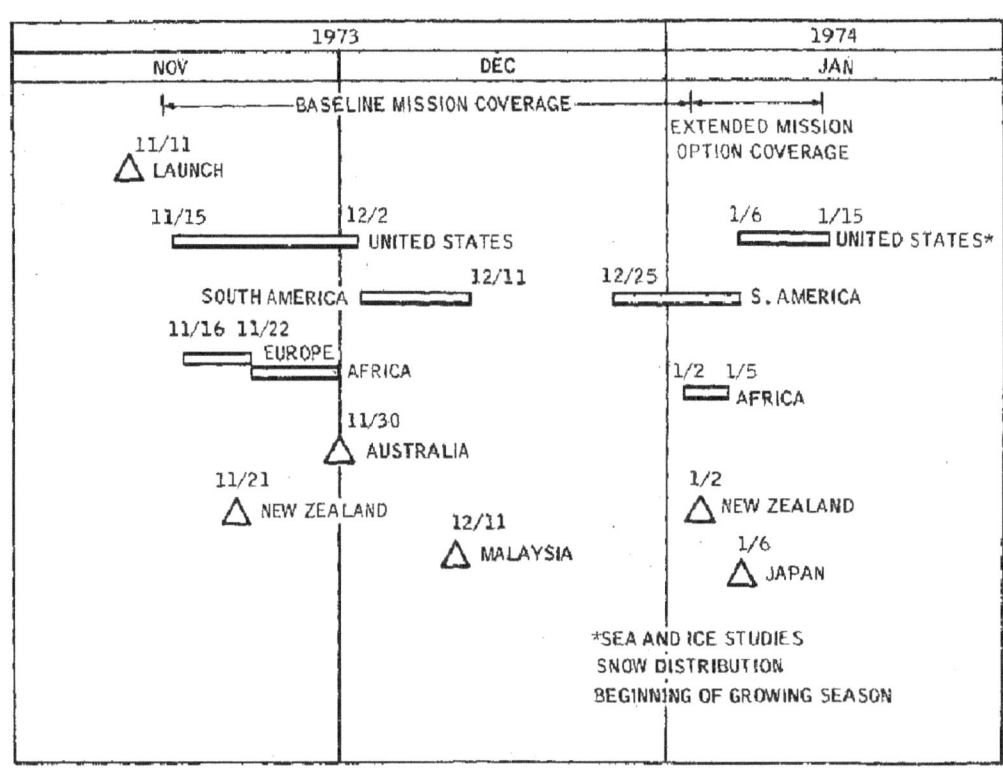

COMET KOHOUTEK AND SKYLAB

An unanticipated major astronomical event has caused revisions in planning for the third and final Skylab manned mission, the SL-4 mission. The passage of the comet Kohoutek was detected early enough in its trajectory to permit scientists to plan ahead for the most promising means to explore its secrets. This will be an extraordinary, opportunity for cometary astronomers, intent on training the best instruments of 1973 on this exciting intruder. The special capability of Skylab rates top priority. Its crew and its instruments are, by good fortune, ready and able to respond, to obtain maximum knowledge about the comet.

Passing inside the Earth's orbit in late November, Kohoutek will travel through the inner solar system during a unique period in the space program, when Skylab and Mariner Venus-Mercury will be in operation and a new NASA C141 Airborne Infrared Observatory is ready for flight.

Surprisingly little hard information is available on the physical nature of comets, despite recorded observations dating back to 467 B.C. Recent work indicates that the clues scientists need exist in the ultraviolet, infrared, and microwave regions of the cometary spectrum.

Comets are generally regarded as samples of primordial material from which the planets formed billions of years ago. Unlike Moon rocks and most meteorites which have experienced melting, the interiors of comet nuclei are believed to have remained in an icy state since their creation.

-more-

A great variety of striking phenomena occur in comets. A few comets actually have disappeared during relatively brief gaps in the observations. One comet split into two comets. A secondary tail apparently formed on another in response to the passage of an interplanetary shock wave. The appearance of a fine spiral pattern in the head of Comet Bennett has been attributed to rotation of the unresolved nucleus. Earlier this year, a flare occurred in a faint comet, increasing its brightness by a factor of almost 10,000 in six days.

The most permanent feature of a comet, the nucleus, is believed to be a sort of dirty ice ball, consisting of frozen gases ("ices") and dust particles. In response to solar radiation as the comet leaves the cold of deep space, the ices sublime and their vapors form an atmosphere, or coma, with a diameter that may reach 100,000 kilometers (60,000 statute miles).

The estimated diameters of cometary nuclei range upward to only a few tens of kilometers or miles. According to one estimate, the nucleus of the Halley Comet loses about 3 meters (10 feet) of surface material each time it passes the Sun. Kohoutek is probably similar in mass to Halley, but going much closer to the Sun, should shed much more.

Separation and ionization due to solar photons and solar wind particles are among the processes which act on gases in the coma, producing "daughter products" - the atoms, radicals, molecules and ions that have been detected spectroscopically in comets. Many astronomers believe that direct detection of the ices and their vapors - the so-called "parent molecules" has not been established. The gases observed thus far are all or mostly daughter products which are unlikely to exist in a solid state under the conditions prevailing in cometary nuclei.

Dust particles liberated from the comet nucleus are impelled in the direction away from the Sun by the pressure of solar radiation. Ions produced in the coma are similarly affected by the charged particles in the solar wind. Thus are formed the dust and plasma tails, which can extend up to 100 million kilometers (60 million miles).

-more-

The dust tails typically look smooth, gently curved and yellow, while the plasma tail is straight, characterized by filaments and an often turbulent appearance and is blue.

If comets condensed from the solar nebula in the region where Jupiter formed, as many astronomers believe, then the parent molecules may be expected to include water, methane and ammonia. On the other hand, if the cometary ices represent aggregated interstellar material, then many more complex substances, including formaldehyde and the other organic molecules that radio astronomers have found in galactic clouds and regions of presumed star formation, should be present.

Hydrogen was first detected in comets a few years ago, thanks to ultraviolet observations with the OAO-2 and OGO-5 satellites. These data showed that the hydrogen atoms occupied an enormous cloud, typically larger than the Sun, surrounding the visible coma.

The origin of comets is unknown. There may be a number of sources. At any rate, comets are occasionally perturbed into the inner solar system where we see them briefly as the long-period comets. Others have been captured in small orbits. These short-period comets include the famous Halley which returns in 1986. Halley was last seen in 1910.

Kohoutek is at least, a long-period comet (10,000 to 80,000 years perhaps) and recent trajectory information raises the possibility that this is the first time that the comet has ever approached the Sun.

Because Kohoutek may be in a relatively undisturbed state, the possibility of obtaining especially valuable scientific information seems clear. To respond to this challenge, NASA has organized "Operation Kohoutek" to obtain physical data on the comet by every suitable means. Dr. Stephen P. Maran of the NASA Goddard Space Flight Center in Maryland is manager of Operation Kohoutek.

The overall objective of Operation Kohoutek is to make a comprehensive investigation of the nature and evolution of the coma and tails as the comet approaches, passes and recedes from the Sun. Among the detailed goals are:

-more-

1. To identify the parent molecules of the gases ("daughter products") observed in comets;

2. To determine the processes that break down the parent molecules and that form the daughter products and excite their radiation spectra;

3. To determine the physical nature and causes of transient events in the comet and their relation to solar activity and phenomena of the interplanetary plasma;

4. To measure the solar wind velocity in the inner solar system;

5. To search for helium, deuterium, molecular hydrogen and other substances that have not yet been found in comets.

Why all the fuss about Kohoutek? From early observations and calculations it appears that Kohoutek is larger than average and will become extremely bright. This will facilitate measurements at very high spectral, spatial and time resolutions, providing maximum scientific data return.

Thanks to Kohoutek's quite small perihelion distance from the Sun, observations of its interactions with the solar wind should reveal new facts about the charged particle environment well within the orbit of Mercury.

As a new or long-period comet in a highly eccentric orbit, Kohoutek may differ substantially from comets such as Encke and Halley that remain within the planetary system, bounded by the orbits of the outer planets. The short-period comets spend a greater fraction of their lives under the influence of solar particles and radiation, and are subject to planetary perturbations. On the other hand, Kohoutek in its present tour of near-solar space should develop a great coma and tails, thanks to the large amount of matter that will be liberated from the frozen nucleus.

Often, the discovery of a major comet comes only a few months before it reaches perihelion, (the closest approach to the Sun) but Kohoutek was found almost 10 months in advance. This early warning permits systematic planning and adequate preparation for a wide variety of coordinated experiments.

-more-

On the other hand, the time involved is far too short to permit development of new spacecraft. Thus the response to the challenge of Kohoutek must make use of existing systems, or ones already well on the way to completion when Kohoutek was found. These are listed in Table 1. Prime among them is Skylab.

Skylab is unique among the spacecraft that will observe Kohoutek, thanks to its capabilities for

- long-term viewing
- near-perihelion viewing
- astronaut response
- payload optimization.

The array of astronomical and solar experiments on Skylab (Table 2) will permit the flight crew to monitor Kohoutek in the UV and visible light ranges regardless of its angular separation from the Sun. This is a critical consideration, because Kohoutek's Sun angle will not exceed $45°$ until January 18th. The unmanned spacecraft are generally constrained to observing at either very large or very small sun angles.

Table 3 indicates the tentative schedule for Kohoutek observations. This is subject to significant up-dating as the individual project offices and experimenters complete and refine their operational plans.

Of particular importance are the ATM instruments on Skylab. They can observe Kohoutek at perihelion when the comet is brightest and receives the most solar energy. At that time, ground-based observations are of very limited scope, due to scattered sunlight in the atmosphere.

White light imagery (S052 experiment) will be performed on ATM at frame rates up to four per minute, much faster than possible with OSO-7, and with higher spatial resolution. Simultaneous mapping of the coma in four UV wavelengths can be accomplished (S055), and high dispersion spectroscopy (S082) may detect the existence of helium and deuterium for the first time in a comet.

The ATM X-ray experiments are not listed in Table 2, since the prospects for detectable cometary radiation in this wavelength range seem poor. However, a major solar flare could induce fluorescence in Kohoutek, leading to a positive result with the S054 instrument.

For a few days just before and just after perihelion, the ATM capabilities will be somewhat reduced due to the larger Sun angles of the comet. During these intervals, however, Kohoutek is too close to the Sun to be observed through the workshop's anti-solar airlock. These are the times for the astronauts to conduct EVA operations. The instruments operated during EVA would be the T025 coronagraph and the new S201B far ultraviolet camera. The T025 observation requires pointing the instrument fairly accurately toward the Sun. For the S201B photography, the Skylab must be maneuvered so that the camera is shadowed by the ATM solar array.

At the airlock, the instruments, including S201B, will operate well before and well after perihelion. An articulating mirror system will be mounted on the airlock and a roll of the spacecraft of up to $90°$ will be made. Implementation of about 24 of these rather major Skylab maneuvers during the mission are being considered. Ordinarily, no more than one would be performed per day.

Comets are known for their unpredictability - for sudden flarings and shape changes. Such transient events are expected to occur in Kohoutek during the Skylab 4 mission and the astronaut crew will react by bringing appropriate instruments into play and increasing the camera frame rates for brief intervals, or taking other special measures.

Only on Skylab, among existing spacecraft, can mission planners change out or modify the instruments to take advantage of an unexpected phenomenon such as the appearance of Kohoutek. Although the stowage list for the Skylab 4 command module is still under review, officials expect to add a new instrument to the orbiting complement. This is the S201B far ultraviolet camera of Dr. T. L. Page and Dr. G. Carruthers (Naval Research Laboratory), which is needed to photograph the hydrogen cloud that will surround the head of the comet. Filters to isolate cometary emissions, a UV-transmitting lens, and extra film to support the desired high frame rates near perihelion are among the other new items expected to be sent up to Skylab.

In addition to the ATM instruments, the following Skylab experiments will be used in the comet study: S019, Ultraviolet Stellar Astronomy; S063, Ultraviolet Airglow Camera; S183, Ultraviolet Panorama Camera; and T025, Multi-filter Coronagraph. The S019 instrument will obtain ultraviolet spectra that will be studied to determine the composition of the comet nucleus and the effects of the solar wind. The S063 camera will obtain ultraviolet and visible color photographs which can help determine the distribution of selected constituents in the coma and tail. The S183 photometric data will help determine the distribution, lifetime and the effect of hydroxyl in the coma. The T025 coronagraph's ultraviolet and visible light photographs should yield information on the particulate production and distribution in the coma and tail.

Some Comet Kohoutek Facts

The orbit of Kohoutek is inclined at 14° to the ecliptic (the plane of the Earth's orbit around the Sun).

Perihelion--the comet's closest approach to the Sun -- will occur Dec. 28 at 21 million km (13 million miles), or 30 solar radii.

Naked-eye visibility should begin in early November prior to sunrise. Eye-balling will switch to after sunset when the comet passes perihelion Dec. 28. Best viewing may come in evening twilight shortly after New Year's Day. Then full Moon will interfere until last third of January. At perihelion, tail will appear short because Earth view will be almost directly along its tail.

Discoverer: Dr. Lubos Kohoutek; discovered photographically March 7, 1973 at Hamburg Observatory in West Germany with the 32-inch Schmidt telescope.

Brightness: Preliminary estimates of Kohoutek's brightness range from visual magnitude -2 to -10. For comparison, the Moon's brightness is -12.7

Designation: Comets bear the names of their discoverers. This is Comet Kohoutek 1973f; the "f" denoting that this is the sixth comet discovered this year. The fifth was Comet Kohoutek 1973e, discovered about a week earlier by the same Dr. Kohoutek.

Observation Systems in Operation Kohoutek

Skylab
Mariner Venus-Mercury
Pioneer Spacecraft
Copernicus (OAO-3)
OSO-7
Sounding Rockets
Airborne Infrared Observatory and Lear-Jet
Far Infrared Balloon Program
Ground-Based Observations

Skylab Experiments for Kohoutek Observations

Name	Description	Principal Investigator	Operation
S052	White Light Coronagraph	R. M. MacQueen High Altitude Observatory	ATM
S055	UV Spectroheliograph	E. M. Reeves Harvard College Observatory	ATM
S082	High Resolution UV Spectrographs	R. Tousey Naval Research Laboratory	ATM
T025	Multi-filter Coronagraph	J. M. Greenberg Dudley Observatory	EVA
S019	UV Objective Prism Spectrograph	K. G. Henize University of Texas and NASA-JSC	Airlock
S063	UV Airglow Camera	D. M. Packer Naval Research Laboratory	Airlock, Windows
S183	UV Panorama Camera	G. Courtes Space Astronomy Laboratory (Marseilles, France)	Airlock
S201B	Far UV Electronographic Camera	T. L. Page Naval Research Laboratory	Airlock, EVA

ORBIT OF COMET KOHOUTEK (1973f), 1973-1974

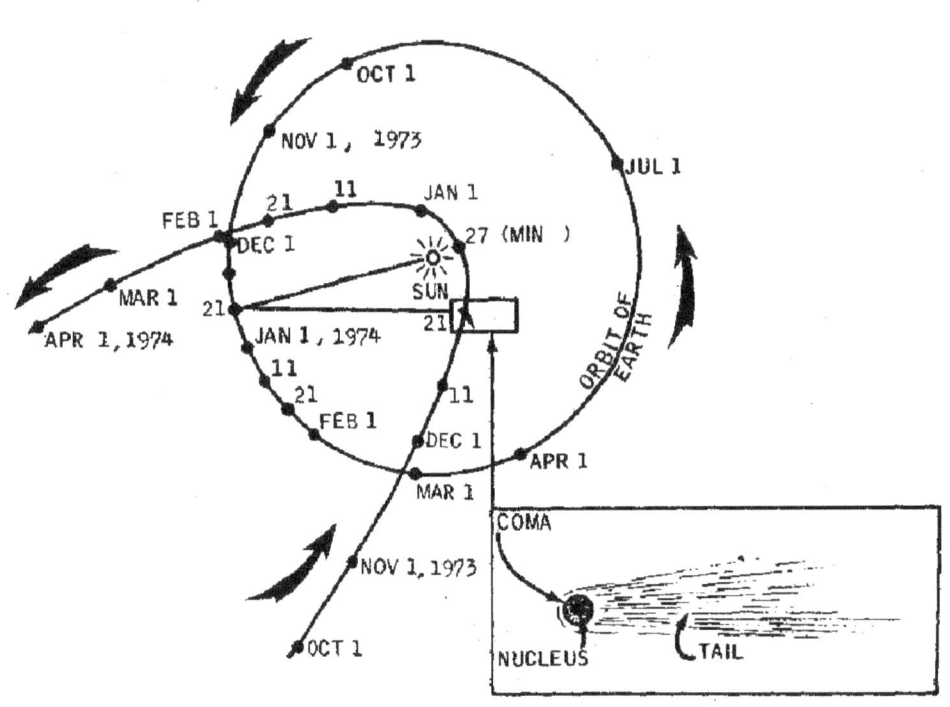

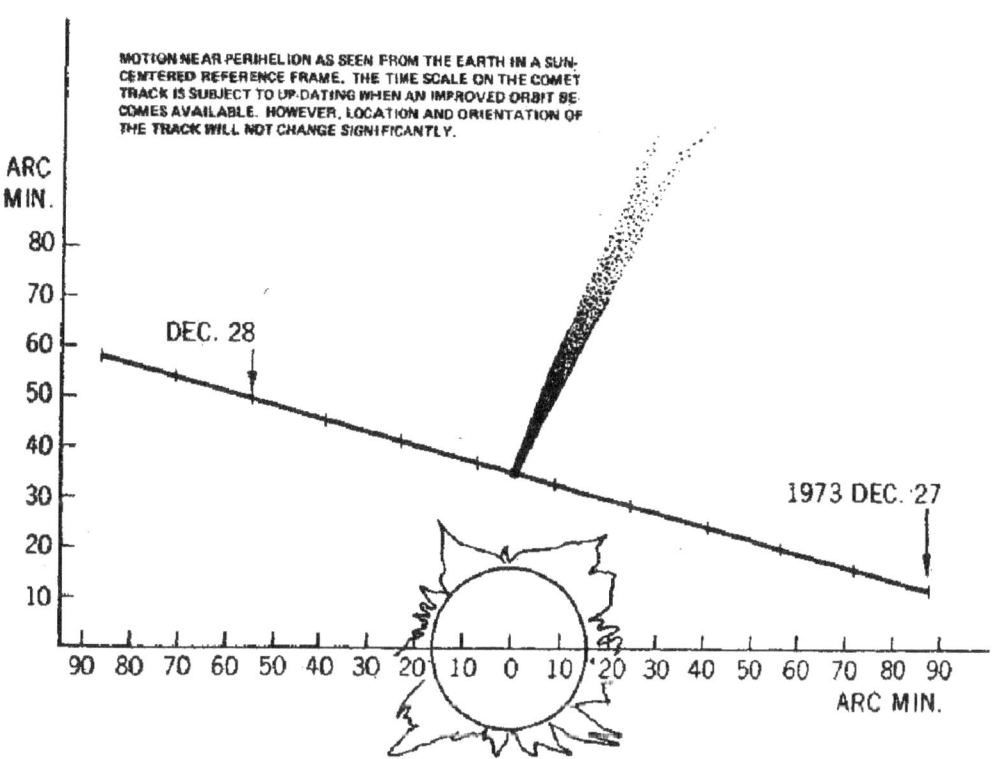

Preliminary Phasing of Kohoutek Operations

Pre-Perihelion
- Late September/October
  - Resume ground-based observations
  - Recover comet
  - Improve orbital definition
- November
  - Launch MVM
  - Launch Skylab 4
  - Begin Lear-Jet flights
  - Begin Skylab airlock observation
  - Launch far infrared balloon
- December
  - Increased priority for airlock observations
  - Comet tail passes Scorpius B-stars
  - Possible STP 72-1 observations

Near-Perihelion
- December 24
  - Halt airlock observations
- December 26-30
  - Intensive ATM Observations
  - EVA for T025, S201B observations
  - OSO-7 observations
  - High dispersion spectroscopy with ground-based solar telescopes

Post-Perihelion
- January
  - Resume airlock observations
  - Pioneer 8 measurements
  - Launch sounding rockets
  - Skylab 4 splashdown
  - Begin MVM observations
  - Prime time for Copernicus observations and ground-based photography
  - Begin C141 flights

Late Post-Perihelion

Spring — Possible reflight of far infrared balloon

SKYLAB 3 ACCOMPLISHMENTS   (July 28 - Sept. 25, 1973)

The second Skylab manned mission not only set a new manned space flight duration record of 59 days and 11 hours but it also accomplished much more scientific experimentation than originally planned.

In two of its main discipline areas--solar observation and Earth resources observation- the Skylab 3 crew was successful in conducting half again as much observation as originally planned.

Bean, Garriott and Lousma, observed the Sun through Apollo Telescope Mount instruments from above the Earth atmosphere 305 hours as compared with the pre-launch plan of 200 hours. Additionally the Sun cooperated with Skylab in presenting an unusual number of active solar events during what was expected to be a quiet period.

In the Earth resources area, the crew was able to conduct 39 passes over selected areas of the Earth to gather data in such areas as forestry, hydrology, oceanography, cartography, geology. Original plans had been to conduct 26 of these passes.

Skylab 3 also exceeded pre-launch plans in the areas of biomedicine, technical and materials processing experiments.

During its 59 days and 11 hours in space, Skylab 3 travelled more than 24 million miles. The mission brought the total United States man-hours in space to 17,831, about the equivalent of nine years work by a man working 40 hours a week.

The crew, after an early experience with motion sickness, adapted well to the weightless environment and was eager for more work assignments as the mission progressed. In fact, during the last portion of the mission, the crew was able to do much more work per day than originally expected. From the 10th to the 15th day of the mission, the crew was able to devote about 19 man-hours a day to scientific experiments. From the 15th day to about the 20th day the rate increased from 27 to 33 man-hours per day in experiment work.

-more-

REAL-TIME FLIGHT PLANNING

Time was when pre-mission space flight plans were followed "by the numbers" with few changes except those caused by systems malfunctions. Skylab flight planning, however, is almost done in real-time, with the pre-mission flight plan serving mainly as a guide to Mission Control Center flight planners. Each day's flight plan is designed to yield the highest experiment data return.

Teleprintered to the Skylab space station early in the morning before the crew wakens, the daily flight plan takes advantage of unique opportunities that enhance data gathering for particular experiments. For example, forcasts of cloud-free EREP sites and ground observatory predictions of unusual solar activity have a bearing upon when EREP passes and ATM runs are scheduled.

The Skylab flight planning cycle begins at midnight Houston time (CST) with a team of flight planners in Mission Control Center drafting a "summary flight plan" for the following crew work day that will start 24 hours after the planning team ends its work shift. This first team is relieved by the so-called "execution" team (day team) of flight controllers concerned only with the existing detailed flight plan for the immediate day. The flight planners on the next, or "swing" shift develop from the summary flight plan a detailed flight plan for the following day, nailing down the activity details first summarized in the early morning hours --- and so on in leapfrog fashion.

Daily flight plans pivot around experiment requirements, spacecraft systems status and optimum crew time usage. Proposed summary flight plans embrace the viewpoints of Skylab systems engineers, experiment principal investigators, flight surgeons, mission management, the flight crew and the weatherman's forecast for potential EREP survey sites. Precedence is given to mandatory operations, ATM, EREP and medical experiments, with other experiments and operations filling the remaining time.

Revised summary flight plans will be reproduced daily and distributed to newspersons at the JSC Newsroom.

DAILY CREW ACTIVITY

The normal Skylab crew workday starts at 6 a.m. and runs until 10 p.m. CST

Breakfast is at 7 a.m., lunch at noon and dinner at 6 p.m. CST --- except for the man on duty at the ATM console during lunch who shifts his meal time so that he can be relieved at the console. Eight hours of sleep are normally scheduled.

- more -

During the mission the astronauts will be operating and monitoring about 60 items of experimental equipment and performing a wide variety of tasks associated with the several hundred Skylab scientific and technical investigations. Depending upon experiment scheduling requirements, Skylab crews have a day off about every seventh day.

About two 15-minute personal hygiene periods are scheduled each day for each crewman and one hour and 30 minutes for physical exercise. Additionally, an hour a day may be set aside for R&R rest and relaxation.

Mission Control Center flight planners fill the remaining eight hours of the crew work day with experiment operations.

Some modifications to flight planning philosophy have been made as a result of experience in the two previous Skylab missions. A marked improvement in crew proficiency was noted after the second week of flight in both crews. Flight plan scheduling has been changed to take advantage of the time gained as crewmen adapt to space station operations.

For example, meal periods have been shortened from one hour to 45 minutes, and the pre- and post-sleep periods have also been shortened. Housekeeping chores, such as trash disposal, filter changing and cleanup which was scheduled in the daily flight plan on the first two missions, will be on the daily "shopping list" for crew option to fit into any slack time.

These changes in flight planning methods have increased the normal experiment day from 22.5 to 28 manhours and are expected to yield more than 200 additional experiment manhours over a 56-day mission.

TYPICAL CREW DAY

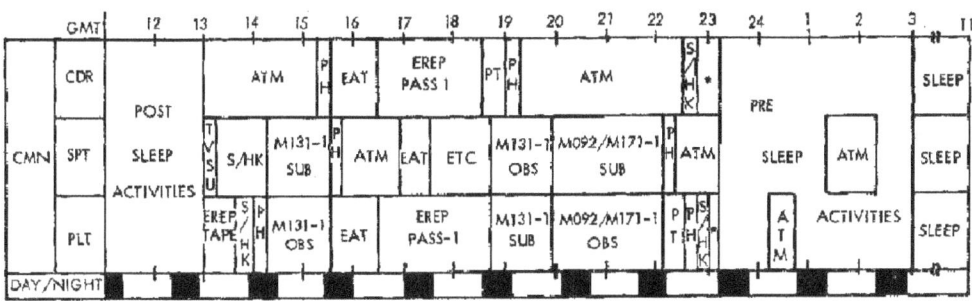

POST SLEEP ACTIVITIES

SYSTEM CONFIGURATION
PH
URINE SAMPLING
T003 EXPERIMENT
BODY MASS MEASUREMENT
BREAKFAST
DINNER PREP
PRD READOUTS
LOAD FILM
REVIEW PADS
STATUS REPORT

S/HK - SYSTEM HOUSEKEEPING

PH - PERSONAL HYGIENE

PT - EXERCISE

TVSU - TV SETUP

* TIME AVAILABLE FOR COROLLARY EXPTS

PRE-SLEEP ACTIVITIES

EVENING MEAL
ATM (1 to 2 PASSES)
MISSION PLANNING
RECREATIONAL ACTIVITIES
CONDENSATE DUMP
TRASH AIRLOCK DUMP
FOOD RESIDUE WEIGHING
STATUS REPORT
T003 EXPERIMENT
SYSTEM CONFIGURATION FOR SLEEP
PH
BREAKFAST PREP

MISSION PROFILE: Launch, Docking and Deorbit

Skylab 4, the third manned visit to space station Skylab, will be launched at 11:41 a.m. EST November 10 from the NASA Kennedy Space Center's Launch Complex 39 Pad B, for a fifth-orbit rendezvous with the space station. The experimental station, designated Skylab 1, was launched into an initial 431x432.9 km (233 by 234 nm) orbit inclined 50 degrees to the equator which is expected to be 427.3 by 432.9 km (231x234 nm) at Skylab 4 rendezvous.

The standard five-step rendezvous maneuver sequence will be followed to bring the astronauts and the Command/Service Module into the space station's orbit---two phasing maneuvers, a corrective combination maneuver, a coelliptic maneuver, terminal phase initiation and braking. The CSM will dock with Skylab's axial docking port at about eight hours after launch.

After verifying that all docking latches are secured, the final Skylab crew will begin activation of the space station but will sleep aboard the Command Module the first night.

As in the frist two manned missions, timekeeping will be on a ground-elapsed-time (GET) basis until GET of eight hours, after which timing will switch over to day of year (DOY), or mission day (MD), and Greenwich Mean Time (GMT or "Zulu") within each day. Mission Day 1 will be the day the crew is launched.

At the completion of the 56-day manned operations period, the crew will return to the CSM, undock and perform two deorbit burns---the first of which will lower CSM perigee to 168.3 km (91 nm) and the second burn will lower perigee to an atmospheric entry flight path. Splashdown will be in the north central Pacific 509 km (310 statute miles) north-northwest of Honolulu, Hawaii. Splashdown coordinates are 25°45'N x 159°15'W. Command Module touchdown, will be at 5:44 p.m. EST January 6, 1974.

(Note: If the mission is extended after the press kit deadline, the JSC Skylab News Center will issue reentry and landing timelines.)

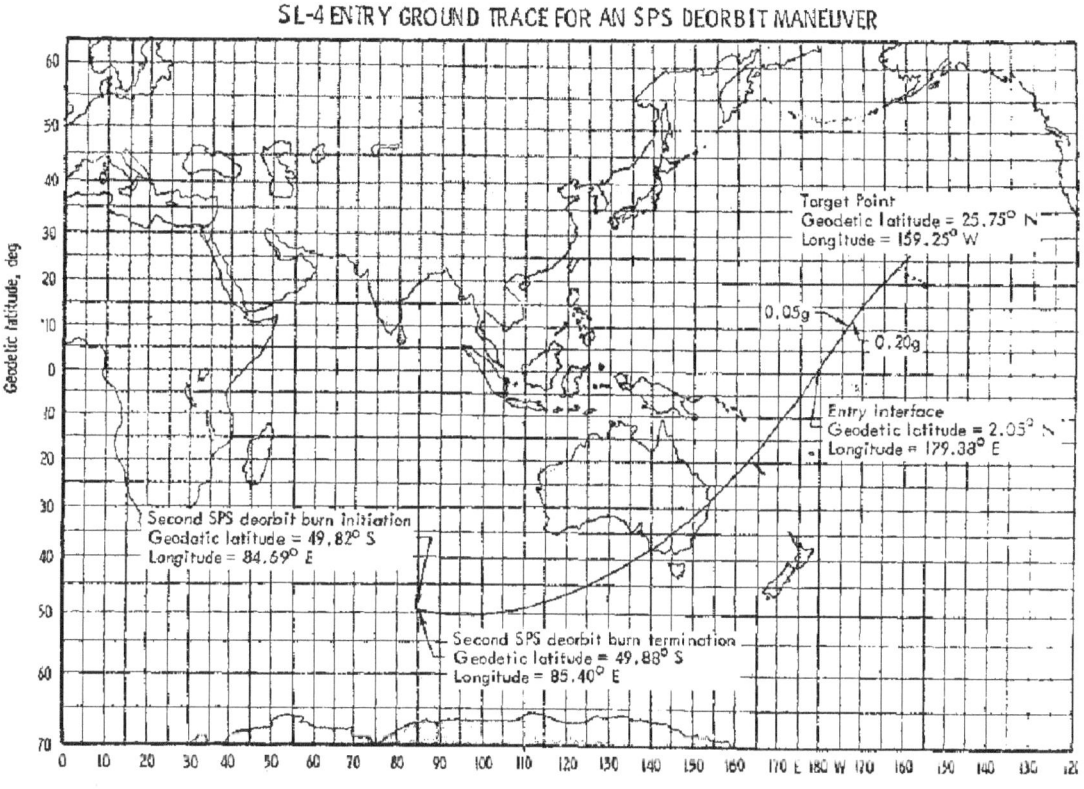

COUNTDOWN

After the July 28 launch of the second crew to man Skylab, the mobile launcher was brought back to the Vehicle Assembly Building at NASA's Kennedy Space Center in Florida. The stages of the next Saturn IB launch vehicle and a boilerplate Command/Service Module (CSM) were erected on the mobile launcher on July 31 and August 1.

Starting August 2, the impact of problems with two of the four control engine quadrants in the docked service module's attitude control system and the possibility of a rescue mission resulted in accelerated processing of the SL-4 launch vehicle and CSM for a possible rescue mission.

Integrated testing of the launch vehicle stages was conducted while the CSM underwent thorough testing - including simulated flights - in the altitude chamber of the Manned Spacecraft Operations Building at KSC.

On August 11, the CSM was moved to the VAB and erected atop the Saturn IB. The vehicles were moved to Complex 39's Pad B on August 14 for pad integration and final tests. A Flight Readiness Test in preparation for the potential rescue mission was conducted September 4-5 and launch preparations went into a "hold" September 11 at a point seven hours prior to the scheduled loading of hypergolic propellants for the Saturn IB's second stage auxiliary propulsion system and for the CSM.

The Skylab 3 mission ended successfully September 25 without the need for a rescue mission and the SL-4 space vehicle was returned to a routine flow on September 25 heading toward a planned launch date of November 11.

As in the previous Skylab launch, SL-4 launch preparations differ from earlier ones in that the Countdown Demonstration Test (CDDT) and the final countdown have been incorporated into a single launch countdown.

The early portion of the countdown will include launch vehicle cryogenic fueling and final countdown activities without astronaut participation.

Following the simulated T-0, the count will go into an operational hold until T-42 hours, 30 minutes, prior to launch. The final recycled count will then proceed to launch. There will be no "dry" test with crew participation in the early portion of the count as was done on earlier missions.

-more-

Key events in the final count, beginning at T-42 hours, 30 minutes, include:

T-36 hours	Begin 8 1/2-hour service module cryogenic fueling and pressurization
T-27 hours	Start CSM mechanical buildup and closeout; to be completed at T-15 hours, 30 minutes.
T-25 hours, 30 minutes	Install launch vehicle batteries.
T-19 hours	Launch vehicle power transfer test.
T-9 hours	Begin clearing pad area
T-8 hours	Replenish RP-1 (first stage fuel)
T-6 hours, 50 minutes	Begin launch vehicle cryogenic propellant load. (Loading takes approximately 3 hours - replenish continues through remainder of countdown)
T-4 hours	Primary damper retracted
T-3 hours, 45 minutes	CSM closeout crew on station
T-2 hours, 40 minutes	Flight crew enters spacecraft
T-1 hour, 51 minutes	Emergency Detection System Tests (to T-1 hour, 21 minutes)
T-58 minutes	LV power transfer test
T-57 minutes	Clear closeout crew from pad area
T-45 minutes	Retract Swing Arm 9 to park position
T-44 minutes	Launch Escape System armed
T-42 minutes	Final launch vehicle range safety checks (to T-35 minutes)
T-35 minutes	Last target update of the Launch Vehicle Digital Computer for Skylab rendezvous

T-15 minutes	Hold for liftoff adjustment - maximum 3 minutes
T-5 minutes	Swing Arm 9 fully retracted
T-3 minutes, 7 seconds	Start automatic sequence
T-50 seconds	Launch Vehicle transfer to internal power
T-3 seconds	Ignition sequence starts
T-0	Liftoff

SKYLAB RESCUE VEHICLE

Preparations for placement of the Skylab Rescue Vehicle, CSM-119, on Pad B at Launch Complex 39 will begin immediately after launch of SL-4. Based upon a November 10 launch, the mobile launcher will be returned to the VAB on November 11 for refurbishment. The erection of the Saturn IB launch vehicle is scheduled for mid-November and the rescue spacecraft - which already has undergone altitude testing - is to be erected atop the two Saturn IB stages and Instrument Unit at the end of November.

The rescue CSM and its Saturn IB are scheduled for transfer from the VAB to the launch pad in early December for pad integration and final tests. Current scheduling calls for SL-R (the designation of the rescue mission) to be in a launch readiness configuration by the end of December.

The countdowns for SL-4 and SL-R are identical from the T-minus 26 hour, 30 minute mark. The SL-4 launch countdown was restructured to match the rescue countdown. Following rescue countdown procedures during an actual launch will enhance confidence and provide a rehearsal for the KSC launch team in the event a rescue mission should become necessary.

The Skylab Rescue Vehicle will remain on Complex 39's Pad B until completion of the SL-4 mission.

-end-

PROJECT MERCURY

FAMILIARIZATION MANUAL

Manned Satellite Capsule

Periscope Film LLC

NASA PROJECT GEMINI

FAMILIARIZATION MANUAL
Manned Satellite Capsule

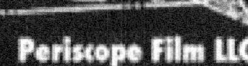

Periscope Film LLC

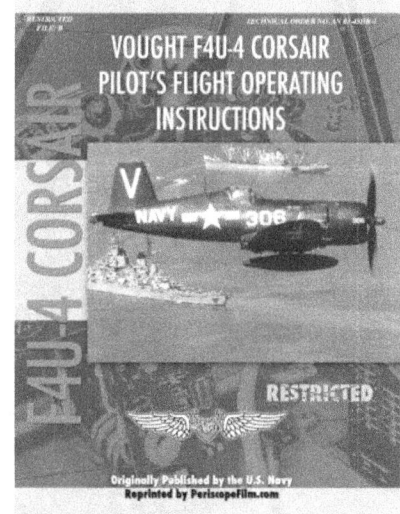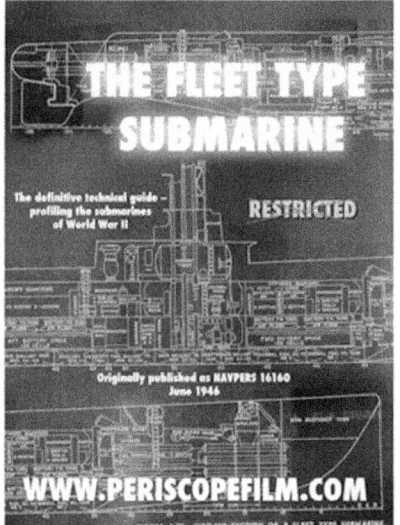

ALSO NOW AVAILABLE FROM PERISCOPEFILM.COM

©2012 Periscope Film LLC
All Rights Reserved
ISBN #978-1-937684-84-6
www.PeriscopeFilm.com

www.ingramcontent.com/pod-product-compliance
Lightning Source LLC
Chambersburg PA
CBHW082019300426
44117CB00015B/2277